FUZZY CONTROL SYSTEMS

FUZZY CONTROL SYSTEMS

Edited by

Abraham Kandel

Department of Computer Science and Engineering
University of South Florida, Tampa, Florida

Gideon Langholz

Department of Electrical Engineering
Florida State University, Tallahassee, Florida
and Tel-Aviv University, Israel

With Foreword by Lotfi A. Zadeh

CRC Press
Boca Raton   Ann Arbor   London   Tokyo

Library of Congress Cataloging-in-Publication Data

Fuzzy control systems / edited by Abraham Kandel, Gideon Langholz ;
  with forward by Lotfi A. Zadeh.
      p.   cm.
   Includes bibliographical references and index.
   ISBN 0-8493-4496-4
   1. Adaptive control systems. 2. Fuzzy systems. I. Kandel,
Abraham. II. Langholz, Gideon.
TJ217.F89 1993
629.8'36--dc20   93-28382
                                                                                   CIP

   This book contains information obtained from authentic and highly regarded sources. Reprinted material is quoted with permission, and sources are indicated. A wide variety of references are listed. Reasonable efforts have been made to publish reliable data and information, but the author and the publisher cannot assume responsibility for the validity of all materials or for the consequences of their use.
   Neither this book nor any part may be reproduced or transmitted in any form or by any means, electronic or mechanical, including photocopying, microfilming, and recording, or by any information storage or retrieval system, without prior permission in writing from the publisher.
   CRC Press, Inc.'s consent does not extend to copying for general distribution, for promotion, for creating new works, or for resale. Specific permission must be obtained in writing from CRC Press for such copying.
   Direct all inquiries to CRC Press, Inc., 2000 Corporate Blvd., N.W., Boca Raton, Florida 33431.

© 1994 by CRC Press, Inc.

No claim to original U.S. Government works
International Standard Book Number 0-8493-4496-4
Library of Congress Card Number 93-28382
Printed in the United States of America   1 2 3 4 5 6 7 8 9 0
Printed on acid-free paper

Grammatici certant,

et adhuc sub iudice lis est

Horatius, *De Arte Poetica,* 78

CONTENTS

FOREWORD Lotfi A. Zadeh   xvii

PREFACE   xix

PART A - GENERAL THEORY   1

Chapter 1   **LEARNING ALGORITHMS FOR NEURO-FUZZY NETWORKS** P. Y. Glorennec   3

       Introduction, *4*
       Neuro-Fuzzy Networks, *5*
       Weight Identification for Φ_2, *9*
       Optimization of Membership Functions, *12*
       Examples, *14*
       Conclusion, *17*
       References, *17*

| Chapter 2 | **TOWARDS A UNIFIED THEORY OF INTELLIGENT AUTONOMOUS CONTROL SYSTEMS** *L. J. Kohout* | 19 |

        Introduction, *20*
        State of the Art and Future Challenges, *20*
        Adaptive Systems and the Dual Control Problem, *22*
        A Systemic Approach to Behavioural Specification
            of Intelligent Systems, *25*
        Behavioural Specification of the Activities of a System
            of Interacting Actors, *28*
        Fuzzification of Formal Models of Adaptive Autonomous
            Control Systems, *32*
        Fuzzy Identification, *36*
        Introducing Control Hierarchies, *38*
        From Abstract Logic Designs to Technological Realizations, *45*
        References, *48*

| Chapter 3 | **REASONING BY ANALOGY IN FUZZY CONTROLLERS** *W. Pedrycz* | 55 |

        Introduction, *56*
        Problem Statement, *56*
        Logic-Based Neurons, *59*
        Analogical Processor - Architecture, *64*
        Learning in the AP, *65*
        Reasoning by Analogy in the AP Structure, *66*
        Analogical Reasoning in Presence of Input Uncertainty, *68*
        Characteristics of the Reasoning Scheme, *69*
        Conclusions, *71*
        References, *71*
            Appendix A, *72*

| Chapter 4 | **INFORMATION COMPLEXITY AND FUZZY CONTROL** *A. Ramer and V. Kreinovich* | 75 |

        Preamble, *76*
        Introduction, *78*
        Uncertainty of Membership Functions, *79*
        Comparison of Uncertainties, *88*
        Closing Remarks, *94*
        References, *95*

Chapter 5   ALTERNATIVE STRUCTURES FOR KNOWLEDGE
             REPRESENTATION IN FUZZY LOGIC CONTROLLERS
             R. R. Yager   99

   Introduction, *100*
   Basic Structures of Fuzzy Logic Controllers, *100*
   Certainty Qualified Antecedents, *107*
   Alternative Formulations for Rule Outputs, *109*
   Chaining of Fuzzy Rules, *114*
   Decoupled Inputs, *121*
   Hierarchical Representation of Rules, *123*
   Conclusion, *134*
   References, *134*

PART B - METHODOLOGIES AND ALGORITHMS   139

Chapter 6   DYNAMIC ANALYSIS OF FUZZY LOGIC CONTROL
             STRUCTURES *A. Garcia-Cerezo, A. Ollero, and J. Aracil*   141

   Introduction, *142*
   Fuzzy Control Structures, *143*
   Dynamic Analysis and Design, *146*
   Fuzzy Controllers with Consequent Functions, *149*
   Example, *152*
   Conclusions, *158*
   References, *158*

Chapter 7   INTELLIGENT FUZZY CONTROLLER FOR EVENT-DRIVEN,
             REAL TIME SYSTEMS AND ITS VLSI IMPLEMENTATION
             J. Grantner, M. Patyra and M. S. Stachowicz   161

   Fuzzy Logic Finite State Machine, *162*
   Algorithm of Creating a Multiple-Input Fuzzy Model, *164*
   Hardware Accelerator, *168*
   VLSI Implementation, *172*
   Conclusions, *179*
   References, *179*

Chapter 8   **CONSTRAINT-ORIENTED FUZZY CONTROL SCHEMES FOR CART-POLE SYSTEMS BY GOAL DECOUPLING AND GENETIC ALGORITHMS** *O. Katai, M. Ida, T. Sawaragi, S. Iwai, S. Kohno, and T. Kataoka* — 181

Introduction, *182*
The Notion of Constraint-Oriented Fuzzy Control, *183*
Construction of Fuzzy Control Scheme by Decoupling the Goals on Cart and Pole, *185*
Construction of Fuzzy Control Systems by Genetic Algorithm Techniques, *188*
Construction of Control Scheme for Cart-Pole Systems by Genetic Algorithms, *191*
Conclusions, *193*
References, *194*

Chapter 9   **A SELF GENERATING AND TUNING METHOD FOR FUZZY MODELING USING INTERIOR PENALTY METHOD AND ITS APPLICATION TO KNOWLEDGE ACQUISITION OF FUZZY CONTROLLER** *R. Katayama, Y. Kajitani, and Y. Nishida* — 197

Introduction, *198*
Self Tuning Method using Interior Penalty Method, *199*
Hybrid Algorithm for Rule Generation and Parameter Tuning, *207*
Application to Knowledge Acquisition of Fuzzy Controller, *218*
Conclusion, *222*
References, *223*

Chapter 10   **FUZZY CONTROL OF VSS TYPE AND ITS ROBUSTNESS** *S. Kawaji and N. Matsunaga* — 225

Introduction, *226*
Fuzzy Control of VSS Type, *226*
Rule Generation of Fuzzy Controller, *232*
Parameter Adjustment of Fuzzy Controller, *234*
Robustness of Fuzzy Controller, *238*
Conclusions, *238*
References, *238*

Chapter 11	**THE COMPOSITION OF HETEROGENEOUS CONTROL LAWS** B. Kuipers and K. Astrom	243

Introduction, 244
Qualitative Descriptions of Incomplete Knowledge, 245
A Heterogeneous Controller for the Water Tank, 248
Guarantees, 250
Simulation Results, 255
Integral Action, 256
Relationship to Fuzzy Logic Control, 259
References, 260

Chapter 12	**SYNTHESIS OF NONLINEAR CONTROLLERS VIA FUZZY LOGIC** R. Langari	263

Introduction, 264
Fuzzy Control Systems, 264
Problem Statement, 267
Discussion, 271
Conclusion, 273
References, 273

Chapter 13	**FUZZY CONTROLS UNDER PRODUCT-SUM-GRAVITY METHODS AND NEW FUZZY CONTROL METHODS** M. Mizumoto	275

Introduction, 276
Min-Max-Gravity Method, 276
Point at Issue of Min-Max-Gravity Method, 278
Product-Sum-Gravity Method, 280
Comparison of Min-Max-Gravity Method and
    Product-Sum-Gravity Method, 282
Realization of PID Controllers by Product-Sum-
    Gravity Method, 285
Fuzzy Control Results, 288
New Fuzzy Reasoning Methods and Control Results, 291
Conclusion, 292
References, 294

Chapter 14 **FUZZY MODELING FOR ADAPTIVE PROCESS CONTROL** *Y. Nakamori* **295**

    Introduction, *296*
    Modeling Methodology, *297*
    Preliminaries, *298*
    Modeling Strategy, *299*
    Hyperellipsoidal Clustering, *301*
    Consequence Modeling, *306*
    Premise Modeling, *307*
    Fuzzy Dynamic Model, *309*
    Predictive Control, *310*
    Conclusion, *313*
    References, *313*

Chapter 15 **FUZZY CONTROLLER WITH MATRIX REPRESENTATION** *M. Nakatsuyama, J. H. Yan, and H. Kaminaga* **315**

    Introduction, *316*
    Fuzzy Control Statements, *317*
    Basic Concept of Matrix Representation, *318*
    Reduction of Matrix Representation, *322*
    Reduction by Simple Self-Tuning, *322*
    Self-Tuning by Modified Simplex Method, *324*
    Neural Networks for Fuzzy Controller, *328*
    Conclusion, *334*
    References, *335*

Chapter 16 **A SELF-TUNING METHOD OF FUZZY REASONING BY GENETIC ALGORITHM** *H. Nomura, I. Hayashi, and N. Wakami* **337**

    Introduction, *338*
    A Conventional Self-Tuning Method, *339*
    Optimization of the Inference Rules by Genetic Algorithm, *341*
    Numerical Examples, *346*
    Conclusion, *353*
    References, *353*

Chapter 17 **HYBRID NEURAL-FUZZY REASONING MODEL WITH APPLICATION TO FUZZY CONTROL** D. Park, A. Kandel, and G. Langholz — 355

Introduction, 356
Fuzzy Models, 357
Hybrid Neural-Fuzzy Reasoning Model, 365
Conclusions, 370
References, 372

Chapter 18 **LEARNING FUZZY CONTROL RULES FROM EXAMPLES** S. G. Romaniuk and L. O. Hall — 375

Introduction, 376
The SC-Net Approach, 377
Learning Fuzzy Motor Control, 381
Summary, 390
References, 394

Chapter 19 **A COMPUTATIONAL APPROACH TO FUZZY LOGIC CONTROLLER DESIGN AND ANALYSIS USING CELL STATE SPACE METHODS** S. M. Smith, B. Nokleby, and D. J. Comer — 397

Introduction, 398
Control Problem, 400
Cell State Space Optimal Control, 402
Fuzzy Logic Controller Tuning, 408
Angular Position Control of an Inverted Pendulum, 411
Extensions to Three and Four Variables, 418
Summary and Conclusions, 425
References, 425

Chapter 20 **AN ADAPTIVE FUZZY CONTROL MODEL BASED ON FUZZY NEURAL NETWORKS** X. Zhang, P. Wang, Z. Shen, and X. Peng — 429

Introduction, 430
Building a Fuzzy Control System, 430
The Implementation of Fuzzy Control, 451
Controllability and Stability of Fuzzy Control, 452
The Comparisons between the Fuzzy Control
    Theory and Modern Control Theory, 454

Conclusion, *455*
References, *456*

PART C - IMPLEMENTATIONS AND APPLICATIONS   457

**Chapter 21   HUMAN FRIENDLY FUZZY TRANSPORTATION
SYSTEM** *T. Iokibe and T. Kimura*   459

Introduction, *460*
System Configuration, *461*
Fuzzy Language Understanding and Fuzzy Image Recognition, *462*
Fuzzy Path Planning, *464*
References, *472*

**Chapter 22   CONTROL OF A CHAOTIC SYSTEM USING FUZZY
LOGIC** *C. L. Karr and E. J. Gentry*   475

Introduction, *476*
The Chaotic System, *478*
Surface-Fitting and the Analytic Solution, *480*
Genetic Algorithm-Designed Fuzzy Logic Controller, *482*
Results, *490*
Summary, *494*
References, *495*

**Chapter 23   APPLICATIONS OF A FUZZY CONTROL TECHNIQUE
TO SUPERCONDUCTING ACTUATORS USING HIGH-Tc
SUPERCONDUCTORS** *M. Komori and T. Kitamura*   499

Introduction, *500*
Superconducting Levitation Mechanism, *500*
Superconducting Radial Bearing, *506*
Superconducting Linear Actuator, *510*
Superconducting Pump Actuator, *515*
Conclusions, *519*
References, *520*

Chapter 24 **A FUZZY LOGIC BASED APPROACH TO MACHINE TOOL CONTROL OPTIMIZATION** *J. R. A. Lopez, E. A. Gutierrez, and L. C. Rosa* — 523

Introduction, *524*
Position Control, *525*
Direct Rule-Based Fuzzy Controller, *526*
Self-Organizing Fuzzy Controller, *530*
The SPB System, *532*
Results Resume, *534*
Fuzzy Position Control of Machine Tools: Conclusions, *536*
A More General Approach: Towards the Concept of Intelligent Machining, *537*
References, *538*

Chapter 25 **FUZZY MANAGEMENT OF CACHE MEMORIES** *M. A. Manzoul* — 541

Introduction, *542*
Organization of Cache Memory, *543*
Fuzzy Cache Management, *544*
Simulation Results, *545*
Conclusion, *546*
References, *549*

Chapter 26 **FUZZY CONTROLLERS ON SEMI-CUSTOM VLSI CHIPS** *M. A. Manzoul* — 551

Introduction, *552*
Logic Synthesis of Fuzzy Controllers, *553*
FPGA Implementation, *558*
Conclusions, *559*
References, *560*

Chapter 27 **GENERAL ANALYSIS OF FUZZY-CONTROLLED PHASE-LOCKED LOOP** *H. N. Teodorescu and A. Brezulianu* — 561

Introduction, *563*
Possibilities of Fuzzy Control in PLLs, *564*
PLL Parameters and Adaptability, *568*
The Case of Synchronous and Coherent Phase-Locked Synchronous Oscillators, *569*
Performance Analysis of an Analog, Fuzzy-Controlled PLL, *571*

The Stability of F-PLL and Chaotic F-PLL, *574*
Advantages and Limits of F-PLL, *575*
References, *576*

**Chapter 28 A FUZZY LOGIC CONTROLLER FOR A RIGID DISK
DRIVE** *S. Yoshida*   579

Introduction, *580*
Seek Control Method, *581*
Seek Employing Fuzzy Logic, *583*
Trial Seek Employing Fuzzy Logic, *584*
Seek Table Reference, *588*
Reversed Seek Table reference, *590*
Correcting Force Unevenness, *595*
Conclusion, *598*
References, *598*

AUTHOR'S BIOGRAPHICAL INFORMATION   601

INDEX   621

FOREWORD

Lotfi A. Zadeh

In examining the contents of the volume edited by Professors Kandel and Langholz I was struck by the vastness of the progress made in the realm of *fuzzy control* since the publication of my 1972 paper "A Rationale for Fuzzy Control." What is particularly impressive is the breadth of applications, which range from machine tool control optimization to the design of human friendly transportation systems. But what is also highly impressive is the progress made in the development of the underlying theory and the construction of effective techniques for its implementation.

To view the developments in fuzzy control in a proper perspective, a bit of history is in order. When I wrote my first paper on *fuzzy sets* in 1965, my expectation was that most of the applications of the theory would be in those fields in which the conventional mathematical techniques are of limited effectiveness. This was, and still is, the case in biological and social sciences, linguistics, psychology, economics and, more generally, in the soft sciences. In such fields, the variables are hard to quantify and the dependencies are too ill-defined to admit precise characterization in terms of difference or differential equations.

It did not take that long, however, to realize that even in those fields in which the dependencies between variables are well-defined, it may be necessary or advantageous to employ fuzzy rather than crisp algorithms to arrive at a solution. The main reason for this state of affairs is that in most real-world settings *precision* is illusory. For example, in the case of the car parking problem, although the kinematics of the car and the geometry of the problem are well-defined, the final position and the orientation of the car are not specified precisely. Thus, it is the imprecision of the goal that makes it possible for humans to park a car without making any measurements or numerical computations. What we see, then, is that humans possess a remarkable innate ability to exploit the tolerance for imprecision to achieve tractability, robustness and low solution cost, whereas the traditional control techniques fail to do so when they employ crisp rather than fuzzy algorithms to arrive at a solution.

In papers published in 1973 and 1974, I have outlined the basic ideas underlying fuzzy control. Among them, the concept of a linguistic variable, fuzzy If-Then rules, fuzzy algorithms, the compositional rule of inference, and the execution of fuzzy instructions. However, it was the seminal work of Mamdani and Assilian in 1975 that showed how these ideas could be translated into a working control system. The contributions of Professor Mamdani and his associates at Queen Mary College, along with the contributions of Professor Van Naute Lemke and his associates at Delft Institute of Technology and those of Professor Sugeno and his associates at Tokyo Institute of Technology, played a key role in the development of fuzzy control and its early evolution.

In the seventies and early eighties, the applications of fuzzy control were centered on industrial systems. A turning point in 1987 was the first consumer product – a shower head conceived and produced by Matsushita. Other consumer products – washing machines, vacuum cleaners, camcorders, cameras, and air conditioners – followed in quick succession. These and other consumer products gave fuzzy logic a much higher visibility in control and stimulated an exploration of its applications in many other fields.

Today, we may be witnessing still another turning point. Specifically, we are observing a paradigm shift from traditional, hard computing to what may be called *soft computing* (SC). As its name suggests, soft computing is concerned with modes of computation which are approximate rather than exact. At this juncture, the principal components of soft computing are *fuzzy logic* (FL), *neural network theory* (NN), and *probabilistic reasoning* (PR), with the latter subsuming *belief networks* (BN), *genetic algorithms* (GA) and the *theory of chaotic systems* (CT). There is substantial overlap between FL, NN, and PR, but in the main, FL, NN, and PR are complementary rather than competitive. For this reason, there are many situations in which FL, NN, and PR may be used to advantage in combination rather than exclusively. A case in point is the rapidly growing use of a combination of FL and NN techniques in the so-called *neurofuzzy* consumer products such as refrigerators, air conditioners, and heaters. Within soft computing, FL is concerned in the main with imprecision and approximate reasoning; NN with learning and curve-fitting; and PR with uncertainty and propagation of belief. In the final analysis, *the role model for soft computing is the human mind.*

The paradigm shift from hard to soft computing is reflected in the orientation and contents of the present volume. With fuzzy control systems as its leitmotif, Professors Kandel and Langholz have assembled a blue-ribbon panel of contributors whose work is at the frontiers of fuzzy logic and soft computing. There is much in the volume that is new and important, both in regard to basic theory and to its applications to the conception and design of sophisticated control and knowledge-based systems.

In my view, *Fuzzy Control Systems* is a must reading for anyone who is interested in acquiring a thorough understanding of fuzzy logic, its role in soft computing, and it applications to control and related fields.

Lotfi A. Zadeh
Computer Science Division
University of California at Berkeley

Berkeley, 1993

PREFACE

The field of fuzzy control systems is one of the most active and fruitful areas of research in which fuzzy sets theory is applied. Fuzzy sets theory was originally proposed by Lotfi Zadeh to formalize qualitative concepts that have no precise boundaries. For example, there are no meaningful landmark values representing the boundaries between *low* and *normal*, or *normal* and *high*. Rather, such linguistic terms are formalized by referring to fuzzy sets of numbers.

A fuzzy controller consists of a collection of control laws whose inputs and outputs are both fuzzy values. All controller rules are fired in parallel and the recommended actions are combined according to fuzzy value combination rules, weighted by the degree of satisfaction of the antecedent. Some process of *defuzzification* is required to convert the resulting fuzzy set description of an action into a specific value for a control variable.

The ability to control a system in *uncertainty* or *unknown* environments is one of the most important characteristics of any intelligent control system. Fuzzy inferencing procedures are becoming, therefore, increasingly crucial to the process of managing uncertainty. Fuzzy sets theory provides a systematic framework for dealing with different types of uncertainty within a single conceptual framework.

In addition, since many applications involve human expertise and knowledge, which are invariably imprecise, incomplete, or not totally reliable, and intelligent control system must combine knowledge-based techniques for gathering and processing information with methods of approximate reasoning. This would enable the control system to better emulate human decision-making processes as well as allow for imprecise information and/or uncertain environments.

This edited volume is divided into three parts. Part A is devoted to the general theory of fuzzy control systems. Part B deals with a variety of methodologies and algorithms used in the analysis and design of fuzzy controllers. The various paradigms considered include fuzzy reasoning models, fuzzy neural networks, fuzzy expert systems, and genetic algorithms. Part C consists of some current implementations and applications of fuzzy control systems in areas such as transportation systems, machine-tool control, management of cache memories, VLSI implementation of fuzzy controllers, and control of chaotic systems, superconducting actuators, phase-locked loops, and hard disk drives.

Many individuals deserve recognition for making this book possible. First and foremost, the contributors to this book for their effort, time, and promptness. We thank Professor Lotfi A. Zadeh for writing the foreword to this book and for his continued encouragement. We are also grateful to the staff of the CRC Press for their advice and commitment to the project, and for a job well done.

This volume is intended for researchers, practitioners, and students involved in the study, research, and development of fuzzy control systems. We hope that they will find this book useful and inspiring. We also hope that the book will serve as an impetus for continued advanced research and development in this exciting field of artificial intelligence endeavor - *fuzzy control systems.*

Abraham Kandel
Gideon Langholz

Florida, 1993

PART A

GENERAL THEORY

Chapter 1
Learning Algorithms for Neuro-Fuzzy Networks

Introduction, *4*
Neuro-Fuzzy Networks, *5*
Weight Identification for Φ_2, *9*
Optimization of Membership Functions, *12*
Examples, *14*
Conclusion, *17*
References, *17*

Neuro-fuzzy networks result from the fusion of neural networks and fuzzy logic. The advantages of both approaches are thus merged. To a linguistic rule-based system, the neural techniques bring supervised learning capabilities for extracting fuzzy rules from a set of numerical data. However, the learning algorithms, which are mostly inspired by the back-propagation algorithm, do not entirely take into account all the specifics of neuro-fuzzy networks. This chapter proposes, therefore, different learning algorithms without back-propagation, which make use of several of the advantages of neuro-fuzzy networks. In particular, the neuro-fuzzy network eliminates the drawback in the design of a conventional fuzzy system whereby the user must tune by trial-and-error the membership functions of the fuzzy sets defined in the input and output universes of discourse.

LEARNING ALGORITHMS FOR NEURO-FUZZY NETWORKS

Pierre Yves Glorennec
Département d'Informatique
Institut National des Sciences Appliquées
35043 Rennes Cedex, France

1 INTRODUCTION

In the design of a conventional fuzzy system, the user must tune by trial-and-error the membership functions of fuzzy sets defined in the input and output universes of discourse. This drawback is eliminated with Neuro-Fuzzy Networks (NFN), which combine the advantages of Neural Networks (NN) and Fuzzy Systems (FS). Thanks to supervised learning methods, it is now possible to optimize both the antecedent and consequent parts of a linguistic rule-based fuzzy system. But almost all the learning methods used for NFN are derived from BackPropagation, which have the well-known following drawbacks :

- slowness, convergence problems, possible local minima;

- difficulty to introduce constraints, for instance about the shape of membership functions;

- all the weights are similarly updated, independantly of their semantic, with a decreasing efficiency from the last layer to the first layer;

- excessive computational hardware for on-chip learning.

We propose a more detailed analyse of NFN, with two different sets of parameters and specific algorithms for each layer. In Section 2, we present the foundations of NFN and the general structure used in this study. In Section 3, we propose three supervised learning algorithms for the consequent part of rules, and, in Section 4, algorithms for optimizing the input membership functions. Examples are given in Section 5.

2 NEURO-FUZZY NETWORKS

2.1 The Conventional Fuzzy Model

For simplicity, we only consider n-input-single-output systems $f : U \subset \mathcal{R}^n \to \mathcal{R}$, where U is the universe of discourse, because the case of m-output systems can be separated into m single-output systems. Let $m_i, i = 1, 2, \ldots, n$, be the number of fuzzy sets defined in the ith input domain. We have :

$$\begin{array}{rll} M &= \sum_{i=1}^{n} m_i & \text{The total number of fuzzy sets} \\ K &= \prod_{i=1}^{n} m_i & \text{The maximum number of fuzzy rules} \end{array}$$

A linguistic rule-based n-input-single-output fuzzy system is composed of a set of rules such as :

$$\text{if } x_1 \text{ is } A_1^{j_1} \text{ and } \ldots \text{ and } x_n \text{ is } A_n^{j_n} \text{ then } y \text{ is } B_i \tag{1}$$

where x_j ($j = 1, 2, \ldots, n$) are the inputs to the fuzzy system, y the output, A_j^k ($k = 1, 2, \ldots, m_j$) and B_i ($i = 1, \ldots, K$) linguistic labels. Thus, the conventional fuzzy model consists of three basic steps : fuzzification, inference process and defuzzification.

Fuzzification is a mapping from the observed input to the fuzzy sets defined in the corresponding universes of discourse. Therefore, fuzzification is a mapping from $U \subset \mathcal{R}^n$ to the unit hypercube $[0, 1]^M$.

Inference Process is a decision making logic which determines fuzzy outputs corresponding to fuzzified inputs, with respect to the fuzzy rules. The designer must specify which implication, conjunction and aggregation operators are used.

Defuzzification produces a nonfuzzy output, using one of the three usual methods : Center Of Area (COA), Max Criterium, or Mean Of Maximum.

There are many different choices within each of these three steps, leading to many different kinds of fuzzy models, see [14] for a review. In this study, we consider the models with :

- the fuzzy rules are all in the form of (1).

- the output membership functions are isocele triangles, gaussian functions or singletons.

- COA defuzzification method (the most commonly-used method) is chosen.

and we consider the set of fuzzy systems which consists of all functions $\Phi : U \subset \mathcal{R}^n \to \mathcal{R}$ of the form

$$\Phi(x) = \frac{\sum_{i=1}^{K} b_i \alpha_i(x)}{\sum_{i=1}^{K} \alpha_i(x)} \quad (2)$$

where $\alpha_i(x)$ is the truth value of Rule i for a given T-norm and b_i is the "center" of B_i, i.e. the point in output space at which the membership function of B_i, μ_{B_i}, achieves its maximum value. If $B_i = \{b_i\}$, i.e. if the fuzzy set B_i is a singleton, then we find Sugeno's method, [19].

2.2 From Fuzzy to Neuro-Fuzzy

As pointed out by several researchers ([2] to [9], [11] to [13], [15], [17]), fuzzy rule-based systems have the underlying structure of a feedforward multilayer neural network[1], with a well defined functionality for each layer, and BackPropagation-like algorithms allow fine tuning of both the membership functions and the parameters in the consequent part of the rules, from a set of input-output data. We can identify a four layered structure, with :

- One input layer feeding the data to the network. The ith input is only connected to the m_i neurons computing the membership values with respect to the fuzzy sets defined in the ith projection of U.

- A first hidden layer, with M Processing Elements (PE), computing the membership degrees with respect to different fuzzy subsets defined on the universe of discourse of each input. The membership functions are triangular in [17], gaussian in [11], [6], sigmoidal in [9]. The outputs are $\mu_{A_i^j}(x_i), j = 1, \ldots, m_i$ and are computed by a real neuron [6], or by a "generalized neuron" [11], [20]. In the first case, the membership function parameters are the synaptic weights.

- A second hidden layer computing the truth value, $\alpha_i(x), i = 1, \ldots, K$, of the fuzzy rules, with one PE (a neuron) for each rule. The chosen T-norm is Min in [17] and [15], product in [9], Lukasiewics's conjunction in [6], or a specific derivable operator in [2], [4], [7].

---

[1] with a judicious choice for parametrisation of the membership functions, of the conjunction and disjunction operators and of the inference method, a fuzzy-rule-based system can be entirely mapped onto an ordinary neural network, [7], [12]. Thus, a NFN is both

1. a Fuzzy System with a neuron-like parametrization for fuzzification, defuzzification and for the sentence connective *And*;

2. an ordinary Neural Network with prior knowledge embedded into the synaptic weights.

- Finally, one output layer giving the system output by (2) or by a more complex way, cf [15]. From (2), we see that the "centers", b_i, of the fuzzy sets B_i are synaptic weights.

This layered structure is shown in Fig. 1, with a two-input-one-output system and two fuzzy sets by input (S for Small and B for Big), for simplicity.

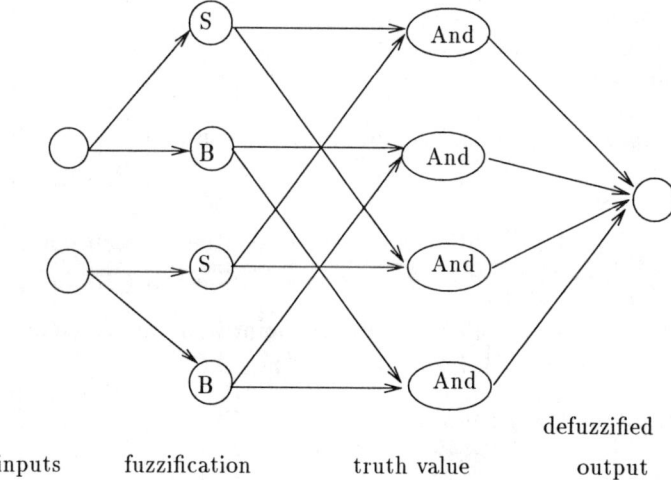

Figure 1: Neural-like Structure of a Fuzzy System

It is convenient to decompose the input-output mapping, Φ, into $\Phi = \Phi_2 \circ \Phi_1 \circ \Phi_0$, where $\Phi_0 : U \to [0,1]^M$ realizes fuzzification, $\Phi_1 : [0,1]^M \to [0,1]^K$ computes the truth values of rules, and $\Phi_2 : [0,1]^K \to \mathcal{R}$ gives the output by (2).

- Φ_0 is a topology preserving mapping, which is, in some cases, injective and/or uniformely continuous. Its parameters define the membership functions for each input. Moreover, the designer may impose constraints about the shape of these membership functions, (e.g. membership functions crossing at the grade 0.5).

- Φ_1 has no adjustable parameters and is entirely defined by the choice of a T-norm.

- Φ_2 is an adaptive Perceptron.

Thus, tuning a NFN consists in tuning two semantically different sets of parameters. To optimize learning, it is quite natural to distinguish between these two sets and to try to tune them specifically. We have developped appropriate learning algorithms for two main purposes :

- To use all the prior knowledge, embedded into the synaptic weights, and to respect possible constraints about the shape of membership functions.

- To avoid BackPropagation-based algorithms, which require excessive computational hardware for on-chip learning, because implementation on an actual neural VLSI chip is now possible, leading to both fuzzy and neural inference by the same hardware [7].

2.3 Initialization

For initialization, we may make the following weak assumptions on the membership functions:

i) The supports of $A_i^1, \ldots, A_i^{m_i}$ form a coverage of the ith input domain, for $i = 1$ to n.

ii) For each fuzzy set, there exists only one point with membership value 1. Let a_i^j be this point for A_i^j, $(\text{Core}(A_i^j) = \{a_i^j\})$.

iii) $\mu_{A_i^k}(a_i^j) = 0$ for $k \neq j$ and $i = 1, \ldots, n$.

These assumptions are easily realized with triangular membership functions crossing at the grade 0.5. Let i be the number of the rule with the premisse

$$x_1 \text{ is } A_1^{j_1} \text{ and } \ldots \text{ and } x_n \text{ is } A_n^{j_n}$$

and we note $\underline{a}_i = (a_1^{j_1}, \ldots, a_n^{j_n})$, $j_k \in \{1, \ldots, m_k\}$

Proposition With these assumptions, we have :

1. With the input vector $x = \underline{a}_i$, only Rule i is fired, with truth value 1. The points $\underline{a}_i, i = 1, \ldots, K$, form a grid on U.

2. $\Phi(\underline{a}_i) = b_i, i = 1, \ldots, K$

Consequences :

1. If we choose a fuzzy partition of each input domain, verifying the previous assumptions, then Φ_0 is entirely determined.

2. Moreover, Φ_2 is entirely determined if we know the desired outputs for the K points $\underline{a}_i$ of the grid.

In this case, prior knowledge is sufficient to determine all the NFN parameters, without learning. In all cases, approximate values for $b_i, i = 1, \ldots, K$, speed up convergence dramatically.

2.4 Training Procedure

We now develop a supervised training procedure to determine conjointly the parameter sets for Φ_0 and Φ_2, given a training set.

1. Fix an initial set of input membership functions and deduce parameters of Φ_0.

2. Compute parameters of Φ_2, using prior knowledge and/or a training method given in Section 3.

3. Compute the NFN error.

4. **If** error $\leq$ given threshold **then** stop
   else

   (a) Modify Φ_0 parameters, using a method of Section 4

   (b) Re-compute the corresponding Φ_2 parameters

   (c) Go to step 3

3 WEIGHT IDENTIFICATION FOR Φ_2

For simplicity, and without loss of generality, we take a two-input-one output network, with fuzzy rules such as :

$$\text{Rule } i : \text{if } x_1 \text{ is } A_1^{j(i)} \text{ and } x_2 \text{ is } A_2^{k(i)} \text{ then } y \text{ is } B_i$$

where $i \in \{1, \ldots, K\}$, $j(i) \in \{1, \ldots, m_1\}$ and $k(i) \in \{1, \ldots, m_2\}$. The function $i \rightarrow (j(i), k(i))$ must be injective. In many cases, operator's knowledge is sufficient to determine a fuzzy partition on the input domains, and, consequently, to deduce Φ_0 parameters. Then, for a given set of fuzzy subsets, we need only to know Φ_2 parameters, i.e. the values in the consequent part of the rules. Let $\mathcal{P} = \{(\mathbf{x}^k, d^k), k = 1, \ldots, P\}$ the training set, where P is the number of training patterns, $\mathbf{x}^k = (x_1^k, x_2^k)$ the kth input pattern, and d^k the corresponding desired output. In this case, the NFN actual response is given as follows, cf eq. (2) :

$$\Phi(\mathbf{x}^k) = y^k = \sum_{i=1}^{K} \frac{\alpha_i^k}{\sum_j \alpha_j^k} b_i \qquad (3)$$

where α_i^k is the truth value of the ith rule for input $\mathbf{x}^k$ for a given T-norm[2], thus $\frac{\alpha_i^k}{\sum_j \alpha_j^k}$ is the relative firing strength of rule i.

In this section, we consider the NFN error, for a given Φ_0 and the pattern $\mathbf{x}^k$:

$$E(\mathbf{x}^k|\Phi_0) = \frac{1}{2}[y^k - d^k]^2 \qquad (4)$$

and the problem is to minimize $\sum_{k=1}^{P} E(\mathbf{x}^k|\Phi_0)$. We have investigated the three following methods : Gradient Descent, Least Square Method and Recursive Least Square Method.

3.1 Gradient Descent

The NFN adjustable parameters are the "centers" $b_i, i = 1, \ldots, K$, of the output fuzzy sets B_i. In gradient descent, these weights are updated by :

$$\Delta b_i = -\epsilon \frac{\partial E(\mathbf{x}^k|\Phi_0)}{\partial b_i} \qquad (5)$$

where ϵ is the learning rate. From equation (2), we have immediatly the expression of $\frac{\partial E(\mathbf{x}^k|\Phi_0)}{\partial b_i}$. A momentum term, η, speeds up the tuning. The weights are updated as follows :

$$\Delta b_i(t) = -\epsilon[y(t) - d(t)]\frac{\alpha_i}{\sum_j \alpha_j} + \eta \Delta b_i(t-1) \qquad (6)$$

The weights are modified proportionally to the network errors and the relative firing strength of the rules.

---

[2]for instance :

$\alpha_i^k = Min(\mu_{A_1^{j(i)}}(x_1^k), \mu_{A_2^{k(i)}}(x_2^k))$ Minimum operator

$\alpha_i^k = \mu_{A_1^{j(i)}}(x_1^k) \cdot \mu_{A_2^{k(i)}}(x_2^k)$ Product operator

$\alpha_i^k = \frac{1}{1+exp[-\beta(\mu_{A_1^{j(i)}}(x_1^k)+\mu_{A_2^{k(i)}}(x_2^k)-1.5)]}$ Smoothed Lukasiewicz's operator [7]

3.2 Least Square Method

With this method, we have to find the optimal vector $\tilde{\mathbf{b}}^* = (b_1^*, \ldots, b_K^*)$ (where the symbol $\tilde{}$ designates vector tranposition) such as :

$$d^k = \sum_i \frac{\alpha_i^k}{\sum_j \alpha_j^k} b_i^* \qquad (7)$$

i.e., $\mathbf{b}^*$ is the solution (if it exists) of the linear system of P equations and K unknowns, denoting $\tilde{\mathbf{d}} = (d^1, \ldots, d^P)$ and $\mathcal{A} = \text{Matrix}(\frac{\alpha_i^k}{\sum_j \alpha_j^k})$:

$$\mathbf{d} = \mathcal{A}\mathbf{b}^* \qquad (8)$$

Therefore, the solution, in the sense of Least Squares, is :

$$\mathbf{b}^* = (\tilde{\mathcal{A}}\mathcal{A})^{-1}\tilde{\mathcal{A}}\mathbf{d} \qquad (9)$$

This method can be used off-line, in order to provide a good initialization of the synaptic weights.

3.3 Recursive Least Square Method

The computation of the optimal vector, $\mathbf{b}^*$, can be made recursively, on-line, updating the weights after each input pattern $\mathbf{x}^k$. For simplicity, the subscript k is omitted. Let

$$\begin{aligned}
\tilde{\alpha} &= (\tfrac{\alpha_1}{\sum \alpha_i}, \ldots, \tfrac{\alpha_K}{\sum \alpha_i}) \quad \text{Relative firing strength vector} \\
\tilde{\mathbf{b}}_i &= (b_1, \ldots, b_K) \quad i^{th} \text{ approximation of } \mathbf{b}^* \\
\mathcal{P}_i &= (\tilde{\mathcal{A}}_i \mathcal{A}_i)^{-1}
\end{aligned}$$

The $(i+1)^{th}$ estimate of $\mathbf{b}^*$ is given by :

$$\mathbf{k}_{i+1} = \mathcal{P}_i \alpha (1 + \tilde{\alpha}\mathcal{P}_i\alpha)^{-1} \qquad (10)$$

$$\mathbf{b}_{i+1} = \mathbf{b}_i + \mathbf{k}_{i+1}(d - \tilde{\alpha}\mathcal{P}_i) \qquad (11)$$

$$\mathcal{P}_{i+1} = \mathcal{P}_i - \mathbf{k}_{i+1}\tilde{\alpha}\mathcal{P}_i \qquad (12)$$

for initialization we can take a first approximation of $\mathcal{P}$ and $\mathbf{b}$ by the Least Square Method or more simply by taking :

$$\begin{aligned}
\mathbf{b}_0 &= 0 \\
\mathcal{P}_0 &= \lambda I \quad I = \text{unity matrix and } \lambda = \text{"big" (e.g. } \lambda = 10^6)
\end{aligned}$$

The computations are not complex and do not need matrix inversion. The term $(1 + \tilde{\alpha}\mathcal{P}_i\alpha)$ is a scalar.

4 OPTIMISATION OF MEMBERSHIP FUNCTIONS

Two methods have been investigated : first order approximation of gradient, and random optimization. A third method, using Genetic Algorithms, is currently in progress.

4.1 First Order Approximation of Gradient

This method applies only if Φ_0 is differentiable (e.g. with gaussian or sigmoidal membership functions and a differentiable T-norm). It is a gradient descent with direct approximation of the gradient by weight perturbation. This method is used for tuning only the parameters of membership functions. We note w_i^j these parameters. If δw_i^j is a small enough perturbation, a first order approximation of gradient is given by :

$$\frac{\partial E}{\partial w_i^j} \approx \frac{E(w_i^j + \delta w_i^j) - E(w_i^j)}{\delta w_i^j} \qquad (13)$$

where $E()$ is the mean square error produced by the NFN on the learning set. This method only requires forward relaxations of NFN and therefore is suitable for VLSI implementations. It has been used in [10] for all the parameters of a Neural Network. Here, we apply it only for Φ_0 parameters. Thus, the number of parameters to be tuned is greatly reduced with regard to the complete network and the computational cost is reduced : we generally have two parameters per membership function, therefore, for M fuzzy subsets, we only have to tune $2M$ parameters by such a method, whereas $2M + K$ are needed in usual methods.

Weight Perturbation Algorithm:

    For each pattern
        Forward Pass
        Compute Error E
        For each Φ_0 parameter
          Weight Perturbation
          Forward Pass
          Compute Error E^*
          Compute Δw_i^j
        Update Φ_0 parameters
        Forward Pass
        Update Φ_2 parameters

We applied this method to the well-known Rosenbrock Valley Problem. A two-dimensional expression of this function is :

$$z = 100(y - x^2)^2 - (1 - x)^2 \tag{14}$$

After a first choice of gaussian input membership functions ([-1,1] was divided into five equal parts), Φ_2 parameters were identified during ten epochs, using a training set of 200 patterns randomly chosen in [-1,1]. Then, Φ_0 parameters were identified by the Weight Perturbation Algorithm. Results are shown in Fig. 2.

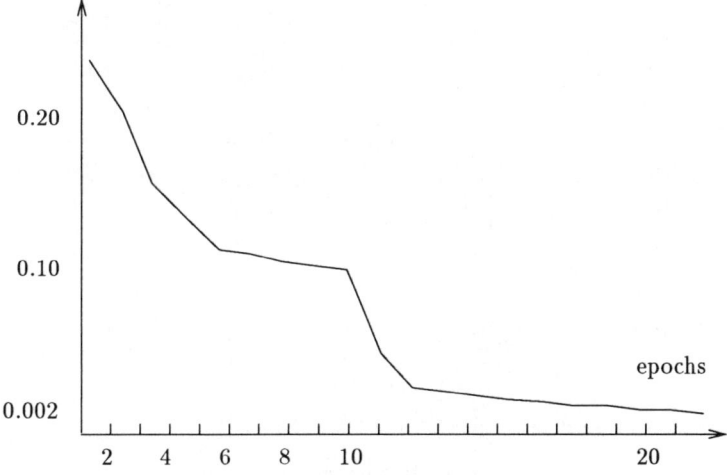

Figure 2: Learning Curve for Rosenbrock's Valley

4.2 Random Optimization

The random optimization method proposed by Matyas [16], is a random walk in a weight space, $X \subset \mathcal{R}^M$. It ensures convergence of an objective function to a global minimum with probability 1 if X is a compact set. An improved method, by Solis and Wets [18], introduces a bias vector which modifies the random walk, exploiting previous results. This method has been used in [1] for learning the weights of Neural Networks. Here, we apply the method only for optimization under constraint of input membership functions. Let $\tilde{\mathbf{w}} = (w_1^{m_1}, \ldots, w_n^{m_n})$ be a Φ_0 parameter vector, $\mathcal{X}$ the set $\{\mathbf{w} \in X$ and $\mathbf{w}$ verifies some given constraints$\}$, $\mathbf{b} \in \mathcal{R}^M$ a bias vector, $\mathbf{g} \in \mathcal{R}^M$ a gaussian random vector and $E : X \to \mathcal{R}^+$ the error function to be minimized.

Random Optimization Algorithm (Solis and Wets, 1981):

1. Select an initial vector $\mathbf{w}^{(0)}$ from prior knowledge, let $\mathbf{b}^{(0)} = 0$ and $k = 0$.

2. Generate a gaussian random vector, $\mathbf{g}^{(k)}$. **If** $w^{(k)} + g^{(k)} \in \mathcal{X}$ **then** go to step 3, **else** go to step 4.

3. **if** $E(w^{(k)} + g^{(k)}) < E(w^{(k)})$
    then $w^{(k+1)} = w^{(k)} + g^{(k)}$ and $b^{(k+1)} = 0.4 g^{(k)} + 0.2 b^{(k)}$
    else if $E(w^{(k)} - g^{(k)}) < E(w^{(k)})$
       then $w^{(k+1)} = w^{(k)} - g^{(k)}$ and $b^{(k+1)} = b^{(k)} - 0.4 g^{(k)}$
    else $w^{(k+1)} = w^{(k)}$ and $b^{(k+1)} = 0.5 b^{(k)}$

4. **if** $k >$ MaxNumberIteration or error $<$ threshold
    then stop
    else $k = k + 1$; go to step 2

In a multiprocessor computer, step 3, which makes two forward passes, can be parallelized :

- processor 1 : compute $E(w^{(k)} + g^{(k)}) - E(w^{(k)})$
- processor 2 : compute $E(w^{(k)} + g^{(k)}) 2 E(w^{(k)})$
    choose the best result
    etc ...

5 EXAMPLES

5.1 Non Differentiable NFN

We take a NFN with triangular membership function and any T-norm. Φ is not differentiable, but we can apply the Random Method of Section 4.2 to identify the parameters of Φ_0 and, for instance, the Gradient Descent of Section 3.1 for the parameters of Φ_2, given a learning set. The parameters determining a triangle are the center, c, the left width, l, and the right width, r, cf Fig. 3. For an isoscele triangle, $l = r$, and we can have l or $r = \infty$. We take, for simplicity, three fuzzy subsets for each input, S for Small, M for Middle and B for Big, with the parameters : $(l_i, c_i, r_i)_{i=1}^3$, with $l_1 = r_3 = \infty$.

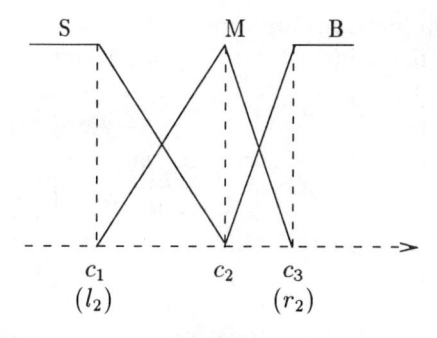

Figure 3: Triangular Membership Functions Crossing at the Grade 0.5

Then, for each input, a set of three fuzzy sets crossing at the grade 0.5 is completely determined by three independant parameters $(c_i)_{i=1}^3$, with

$$c_i = c_{i-1} + l_i = c_{i+1} - r_i \qquad (15)$$

$$c_1 < c_2 < c_3 \qquad (16)$$

From the $(c_i)_{i=1}^3$ parameters and the two previous equations, we can rebuild Φ_0. Therefore, we can apply the Random Optimization method in the following manner :

- produce a new vector $(c_i)_i$ for each input
- build the corresponding NFN
- apply the method
  Update $(c_i)_i$ for each input
  Update Φ_2 parameters

5.2 Differentiable NFN

We define the NFN in the following manner :

- The And operator is a neural smoothed form of the Lukasiewicz's And [7].

$$\alpha_i^k = \frac{1}{1+exp[-\beta(\mu_{A_1^{j(i)}}(x_1^k)+\mu_{A_2^{k(i)}}(x_2^k)-1.5)]}$$

with the notations of Section 2.

- The membership functions are expressed by two sigmoids, a sigmoid being defined by two parameters c (for "center") and l (acting as the half left width), see Fig. 4 :

$$sigm_{c,l}(x) = \frac{1}{1 + \exp -\alpha \frac{x-c-l}{l}} \qquad (17)$$

with

- α is computed so that the residual value of the sigmoid at x = c - 2l is a "small" ϵ ($\alpha = \text{Log } \frac{1-\epsilon}{\epsilon}$)
- $sigm_{c,l}(c) = 1 - \epsilon$ and $sigm_{c,l}(c - 2l) = \epsilon$

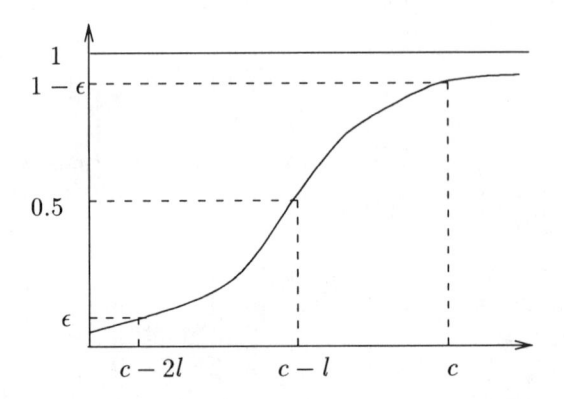

Figure 4: Parametrization of a Sigmoid

As in the previous subsection, we define three membership functions S, M and B by input :

$$\mu_S(x) = 1 - sigm_{c_1,l_1}(x) \qquad (18)$$

$$\mu_M(x) = sigm_{c_1,l_1}(x) - sigm_{c_2,l_2}(x) \qquad (19)$$

$$\mu_B(x) = sigm_{c_2,l_2}(x) \qquad (20)$$

With the constraints $c_i = c_{i-1} + 2l_i$, for i=1 to 3, we only have 3 parameters by input to optimize with the Random Optimization Method or the Weight Perturbation Method. The parameters of Φ_2 are optimized by Gradient Descent or a Least Square Method.

6 CONCLUSION

NFN exhibits a massively parallel and layered feedforward structure, with a different semantic for each layer. Prior knowledge can be embedded into the synaptic weights, reducing the training time dramatically. The global parameter set can be broken into two subsets, with appropriate learning algorithms for each subset. Finally, since the proposed algorithms avoid backward passes, implementation on neural-like VLSI chips is easier.

References

[1] Baba N., "A new approach for finding the global minimum", *Neural Network*, Vol 2, 1989.

[2] Berenji H.R., Khedkar P., "Fuzzy Rules for Guiding Reinforcement Learning", *Proc. of Int. Conf. IPMU'92*, Mallorca, Spain, 1992.

[3] Brown M., Harris C., " A nonlinear Adaptive Controller", *IMA J. of Math. Control*, 1991.

[4] Feldkamp L.A., Puskorius G., " Architecture and Training of a Hybrid Neural-Fuzzy System", *Proc. of Iizuka'92*, Iizuka, Japan, 1992.

[5] Glorennec P.Y., "Adaptive Fuzzy Control", *Proc. of 4th IFSA congress*, Brussels, 1991.

[6] Glorennec P.Y., "Un réseau neuro-flou évolutif", *Proc. of Neuro-Nimes'91*, Nimes, France, 1991.

[7] Glorennec P.Y., "A neuro-fuzzy inference system designed for implementation on a neural chip", *Proc. of Iizuka'92*, Iizuka, Japan, 1992.

[8] Hayashi Y., Czogala E., Buckley J., "Fuzzy Neural Controller", *Proc. of FUZZ-IEEE'92*, San Diego, 1992.

[9] Horikawa S.I., Furuhashi T., Okuma S., Uchkawa Y., "A fuzzy controller using a neural network", *Proc. of Iizuka'90*, Iizuka, Japan, 1990.

[10] Jabri M., Flower B., "Weight Perturbation : An Optimal Architecture and Learning Technique for Analog VLSI Feedforward and Recurrent Multilayer Networks", *IEEE Trans. on Neural Networks*, Vol. 3, No. 1, Jan. 1992.

[11] Jang J-S.R., "Rule extraction by generalised neural Network", *Proc. of 4th IFSA congress*, Brussels, 1991.

[12] Kawamura A., Watanabe N., Okada H., Asakawa K., "A Prototype of Neuro-Fuzzy Cooperation System", *Proc. First IEEE Conf. on Fuzzy Systems*, pp. 1275-1282, San Diego, 1992.

[13] Keller J., Yager R., Tahani H., "Neural network implementation of fuzzy logic", *Fuzzy Sets and Systems*, Vol. 45 no. 1, jan. 92

[14] Lee C.C., "Fuzzy Logic in Control Systems, parts I and II", *IEEE Trans. on SMC*, Vol. 20, No. 2, March/April 1990.

[15] Lin C.T., lee C.S., "NN-based fuzzy logic control and decision system", *IEEE trans on computers*, Vol. 20, No. 12, dec. 1991.

[16] Matyas J., "Random Optimization", *Automation and Remote Control*, Vol. 26, pp. 246-253, 1965.

[17] Nomura H., Hayashi I., Wakami N., "A self-tuning method of fuzzy control", *Proc. of 4th IFSA congress*, Brussels, 1991.

[18] Solis F., Wets J., "Minimization by random search techniques", *Mathematics of Operation Research*, Vol. 6, 1981.

[19] Sugeno M., "An introductory survey of fuzzy control", *Information Sciences*, Vol.36, pp. 59-83, 1988.

[20] Wang L.X., "Fuzzy systems are universal approximators", *Proc. First IEEE Conf. on Fuzzy Systems*, pp.1163-1169, San Diego, 1992.

Chapter 2
Towards a Unified Theory of Intelligent Autonomous Control Systems

Introduction, *20*
State of the Art and Future Challenges, *20*
Adaptive Systems and the Dual Control Problem, *22*
A Systemic Approach to Behavioural Specification of Intelligent Systems, *25*
Behavioural Specification of the Activities of a System of Interacting Actors, *28*
Fuzzification of Formal Models of Adaptive Autonomous Control Systems, *32*
Fuzzy Identification, *36*
Introducing Control Hierarchies, *38*
From Abstract Logic Designs to Technological Realizations, *45*
References, *48*

To make a significant advance in the field of intelligent autonomous control systems requires an integrated approach exploring adequately the advances of the disciplines relevant to this topic. This chapter provides foundations for descriptive and computational approach that puts on equal footing the mathematical control-theoretical part and the AI-based symbolic part needed for such a unification. An intelligent system consisting of multiple centers can be viewed as *a community of actors* placed in a specific environment, with each actor having the *disposition* for communication with other actors and for coordination of activities with respect to some rules of that guaranteed correct interaction. The basic construct *formally representing* a cognitive/action model of an agent is a pair of mathematical structures called *abstract logic*, consisting of an abstract algebra and a generalized closure system. These are used to represent processes realizing the activities of actors, the dynamics of the environments, and the dynamics of interaction of actors and environments. To capture uncertainties of the real world, the structures are fuzzified. Formal logic representations of hierarchical autonomous systems are supplemented by the guidelines of how these can be realized in modern technology and further references to the literature are appended.

TOWARDS A UNIFIED THEORY OF INTELLIGENT AUTONOMOUS CONTROL SYSTEMS

Ladislav J. KOHOUT

Department of Computer Science B-173 and The Institute for Cognitive Sciences, Florida State University, Tallahassee, Florida 32306, USA.

1 INTRODUCTION

To make a significant advance in the of field Intelligent Autonomous Control Systems requires an integrated approach that would explore the advances of various disciplines relevant to this topic. Such a unification requires a formal descriptive and computational approach that would put on equal footing the mathematical control-theoretical part and the AI based symbolic part. The aim of this paper is to contribute towards the development of formal techniques as well as of a methodological foundation that could deal with the outlined challenges. The starting point of this endeavour is to abstract the relevant concepts and logical forms from the field of adaptive control and combine these with adequate formalisms based on fuzzy relational methods that have already been applied in Knowledge Engineering and are backed by powerful algorithms supporting fuzzy relational computations.

2 STATE OF THE ART AND FUTURE CHALLENGES

2.1 Three views of what are the most important robotic issues

Let us briefly look at three different views of robots and their functions: the view of a control engineer involved in robotics, the view of a computer scientist engaged in applying symbolic AI in robotics, and the view of a designer of a Computer Aided Manufacturing (CAM) system. We shall see that each view emphasises different aspects of robotics and presents different problems and challenges to the field. All three viewpoints have to be addressed adequately by any theory or methodology claiming to be capable of integrating these three domains in a unified framework.

A control engineering viewpoint. A robot is a programmable device with varied degree of adaptivity, capable of sensing and actively acting upon its environment. Advanced robot systems are hierarchically structured. Each level of this hierarchy usually forms a subsystem composed of a task decomposition block, a world model block and a feedback block [1]. This is a typical structure of a robot manipulator [2].

Computer science and AI viewpoint. The hardware of a typical robot can be split into sensors and actuators, special-purpose controllers and general-purpose computers. A robot possessing this hardware structure should be different from the special-purpose devices that were typical of automation of the 1960s in that they are required to handle a range of industrial tasks with no change in hardware and only modest changes in programming. This requires considerable symbolic problem solving ability. The activities that such a robot conducts have to deal with the objects and processes of the robot's environment in terms of adequate generality. A fitting example can be quoted [3]: "you program the robot to understand the assembling and disassembling in somewhat general terms, and then have the robot figure out how to solve the particular problem of changing the oil filter on a 1987 International Harvester model 4A by applying its general assembly/disassembly knowledge". This of course requires application of some AI techniques.

Computer aided manufacturing viewpoint. Automated manufacturing systems involve integration of management and control aspects in distributed large scale systems, which require decision architectures for event-driven complex dynamic systems including robots. These have to be integrated in a unified framework for solving production and planning problems in an automated manufacturing environment. The complexity of the systems dictate decentralization of the control. Occurrences of unknown events enforce the requirement of control adaptivity. The architectures developed in the framework of optimal control theory have not yet offered such a combination of features.

2.2 Some Pros and Cons of the Available Methodologies

Optimal control theory provides a mechanism for selecting from all the available trajectories of a dynamic system those that optimize a given criterion, a functional on the space of all such trajectories. Classical optimal control theory implicitly assumes that:

1. There exists a central agent that controls the system.

2. The agent is an ideal anticipatory agent, i.e. it has a perfect knowledge of the future (this is implied by assuming the knowledge and global applicability of the optimization criteria).

3. The optimal trajectories are chosen once and for all at the commencement of the system's activity.

A problem appears if such a functional is not applicable to all the available trajectories. In that case the space of optimal solutions is only partial. Another problem appears, if the functional, or even the mathematical model of the controlled system, are not available, or are only approximate. This necessitates the use of some adaptive features as well as some kind of fuzzy approximation technique.

The symbolic problem solving ability of the AI approach requires an integration of adequate software engineering methodologies to deal with the integration of hardware and software requirements. These methodologies should, however, be compatible with control engineering techniques. This is not the case at present.

With respect to the goals of computer aided manufacturing, pros and cons of current approaches are best demonstrated by contrasting two different important design methodologies of the field. The "intelligent control" of Saridis [4],[5] and "knowledge-based control" of Meystel [6] are as design methodologies both directed towards obtaining a multi-layer control architecture that is directly originated by the system to be controlled, and then satisfies the system's integrator requests. The former method is mainly oriented towards analysis of the integration of the participating modules. The latter, however, primarily deals with organization of control modules and formal representation of their tasks. Industrial control problems require a good *balance between theoretical formalizations* of problem that are used as mathematical models of the controlled plant that assure theoretically optimal properties of the proposed solutions, and the *constraints* that can be handled with *difficulty* in mathematical terms. Villa originated Hybrid Control System (HCS) methodology and architecture [7] which should provide a better balance between formal methodologies and knowledge-based reasoning. The control engineering part of Villa's method is based on formal mathematical approaches while the symbolic AI part is mainly heuristic.

To make a significant advance in the field requires a formal approach that would put on equal footing the mathematical control-theoretical part and the AI based symbolic part. In particular, in CAM systems, tasks of information retrieval in the system information management part and the tasks of control engineering part strongly interact. This in turn requires a unified approach that can successfully clarify the ways of these interactions. The aim of this paper is to contribute towards the development of a methodological foundation that could deal with the outlined problems. The starting point of this endeavour is to abstract the relevant concepts and logical forms from the field of adaptive control and combine these with adequate formalisms based on fuzzy relational methods.

3 ADAPTIVE SYSTEMS AND THE DUAL CONTROL PROBLEM

3.1 Identification vs. Optimization in Changing Environments

The optimal control approaches have substantial merits, but also some drawbacks when viewed from the perspective of adaptive systems. Optimal Control Theory provides mechanisms for selecting trajectories in the state-space of the control system that optimize a given criterion represented by a functional on the space of such trajectories. This presupposes having a *precise mathematical model* of the control plant that is used by the optimal controller. If the controlling agent takes actions that *change the structure* of the controlled plant or its environment, then the control problem, in order to be fully described also requires the introduction of a model that captures interaction of the plant with the environment. In some cases this might not be enough and a mathematical model of *possible actions* of the controlling agent has to be added as well.

An autonomous controller might perform actions that change the structural parameters of the plant the controller is interacting with. This causes mismatch with the fixed mathematical model that the controller would have to possess. This in turn introduces the *identification problem*. For example, in the case of a continuously distributed system, this is reflected in the time-dependent changes of coefficients of the partial differential equations that describe such a system. If there is no mechanism available that performs the update accordingly, the model that the controller possesses would soon become obsolete, as the actions modifying the structural properties of the environment gradually take effect. *Repeated attempts at identification* of the new properties of the control plant and its environment have to be made by the controller leading to what is called 'identification problem'.

In the early development of adaptive control theory the practice was to split the overall control problem into two parts: system *identification* problem and system *optimization* problem. The former was concerned with determining the characteristics of the system to be controlled (plant), the latter determining the best controller with respect to some performance criteria. This separation lead to powerful theories of identification and optimization being developed separately from one another. Unfortunately, in any really adaptive control situation with a versatile enough controlling agent which can change the structural parameters of the controlled system or its environment such a separation is impossible. Even for such a simple problem as is the control of a uncomplicated linear system, it has been shown by Sworder [8] that the optimal identification control coupled to the optimal control algorithm does not lead to the optimal controller. In more complex systems, there may be a conflict between the requirement to control the system and the requirement to identify the relevant characteristics of the system in order to control it optimally. This so called "dual control problem" has been extensively analyzed by Feldbaum [9], [10], who succintly summarized it:

> In dual control systems, there is a conflict between the two sides of the controlling process, the investigational and the directional. An efficient control can only be effected by a well-timed action on the object. But the control can only be effective when the properties of the object are sufficiently well known. One needs, however, more time to become familiar with them".

Although Sworder and Feldbaum were mainly concerned with optimizing an overall performance criterion using an integrated strategy, similar problems also hindered the attempts to combine *independent* identification and optimization algorithms into a *single learning algorithm*. Gaines [11] pointed out:

In general, it is clear that overall optimization is not possible, and there will always be a conflict between strategies designed to learn about the environment, and strategies designed to control it... One obvious implication of this consideration is that there can be no *universal* learning machine, for example, one that is eventually as good or better in its steady-state performance than any other uniformly over all tasks. Different machines, with different relationship between exploration and control, have different 'personalities', and a machine temperamentally suited to one environment will totally be unsuited to another. It is from consideration of the relationship between types of environment relative to types of machine that much progress may be expected in the understanding of learning systems in the future.

3.2 The Need for Hierarchies in Adaptive Control Systems

The fundamental building block of most forms of adaptive controller is a *two-level hierarchy* in which the lower level implements one of the class of control policies, whilst the upper level selects the class to be implemented. In the case of multilevel control, this pair of units is replicated at various levels of abstraction within the hierarchical or heterarchical control structure [7], [12]. Similar twin structures (of virtual machines) appear in the abstract schemes describing adaptive computer architectures ([13], section 5.6). Also in the schemes capturing adaptive behaviour of neuro-physiological centres of the brain, this building block of twin structures plays an important role in the teleological explanation of dynamics of various activities and of their interaction in the normal and pathological situations [14],[15],[12].

It has been emphasised by Gaines [11] that this conceptualization of adaptive control based on 2-level building primitives as described above, introduces relativity into the definition of the term 'adaptive' from a structural point of view. This is so because any particular controller may be split up in many ways according to the definition of a class of possible control policies; a similar relativity in the behavioural definition of adaptive has been noted in [16] due to the variety of possible ways in which an interaction may be partitioned into 'tasks'. Gaines further suggests [11] that "the binary relations over tasks, generated by consideration of a controller's adaptive behaviour, may be used as the basis for a taxonomy of environments according to the problems involved in adapting to them".

This method can be reversed, starting from a family of environments or contexts in which an adaptive system is to operate and using this for the explanation of adaptive behaviour of various substratum structures (e.g. physiological centres of the brain [17]). It can also be used to advantage in the design of functional structures of behaviour from which the substratum of the adaptive multi-level controller can be synthesised [18]. This also carries over into the design of multi-environment KBS and other adaptive computer architectures [19], [12], thus indicating that a unified methodology is possible. The foundations on which such a methodology can be based, are outlined in the sections that follow.

4 A SYSTEMIC APPROACH TO BEHAVIOURAL SPECIFICATION OF INTELLIGENT SYSTEMS

4.1 A Conceptual Framework for Describing and Synthesizing Actions of an Intelligent System

In a very general way, an artificial or natural Intelligent System [12] composed of a *substratum* (e.g. hardware, brain) consisting of multiple centres, can be viewed as a *community of actors* placed in a specific environment (Kohout [20]). Each actor has to have the *disposition* for communication with other actors as well as for co-ordination of activities, and has to adhere to some rules of behaviour that guarantee correct interaction.

If a particular family of neural or other modules is taken as such an actor, a suitable methodology capable of dealing with the synthesis of behaviour of artificial actors becomes the corner-stone of the approach. For this purpose a methodology of Activity Structures has been developed [13],[12]. Adopting this methodology we have at our disposal a design approach which is powerful enough to reconcile the diversities of various design methods as well as the differences in the implementation technology of individual centres. Thus, our Activity Structures based unified approach can be used to deal with different types of implementation substrata, e.g. with neural nets modules, the conventional AI modules based on discrete (digital) symbolic computations, and also with analogue and/or digital conventional or adaptive control engineering modules. The method allows us capture the *mutual interactions* of different types of modules. Below, we outline the basic concepts that are used in this application of Activity Structures methodology.

Actors and Agents   An autonomous organism is a system or subsystem of type "agent" which must exhibit behaviour that possesses some basic competence attributes in order to belong to this class. This basic competence consists of fulfilling the following basic aims (Kohout [12],[21]):

a) The aim of stable existence or more generally of surviving;
b) The aim of optimising its performance in a particular environment or a class of environments.

In order to succeed in (a), (b) the autonomous organism has to possess the aptitude for controlling its own behaviour as well as the aptitude for symbolic communication [15].

In the sequel, the term 'actor' will denote an agent with the dispositions

1. *communicate*,

2. to *co-ordinate* actions and

3. to adhere to some rules of behaviour that *guarantee correct interaction.*

An actor has to possess the aptitude to adapt and learn in order to operate in a stable mode and survive functionally or indeed stucturally, in an environment that is not constant but changes its characteristics. The aims of an actor, agent or a complete autonomous system are realised by reaching appropriate final goals. That can be achieved by successfully performing appropriate *tasks*. A task is defined as a concurrent family of activities that are generated by appropriate families of concurrent processes (Kohout [22],[21],[12]). In order to guard against undesirable side effects of concurrent activity, some protection rules are essential. A structure ensuring dynamic protection becomes necessary for achieving correct stability of the inferential activities and of decision making when dealing with a dynamically changing world.

Further, it is not assumed that the aptitudes (a), (b), of the actors/modules composing an Intelligent System are universal, but it has to be assumed that these are effective only in some environments [23]. The consequence of this assumption is that the competence of agents/modules *with respect* to a particular **class of environments** has to be distinguished. This is in accordance with all the current experience of humankind with the known adaptive life forms as well as with the existing technological artifacts.

In a knowledge-based system intended for computer-aided manufacturing, each environment corresponds to a different industrial or manufacturing process to be controlled. In robotics, again, each modality of the robot's activity represents a separate environment and is also executed in a multiplicity of contexts. The fact that multiple environments and context are involved [23] necessitates introducing the notion of relevance models. The intentional use of cognitive models by the actor/agent requires the use of the notion of *purpose*. The domain of applicability of a particular cognitive model is then determined by the pair <Purpose, Relevance>. (cf. Kohout [15],[24]).

An actor, module or a centre of a specific activity interacts with its environment. This environment consists of the relevant part of the world in which the actor is placed and of other actors that mutually interact. For description of this scenario, it is essential to represent adequately the general logical structure as well as the particular computational mechanisms (processes), by which reasoning over the inner representation of the relevant parts of the external world is performed by the actors. This can be achieved by a detailed elaboration of the meta-structure that captures the situations of the world together with their image in the cognitive models which the actor creates in order to represent the relevant features of the world (Kohout [12],[20]). Also the nature, character and dynamics of the intentional acts of the agent/actor need to be described. This topic is further discussed in the sequel.

Interaction of an agent with its environment is mediated by means of processes dealing with afferent and efferent activities. These processes are executed within the actor's substratum (hardware or software virtual processors). *Afferent activities* use some input information and utilise its relevant portions by appropriate filtration; *efferent activities* realize a multitude of environment-directed intentional acts (e.g. actions, communications, etc.). Thus the perceived reflection of the external world is decomposed into a generating family of relevance cognitive models [20], [25],[12] within the system's information processing structures factored by <PURPOSE, RELEVANCE>, thus generating cognitive heterarchies. For further details see Section 8 below. This also requires introducing finer notions of hierarchies or heterarchies of knowledge representation and control centres within the structure of an autonomous intelligent system, as discussed in the sequel.

Issues of concurrency of processes and tasks   An actor/module is not usually isolated, but is an actor in *interaction*. This interaction may be co-operative and/or competitive, involving other agents/actors as well. The requirement for an actor, to consider and deal successfully with the activities and multiple interactions of agents/modules surrounding it, leads to the necessity of introducing into the overall framework an additional systemic construct, a *norm* that would specify what is considered to be successful interaction with other active agents.

Synthetizing concurrent activities of a collection of actors is concerned with sharing information, with many units producing and consuming information. This sharing can be co-operative, competitive or mixed, having both the former and the latter aspects. More specifically, it can be seen that in an Intelligent System with distributed structure, a task can be decomposed into a family of functional [15], mutually interacting processes. These functional structures are intensional descriptions of the activities of the substratum modules [12] of which the system realising a given task is composed.

Interaction of processes which can be co-operative or competitive, requires that the participating processes strictly adhere to some prescriptive (imperative) rules of behaviour that guarantee the correct form of interaction. To deal with the activities of this kind in a systematic way, the theory and methodology of Activity Structures can be used (Kohout [13],[21],[24],[12]). In this methodology, a careful distinction between the purpose, aim and goal was made, and the concept of protection and intention structures was introduced.

Activity Models of the World in which an Agent Operates   An appropriate activity model of the world in which the agents act is required in order to achieve the *correct* synthesis of the activities of a particular actor. An actor behaves in a *correct way*, if it selects the bf adequate intentions and executes *correct actions*. Use of this model is conditioned by the relationship of the knowledge of the present state with the predicted future dynamic state of the world in which the actor might act. An action oriented cognitive model the actor uses for acquiring knowledge about dynamically changing world has to be included in the total activity model of that actor. This model ought to accommodate not only information concerning the activities of other actors/agents, but should also contain information about the consequences of activities of the very actor that possesses and uses this model [26],[12].

The descriptive apparatus of the model should allow for capturing not only the relevant systemic structures but also the dynamics of their change. The later is indispensable for depicting adaptivity and learning of the actor/module and time-variability of the relevant environments.

5 BEHAVIOURAL SPECIFICATION OF THE ACTIVITIES OF A SYSTEM OF INTERACTING ACTORS

Each actor/module exhibits individual activities, using these for its own purpose. These activities, however, may also have a specific purpose in the wider context of the whole distributed system. In order to perform adequately the activity, the actor must extract the relevant input information which in turn will be used to direct subsequent future activities. Thus the following problems arise:

1. How to describe a goal oriented activity of an agent/actor/module in a way that will be useful for the behavioural synthesis.

2. How to describe the total activity of the whole distributed system together with the activities of its environments, and the effects of their interaction.

3. How to make the description of (2) sufficiently refined, so that it would capture all the relevant aspects of mutual interaction of agents, but leave out all irrelevant details.

In order to deal with problem (1) *task descriptors* were introduced [21],[22] (see below, Sec. 8.1). Problems (2) and (3) can be dealt with by means of structures called *Abstract Logics* [27],[20] which combine the description by algebras, of the structures generating behaviour [21],[12] with the description of dynamic consequences of this behaviour within the environment by means of generalized topologies [28],[12].

5.1 Models of Actors interacting with their Environments

In order to construct efficient artificial actors, it is essential to depict adequately not only the general logical structure and particular computational mechanisms (processes) realizing the activities of the actors, but also the dynamics of the **environments**, and the **dynamics of interaction** of the actors and environments. This can be achieved by detailed elaboration of meta-structures that can capture the situations of the world and their reflections in the cognitive models which the actor uses to represent relevant features of the world. The meta-structure must also model the conceptual nature and the dynamics of the intentional acts of the actor within the context of the environments. For this purpose Abstract Logics are used, composed of a systemic structure W and a consequence structure (Kohout [20],[21],[12]). The systemic structure consists of object-aggregates W and events operating on the object-aggregates. The object-aggregates themselves are some general relational structures, and the operations are generalised morphisms (bundles of operators). The consequential structure consists of closures of the carrier W induced by the actions of the algebra-model. As this abstract representation can capture both a continuous and discrete topology, it can unify in a coherent framework the discrete symbolic and the continuous analogue, control structures. This conceptual system for dealing with multiple-agents has been supported by the formal apparatus of considerable expressive power, that permits not only formal mathematical development, but is also well suited for direct computer representation (Kohout [13],[12]).

In order to realize the full descriptive potential of the formal apparatus we shall examine the nature of the actors' inputs. The agent's cognitive activity contains inductive and abductive acts employed to facilitate successful execution of an action or activity in the agent's environment. These agent's intentional acts ought to be considered within an adequate conceptual framework. Thus we introduce an **Intentional Framework for Directing a Purposeful Activity** (IFDPA) which consists of (Bernštejn [26], Kohout [12]):

1. perception and evaluation of the situation;

2. determining what the situation must became instead of what it is;

3. deciding what must be done;

4. determining how it must be done and what are the available resources to achieve this end.

This action paradigm provides the conceptual meaning for the components of the abstract model represented by Abstract Logics. For effective application of (1), (2), (3) a good cognitive model of the agent's surroundings is needed.

A **direct percept** ϕ is a relation from the world elements $\mathcal{P}(W)$ into the elements W_a of the cognitive structure of the agent a:

$$\phi : \mathcal{P}(W) \to W_a$$

This general relation is realised as a concrete afferent projection by a bundle of morphisms.

The basic construct $\mathcal{L}_{W_a}$ used to *formally represent* a cognitive/action model of an agent is a pair of mathematical structures called the abstract logic (Brown and Suszko [27], Kohout [12], [20]). Thus for an agent a, the pair

$$\mathcal{L}_{W_a} = (\mathcal{W}_a, \mathcal{C}_{W_a})$$

consisting of an abstract algebra $\mathcal{W}_a$ and a generalised closure system $\mathcal{C}_{W_a}$ is the basic representational construct. The first formal structure, *abstract algebra* $\mathcal{W}_a$ is itself a system composed of two parts:

$$\mathcal{W}_a = (W_a, \mathcal{F}_{W_a}),$$

where the set W_a is the so called carrier of the algebra, and $\mathcal{F}_a$ is the family of operations over the carrier set of the algebra. The carrier W_a represents the set of possible states of the world as it is perceived by the agent a, and is in fact a reflection (image) of the world W in which the agent acts. The set of operations on the carrier represents the actions (or, more generally, events) that can cause the change of the states in W. Although the majority of the elements of $\mathcal{F}_{W_a}$ usually have the counterpart in the reality (the world W), there may be some actions or events that may not have corresponding elements in the world W. This captures the situation in which some intended actions of an agent (intentions) although conceivable by the agent are **not realisable** for one or another reason.

The second formal structure, a *generalised closure system* $\mathcal{C}_W$ is a family of subsets of W that is generated by a *generalised closure operator* **Cn**. Let $\mathcal{P}(W)$ designate the power set of W, that is the set of all subsets of W. Then the generalised closure operator itself is defined as a meta-relation:

Cn: $\mathcal{P}(W) \longrightarrow \mathcal{P}(W)$ that must satisfy the following closure axioms:

I: if $A \subseteq W$, then $A \subseteq \mathbf{Cn}(A)$;
M: if $A \subseteq B \subseteq W$ then $\mathbf{Cn}(A) \subseteq \mathbf{Cn}(B)$;
O: $\mathbf{Cn}(0) = 0$, where 0 is the empty set.
U: For any $A \subseteq W$, $\mathbf{Cn}(A) = \mathbf{Cn}(\mathbf{Cn}(A))$.

Further technical details concerning generalised closure systems are contained in Kohout ([28], [12]). This system, satisfying **I, M, O, U** axioms, captures adequately all the aspects of consequences of reasoning. When we deal with stability and instability induced by some time-induced changes, it might be necessary to relax the **U** axiom in some instances (Kohout [28]).

When a cognitive model is used by an agent, the question of its correctness arises. It is not only the question of correct representation of the agent's knowledge but also the question of its correct use. Two new research issues emerge here:

1. Are the consequences (the information/prediction) generated by the model correct?

2. In what interactions and/or in which environments and contexts are these correct?

As the abstract logic can formally capture the general framework in which these issues can be concretely studied, questions (1) and (2) provide a justification for using the pair of abstract algebra and generalised topology in abstract logic setting for modelling the knowledge structures of interacting agents.

There are two important issues involved in investigating the credibility of agent's models:

1. the correspondence of the models to the modelled reality,

2. the credibility of the consequences generated by the models, when these are used by the agents in interaction with *environments that change*.

Each cognitive act, the purpose of which is to acquire some relevant knowledge utilized in subsequent control acts, leads to a particular model which is usually not a complete model but a only partial one. Combination of the partial models has to be done according to some coherence criteria that these models must satisfy in order to form an adequate total model.

Now, taking two abstract logic systems,

$$\mathcal{L}_W = (\mathcal{W}, \mathcal{C}_W) \text{ and } \mathcal{L}_{Wa} = (\mathcal{W}_a, \mathcal{C}_{Wa}),$$

one simulating the external world W and the other the internal representation of the agent W_a, allows us to define important notions of correctness, completeness and faithfulness of the agent's representation of the external world, thus evaluating various aspects of credibility of models of the external world (Kohout [20], [12]).

We can recall [12] that for each agent a, a mapping ϕ from the world W to the carrier W_a representing the internal representation possessed by the agent a is defined as:

$$\phi : \mathcal{P}(W) \longrightarrow W_a,$$

where $\mathcal{P}(W)$ is the power set of the carrier W. Thus, let ϕ be a homomorphism of structures, $\phi \epsilon Hom(\mathcal{W}, \mathcal{C}_{Wa})$, and let $\mathcal{S}_{Wa} = (\mathcal{W}_a, \mathcal{C}_{Wa})$ be the closure spaces associated with the abstract logics. Then the agent's model $\mathcal{L}_{Wa}$ is:

(i) CORRECT if $\phi^{-1}(\mathbf{Cn}_{Wa}(\phi(X))) \subseteq \mathbf{Cn}_W(X)$

(ii) COMPLETE if $\phi(\mathbf{Cn}_W(X)) \subseteq (\mathbf{Cn}_{Wa}(\phi(X)))$

(iii) FAITHFUL if it is *complete* and *correct*.

In order to deal adequately with the description of purposeful behaviour of actors/modules, general meta-logical properties of the activity models used to represent these have to be determined. From the assumption of locality of relevance it follows that any model consists of a family of submodels, which form a functional heterarhy (or, as a special case, a hierarchy) of partial submodels [20],[12]. This will be discussed in greater depth in Sec. 8 bellow, which is concerned with advantages and problems of multi-level control of actions performed by an autonomous intelligent system.

6 FUZZIFICATION OF FORMAL MODELS OF ADAPTIVE AUTONOMOUS CONTROL SYSTEMS

6.1 The Effects of Imprecision and Uncertainty

It cannot be assumed that the above described formal models can be built accurately because imprecision enters into the act of cognition of an actor as well as into the act of our descriptions of the actors' intentional acts. In this context, imprecision of two kinds was identified (Kohout [20], p.705):

1. imprecision of accumulated data that is used in the process of inductive inference leading to the construction of a model;

2. imprecision caused by the fact that the theories which are inbuilt into the individual agents/modules (be it by a priori design or by enforced or autonomous learning); the models may be only partially true, approximating better or worse the ideal theory.

In order to deal adequately with these problems, fuzzy power sets and closures of various kinds were developed [29],[30],[31],[32].

Closures in abstract logics (cf. Sec. 5 above) are the essential tools for investigation and proper understanding of interactions between the parts of knowledge structures employed by agents. As the formal apparatus of generalised topologies referred to in Sec. 5 uses classical set theory, the crisp closures of these theories *have to be extended* to capture the essential multi-valuedness of the degrees of plausibility of consequences in abstract logics models.

Such a desirable expansion of technical tools has been achieved by developing many-valued logic based extensions (Kohout [33],[34]) of the classical sets which allow us to deal adequately with imprecision introduced by (a) above. Fuzzy power sets and closures of various kinds developed for this purpose are available (Bandler and Kohout [30],[31],[29]). On the other hand, in order to deal with imprecisions introduced in (b), more sophisticated tools are needed. These were provided in the form of many-valued extensions of mathematical relations (Bandler and Kohout [35],[36]) and of new products of relations (Bandler and Kohout [37],[38]). These new, so called *triangle* and *square* relational products have been used extensively to deal with *knowledge representation under uncertainty* and for dealing with information *containing imprecision* in various AI and computer science applications. The best starting point for a reader interested in the new products, fuzzy closures and interiors is the Computer and Control Encyclopedia entry [35] together with our two recent survey articles [36],[39] which also contain a selected bibliography of further references.

6.2 Fuzzification of Power Sets and Relational Systems

The meta-operations of set inclusion and closure composition as expressed above represent general logical types of structures, with type composition laws equally applicable to fuzzy systems. It is sufficient to exchange the Boolean *AND, OR* and → (*Implication Operator*) connectives for a suitable many-valued logic type according to the meta-properties required for a particular fuzzy system. This choice, of course depends on the particular field of application. The papers of Bandler and Kohout [30],[31] have opened a new avenue in extending set-theoretical methods of dealing with imprecision and uncertainty in this direction. The key techniques explored in depth in [31] paper are based on the following observations:

1. the degree to which A is a subset of B *is equal* to the degree of membership of A in the power set $\mathcal{P}(\mathcal{B})$ of B;

2. A is a subset of B if and only if $x \epsilon A$ implies $x \epsilon B$ for all elements x.

These were expressed in an earlier 1978 paper of Bandler and Kohout [30] as meaning postulates in the following form:

1. Connection between $\subseteq$ *and* →:
   $A \subseteq B$ means $(b \in A \rightarrow b \in B)$.

2. Subset relation as a construct build of $\in$ and $\mathcal{P}(B)$:
   $A \subseteq B$ means $A \in \mathcal{P}(B)$.

3. Binding two meanings of $A \subseteq S$ yields the third one:
   $A \in \mathcal{P}(B)$ means $(b \in A \rightarrow b \in B)$.

These postulates are powerful enough to provide adequate fuzzification of the logic constructs presented in Sec. 5.

Construction of general fuzzy possibilistic families of power sets, with the "Theory of Possibility" as a special case If these "meaning postulates" are used as the restricting conditions in Zadeh's calculus of fuzzy restriction [40], this yields a generic possibilistic specification of a large variety (family) of power set theories [30], [31] of which, of course, the crisp power set is a special instance.

A family of fuzzy power set systems is defined by Bandler and Kohout [30], [31] as the linguistic proposition p:= $< A \subseteq B$ is $A \in \mathcal{P}(\mathcal{B}) >$, the meaning assigned to it being determined by the *possibilistic* postulate (1) and the *restriction* postulate (2) of the following form:

1. $\pi(Q \subseteq S) = \mu_{\mathcal{P}(S)}Q$

2. $R_{\subseteq}(A(\mu_Q x, \mu_S x)) = \mu_{\mathcal{P}(S)}Q = \bigwedge_{x \in U}(\mu_Q x \rightarrow \mu_S x)$.

It is obvious, that the possibilistic postulate (1) above is a particular instance of the linguistic sentence *sent* := X *is* F, making a statement about the relationship of the whole component *sent* to the assumed parts of it, the individual names X and F. Clearly, X is the name of a construct (object) and F is the label of a fuzzy subset of the universe of discourse U. The restriction postulate (2) above is a particular instance of Zadeh's relational assignment equation. In its general form, the assigned meaning is defined by the relational assignement equation $\phi(sent)$: $R(A(X)) = F$; where A is an *implied attribute* of X, i.e. an attribute which is implied by X and F. R denotes a fuzzy restriction on $A(X)$ *to which* the value of F is assigned by $\phi(sent)$. In the possibilistic construction of power sets by Bandler and Kohout, F is not just a fuzzy subset (a first order construct), as in the definition by Zadeh [41] of the possibility distribution and the associated possibility measure. Instead, it is a functor over a 2-ary fuzzy relation $\bigwedge_{x \in U}(\mu_S x \to \mu_Q x)$, that is a *second order* logic construct. In its full generality, the relations in Zadeh's calculus of restriction must be given by their intension and their reducibility to the *satisfaction graphs* (cf. Bandler and Kohout [35]) must be proved. For further discussion of related issues see [42].

As seen from the postulates (1) and (2) above, this generalisation has played an important role in the possibilistic construction of a variety of fuzzy power sets in Bandler and Kohout ([30], [31]), of which, as we said before, the so called "possibility theory" was just one special instance. The specific concrete power set system was selected from the family specified by the possibilistic (1) and restriction (2) postulates above by means of the choice of an appropriate many-valued logic implication operator. It is important to realize that (1) and (2) above yield a wide variety of useful possibilistic systems, depending on the choice of $\bigwedge$ and $\to$, not just the standard ones [42]. This can be put to advantage in design of multi-level fuzzy controls with a priori prescribed dynamic properties.

The power set structures capturing the dynamic consequences generated by the algebraic part of the abstract logic model had the structures of generalized topologies defined by means of set inclusions. These set inclusions were viewed as general logic types, **open concepts** that were made specific by the choice of the base many-valued logic. The same applies to the relational compositions that are required for defining *equality of powersets* and other operations used in fast fuzzy relational identification algorithms that are of direct practical importance [32],[29]. Thus the logical types of relational compositions in Def. 1 are *instantiated* and *particularized* by the choice of specific many-valued connectives in the logically equivalent Definition 2.

Definition 1:

Circle product: $\pi\{x(R \circ S)z\} \Leftrightarrow \pi\{xR \text{ intersects } Sz\}$
Triangle Subproduct: $\pi\{x(R \triangleleft S)z\} \Leftrightarrow \pi\{xR \subseteq Sz\}$
Triangle Superproduct: $\pi\{x(R \triangleright S)z\} \Leftrightarrow \pi\{xR \supseteq Sz\}$
Square product: $\pi\{x(R \square S)z\} \Leftrightarrow \pi\{xR = Sz\}$

where π denotes the degree of possibility of the corresponding formulas.

Definition 2:

In the formulas, R_{ij} represents the fuzzy degree to which the statement $x_i R y_j$ is true.

$\pi(R \circ S)_{ik} = \bigvee_j (R_{ij} \wedge S_{jk})$ $\pi(R \triangleleft S)_{ik} = \bigwedge_j (R_{ij} \to S_{jk})$

$$\pi(R \triangleright S)_{ik} = \bigwedge_j (R_{ij} \leftarrow S_{jk}) \qquad \pi(R\square S)_{ik} = \bigwedge_j (R_{ij} \leftrightarrow S_{jk})$$

It is important to distinguish from what we call the *harsh* products (defined above), the family of the *mean* products. Given the general formula

$$(R@S)_{ik} ::= \#(R_{ij} * S_{jk})$$

the outer connective denoted $\#$ is replaced by $\sum$ and the resulting product normalized appropriately. The mean products are very important in some applications [43].

The customary logical symbols for the logic connectives *AND, OR*, both *implications* and the *equivalence* in the above formulas represent the connectives of some many-valued logic, **chosen** according to the properties of the products required. Lack of space does not permit us to go into details of the available repertory of various implication operators here, the interested reader will find the definitions of the most useful implication operators used in the triangle products in Bandler and Kohout [44], [45].

As said before, by a suitable choice of the *multi-valued logic implication operator* (Bandler and Kohout [46]) in the definition of a subset, a whole *repertory of extended alternative set theories* is built. It is of course assumed that for the limit values of subset membership (that is, the values 0 and 1) all thus defined many-valued set theories will default into its special case, namely the crisp set theory. The extensions of the classical theory of sets and relations thus defined are important for at least two reasons:

1. They provide a viable choice of alternatives for dealing with uncertainty and imprecision, supplying a whole family of set systems, each of which has some *well-defined specific* and *desirable properties* that may be needed in a particular use.

2. The *unifying logical framework* they provide offers the formal link between various alternative systems, such as probability theory, fuzzy set theory of Zadeh [47], and other different many-valued logic based set theories (Klaua [48],[49], Zadeh [41], Bandler and Kohout [31],[30]).

Opening a new field and dealing with a whole spectrum of alternative fuzzy set theories and computational algorithms poses, of course, new questions. One should ask:

1. What are *the essential differences* between the newly available alternatives, in both their theoretical aspects, and in their applications?

2. How do we determine which system is *the best* in a particular theoretical use or an empirical application; and what are the optimality criteria for defining appropriately what is "the best"?

Because the algorithmic and computational aspects are very important in any new mathematics, our general framework of many-valued logics based power sets, mathematical relations and *relational products* (Bandler and Kohout [44],[50],[45],[35]) has been supplemented by the development of new *fast relational computational algorithms* (Bandler and Kohout [29],[32]) and of a new methodology for empirical data evaluation [43],[51],[52],[38],[53],[54]). This leads to the notion of **identification of emergent properties** [42] of the information coming from the environment to the intelligent control system, providing new algorithmic techniques that can be used in *fuzzy identification problem*, as discussed in the next section.

7 FUZZY IDENTIFICATION

The role of control concepts as the original impetus for developing fuzzy mathematics is not sufficiently stressed in the current literature. One important part of Zadeh's epistemology is the *identification problem* – the problem of identifying the characteristics of a black box from the input/output relationship by experimental means. In a paper of 1956, Zadeh [55] formulates the problem as follows:

> Given 1) a black box, B, whose input-output relationship is not known a priori; 2) the input space (domain) of B, i.e. the class of all time functions on which operation with B is defined; 3) a **class** of black boxes, $\mathcal{A}$, which on the basis of *a priori* information about B is known to contain B. Determine – by observing the response of B to various inputs – a member of $\mathcal{A}$ which is equivalent to B in sense that its responses to all time functions in the input space of B are identical with those of B.

Here we already have a **family of models** to which we approximate. In order to interpret this starting point correctly, we must realise that "identical" may mean stochastic identity.

Zadeh introduced the term identification in 1956 and over the next 20 years the identification problem became an important area of study for modern (state-space based) control theory, justifying continuous series of symposia on the topic, publication of major books, and the development of important applications including adaptive filtering techniques applied in commercial systems. The spectrum of the problems and applicability is however wider, including biological models of behaviour, irreversible models of human operators, when one crosses over into the domain of grammatical inference.

Although the main part of Zadeh's paper was concerned with continuous nonlinear system identification, he recognized that the framework was much wider and formulated the problem in general systemic terms. Interesting discussion of Zadeh's identification problem was provided by Gaines [56]. Gaines pointed out that (a) identification may conflict with other objectives; (b) it may not necessarily have a unique solution, but a number of solutions depending on other factors. Starting from these additional assumptions, Gaines developed concrete identificational techniques of considerable practical importance, including those for identification of fuzzy grammars [57].

Theoretical and practical work concerning this topic demonstrated that even for stochastic systems there was no unique solution and the "solution" had to be defined on an *admissible subset of models* [56]. This establishes a Galois connection between the orders of approximation and complexity in this admissible subset. This was elegantly incorporated by Ralescu into a category theoretical formulation, unifying in a common framework the identification problem for deterministic, stochastic and fuzzy systems, as an adjunction functor. Unfortunately, this important result of Ralescu is very little known.

Zadeh's definition of identification possesses two key features of the general identification problem:

1. the class of possible models must be determined in advance;

2. identification is an active process of testing hypotheses by interaction with the system being modelled / represented.

Gaines remarked that (1) "is an example of precedence of ontology over epistemology – the philosophical problems involved do not seem to have worried system theorists a great deal. ..... and this is an error." (p. 94, in Gaines [56]).

Bandler and Kohout were concerned with this defect of identification. They developed a methodology that provides the techniques, by means of which the class of possible fuzzy relational structures not predetermined in advance (possible models) can be built up from the empirical data. In order to deal adequately with the issues of *relevance* and *context*, it was necessary to find adequate general structures for representing knowledge [30], [43], [37], [35] and also to develop suitable algorithmic procedures for identification of various relational properties of fuzzy relational complexes [32], [29]. These complexes are the structures that during the identification process emerge dynamically through the mutual interaction of observers and the observed [44], [58]. Such a methodology is useful for recognizing/extracting structures from the empirical situations/ materials/ texts and for identification of emerging fuzzy dynamic structures without any a priori assumptions [36].

Bandler and Kohout took the view that the right way in which the class of possible models should be identified is to allow the possible structures to emerge within a given *normative* meta-context, so that an a priori explicit postulation of the class of models is no longer needed. They developed a methodology satisfying this substantial requirement. To deal with the locality of models was however not possible within the previously existing relational methods, be they crisp, fuzzy or probabilistically based. The previously existing methods did not take into consideration the locality of relational properties of the attributes of objects to which concepts were assigned. This necessitated introducing *two important revisions* (Bandler and Kohout [32], [29]) concerning the theory of fuzzy (as well as crisp) mathematical relations. These extensions of the mathematics of relations consist of introducing:

- local reflexivity [32], [35], [29];

- 3 types of partial homomorphisms [53] of mathematical models and structures (week, strong, very strong homomorphisms).

This has been applied in a number of situations [19]. The successful application, however, crucially depends on the proviso that some normative criteria are used, by which one can judge how faithfully local models reflect the reality [20], [21] and to what extent the models distort it [21], [12]. Over the last 10 years we have developed appropriate fuzzy theory of relations [35], algorithms [32], [29], and supporting methodology [19], [36] for achieving this objective.

8 INTRODUCING CONTROL HIERARCHIES

With basic abstract conceptual structures, backed by computational techniques using fast fuzzy relational algorithms, one is in the position to develop new, better structures for adaptive control systems incorporating the advantages of fuzzy approximation. What is needed is an algebraic relational structure that can represent the correct conceptual model, and is compatible with the abstract logic representation of the problem and can be adequately fuzzified. Such a relational representation of control hierarchies and heterarchies will be provided in this section.

8.1 The Behaviour of an 'Intelligent' Autonomous System

The behaviour of an autonomous intelligent system is parametrically characterised by the quadruple (input, output, situation, purpose). This paradigm is presupposed in our previous work [12],[19]. As shown in Chapter 9 of [12], one structure (i.e. subsystem, or actor/agent – participant in a distributed system) can exhibit several distinct behaviours determined by the parameters: situation and purpose. This gives a scope for introducing useful redundancy, namely that the same behaviour can be executed by several distinct structures (subsystems, cooperating agents).

The model of Sec. 5 describing how the autonomous intelligent system views the external world does not take into consideration that, in a large majority of cases an artificial or natural autonomous organism is a multi-centre structure. Each distinct part of that complex structure may receive some information from the outside world and interact with other parts and/or with the environment, to provide some information, to control, initiate or execute an action or a complex activity. In this case the model $\mathcal{L}_{Wa}$ of Sec.4 may be decomposed into a family of sub-models where each sub-model belongs to a different part of the organism's structure. The decomposition of the global model $\mathcal{L}_{Wa}$ into a family of sub-models should preserve all properties of the global model $\mathcal{L}_{Wa}$. In particular, the family of sub-models $\{\mathcal{L}_{a,Mi}\}$ should be a faithful representation of $\mathcal{L}_{Wa}$. If the family $\{\mathcal{L}_{a,Mi}\}$ possesses the properties specified in the definition below, it is called a *generating family* [12].
Definition:
A system of abstract logics is called a generating family if and only if:

1. Mappings $\rho_i : W_a \to M_i$ are defined for all i and $\rho_i \in Hom(\mathcal{O}, \mathcal{M}_\}))$.

2. Each model $\{\mathcal{L}_{a,Mi}\}$ is a faithful representation of $\mathcal{L}_{Wa}$.

It is plausible to assume that each sub-model in the family $\{\mathcal{L}_{a,Mi}\}$ should be a faithful representation of some aspects of the world model $\mathcal{L}_{Wa}$. As we cannot go into the mathematical consequences of this assumption which leads to the so called projective family of $\mathcal{L}_{Wa}$, the interested reader is referred to [12] for further details.

We have seen that steps (1) and (2) of the Intentional Framework for Directing a Purposeful Activity (IFDPA) introduced in section 5 had to deal with evaluation of the current situation and prediction of the future intended situation. For this, a model of past, present and future situations has to be formally represented. In order to make this model, a suitable description of the environment is needed. Let a set of descriptors (sentences) $\{a_1, a_2, ...a_n\}$ describe the environment. Taking this set as a state-vector yields the state-space of the 'world' model. From the current state given by the world model the next potential state of the world is computed by a predictive transformation.

The notion of predictive transformation is utilized in this context, because it is demanded by step (3) of IFDPA requiring the knowledge of 'what must be done'. What must be done depends on the aim or purpose of the actor/organism and can be characterised and formalized by introducing the concept of *task descriptor* [12].

Task descriptors

Definition 3: *task descriptor*

Let γ be an activity leading to a certain aim or having a particular purpose. A task descriptor T_γ, associated with the activity γ, performing a set of tasks T, is a partition of T into two disjoint subsets

$$T_\gamma = (T_p, T_f)$$

where T_p is the subset of *permitted* tasks and T_f the subset of *forbidden* tasks, of T.

The knowledge of 'What are available resources' is demanded by step (4) of IFDPA. This question of resources has two facets: the resources can be characterised by a functional classification of the processes available to generate the desired activities, or by the substratum modules that embed these dynamic processes in the form of some virtual machines. This provides a link with Activity structures methodology. Given a set of processes X, an Activity Structure gives a description of the family of classes of X that are needed in order to perform an activity leading to a certain aim ([21],[12],[18]). An activity specified this way is embedded in a chosen substratum, thus leading to a physical realization of the mechanism that is potentially capable of generating this activity when required. This approach forms a basis for the synthesis of virtual architectures [14],[12] from the teleological dynamic description of activities provided by the task descriptors.

Care should also be taken to insure that forbidden tasks are not realized. This leads to the notion of *protection* and *permission* structures ([12], ch. 5,6 and 7). Permission structure is a mechanism for avoidance of disastrous consequences. The role of this structure is twofold:

1. If the predicted state belongs to the forbidden region, a rapid action should be taken, which brings the 'intelligent' system into the permitted region of the world model state space.

2. Only such intentions should be converted into actions, which do not cause that the system is taken into a state that belongs to the forbidden region.

The problem of protection against the undesirable actions and activities of individual agents-participants is also closely connected with the purposefulness of behaviour of these agents-participants. Thus our definitions are completed by introducing the terminology explicitly defining *purposefulness*: An *aim* is an a priori specified (required) result of a sequence of actions that form a particular definite activity. Aim controllable by a group of subjects X is an aim which can be achieved by a sequence of actions exclusively performed by the group X. An aim that is protectable by a group of subjects X is an aim that cannot be achieved by an activity outside X without the specific permission of the group X. Further elaboration of these notions and an outline of the way in which the concepts of *aim, goal* and *purpose* are related to *task descriptors* is contained in Kohout ([13]). How these concepts are relevant to a multi-center control autonomous system we shall see in the sequel.

8.2 From Two-level to Multi-level Control

In section 3.2, we have discussed the need for structuring of control, and also provided the basic motivation for building multi-level systems out of two-level hierarchical units. Here, we shall look first at the two level hierarchical control module in some detail. We shall examine what is the *minimal abstract structure* that such a module should consist of, and how it can be placed within a multi-level system. Then, we shall look more closely at the ways multi-level (hierarchical or heterarchical) systems can be further structured. We have to examine these issues from two different points of view: firstly, the point of view of dynamics of control behaviour of the system and its decomposition into functional hierarchies of controls; secondly, from the point of view of decomposing the substratum of the autonomous adaptive system into a collection of virtual physical blocks (modules). After describing its abstract structure and the computational dynamic rules generating its control behaviour, we shall briefly discuss the inherent problems induced by the ways in which a collection of such units is combined into a coherent co-operating family.

The dynamics and conceptual structures of an autonomous intelligent system have been discussed in rather general terms of abstract logics and of behavioural sequents. In the design of specific autonomous systems it is possible to make some choices of representation and come up with more specific computational structures of individual units of the hierarchy. In this section I shall present a particular formalization that provides the computational structures for concrete application of fuzzy relational structures.

This section presents a formal description of an i-th level of a multi-level control system (cf. [12] ch.9) that is compatible with the decomposition paradigm of the knowledge models outlined in section 8.1 and previously elsewhere ([20],[21],[12] chapter 3). The present formalization is compatible with Bernštejn's theories of hierarchical neurological control of human movement. The formulation of my formal model was motivated by Bernštejn's important but neglected monograph [17] but also by the little known yet important monograph of Laufberger on neural nets [59].

Figure 1 depicts a pair of substratum units and their coupling with the environment. Each unit interacts with the environment by means of three types of interfaces: *afferent* paths, *efferent* paths and the *upper control* paths (cf. [12], ch. 9). Let X be the class of all perceps that are extracted at a certain predefined resolution level as the atomic elements of the perceptual field of the organism/multi-level autonomous system. The afferentation at the i-th level of the functional hierarchy is combined with the set of internal controls H_i which come from the other, superior levels of the control hierarchy. E_k is the set of all efferent controls provided by this level to its environment. The collection of relevant features Y of the class Y of all the perceptual elements depictable by an organism/autonomous system is transformed by a perceptual transformation π_i into the set of inputs entering the afferent relational structure A_i. The relational control structure $\mathcal{F}_i$ (the power set of the control command set AC_i) is combined with the afferent relational structure into the action relational structure ('action frame' of the i-th level). The overall strategy scheme ρ_i does depend not only on the action frame, but also on the family of the inner controls which contain the control parameters incoming into the i-th level from other, superior levels of the hierarchy. A formalization in terms of MVL-relational transition rules is also presented in Figure 1.

The next question to ask is, how the activities of the i-th level unit fit into the overall scheme of task descriptors and into the Intentional Framework for Directing a Purposeful Activity (IFDPA) introduced in section 4. The afferent information π_i obtained from the sensory input by a perceptual transformation *determines* the initial situation ('what is'). This has to be evaluated (cf. [12], ch. 9.5) in order to determine the desired situation, say, ϕ^{fin} (the goal, 'what it must become').

The next step of the control strategy determines 'what must be done'. That is, what are the intermediate subsets of situations $\{\phi^i\}$ that can transform the initial ϕ^0 into the final ϕ^{fin}. A trajectory $\Phi(\phi^0, \phi^{fin})$ is a sequence of immediate actions $\phi^1 \phi^2 \phi^3 ... \phi^k$ which transforms ϕ^0 into ϕ^{fin}. Note that there may be possible to find more than one sequence of trajectories transforming ϕ^0 into ϕ^{fin}, a family of sequences $\Phi(\phi^0, \phi^{fin}) = \{\Phi^1(\phi^0, \phi^{fin}), \Phi^2(\phi^0, \phi^{fin}), ..., \Phi^s(\phi^0, \phi^{fin})\}$. From $\Phi(\phi^0, \phi^{fin})$, a set of **possible control strategies** and associated **possible action commands** of i-th level is determined. All this is done by an appropriate refinement of those task descriptors

Figure 1: MULTILEVEL CONTROL: Interaction of two Levels.

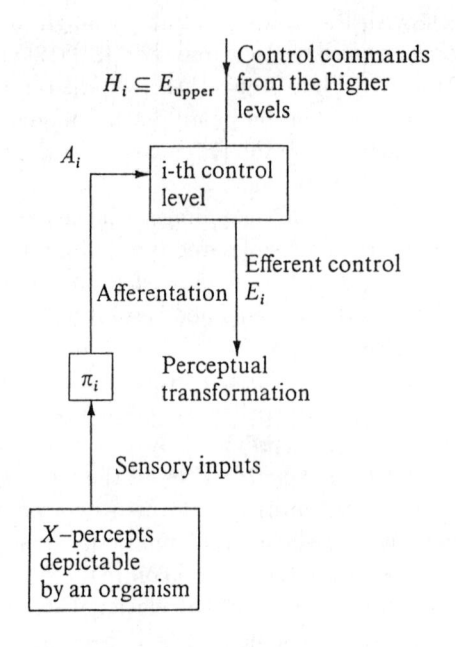

$H_i \subseteq E_{upper}$ | Control commands from the higher levels

A_i → i-th control level

Afferentation | Efferent control E_i

π_i | Perceptual transformation

Sensory inputs

X–percepts depictable by an organism

E_{upper}–all potential efferent outputs of the higher levels that can affect the i-th level;

Upper control of the i-th level:

$$\mathscr{X}_i \subseteq \mathscr{P}\left(\bigcup_j^r E_j\right), \quad j > i$$

(j goes over the indexes of all levels higher than i).

AC_i–action commands of the i-th level

The descriptors of the i-th level:
Control frame of the i-th level
Control strategy σ_i
Control state-space ρ_i
Control variables

$\mathscr{F}_{c_i} \subseteq \mathscr{P}(AC_i)$
$\sigma_i \subset A_i \times \mathscr{P}(\mathscr{F}_{c_i})$
$\rho_i \subset \sigma_i \times \mathscr{X}_i$
$\mathscr{S}_a \subset A_i$ afferent information
$\mathscr{S}_e \subset E_i$ efferent control
$\mathscr{S}'_c \in \sigma_i$ new control strategy
$\mathscr{S}_u \in \mathscr{X}_i$ upper control

Functional rewriting rules of the i-th level:

$$\frac{\mathscr{S}_a, \mathscr{S}_u, \mathscr{S}_c}{\mathscr{S}_e, \mathscr{S}'_c}$$

(i.e. this rule should be viewed as a mapping $\mathscr{P}(A_i) \times \mathscr{X}_i \times \sigma_i \to \mathscr{P}(E_i) \times \sigma_i$)

which lead to the realization of that particular final goal.

Not all the possible action commands are permissible (cf. [12], ch. 6), some intermediate situations are forbidden relations. In order to satisfy a protection criterion, these situations have to be excluded, otherwise they will cause the potential violation of a protection criterion, which of course may be defined as either crisp or fuzzy [60].

8.3 Organization and Strategies of Multi-Level Control

When a more complex than two-level system is considered, the global aims of the autonomous system/organism have to be linked properly with the local resources of each level and centre Although the individual centres will cooperate, it is usually necessary to assign to some centre a leading role, or at least to use it as an arbiter of possible conflicts. In a two level system, the top level has the power to adjust the parameters, and restrict the strategies of the lower level. In an extension of the control strategies to multi-level control, the top level, the so called functional *leading level*, will determine the *aim* and hence the strategy of the control act. The plan of control strategies must be compatible with the decomposition criterion of the generating family of models as described above (cf. section 8.1). In addition, the possible trajectories of each subordinate level must not violate the protection criteria specified for that level.

Selection of the resources from the pool of available resources will of course become more intricate than is in the case of a single level or a two level hierarchy of control. Choice of a strategy for the leading level depends on the situation of the environment and the intentions of the autonomous system/organism. This choice is subject to the following filtration criteria:

1. All candidate strategies that lead to a chosen aim are pre-selected.

2. Restriction to permitted strategies is enforced.

3. The choice of other participating subsystems is made according to the functions that are needed for realization of the permitted strategies.

4. Choice of a suitable equivalence class of strategies of lower levels is made.

5. The most suitable class of strategies of those produced by (2) is selected according to some cost or performance criteria.

The equivalence class of most suitable strategies is in fact a substructure of the decomposition of the system's behaviour into the generating family $\mathcal{M}_i$ (cf. section 8.1). Decomposition of the behaviour into a model consisting of generating families represents a functional decomposition of behaviour of the organism/autonomous system; that is a decomposition into *Functional Activity Structures* (FAS). Decomposition of the control hierarchy into individual modules over which the FAS are distributed (localized) on the other hand represents a decomposition of *Substratum Structures* (Sub)[1] into individual substratum modules – virtual machines or physical subsystems; i.e. hardware in a computer, or neural units (system's 'anatomy') in an artificial or natural neural network.

Functional rewriting rules $\mathcal{R}_i$ of the i-th level (see Figure 1) determine the family of control parameters that are produced by this level together with the new control strategy ϕ'_c of this level. It can be seen that the formal model presented here represents a specialisation of the model given in chapter 3 of [12]. In fact, to be of practical use in design of general adaptive control, the decomposition of the whole level hierarchy of the structures described here should satisfy the general structural axioms of faithful decomposition of static and dynamic structures of the model hierarchy as defined in the most general case of generating family (cf. section 8.1 above).

The class of suitable strategies is contained in the closure space of the generating family model $\{\mathcal{L}_{a,Mi}\}$. Expressed more formally, the inclusion Φ (ϕ^0, ϕ^{fin}) $\in \{\mathcal{C}_{a,Mi}\}$ holds, crisply or fuzzily. Expressed in terms of Functional Activity Structures (FAS), Φ (ϕ^0, ϕ^{fin}) contains the information as to 'what can be done' (expressed in behavioural terms), taking into consideration the intentions of the organism/autonomous system, and the potential, of both the system and of the present and future environmental situations.

The selection of suitable substratum modules to realize one of the chosen potential strategies represents the localization of function. A procedure achieving this can be constructed, using the general procedure for embedding the FAS into substratum (cf. [12], ch. 10.6):

1. Combine intention and potentiality sub-structures of $\mathcal{M}_i$.

2. Resolve conflicts.

3. Select a suitable realization level in Sub.

4. Combine with the information of (1) above.

5. Resolve conflicts.

It should, however be understood that in the case of a hierarchical autonomous system the localization of functions is partially performed decomposition into the family $\mathcal{M}_i$. It is impossible to discuss further this involved question here. It suffices to say that the redundancy of FAS may allow for compensation of impairment of some Substratum modules in a complex hierarchical control autonomous system by re-localisation of some strategies and functions (refer to [25],[12] ch. 9 for a detailed example).

---

[1]Here we shall employ just bare essentials of the part of Activity Structures methodology concerned with the interplay of FAS and Substratum structures that are needed to carry through the argument. The reader is referred for further details to [12] and the wealth of papers referenced therein. Of these, the most relevant to the present discussion are [20],[14], [15],[24],[13] and [12] ch. 4 and 10.

The decomposition into hierarchies is not arbitrary. One functional structure may control details, the other structure a family of curves with the unessential details left out. The parameters have to be decomposed in such a way that one subsystem selects the equivalence classes and the other subsystems tune the parameters according to some local requirements. For example, in the model of human movement control [12] (chapter 9, Fig. 9.10) the former is S.EXEC and the latter STRANSF. It can be seen from this example, that in order to make the system workable, it is necessary to impose certain assembly preference metrics onto the component tasks which jointly generate the resulting behaviour of the system. Wrong partial preference order may make the tuning inherently unstable. The question of stability can by studied by means of adaption automata (Gaines [16]) on which suitable topologies are induced by the preference metrics. It was indicated by Kohout in [61] that the adaption automaton (which in the case of our example would be formally capturing the process of tuning) will be globally stable with respect to tolerance topology if the input-induced preordered topology of this adaption automaton *"is a refinement of the tolerance topology. In other words, the automaton is globally stable if the U-modification* (cf. [12] Appendix 1) *is finer than the topology induced by the inputs. In the terms of measures, the necessary condition for the global stability is that the tolerance semi-pseudo-metric is also a pseudometric."* It is outside the scope of this paper to go into further details concerning this matter. It suffices to say here that this formulation in metric terms [61] equally applies to fuzzy or other many-valued logic based topologies as it does to crisp topologies and ought be studied further, at least for the fuzzy power set systems that have been found most useful in practical applications [30],[31].

9 FROM ABSTRACT LOGIC DESIGNS TO TECHNOLOGICAL REALIZATIONS

Formal logic representations of hierarchical autonomous systems has to be supplemented by the description of the way these can be realized in modern technology. Some guidelines and further references are provided in this section.

The architectural shell of the multi-environmental system is formed of a collection of mutually linked substratum modules. Each module is represented in its most general form by a generalised virtual machine module as briefly described below (see also [13], section 5.6).

9.1 Generalised Virtual Machines

The Generalised Virtual Machine (GVM) has three communication interfaces: the *upper interface*, the *upper control interface* and the *lower interface*. The upper interface accepts protocols in the main language $\mathcal{L}_i$, the upper control interface accepts protocols in the upper control language $\mathcal{L}_{i_u}$ and the lower interface emits protocol in the lower interface language $\mathcal{L}_{i_e}$.

The condition enforcing *correct coupling* of VM's requires that the output language $\mathcal{L}_{k_e}$ of the target machine VM_k must be *acceptable* by the input of the host machine VM_j. In other words, the language $\mathcal{L}_{k_e}$ must be be a *sublanguage of* (i.e. it has to be contained in) the language $\mathcal{L}_j$ of the host VM.

A GVM_i *accepts* the protocol in the language $\mathcal{L}_i$ and *translates selected portions* of it into the protocol in the language $\mathcal{L}_{i_e}$.

A GVM_i *executes* a portion of a protocol by the following algorithm:

1. it segments the protocol,

2. stores part of it in its inner state vector,

3. performs transformation and translation,

4. communicates with other modules (GVMs) to obtain necessary computations,

5. recomposes the result,

6. makes the result available through the upper interface.

As noted previously, conventional virtual machine is a special case of GVM because it has only the upper interface that accepts the language $\mathcal{L}_i$, and the lover interface that produces language $\mathcal{L}_{i_e}$. Thus we can see that the conventional virtual machine has no upper control interface.

To make the notion of virtual machine useful, we have to have at least a pair of machines, mutually interconnected. We have already seen that the basic adaptive control unit consists of two virtual machines. Thus let us consider a pair of *coupled* virtual machines VM_j and VM_k, where the coupling is done by interconnecting the lower interface of VM_k to the input of VM_j. With this type of coupling, the VM_k becomes the upper VM and VM_j the lower VM. The former is usually called the target machine and the latter the host machine. A given pair of virtual machines <Upper VM, Lower VM> is sufficient to define a translation from one machine of the pair to the other machine, of the texts (programs, protocols) that these machines execute. Well known computer science concepts 'compilation' and 'interpretation' represent special instances of such a translation within a system with discrete data structures. The protocols however may not be restricted to the structures with a discrete topology, but may also contain continuous control structures.

9.2 A Reconfigurable Architectural Shell with General Knowledge Structures

Interconnections of the modules have a dynamic nature and might change during the operation of the system. Each virtual module has a variable structure which can be changed by the tuning input that comes from a higher level. Each module operates on a dynamic stream of information that comes from the perceptual input. Each module has an efferent output which can be linked either to a perceptual input or a tuning input of another module.

The mutual interconnections of the individual modules are formed by a generator of bindings. It consists of

- an interaction controller
- activity selector.

The interaction controller binds dynamically the virtual modules. The activity selector chooses the appropriate activities. Each activity is composed of some functions selected from the repertory of teleological functions by the activity selector.

Dynamic information that the generator of bindings has to provide consists of the following:

1. information specifying which VMs are to be bound;

2. what to bind, namely:
   a) what to accept from the environment of each module,
   b) what to affect in the environment.

Substituting a many-valued (fuzzy) logic for the conventional crisp valuations of truth or falsity in individual operations in each Virtual Machine achieves information-flow triggered binding of the individual modules. This leads in most cases to the elimination of procedural (explicit) generators of bindings, or at least reduces its complexity. The structure of the binding effected by the information-flow triggering can also be modified. This is done by the dynamic changes of the alpha-cuts on the valuations of those expressions in each virtual module that effect decision making or activation of other Virtual Machines.

In addition to the substratum shell structures, some functional structures have to be provided as well in such a general shell. In order to provide these functional structures, the levels of control will have to be split into several categories, according to the class of the control functions they will have to provide. For example in systems discussed in [12] ch. 9 and in [18] there were five control hierarchies: A – of kinetic regulations, B – of synergies, C – of spatial field control, D – of object oriented manipulation, E – of symbolic processing and communication.

The general kind of architecture for a robotic system, the functional refinements of which are given by the information of the type similar to that of Table 13.1 printed in [18] can provide the following functions:

1. Support knowledge structures required for dynamic simulation of movement, which can also be used to drive some robot hardware.

2. Capture the knowledge concerning not only the norm but the possible malfunctions of its behaviour.

3. Provide the knowledge about their internal substratum (e.g., hardware) structure.

4. Use this information (cf. 3) for restructuring its behaviour, when the part of its structure is impaired, is malfunctioning, and has not yet been repaired.

5. Support the knowledge needed for object-oriented actions.

6. Support the knowledge structures needed for symbolic co-ordination and planning.

It is not difficult to see that all these functions may be needed in a well co-ordinated manufacturing process. Then of course, much more is required from the system than just providing all the "robotic" functions. This is the reason why more knowledge has to be added to the scheme discussed so far. The conceptual modification of functional levels for the purpose of CAM have been described in [19] pp. 238–240 where the technical design of the minimal knowledge structure shell for a KBS for the support of manufacturing activities by a Computer Assisted Manufacturing (CAM) is discussed.

REFERENCES

1. Albus, J.S., Barbera, A.J., and Fitzgerald, M.L.
   Hierarchical control for sensory interactive robots. In *11th International Symposium on Industrial Robots*, pages 497–505. Tokyo, 1981.

2. Palm, R.
   Fuzzy controller for a sensor guided robot manipulator. *Fuzzy Sets and Systems*, 31:133–149, 1989.

3. Dean, T.
   Robot problem solving. In Adeli, H., editor, *Knowledge Engineering (vol. 2, Applications)*, pages 84–115. Mc-Graw Hill, New York, 1990.

4. Saridis, G.N.
   Identification methods for intelligent control systems. In *Proc. 5th IFAC Symposium on Identification and System Parameter Estimation (Darmstad, Germany)*, pages 145–149. IFAC, 1979.

5. Saridis, G.N. and Valavanis, K.P.
   Mathematical formulation of the organization level of an intelligent machine. In *Proc. 1986 IEEE Internal. Conf. on Robotics and Automation*, pages 257–272. IEEE, New York, 1986.

6. Meystel, A.
   Knowledge representation for intelligent control systems (tutorial on intelligent control). In *Proc. 1987 IEEE Internal. Conf. on Robotics and Automation (Rayleigh, NC)*. IEEE, New York, 1987.

7. Villa, A.
   Hybrid Knowledge-Based Analytical Control Systems in Manufacturing. An Abacus Book, Gordon and Breach, London, 1991.

8. Sworder, D.D.
   A study of the relatinship between identification and optimization in adaptive control problems expert systems.
   J. Franklin Inst., 281:198–213, 1966.

9. Feldbaum, A.A.
   Dual control theory i, ii. *Avtomatika i Telemechanika*, 21:1240–49, 1453–64, 1960.

10. Feldbaum, A.A.
    Dual control theory ii, iv. *Avtomatika i Telemechanika*, 22:3–16, 129–142, 1961.

11. Gaines, B.R.
    The Human Adaptive Controller. Ph.D. Thesis, Cambidge University, Cambridge, U.K., 1972.

12. Kohout, L.J.
    A Perspective on Intelligent Systems: A Framework for Analysis and Design. Chapman and Hall & Van Nostrand, London & New York, 1990.

13. Kohout, L.J.
    Activity Structures: A methodology for design of multi-environment and multi-context knowledge-based systems. In Kohout, L.J., Anderson, J., and Bandler, W., editors, *Knowledge-Based Systems for Multiple Environments*, chapter 5. Ashgate Publ. (Gower), Aldershot, U.K., 1992.

14. Kohout, L.J.
    The functional hierarchies of the brain. In Klir, G.J., editor, *Applied General Systems Research: Recent Developments and Trends*, pages 531–544. Plenum Press, New York, 1978.

15. Kohout, L.J.
    On functional structures of behaviour. In Kohout, L.J. and Bandler, W., editors, *Knowledge Representation in Medicine and Clinical Behavioural Science*, chapter 7, pages 69–94. an Abacus Book, Gordon and Breach Publ., London and New York, 1986.

16. Gaines, B.R.
    Axioms for adaptive behaviour. *International Journal of Man-Machine Studies*, 4:169–199, 1971.

17. Bernštejn, N.A.
    O Postrojenii Dviženij. Nauka, Moscow, 1947.

18. Kohout, L.J.
Distributed architectures for computer aided manufacturing (CAM) and other embedded robotic systems. In Kohout, L.J., Anderson, J., and Bandler, W., editors, *Multi-Environmental Knowledge-Based Systems*, chapter 13 Ashgate Publ. (Gower), Aldershot, U.K., 1992.

19. Kohout, L.J., Anderson, J., and Bandler, W.
Knowledge-Based Systems for Multiple Environments. Ashgate Publ. (Gower), Aldershot, U.K., 1992.

20. Kohout, L.J.
Representation of functional hierarchies of movement in the brain. *Internat. Journal of Man-Machine Studies*, 8:699–709, 1976.

21. Kohout, L.J.
Methodological Foundations of the Study of Action. Ph.D. Thesis, University of Essex, U.K., January 1978.

22. Kohout, L.J.
Analysis of computer protection structures by means of multiple-valued logics. In *Proc. of the 8th Internat. Symposium on Multiple-Valued Logic*, pages 260–268, New York, 1978. IEEE.

23. Kohout, L.J. and Mohamad, S.M.A.
Development of support tools and methodology for design and validation of multi-environmental computer architectures. In Kohout, L.J., Anderson, J., and Bandler, W., editors, *Knowledge-Based Systems for Multiple Environments*, chapter 17. Ashgate Publ. (Gower), Aldershot, U.K., 1992.

24. Kohout, L.J.
Activity structures as a tool for design of technological artifacts. *Systems and Cybernetics: An International Journal*, 18(1):27–34, 1987.

25. Kohout, L.J.
Control of movement and protection structures. *Internat. Journal of Man-Machine Studies*, 14:394–422, 1981.

26. Bernstein, N.A.
The Co-ordination and Regulation of Movement. Pergamon Press, Oxford, 1967.

27. Brown, D.J. and Suzsko, R.
Abstract logics. *Dissertationes Mathematicae*, 102:5–42, 1973.

28. Kohout, L.J.
Generalised topologies and their relevance to general systems. *Internat. Journal of General Systems*, 2:25–34, 1975.

29. Bandler, W. and Kohout, L.J.
    Special properties, closures and interiors of crisp and fuzzy relations. *Fuzzy Sets and Systems*, 26(3):317–332, June 1988.

30. Bandler, W. and Kohout, L.J.
    Fuzzy relational products and fuzzy implication operators. In *International Workshop on Fuzzy Reasoning Theory and Applications*, London, September 1978. Queen Mary College, University of London.

31. Bandler, W. and Kohout, L.J.
    Fuzzy power sets and fuzzy implication operators. *Fuzzy Sets and Systems*, 4:13–30, 1980.

32. Bandler, W. and Kohout, L.J.
    Fast fuzzy relational algorithms. In Ballester, A., Cardús, D., and Trillas, E., editors, *Proc. of the Second Internat. Conference on Mathematics at the Service of Man*, pages 123–131, Las Palmas, 1982. (Las Palmas, Canary Islands, Spain, 28 June - 3 July), Universidad Politechnica de las Palmas.

33. Kohout, L.
    The Pinkava many-valued complete logic systems and their application to the design of many-valued switching circuits. In Rine, D.C., editor, *Proceedings of 1974 Internat. Symposium on Multiple-Valued Logic*, pages 261–284, New York, 1974. (West Virginia University, May, 1974), IEEE 74CH08945.

34. Kohout, L.J.
    Application of multi-valued logics to the study of human movement disorders. In *Proc. of the Sixth Internat. Symposium on Multiple-Valued Logic*, pages 224–232, New York, May 1976. IEEE.

35. Bandler, W. and Kohout, L.J.
    Relations, mathematical. In Singh, M.G., editor, *Systems and Control Encyclopedia*, pages 4000 – 4008. Pergamon Press, Oxford, 1987.

36. Kohout, L.J. and Bandler, W.
    Fuzzy relational products in knowledge engineering. In Nov'ak, V. et al., editor, *Fuzzy Approach to Reasoning and Decision Making*, pages 51–66. Academia and Kluwer, Prague and Dordrecht, 1992.

37. Bandler, W. and Kohout, L.J.
    Fuzzy relational products as a tool for analysis and synthesis of the behaviour of complex natural and artificial systems. In Wang, P.P. and Chang, S.K., editors, *Fuzzy Sets: Theory and Applications to Policy Analysis and Information Systems*, pages 341–367. Plenum Press, New York and London, 1980.

38. Kohout, L.J. and Bandler, W.
    Relational-product architectures for information processing. *Information Science*, 37:25–37, 1985.

39. Kohout, L.J. and Bandler, W.
    Use of fuzzy relations in knowledge representation, acquisition and processing. In Zadeh, L.A. and Kacprzyk, J., editors, *Fuzzy Logic for the Management of Uncertainty*. John Wiley, New York, 1992.

40. Zadeh, L.A.
    Calculus of fuzzy restrictions. In Zadeh, L.A., Fu, K.S., Tanaka, K., and Shimura, M., editors, *Fuzzy Sets and Their Applications to Cognitive and Decision Porcesses*, chapter 1, pages 1–39. Academic Press, New York, 1975.

41. Zadeh, L.A.
    Fuzzy sets as a basis for a theory of possibility. *Fuzzy Sets and Systems*, 1:3–28, 1978.

42. Kohout, L.J.
    Quo vadis fuzzy systems: A critical evaluation of recent methodological trends. *Internat. J. of General Systems*, 19(4):395–424, 1991.

43. Bandler, W. and Kohout, L.J.
    The use of new relational products in clinical modelling. In Gaines, B.R., editor, *General Systems Research: A Science, a Methodology, a Technology (Proc. 1979 North American Meeting of the Society for General Systems Research).*, pages 240–246, Louisville KY, January 1979. Society for General Systems Research.

44. Bandler, W. and Kohout, L.J.
    Semantics of implication operators and fuzzy relational products. *Internat. Journal of Man-Machine Studies*, 12:89–116, 1980.
    Reprinted in Mamdani, E.H. and Gaines, B.R. eds. *Fuzzy Reasoning and its Applications*. Academic Press, London, 1981, pages 219-246.

45. Bandler, W. and Kohout, L.J.
    A survey of fuzzy relational products in their applicability to medicine and clinical psychology. In Kohout, L.J. and Bandler, W., editors, *Knowledge Representation in Medicine and Clinical Behavioural Science*, pages 107–118. an Abacus Book, Gordon and Breach Publ., London and New York, 1986.

46. Bandler, W. and Kohout, L.J.
    Fuzzy implication operators. In Singh, M.G., editor, *Systems and Control Encyclopedia*, pages 1806–1810. Pergamon Press, Oxford, 1987.

47. Zadeh, L.A.
    Fuzzy sets. *Information and Control*, 8:338–353, 1965.

48. Klaua, D.
Über einen Ansatz zur mehrwertigen Mengenlehre. *Monatsb. Deutsch. Akad. Wiss. (Berlin)*, 7:859–867, 1965.

49. Klaua, D.
Einbettung der klassischen Mengenlehre in die mehrwertige. *Monatsb. Deutsch. Akad. Wiss. (Berlin)*, 9:258–272, 1967.

50. Bandler, W. and Kohout, L.J.
On the general theory of relational morphisms. *International Journal of General Systems*, 13:47–66, 1986.

51. Bandler, W. and Kohout, L.J.
Activity structures and their protection. In Ericson, R.F., editor, *Improving the Human Condition: Quality and Stability in Social Systems (Proc. Silver Anniversary 1979 International Meeting of the Society for General Systems Research).*, pages 239–244, Louisville KY, 1979. Society for General Systems Research, Distributed by Springer Verlag, New York.

52. Kohout, L.J. and Bandler, W.
Approximate reasoning in intelligent relational data bases. In Parslow, R.D., editor, *BCS'81: Information Technology for the Eighties (Proc. Conf. The British Computer Society, London, 1-3 July 1981)*, pages 483–485. Heyden & Son, London, 1981.

53. Bandler, W. and Kohout, L.J.
On new types of homomorphisms and congruences for partial algebraic structures and n-ary relations. *International Journal of General Systems*, 12:149–157, 1986.

54. Kohout, L.J. and Bandler, W.
The use of fuzzy information retrieval techniques in construction of multi-centre knowledge-based systems. In Bouchon, B. and Yager, R.R., editors, *Uncertainty in Knowledge-Based Systems (Lecture Notes in Computer Science vol. 286)*, pages 257–264. Springer Verlag, Berlin, 1987.

55. Zadeh, L.A.
On the identification problem. *IRE Transaction on Circuit Theory*, CT-3:277–281, 1956.

56. Gaines, B.R.
General system identification: Fundamentals and results. In Klir, G.J., editor, *Applied General Systems Research: Recent Developments and Thrends*, pages 91–104. Plenum Press, New York, 1978.

57. Gaines, B.R.
Systems identification, approximation and complexity. *International Journal of General Systems*, 3:145–174, 1977.

58. Kohout, L.J.
    Observers, constraints and possibilistic systems. In *3rd Internat. Workshop on Fuzzy Reasoning*, U.K., September 1978. Queen Mary College, University of London.

59. Laufberger, V.
    Vzruchová Theorie (A Theory of Activation). Prague, 1947.

60. Kohout, L.J. and Bandler, W.
    Computer Security Systems: Fuzzy Logics. In Singh, M.G., editor, *Systems and Control Encyclopedia*. Pergamon Press, Oxford, 1987.

61. Kohout, L.J.
    Automata and topology. In Mamdani, E.H. and Gaines, B.R., editors, *Discrete Systems and Fuzzy Reasoning*, pages 88–123. Queen Mary College, University of London (Workshop Proceedings), EES-MMS-DSFR-76, 1976.

Chapter 3
Reasoning by Analogy in Fuzzy Controllers

Introduction, *56*
Problem Statement, *56*
Logic-Based Neurons, *59*
Analogical Processor - Architecture, *64*
Learning in the AP, *65*
Reasoning by Analogy in the AP Structure, *66*
Analogical Reasoning in Presence of Input Uncertainty, *68*
Characteristics of the Reasoning Scheme, *69*
Conclusions, *71*
References, *71*
Appendix A, 72

Reasoning by analogy in the context of fuzzy controllers can be viewed as a mechanism for providing conclusions about fuzzy control based upon a level of similarity achieved between facts available in the collection of *if-then* rules and a new piece of data for which an appropriate control action has to be determined. The essential feature of this scheme is that, instead of transforming fuzzy sets via a certain fuzzy relation, it operates on the levels of matching between the conditions and the related levels of matching for the conclusions. It is shown in this chapter that this has a profound impact on expressing relevancy of the results of reasoning, and on handling incomplete and/or imprecisely defined input data available for the process of reasoning. Both of these aspects contribute as novel design features of the fuzzy controller. The scheme of reasoning by analogy is of particular interest in processing incomplete input information. The results of inference derived in this case clearly reflect the deteriorated character of the available data by giving rise to interval-valued fuzzy sets. The higher the uncertainty in the input data, the broader the derived intervals of possible grades of membership functions of conclusions. In a similar fashion one can detect inconsistent rules as they again produce interval-valued fuzzy sets. An interesting problem of determining dominant (significant) subsets of control rules which can be used as a suitable representatives of the overall control protocol is also discussed.

REASONING BY ANALOGY IN FUZZY CONTROLLERS

Witold Pedrycz
Electrical & Computer Engineering
University of Manitoba
Winnipeg, Manitoba, Canada R3T 2N2
pedrycz@eeserv.ee.umanitoba.ca

1. INTRODUCTION

Reasoning by analogy has been exploited in many areas of Artificial Intelligence, cf [4]. The relevant studies involving fuzzy sets have been also reported in the existing literature where this mechanism of reasoning was widely applied. This refers in particular to models of fuzzy reasoning and fuzzy controllers. The scheme accommodated in pattern recognition as reported in [6] [9] uses reasoning by analogy as a classification mechanism while [3] deals with the problem of concept formation. Reasoning by analogy has been also considered as a tool for implementing reasoning in frame-based systems, cf. [2]. In [1] a cognitive system being a certain type of the scheme of reasoning by analogy has been presented along with its VLSI implementation.

In this study we will propose a new architecture based on some principles of the reasoning by analogy in the presence of fuzzy sets that have been developed in the previous studies [1] [3]. The current architecture with the aid of which this reasoning will be implemented exploits a logic-based structure named Analogical Processor (AP). The AP units will be designed through a collection of AND, OR, and OR/AND neurons. It will be clarified how a fuzzy controller can be constructed as a family of interconnected Analogical Processors. The study on handling uncertainty associated with the input information to be used in the reasoning scheme will be carried out. The introduced synthetic characteristics of the APs' modules are particularly interesting in designing fuzzy controllers.

The organization of the material is the following. The general problem statement together with the scheme of reasoning by analogy and an analysis of its main functional requirements will be exposed in Section 2. Section 3 introduces to the several classes of logic-based neurons. The architecture of the Analogical Processor will be given in Section 4. The learning issues of the processor will be studied in Section 5. Then we will be concerned with analogical reasoning realized within the introduced structure such as functioning of the AP units, characteristics of the reasoning process and reasoning in the presence of the uncertainty factor biasing the available information. Finally, conclusions will be summarized in Section 7. Numerical examples illustrating the material will be distributed throughout the entire text.

2. PROBLEM STATEMENT

From a formal point of view, the models of reasoning by analogy can be conveniently discussed in the following conceptual framework. Let be given a family of cases (situations) where each of them is composed of a collection of fuzzy sets defined in finite universes of discourse. This family is furnished as

$$r_1: \{a_{11}, a_{12}, ..., a_{1n}\}$$

$$r_2: \{a_{21}, a_{22}, ..., a_{2n}\}$$

$$\vdots$$

$$r_N: \{a_{N1}, a_{N2}, ..., a_{Nn}\}$$

where a_{ij} are discrete fuzzy sets, namely $a_{ij} \in [0,1]^{nj}$, $j = 1,2,...,n$, $i = 1,2,...,N$. The family of the available cases will be treated as a source of knowledge about the problem. The problem itself may be interpreted in many different ways. For instance:
1. a_{ij} are fuzzy sets forming entries of the slots of the instance frames $r_1, r_2, ..., r_N$.
2. $r_1, r_2, ..., r_N$ are control rules of the fuzzy controller. Then some elements in r_k represent conditions of the rule while the others deal with its conclusions (actions).
3. In pattern classification one treats $r_1, r_2, ..., r_N$ as patterns available for training purposes. Then all but one component of the situation pertain to the features of the pattern. The remaining component characterizes its class membership (or multimembership class assignment), see [3] [9].

Bearing this in mind the variety of applications one can envision a general scheme of reasoning by analogy will be realized within a functional structure visualized in Fig.1.

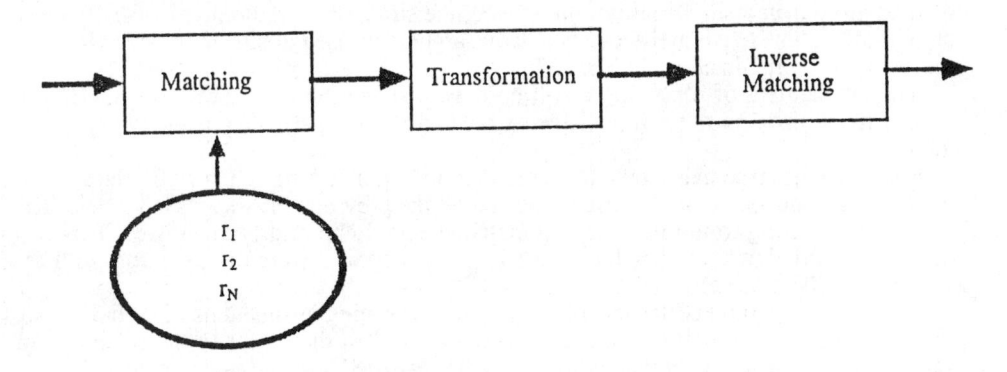

Fig.1. Reasoning by Analogy - Main Functional Blocks

It consists of three clearly visible blocks:
- **matching**. This is the place where the cases r_k's and a new case r are compared by considering all the available objects existing there. The block returns a vector (or a scalar) result of matching. It describes how the available pieces of knowledge match r. We will be also referring to this procedure as an act of matching realized in the space of known objects.
- **transformation**. At this level the obtained results of matching are transformed (mapped) from the space of the known objects into the space of objects to be determined.
- **inverse matching**. This block produces characteristics (grades of membership) of the unknown objects. According to the level of matching achieved at the first block, the results derived now could be less precise than those existing for the original objects in r_k. Depending on the nature of the objects the general scheme of reasoning by analogy takes on quite diverse interpretations. For instance:

- In fuzzy controllers, the objects representing fuzzy sets of conditions (e.g., error and change of error) are matched with the current conditions of the system under control. The inverse matching procedure yields an object particularizing a relevant fuzzy set of control or its interval-valued representative.
- In the classification problem, the features become available while the reasoning by analogy provides information about the possible range of grades of membership of this object in the classes discussed for the patterns.

The reasoning by analogy is completed as follows. Let a new situation

$$r = \{\mathbf{a}_1, \mathbf{a}_2, ..., \mathbf{a}_n\}$$

be incompletely specified, namely the characteristics (membership values) of one or several objects ($\mathbf{a}_i$) are not available. They have to be determined based on the available and fully specified cases $r_1, r_2, ..., r_N$. The key idea of reasoning by analogy in its qualitative setting is to derive unknown elements of r by finding some similarities between one or several situations given previously and the objects specified in r. If the matching is *high* with respect to a certain situation then the unknown object(s) should take entries *similar* to those existing in the corresponding case. A sort of general "continuity" property should be also preserved in this process of reasoning: The higher the matching, the higher the similarity of the derived case to that one that has been established as the best candidate. Since there could be more than a single case of this nature, the situation of reasoning is usually more complicated. In this regard the above qualitative arguments used in this intuitive explanation require further formalization.

The mechanism of reasoning could be developed in many possible ways. Before proceeding with one among possible constructions it is worth posing two general requirements:

- The result of reasoning should reflect the strength of matching achieved for the available components of the object under discussion and all the previous fully specified cases. The uncertainty resulting from this imperfect matching should be clearly quantified. This will make the entire algorithm self-flagging and augment the results of reasoning with the associated certainty levels.
- The scheme should be capable of carrying out reasoning in situations of partially incomplete input information about the objects viewed as the inputs to the scheme. For example, in pattern classification, the pattern to be classified may have one or some features which are provided as *unknown*. This phenomenon of incomplete information should be reflected in the format of the produced results contributing to their higher uncertainty or imply a lowered level of confidence attached to the results of reasoning carried out under these circumstances.

Viewing all the discussed objects as fuzzy sets, the uncertainty related with the inferred object should be incorporated into a concept of a higher order than a fuzzy set itself. The obvious generalization can be accomplished by studying interval-valued fuzzy sets, see [10]. Let us briefly remind that the generic idea of interval-valued fuzzy sets is to depart from single numerical values of the membership degrees and extend them to subintervals of the unit interval. An interval-valued fuzzy set $\tilde{\mathbf{a}} = [\mathbf{a}_-, \mathbf{a}_+]$ $\mathbf{a}_-, \mathbf{a}_+ \in [0,1]^n$ contains a family of fuzzy sets $\mathbf{a} \in [0,1]^n$ such that

$$\mathbf{a}_- \leq \mathbf{a} \leq \mathbf{a}_+$$

(where the above inequality is satisfied coordinatewise). Thus the uncertainty associated with $\tilde{\mathbf{a}}$ invokes all fuzzy sets satisfying the above inequality. All of them are equally possible. The higher the uncertainty level, the broader the bounds of $\tilde{a}$. Further constructs modelling this character of uncertainty applied here could result in fuzzy sets of higher order or probabilistic sets.

3. LOGIC-BASED NEURONS

In this section we will introduce and study basic properties of neurons developed with the aid of logic operations (fuzzy set connectives). The two first neurons are used to carry out generalized AND and OR operations. Consider a collection of inputs x_i, $i = 1,2,\ldots,n$, being arranged in a vector form $\mathbf{x} \in [0,1]^n$. The OR neuron is described formally as

$$y = \text{OR}(\mathbf{x}; \mathbf{w}) \qquad (1)$$

namely

$$y = \text{OR}[x_1 \text{ AND } w_1, x_2 \text{ AND } w_2, \ldots, x_n \text{ AND } w_n]$$

where $\mathbf{w} = [w_1, w_2, \ldots, w_n] \in [0,1]^n$ is a vector of connections (weights) of the neuron. The standard implementation of the fuzzy set connectives standing in the description of the neuron takes advantage of triangular norms (s- and t-norms),

$$y = \underset{i=1}{\overset{n}{S}} [x_i \; t \; w_i] \qquad (2)$$

In the AND neuron the OR and AND operators are interchanged. We obtain

$$y = \text{AND}(\mathbf{x}; \mathbf{w}) \qquad (3)$$

which in the notation of the triangular norms reads as

$$y = \underset{i=1}{\overset{n}{T}} [x_i \; t \; w_i] \qquad (4)$$

One can look at (2) as a kind of a standard Σ-neuron, say

$$y = f\left(\sum_{i=1}^{n} w_i x_i\right) \qquad (5)$$

where "f" denotes a nonlinear transformation (e.g., implemented as a sigmoid function). The second type of the neuron described by (3) reminds somewhat a so-called Π-neuron, where

$$y = \prod_{i=1}^{n} (w_i + x_i)$$

Nevertheless, there exists a significant difference: (5) has no logical meaning and no transparent logical interpretation is available. Neurons (1) or (3) have an obvious logic-oriented interpretation which can be inferred directly from the expressions.

The OR/AND neuron is constructed by fusing several AND and OR neurons into a single structure as shown in Fig.2. The main motivation behind combining several neurons and considering them as a single structure lies in an ability of this neuron to synthesize an intermediate logical behavior which is placed inbetween the pure AND and pure OR characteristics of the previous neurons. The influence coming from the OR (AND) part of the neuron is properly balanced by selecting suitable values of the connections λ and μ, see Fig.2. When $\lambda = 1$ and $\mu = 0$, a pure AND character of the neuron becomes evident. In the second extreme when $\lambda = 0$ and $\mu = 1$, the structure functions as a pure OR neuron. We will use the notation

$$y = OR/AND(x; w, \lambda, \mu)$$

to emphasize the nature of the characteristics produced by the neuron.

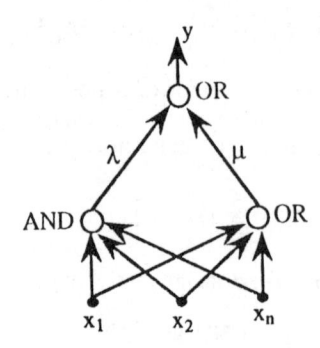

Fig.2. Topology of the OR/AND Neuron

The characteristics of the OR, AND, and OR/AND neurons for some selected values of the connections are shown in Figs.3, 4, and 5.

In comparison to the AND, OR, and OR/AND neurons realizing operations of the aggregative character, the MATCH neuron accomplishes reference computations. Its functional behaviour is described as

$$y = MATCH(x; r, w) \qquad (6)$$

or equivalently

$$y = \overset{n}{\underset{i=1}{S}} \left[w_i \, t \, (x_i \equiv r_i) \right]$$

where $r \in [0,1]^n$ stands for a reference point defined as an element of the unit hypercube.

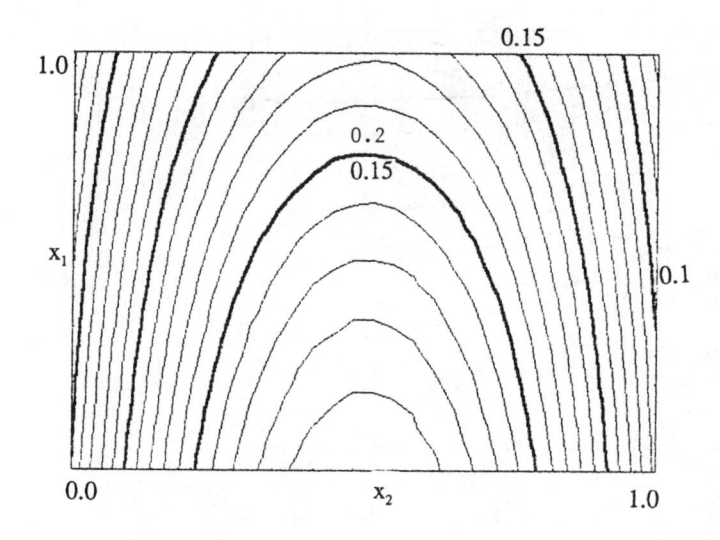

Fig.3.Characteristics of the AND Neuron,
   y=AND([x_1 x_2 $\bar{x}_1$ $\bar{x}_2$],[[0.80 0.10 0.56 0.135])

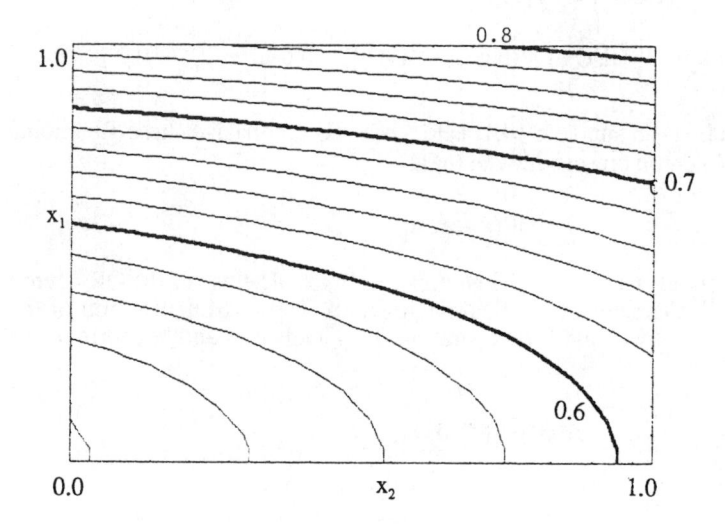

Fig.4.Characteristics of the OR Neuron, y=OR([x_1 x_2 $\bar{x}_1$ $\bar{x}_2$], [0.75 0.31 0.43 0.15])

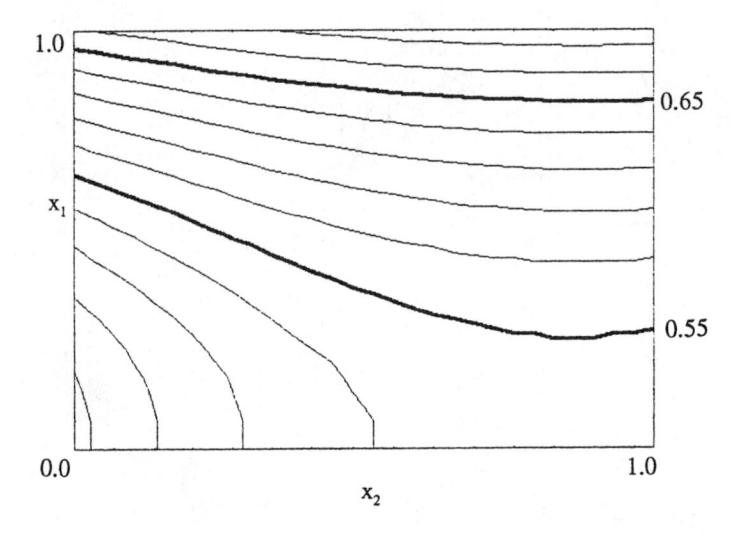

Fig.5.Characteristics of the OR/AND Neuron, $y=$ OR($[[z_1\ z_2]$, $[0.43\ 0.85]$)
$z_1=$ AND ($[x_1\ x_2\ \bar{x}_1\ \bar{x}_2]$, $[0.81\ 0.1\ 0.56\ 0.135]$)
$z_2=$ OR($[x_1\ x_2\ \bar{x}_1\ \bar{x}_2]$, $[0.75\ 0.31\ 0.43\ 0.15]$)

This neuron functions as follows: First the inputs summarized in **x** match **r** using matching operator "≡" and afterwards the results of this operation are aggregated by the OR neuron. The matching operator will be defined as [7]

$$a \equiv b = \frac{1}{2}\left[(a\ \varphi\ b) \wedge (b\ \varphi\ a) + (\bar{a}\ \varphi\ \bar{b}) \wedge (\bar{b}\ \varphi\ \bar{a})\right]$$

where: $\wedge$-minimum and $a\ \varphi\ b$ sup $\{c \in [0,1]\ |\ atc \leq b\}$. To emphasize these functional features of the MATCH neuron one can rewrite (6) as

$$y = OR(x \equiv r;\ w)$$

which stresses the referential character of the neuron, see Fig.6. The use of the OR neuron indicates an "optimistic" character of the final aggregation. The pessimistic form of the aggregation can be realized by using AND operation. This leads into another form of the MATCH neuron:

$$y = AND(x \equiv r; w)$$

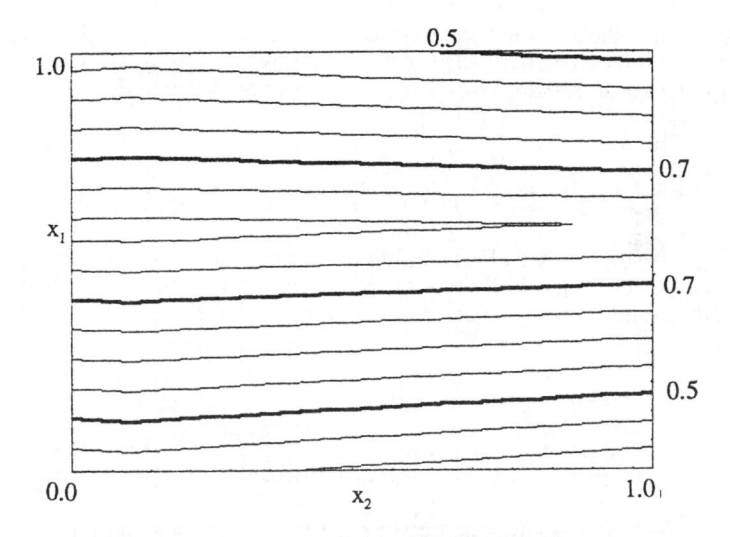

Fig.6. Characteristics of the MATCH Neuron,
$y = $ OR $([x_1 \equiv 0.56 \ x_2 \equiv 0.35], [0.8 \ 0.1])$

The final AND-type used in the MATCH neuron

$$y = \text{AND} ([x_1 \equiv 0.56 \ x_2 \equiv 0.35], [0.8 \ 0.1])$$

produces the results shown in Fig.7.

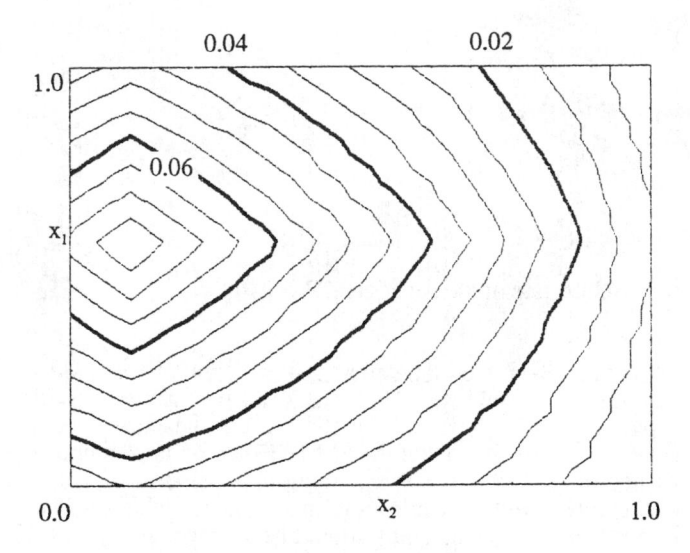

Fig.7. Characteristics of the MATCH Neuron,
$y = $ AND $([x_1 \equiv 0.56 \ x_2 \equiv 0.35], [0.8 \ 0.1])$

These neurons will be used in the construction of a generic neural network called Analogical Processor (AP). The AP realizes a basic process of reasoning by analogy by supporting the phase of matching and the transformation of analogy from the condition into the conclusion space.

4. ANALOGICAL PROCESSOR - ARCHITECTURE

The Analogical Processor forms a class of logic-based neural networks with many inputs and a single output. To focus our discussion, let us consider an "if-then" statement with two subconditions **a** and **b** specified as fuzzy sets defined in discrete universes of discourse,

$$\mathbf{a} = [a_1, a_2, ..., a_{n1}] \quad \mathbf{b} = [b_1, b_2, ..., b_{n2}]$$

and a single conclusion formed by fuzzy set $\mathbf{c} = [c_1, c_2, ..., c_m]$. Since we will be working with a single component of **c**, its index can be ignored and we will be simply referring to it as c.

Now let a certain state of the environment be expressed by fuzzy sets **a**' and **b**'. The role of the network is to produce a level of similarity y at the output node (representing element c) for a certain collection of the results of matching observed at the input. The AP architecture consists of three layers, see Fig.8.

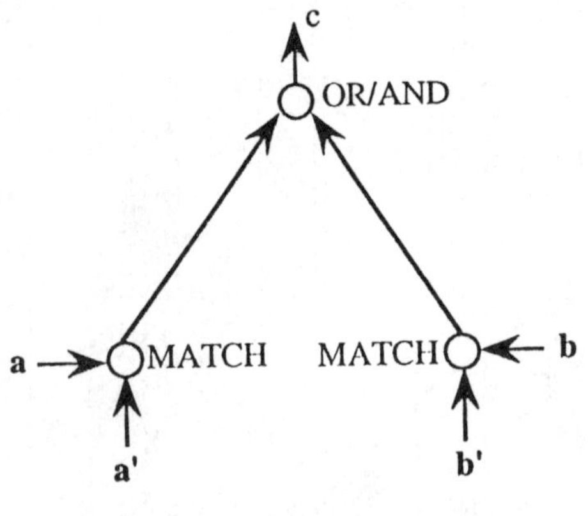

Fig.8. General Architecture of the Analogical Processor

The role of the first layer is to distribute the inputs **a**' and **b**' which afterwards are sent to the MATCH neurons constituting a hidden layer. Note that **a** and **b** constitute the reference points for **a**' and **b**', respectively. The results of matching are then collected by the OR/AND neuron (output layer). While the role of the MATCH neuron is well understood, the use of the OR/AND neuron deserves some explanation. This form of the neuron provides enough representation flexibility and allows for a smooth switching between the OR-type of aggregation (requiring that the matching should be achieved for at least **a** and

a' or **b** and **b'**) and the AND-type of aggregation. The character of this aggregation is not available in advance and usually is far enough from the pure types conveyed by the AND or OR characteristics.

The detailed formulas describing the network are written now as

hidden layer:

$$z_1 = \text{MATCH}(\mathbf{a'}; \mathbf{a}, \mathbf{w}_1)$$
$$z_2 = \text{MATCH}(\mathbf{b'}; \mathbf{b}, \mathbf{w}_2)$$

output layer:

$$y = \text{OR/AND}(\mathbf{z}; \mathbf{u}, \lambda, \mu)$$
$$\mathbf{z} = [z_1, z_2]$$
$$\mathbf{u} = u_1, u_2$$

The AP can be immediately extended to handle any number of subconditions in the case (rule) like $\mathbf{a}_1, \mathbf{a}_2, ..., \mathbf{a}_c$ where $\mathbf{a}_i \in [0,1]^n$, $i = 1, 2, ..., c$. To handle the entire m-dimensional space we can follow two ways:
- "m" single output APs are combined,
- form the AP structure with m-outputs.

As it will be discussed later these two topologies have a certain impact both on the learning of the network and the overall architecture. The first topology is highly modular: it can be easily extended to any size of the output space. The learning in this case can be also faster since no re-training of the already existing modules is required. The second m-output version of the AP is more economical as it results in more condensed structures. This version is recommended for a higher dimensionality of the output space. Unfortunately, the structural modularity cannot be sustained within this structure.

5. LEARNING IN THE AP

We will now address learning in the AP as a central issue determining an efficient use of the network. The learning is based on available cases and is carried out in a supervised mode. The learning set consists of "N" cases and proceeds in a referential format, including all different pairs of the cases. These cases used for training can be conveniently arranged in a tabular form

$$\begin{bmatrix} (r_1, r_1) & (r_1, r_2) & (r_1, r_N) \\ (r_2, r_1) & (r_2, r_2) & (r_2, r_N) \\ (r_N, r_1) & (r_N, r_2) & (r_N, r_N) \end{bmatrix}$$

Each pair (r_i, r_j) represents an individual training situation in which one of the cases, say r_i, is viewed as a reference case.

By following this way the AP is exposed to the levels of matching achieved between the input and output objects of the cases. Note that the matrix is symmetrical so only a single element out of each pair (r_i, r_j) and (r_j, r_i) will be included in learning set. Furthermore, the elements that are situated on a main diagonal (pairs (r_i, r_i)) always expose the AP to the level of matching equal identically to one both at the input and the output of the network. In total, the set of "N" cases produces

$$N(N+1)/2$$

pairs to be utilized in this mode of learning.

The learning set created in the above way is used to adjust the connections of the neurons in the structure so that a certain performance index Q is minimized,

$$\min_{\text{connections}} Q$$

The standard gradient-based method yields the updates proportional to $-\frac{\partial Q}{\partial \text{ connections}}$. The resulting learning scheme is standard to a high extent; the reader is referred to [9] with respect to more specialized computational details as well as some improvements which could be particularly essential for some types of triangular norms being used to implement the network. The learning scheme described in Appendix A serves as an illustration on how the details of the algorithm can be worked out.

6. REASONING BY ANALOGY IN THE AP STRUCTURE

We will discuss in detail how the reasoning process is realized applying the structure given in Fig.1. The basic idea is to cycle through all the cases (rules) and summarize outcomes of matching at the output node. In sequel one can study an architecture where the cases are analyzed in parallel; in contrast to that, the current mode can be thus viewed as a serial mode of reasoning by analogy.

Let us start from the first case composed of $\mathbf{a}_1$ and $\mathbf{b}_1$. The input of the network is furnished with $\mathbf{a}$ and $\mathbf{b}$. The output of the network is γ_1 which describes the transformed result of matching achieved for $(\mathbf{a}, \mathbf{b})$ and $(\mathbf{a}_1, \mathbf{b}_1)$. Cycling through all the cases used in the learning phase we derive a collection of the levels of matching for the unknown output object and the objects included in the cases. They are summarized as the pairs:

$$c_1, \gamma_1$$
$$c_2, \gamma_2$$
$$\vdots$$
$$c_N, \gamma_N$$

The maximal value of γ_i contributes to the formation of the result of reasoning emerging in this situation. Let us first assume that there exists only a single maximal element in the entire collection of γ_i's,

$$\exists!_{i_0} \gamma_{i_0} = \max_{i=1,2,\ldots,N} \gamma_i$$

This highest level of γ_{i_0} is used in building the corresponding range of possible grades of membership. Since γ_{i_0} and c_{i_0} are provided, we can look at this problem as inverse matching [7] in which one has to generate all c's which satisfy the requirement:

the level of matching of c and c_{i_0} is at least γ_{i_0}

This translates into the inequality

$$c \equiv c_{i_0} \geq \gamma_{i_0}$$

The solution to this inequality forms always a subinterval of [0,1], namely $[c_-, c_+]$, see [7]. This subinterval of [0,1] reduces to a single point (a single membership grade) if and only if $\gamma_{i_0} = 1$. In all the remaining cases its width is a nondecreasing function of γ_{i_0}. Lower values of γ_{i_0} imply higher uncertainty, and consequently, broader ranges of the feasible membership values.

In the case of the fuzzy controller working in a numerical mode (viz. generating a single numerical response) the interval-valued fuzzy set $[\mathbf{c}_-, \mathbf{c}_+] = [[c_{1-}, c_{1+}], [c_{2-}, c_{2+}] \ldots [c_{m-}, c_{m+}]]$ defined in space $\mathbf{Y} = [y_1, y_2, \ldots, y_m]$, $y_j \in \mathbf{R}$ has a straightforward consequence in producing a range of possible numerical values.

The transformation procedure applied to a fuzzy set of control $[c_1, c_2, \ldots, c_m]$ yields y_o as a type of weighted average

$$y_o = \frac{\sum_{j=1}^{m} c_j y_j}{\sum_{j=1}^{m} c_j}$$

For the interval-valued fuzzy set the same expression produces bounds for the numerical control, say $[y_{o-}, y_{0+}]$

$$\tilde{y} = \frac{\sum_{j=1}^{m} [\delta_j c_{j-} + (1 - \delta_j) c_{j+}] y_j}{\sum_{j=1}^{m} [\delta_j c_{j-} + (1 - \delta_j) c_{j+}]}$$

where $\delta_i \in \{0,1\}$. Depending on its value, δ_j selects element from the interval. The above expression is thus a function of $\delta = [\delta_1, \delta_2, \ldots, \delta_m]$. As such, when minimized or maximized over δ, it generates an interval of admissible numerical values of control.

$$\min_{\delta} \tilde{y} \to y_{o-} \quad \max_{\delta} \tilde{y} \to y_{o+}$$

The second special case of inverse matching that should be dealt with separately occurs when several values of γ_i's are the same. Denote the set of indices for which γ_i attains maximum by I:

$$I = \{i \mid \max_{j} \gamma_j = \gamma_i\}$$

The previous procedure has to be slightly modified to take care of all subintervals generated by those c_i's indicated by I. Denote these intervals by $[c_{i-}, c_{i+}]$, $i \in I$. The pessimistic way of aggregation of the uncertainty about the grades of membership resulting from these cases is to build the interval

$$[\min_{i \in I} c_{i-}, \max_{i \in I} c_{i+}]$$

The situation described above where more than a single case becomes highly activated, card (I) > 1, occurs for instance, if **a** is biased by an initial uncertainty or several competing cases have been included in the data set. This topic will be studied in more detail in the next section.

7. ANALOGICAL REASONING IN PRESENCE OF INPUT UNCERTAINTY

The uncertainty occurring in the input information provided to the AP can be easily modelled and delineated in the results of reasoning. To narrow a scope of our discussion let the input **x** of the AP described as

$$\gamma = AP(\mathbf{x}; \mathbf{a}, \mathbf{W})$$

be biased by the factor of uncertainty (in the above notation **W** is used to concisely summarize all the connections of the processor). In other words **x** is viewed as "unknown". By stating this we acknowledge that the values of membership are not given at all. In other words, any membership value is admitted, so x_i expands to the unit interval and cannot be handled as a single numerical value (grade of membership). For a total ignorance one considers that $x_i \in [0,1]$[1]. The consequence of this fact is that the coordinates of **a** match the unit interval so the upper numerical limit of matching achieved in this situation is always 1. Thus since there is always an element in the interval equal to a_i, this implies

$$a_i \equiv [0,1] = \sup_{\xi \in [0,1]} (a_i \equiv \xi) = 1$$

The limited uncertainty level about **x** can be represented as a subinterval of [0,1], for instance its i-th coordinate can be modelled as $[x_{i-}, x_{i+}] \subset [0,1]$. Then the result of matching of this interval and a_i returns a value

$$a_i \equiv [x_{i-}, x_{i+}] = \sup_{\xi \in [x_{i-}, x_{i+}]} (a_i \equiv \xi)$$

The straightforward consequence of these computations is the elevated level of matching.

Denote by γ_k the output of the AP obtained for the analogical reasoning completed for **x** and $\mathbf{a}_k$. If **x** is associated with the maximal level of uncertainty (the unit interval) then the results of reasoning are the same for all the cases, k = 1,2,..., N. This yields

---

[1]One should stress that a way of modelling uncertainty by a fuzzy set equal identically 1 (identity membership function) over the entire universe of discourse, as commonly found in the literature, is not suitable here: the uncertainty should result in constructs of an order higher than fuzzy sets (such as e.g., interval-valued fuzzy sets).

$$\gamma_k = \gamma = AP(\mathbf{x}; \mathbf{a}_k, \mathbf{W})$$

When converted into membership values, they produce intervals,

$$\gamma = c_1 \equiv c \quad [c_{1-}, c_{1+}]$$
$$\gamma = c_2 \equiv c \quad [c_{2-}, c_{2+}]$$
$$\vdots$$
$$\gamma = c_N \equiv c \quad [c_{N-}, c_{N+}]$$

which have to be aggregated to represent uncertainty visible at the output level of the reasoning scheme. Considering a conservative mechanism of uncertainty aggregation we come up with the interval

$$[\min_{k=1,2,\ldots,N} c_{k-}, \max_{k=1,2,\ldots,N} c_{k+}]$$

An interesting situation arises where all γ_k's are equal to 1, namely $AP(\mathbf{x}; \mathbf{a}_k, \mathbf{W}) = 1$, $k = 1, 2, \ldots, N$. Then the AP generates successively c_k's appearing in the corresponding cases. The same aggregation as applied above yields an interval covering all results of reasoning

$$[\min_{k=1,2,\ldots,N} c_k, \max_{k=1,2,\ldots,N} c_k]$$

Hence the complete uncertainty at the input of the reasoning scheme triggers all the output objects of the cases. The limited level of uncertainty tends to narrow down the range of possible output objects being considered as the corresponding results of reasoning.

7. CHARACTERISTICS OF THE REASONING SCHEME

The APs can be characterized by describing results of reasoning. Two types of obtained results will be of interest: (i) attraction regions of the cases and (ii) hierarchy of cases. Both of them allow us to get a better insight into the way how the reasoning is accomplished. The attraction regions of the cases are built by determining all the grades of membership of the object for which a certain case r_{k_o} generates the highest output of the AP. These regions can vary for different coordinates of the object being determined. Let

$$\gamma = AP(\mathbf{x}; \mathbf{a}_k, \mathbf{W})$$

be the output of the AP produced by the k-th case. Then the attraction region, $Attr(r_k)$, denotes all $\mathbf{x}$'s such that

$$Attr(r_k) = \{\mathbf{x} \mid AP(\mathbf{x}; \mathbf{a}_k, \mathbf{W}) = \max_{\substack{j=1,2,\ldots N \\ j \neq k}} AP(\mathbf{x}; \mathbf{a}_j, \mathbf{W})\}$$

The attraction regions can be additionally quantified by adding a threshold level $\beta \in [0,1]$. Its role is to identify $\mathbf{x}$'s yielding a sufficiently high level of similarity produced by the AP,

$$\mathrm{Attr}(r_k, \beta) = \{ x \mid \mathrm{AP}(x; a_k, W) = \max_{\substack{j=1,2,\ldots N \\ j \neq k}} \mathrm{AP}(x; a_j, W) \text{ and } \mathrm{AP}(x; a_k, W) \geq \beta \}$$

By studying the distribution of the attraction regions we can gain a better understanding about the performance of the cases incorporated in the reasoning scheme. Consider two situations shown in Fig.9 and illustrating distribution of $\mathrm{Attr}(r_k, \beta)$ for several values of "k" and the given level of β. The first drawing, Fig.9(a), illustrates the cases that are distributed fairly sparsely in the space of matching: note that there is no overlap between the attraction regions, however, some area is not covered by these regions at all. In Fig.9(b), the distribution of the attraction regions indicates their significant overlap. This situation is characteristic of cases which are conflicting (contradictory). Another reason for this overlap could be that there have been too many cases accommodated in the reasoning scheme during its learning phase.

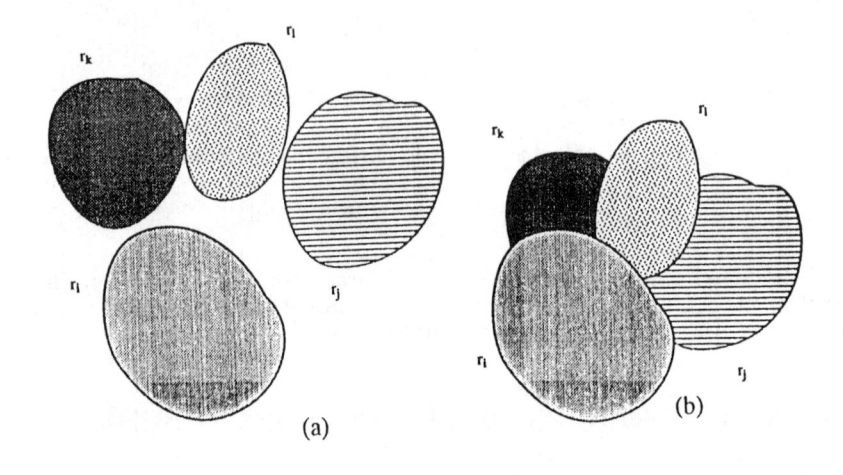

Fig.9. Distribution of Attraction Regions

This performance index allows us to look at the reasoning scheme as a sort of an associative memory in which the quality of recall depends heavily on the number of the memorized items and the relationships between them.

The studies of the attraction regions could be significantly helpful in the analysis of fuzzy controllers. Let us remind ourselves that the cases become just control rules:

r_k: if A_k and B_k and ... then U_k

$k=1,2,\ldots,N$. The attraction regions can partition the space of conditions as shown in Fig.9. In the context of reasoning applied to fuzzy controllers the interpretation of the attraction regions refers to the nature of the control rules:

-if there is no overlap between the regions, the rules are nonconflicting and each region identifies situations for which a control action can be inferred with a high degree of confidence (at least β),

-the overlap indicates that some of the control rules are conflicting.

The hierarchy of cases is constructed by studying their representation capabilities. The idea is to quantitatively specify how a certain case "k_o" represents all the remaining cases. For a single output AP these capabilities are characterized by the following sum obtained for the AP when being exposed to a_k's, namely

$$c_{k_o} = \sum_{k=1}^{N} AP(a_{k_o}; a_k, W)$$

The higher the value of c_{k_o}, the higher the capabilities of this case to represent other situations.

This index introduces a linear ordering among the cases. The most significant can be thus extracted from the entire collection. This in turn could contribute to the reduction of the architecture necessary to implement reasoning.

8. CONCLUSIONS

We have proposed an idea of reasoning by analogy when available cases consist of objects viewed as fuzzy sets and formulated in finite dimensional hypercubes. The general concept is then realized in the form of Analogical Processors. Their architecture as well as the learning aspects have been discussed. The uncertainty associated with the results of reasoning is quantified in the form of interval-valued fuzzy sets. This feature constitutes an evident advantage of the proposed architecture in comparison to some other implementations of this type of reasoning. The associated characteristics of the AP provide a detailed look into the results of reasoning depending on the cases available during the training phase and the nature of the case for which the reasoning has to be performed.

REFERENCES

1. Diamond,J., McLeod,R.D., Pedrycz,W., "A fuzzy cognitive structure: Foundations, applications and VLSI implementation", *Fuzzy Sets and Systems*, 47, 49-64, 1992.

2. Di Nola,A., Pedrycz,W., Sessa,S., "Knowledge representation and processing in frame-based structures", In: *Fuzzy Engineering Toward Human Friendly Systems*, vol.1, Proc. Int. Fuzzy Engineering Symp. '91, Yokohama, pp. 461-470, 1991.

3. Hirota K., Pedrycz,W, "Concepts formation: representation and processing issues", *Int. J. of Intelligent Systems*, 7, 3-13, 1992.

4. Kling, R.E., "A paradigm for reasoning by analogy", *Artificial Intelligence*, 2, 147-178, 1971.

5. Koczy,L.T., Hirota,K., "Reasoning by analogy with fuzzy rules", *IEEE Conf. on Fuzzy Systems*, March 8-12, 1992, San Diego, pp.263-270.

6. Pedrycz,W., "A fuzzy cognitive structure for pattern recognition", *Pattern Recognition Letters*, 9, 305-313, 1989.

7. Pedrycz,W., "Direct and inverse problem in comparison of fuzzy data", *Fuzzy Sets and Systems*, 34, 223-235, 1990.

8. Pedrycz,W., "Neurocomputations in relational systems", *IEEE Trans.on Pattern Analysis and Machine Intelligence*, 13, 289-296, 1991.

9. Pedrycz,W., Bortolan,G., Degani,R., "Classification of electrocardiographic signals:a fuzzy pattern matching approach", *Artificial Intelligence in Medicine*, 3, 331-46, 1991.

10. Sambuc,R., "Functions φ-flous:Application de l'Aide a Diagnostique en Pathologie Thyroidienne", These Univ.de Marseille, Marseille, 1975.

11. Turksen,I.B., Tian,Y., "Bounds on multiple antecedent fuzzy s-implications and reasoning", In: *Fuzzy Engineering Toward Human Friendly Systems*, vol.1, Proc. Int. Fuzzy Engineering Symp.'91, Yokohama, pp. 185-196, 1991.

APPENDIX A

This appendix summarizes all details regarding parametric learning in the many input-single output AP. The formulas describing the processor are:

the output layer consists of a single OR/AND neuron

$$y = OR(\mathbf{u}; [\lambda\, \mu]) \quad \mathbf{u} = [u1, u2]$$

$$u_1 = AND(\mathbf{z}; \mathbf{w}_1) \quad u_2 = OR(\mathbf{z}; \mathbf{w}_2)$$

$\mathbf{z}, \mathbf{w}_1, \mathbf{w}_2 \in [0,1]^c$

the hidden layer consists of "c" MATCH neurons

$$z_i = MATCH(\mathbf{x}_i; \mathbf{r}_i, \mathbf{v}_i) = AND\, (\mathbf{x}_i \equiv \mathbf{r}_i; \mathbf{v}_i)$$

$$\mathbf{x}_i = [x_{i1}, x_{i2}, ..., x_{in_i}] \in [0,1]^{n_i}$$

$$\mathbf{r}_i = [r_{i1}, r_{i2}, ..., r_{in_i}] \in [0,1]^{n_i}$$

$i = 1,2,...,c$

The learning scheme will be described in its on-line version: the input

$$\mathbf{x} = [\mathbf{x}_1, \mathbf{x}_2, ..., \mathbf{x}_c]$$

$\mathbf{x} \in [0,1]^{n_1+n_2+...+n_c}$ and the corresponding target value t are given as a training pair $(\mathbf{x}, t)$. The objective is to minimize the performance index

$$Q = [t - AP(\mathbf{x}; \textbf{parameters})]^2$$

with respect to all the parameters of the processor. The gradient-type method gives rise to the following general expression

$$\text{new\_\textbf{parameters}} = \textbf{parameters} - \alpha \frac{\partial Q}{\partial \textbf{parameters}}$$

where $\alpha \in [0,1]$. Furthermore

$$\frac{\partial Q}{\partial \textbf{parameters}} = -2(t - AP(\mathbf{x}; \textbf{parameters})) \frac{\partial AP}{\partial \textbf{parameters}}$$

The increments (adjustments) of the parameters (connections) of the AP given below have been derived for the t-norm viewed as the product operation (ab) and the s-norm delineated as the probabilistic sum (a s b = a + b - ab). We obtain

$$\frac{\partial AP}{\partial \lambda} = u_1(1 - u_2\mu) \qquad \frac{\partial AP}{\partial \mu} = u_2(1 - u_1\lambda)$$

$$\frac{\partial AP}{\partial w_{1i}} = \frac{\partial AP}{\partial u_1} \frac{\partial u_1}{\partial w_{1i}}$$

$$\frac{\partial AP}{\partial u_1} = \lambda(1 - \mu u_2)$$

$$\frac{\partial u_1}{\partial w_{1i}} = A(1 - z_i), \quad A = \prod_{\substack{j=1 \\ j \neq i}}^{c} (w_{1j} + z_j - w_{1j}z_j)$$

$i = 1,2,...,c$

$$\frac{\partial AP}{\partial w_{2i}} = \frac{\partial AP}{\partial u_2} \frac{\partial u_2}{\partial w_{2i}}$$

$$\frac{\partial AP}{\partial u_2} = \mu(1 - \lambda u_1)$$

$$\frac{\partial u_2}{\partial w_{2i}} = z_i(1 - B), \quad B = \mathop{S}_{\substack{j=1 \\ j \neq i}}^{c} (w_{2j}z_i)$$

$$\frac{\partial AP}{\partial v_{ji}} = \left[\frac{\partial AP}{\partial u_1} \frac{\partial u_1}{\partial z_j} + \frac{\partial AP}{\partial u_2} \frac{\partial u_2}{\partial z_j} \right] \frac{\partial z_j}{\partial v_{ji}}$$

$$\frac{\partial u_1}{\partial z_j} = (1 - w_{1j})A_1, \quad A_1 = \prod_{\substack{l=1 \\ l \neq j}}^{c} (w_{1l} + z_l - w_{1l}z_l)$$

$$\frac{\partial u_2}{\partial z_j} = w_{zj}(1 - B_1), \quad B_1 = \mathop{S}_{\substack{l=1 \\ l \neq j}}^{c} w_{2l}z_l$$

$$\frac{\partial z_j}{\partial v_{ji}} = B_2(1 - (x_i \equiv r_{ji})), \quad B_2 = \prod_{\substack{l=1 \\ l \neq i}}^{n_j} (v_{jl} + (x_l \equiv r_{jl}) - v_{jl}(x_l \equiv r_{jl}))$$

$j = 1, 2, ..., c, \quad i = 1, 2, ..., n_j$

Chapter 4
Information Complexity and Fuzzy Control

Preamble, *76*
Introduction, *78*
Uncertainty of Membership Functions, *79*
Comparison of Uncertainties, *88*
Closing Remarks, *94*
References, *95*

This chapter analyzes the complexity of extracting information from fuzzy data. This process is also considered as contribution of information, hence also alleviating uncertainty about the result of the process itself. The estimate of complexity serves as a measure of information, and of uncertainty, of the underlying fuzzy membership assignment. The chapter mainly studies the complexity of element identification through binary search, based on progressively more complex fuzzy domains. The use of the expedient of function *rearrangement* permits handling fuzzy sets based on arbitrary measurable functions. Also dealt with is the relationship of complexity to the choice of fuzzy operators. Such operators correspond to binary connectives of fuzzy logic and their choice plays significant role in the performance of fuzzy control processes. Choices corresponding to maximum and minimum values of information are identified. In terms of fuzzy control, they represent either the most stable or the smoothest control strategies. Also explores are ties to Shannon entropy (the probabilistic measure of information) and to U-uncertainty (the possibilistic measure of uncertainty).

INFORMATION COMPLEXITY AND FUZZY CONTROL

Arthur Ramer
School of Computer Science
University of New South Wales
Kensington, N.S.W. 2033, Australia

Vladik Kreinovich
Computer Science Department
University of Texas
El Paso, TX 79968

1 PREAMBLE

Our article analyzes complexity of extracting information from fuzzy data. Such process can be properly viewed as contributing information, hence also alleviating uncertainty about the result of the process itself.

Notion of information has been utilized frequently, both formally and informally. In the formal vein, Shannon entropy of probabilistic distributions is foremost, followed by host of less perfect variants. In the theory of possibility [3], based on fuzzy sets, there is U-uncertainty and several related functions [7, 16]. Both these concepts can be introduced axiomatically, using general rules of combining *information* [7]. Such rules deal with combining independent events, restricting ranges of possible outcomes, and other operations on distributions. Specialized to either probability or possibility theories, they become axioms, which can be shown sufficient to characterize the respective information measures uniquely [16].

Independent of those methods, a question of *complexity* of computing numerical answers to analytical queries has been studied [18, 19]. Here the query is associated with a function about which we have only a partial information. A typical example is finding a root or the integral of a function, for which available are only the values at a discrete set of points. A particularly important example, which also serve as the point of departure in [18], is the *binary search*.

The very same problem arose in analysis of complexity of securing information from fuzzy membership functions. This problem was introduced in [10, 11], where the resulting information gain was termed *entropy*. However, it is clearly most closely related to the information-based complexity [18, 19], and we shall use a neutral term *information* throughout this article. This usage obviates the need of axiomatic characterization, which is not germaine to this concept. At the same time there are certain ties to Shannon entropy, which we discuss in the closing section [14].

Other ties, perhaps more significant, are to the possibilistic information on continuous domains. This relationship can be made formal as we point out in the closing section. In particular, the main information expression is virtually identical both in our information complexity study, and in possibilistic uncertainty.

Here we begin by studying the complexity of element identification through binary search, based on progressively more complex fuzzy domains. (Obviously, for crisp domains our results parallel those in [18].) Using the expedient of function *rearrangement* we can apply it to fuzzy sets based on arbitrary measurable functions. The second part deals with relationship of complexity to the choice of fuzzy operators. Such operators correspond to binary connectives of fuzzy logic; their choice plays significant role in performance of fuzzy control processes. We show how to select the operators corresponding both to the maximum and to the minimum values of information. In terms of fuzzy control they represent either the most stable, or the smoothest control strategies.

Numerical quantification of *information* can be always viewed equally as quantification of *uncertainty*. Any analysis, experiment, or observation which *gains* information, conveys *removing* uncertainty. In all the frameworks discussed analytically, rather than epistemically, these notions can be used interchangeably. We shall do so here, mostly for stylistic reasons.

In our notation we follow the possibilistic interpretation of fuzzy sets and membership grades. In particular we use $\pi(x)$ to denote the membership on domain X, thus making $\pi(x)$ the *possibility* of $x \in X$. As an added convenience we can use $\mu(x)$ to mean the Lebesgue measure, which we assume defined on all 'continuous' domains of discourse.

2 INTRODUCTION

Traditional control theory is applicable in the situations when the analytical behavior of the system to be controlled, is known. However, in many situations, e.g., in space exploration, we do not have this knowledge, but we still have to make control decisions. In such situations, a reasonable idea is to find a human operator who is good in this kind of control, and translate his control experience into a precise formula. This control experience is usually formulated in terms of the natural-language rules like *if x is small, then control must also be small*. The methodology of translating these rules into an actual control strategy, due to its relationship to fuzzy sets and logic, became known under the name of *fuzzy control* [1, 13, 12].

Three main steps are necessary to specify this translation. Firstly, we must determine membership functions that correspond to natural language terms (like *small* or *big*) that appear in the rules. Secondly, we must choose operations that correspond to & and ∨. As a result we get a membership function $\pi_C(u)$ for a control; then we need a method to transform this function $\pi_C(u)$ into a single control value (*defuzzification method.*)

Different choices may lead to dramatically different control strategies, making such choice very important [10, 11]. We begin by considering the problem of interpreting &− and ∨−operations. As making a choice restricts the set of possible control strategies, and wrong choice may lead to a poor control strategy, it is reasonable to preserve as many possibilities as feasible. In other words, we should choose &− and ∨−operations in such a way that the uncertainty corresponding to $\pi_C(u)$ is the greatest possible. This methodology is well known in probability theory under the name of *maximum entropy formalism*, and has been applied widely to a variety of problems, ranging from pattern recognition to processing uncertainties in expert systems [5, 8, 2, 9].

Just like in the probabilistic case, we want to evaluate the uncertainty of a membership function as the average number of binary questions that one needs to ask in order to determine the value. In the present paper, we propose the formulas that compute this uncertainty and determine operations for which this uncertainty is maximized. We prove that the desired maximum uncertainty is attained when we use $min(a+b,1)$ for $\vee$, and $min(a,b)$ for $\&$.

In control theory terms, maximum entropy leads to the maximally stable strategy. This result is an intuitive one—we minimized the lost opportunities, and hopefully ended up with the best possible control.

The above arguments are reasonable only if we are ready to apply various defuzzification techniques to extract the best control from $\pi_C(u)$. However, in industrial applications, a defuzzification rule is usually fixed. Since this rule is not necessarily the most appropriate [20, 11], it is reasonable to try to depend on it to the least extent. It suggests choosing $\&$ and $\vee-$ operations based on condition that the uncertainty related to $\pi_C(u)$ is the *least* possible. We find that the operations for which the resulting uncertainty is the least are $max(a,b)$ for $\vee$ and ab for $\&$. In control terms, minimum entropy leads to the maximally smooth strategy. This result is also quite intuitive—since we are extremely cautious, we end up with a very smooth control.

3 UNCERTAINTY OF MEMBERSHIP FUNCTIONS

3.1 Motivation

Let us recall where the values $\pi(x)$ of a membership function come from. If $\pi(x)$ corresponds to, say, *small*, then $\pi(x)$ is our degree of belief that x is small. One of the most natural ways to evaluate numerically this degree of belief is to poll several experts, asking whether they consider x small or not, and after M out of N answer "yes", take M/N as $\pi(x)$ [7]. This approach permits interpreting the value $\pi(x)$ as either frequency or subjective probability that x is small.

With this interpretation in mind, let us estimate uncertainty $U(\pi)$ that corresponds to function $\pi(x)$, and suppose that a certain notion like *small* is being described by the function $\pi(x)$. If the only thing we know about some real value x is that it satisfies this property, then how many binary questions do we have to ask to determine x? As it will be become apparent, we do not just want to compute the simple complexity of answering such query, but also to recognize the maximum values that related possibilities may assume.

The analysis is best organized as a sequence of progressively more general cases. We begin with finitely many alternatives, when it is easy to estimate uncertainty, and proceed towards a general membership function. The discussion places stress on the background and motivation, leaving some mathematical details to the last part of the article.

3.2 Information of crisp assignments

We want to evaluate uncertainty according to the number of binary questions that we have to ask to get the complete knowledge. This number is easy to dfine and to understand in the discrete case, when we have finitely many alternatives $x_1, \ldots, x_n$. In this case finitely many questions suffice to determine the actual alternative and our uncertainty can be estimated as the smallest number of binary *yes-no* questions that we have to ask to determine x_i. As x_i come from a numerical domain—they represent possibility values—they can be considered as ordered, and binary search employed. For n alternatives we need $Q = \lceil \log_2(n) \rceil$ questions, thus proving

PROPOSITION 1 *The uncertainty value of n alternatives is* $\log_2(n)$.

Now, let us start describing uncertainty for membership functions, starting with the simplest one, when the fuzzy set is crisp. First, let us recall basic definitions.

By a *membership function* we mean a function $\pi : R \to [0,1]$ that is not identically 0. If for every x, $\pi(x) \in \{0,1\}$, the function is called *crisp*. Such functions are *characteristic* functions of sets $S \subset R$, permitting us to term such standard sets as *crisp*.

We are facing the problem of estimating uncertainty when membership function $\pi(x)$ is the characteristic function of an interval $[a,b]$, expressing the property that $x \in [a,b]$. Now the set of alternatives coincides with the set of all points on an interval and is infinite. If we ask finitely many questions, then we will have only finitely many possible answers. Therefore, to find the precise value of x, we need to ask infinitely many questions.

The natural solution of this problem is related to the fact that in real life, we never know precise values of physical quantities. We can measure them, but all the measurements have finite precision. The only information that we can get from a measurement procedure is a value $\tilde{x}$ such that the actual value x satisfies the inequality $|x - \tilde{x}| \leq \varepsilon$, where $\varepsilon > 0$ is the precision of this measurement. In other words, the actual value x belongs to the interval $[\tilde{x} - \varepsilon, \tilde{x} + \varepsilon]$. After we know such an $\tilde{x}$, we say that we know x with precision ε.

Therefore, to estimate of the information that $x \in [a, b]$, we fix some ε, and estimate the number of binary questions that have to be asked in order to find x with precision ε.

DEFINITION 1 *Let S be a set of real numbers, and $\varepsilon > 0$. ε−alternatives are a set $\{x_1, \ldots, x_n\}$ of real numbers such that*

$$S \subset \bigcup_{i=1}^{n} [x_i - \varepsilon, x_i + \varepsilon].$$

They are called ε−alternatives because if we know which of the values x_i is ε−close to the actual x, then we know x with precision ε. They are considered *alternatives*, because according to the above definition, for each possible $x \in S$ there exists an x_i, for which $|x - x_i| \leq \varepsilon$. To express the *uncertainty* of this set of alternatives we take $\log_2(n)$.

For a given set S we can have several different sets of ε−alternatives. Different sets of ε−alternatives may have different number of elements, and thus lead to different values of uncertainty. It is natural to consider the smallest of these values as the measure of uncertainty of S. We introduce a convenient notation.

DEFINITION 2 *Let S be a set of real numbers, and $\varepsilon > 0$. By ε−uncertainty $Q(\pi_S, \varepsilon)$ of S, we mean the uncertainty of the smallest possible set of ε-alternatives.*

This notation may look somewhat clumsy: why not simply $Q(S, \varepsilon)$? The reason is that later on, we will define $Q(\pi, \varepsilon)$ for membership functions π that do not necessarily describe crisp sets.

A straightforward argument establishes the next result.

PROPOSITION 2 $Q(\pi_{[a,b]}, \varepsilon) = \log_2(\lceil (b-a)/(2\varepsilon) \rceil)$.

To prove it we observe that the smallest family of 2ε-length intervals, whose union covers $[a, b]$ has $\lceil (b - a)/(2\varepsilon) \rceil$ elements. Taking its logarithm gives the proposition.

By explicitly mentioning a precision ε when we described uncertainty, we ended up with a finite value of uncertainty. The drawback is that instead of one value that described uncertainty in a finite case, we now have a function that correspond to different $\varepsilon > 0$. It would be desirable to isolate the components contributed by the interval $[a, b]$ itself, and by the precision factor ε.

As we are interested primarily in behavior of $Q(\pi, \varepsilon)$ for small $\varepsilon > 0$ the following definition is in order.

DEFINITION 3 *Two functions $f(\varepsilon)$ and $g(\varepsilon)$ are asymptotically equivalent, denoted $f(\varepsilon) \sim g(\varepsilon)$ if $\lim_{\varepsilon \to 0} f(\varepsilon) = \infty$, and $\lim_{\varepsilon \to 0} (f(\varepsilon) - g(\varepsilon)) = 0$.*

PROPOSITION 3 *For $\varepsilon \to 0$*

$$Q(\pi_{[a,b]}, \varepsilon) \sim \log_2(b - a) - \log_2(2\varepsilon).$$

From it follows that when we consider intervals, the behavior of the function $Q(\pi_{[a,b]}, \varepsilon)$ for small $\varepsilon > 0$ is uniquely determined by a single number $\log_2(b-a)$. We will see that the same is true for more complicated sets. However, for fuzzy (non-crisp) membership functions π, we will need two numbers to describe the behavior of $Q(\pi, \varepsilon)$: the coefficients at 1 and $\log_2(2\varepsilon)$. Therefore, we consider these coefficients as the measure of uncertainty that corresponds to a membership function π.

DEFINITION 4 *Let π be a membership function. We say that for π, uncertainty is dfined if $Q(\pi, \varepsilon) \sim u - m \log_2(2\varepsilon)$ for some real numbers u and m. The pair (u, m) is called the uncertainty of a fuzzy set π. The value of u will be denoted by $U(\pi)$, and the value of m by $m(\pi)$.*

The following is immediate.

PROPOSITION 4 *For the interval $[a, b]$, $U(\pi_{[a,b]} = \log_2(b - a)$ and $m(\pi_{[a,b]}) = 1$.*

To avoid possible misunderstanding, let us note that the previously defined uncertainties stated the number (or an average number) of binary questions needed to to determine the correct alternative (uniquely or with the precision ε.) As the number of questions is always non-negative, all previously defined measures of uncertainty were themselves non-negative.

However, it can be easily seen, that the expression $U(\pi_{[a,b]}) = \log_2(b-a)$ becomes negative if $b - a < 1$. There is not contradiction in there, because $U(\pi_{[a,b]})$ is not equal to the (non-negative) number of questions. To find the number of questions $Q(\pi_{[a,b]}, \varepsilon)$, one has to select ε, and combine $U(\pi_{[a,b]})$ with the term $-\log_2(2\varepsilon)$. For a small ε, this second term is positive and tends to $+\infty$ as $\varepsilon \to 0$. Therefore, for sufficiently small ε the resulting sum is positive.

There is an analogy between this situation and the entropy of continuous distributions. For discrete variables, that take only finitely many values x_i with probabilities p_i, the entropy is defined as $-\sum_i p_i \log_2(p_i)$. This sum is always non-negative. For continuous variable with density $\rho(x)$ the usual analog of entropy is the integral $S = -\int \rho(x) \log_2(\rho(x))\, dx$. Unlike discrete entropy, this integral can be negative: for example, if we take a uniform distribution on an interval $[a, b]$, i.e., a function $f(x) = 1/(b-a)$ for $x \in [a, b]$ and 0 outside this interval, then this entropy is equal to $\log_2(b-a)$, and this value is negative for $b - a < 1$. The reason is the same as for our non-statistical dfinition: Shannon's entropy $-\sum_i p_i \log_2(p_i)$ can be interpreted as an average number of questions, and therefore, is always non-negative. But its continuous analog is only indirectly related to the number of questions: namely, $S - \log_2(2\varepsilon)$ is the average number of questions that we have to ask to determine x with precision ε.

3.3 Information of fuzzy sets

We start with a crisp set, whose nontrivial part $\{x : \pi(x) > 0\}$ is a collection of intervals.

PROPOSITION 5 *For an arbitrary set S which is a union of disjoint intervals, uncertainty is defined for π_S, $U(\pi_S) = \log_2(\mu(S))$, and $m(\pi_S) = 1$.*

We use $\mu(S)$ to denote Lebesgue measure, here simply the sum of lengths of the intervals. A much more important is the case of $\pi(x)$ piecewise constant. First, a couple of definitios.

DEFINITION 5 *Membership function $\pi(x)$ is normalized if $\sup_x \pi(x) = 1$.*

DEFINITION 6 *Membership function is piecewise constant if there exist values* $x_1 < x_2 < \cdots < x_n$ *such that* $\pi(x) = 0$ *for* $x < x_1$ *and* $x > x_n$, $\pi(x) = const$ *on each of the intervals* (x_i, x_{i+1}), *and for each* i, $\pi(x_i)$ *coincides either with the value of* $\pi(x)$ *for* $x < x_i$, *or with the values of* $\pi(x)$ *for* $x > x_i$.

The estimate of the number of binary questions is complicated by the need to recognize the relative weight of each possibility value. The function being piecewise constant, it takes only finitely many different values. Let us order them $h_0 = 0 < h_1 < h_2 < ... < h_k$ and put $\sup \pi(x) = h_k$. In this section, we consider only normalized functions, thus $h_k = 1$.

Let us first consider an example. Suppose that a membership function $\pi(x)$ is equal to 1 for $x \in [-a, a]$, is equal to 0.6 for $x \in [-2a, -a]$ and $x \in [a, 2a]$, and $\pi(x) = 0$ for $x \notin [-2a, 2a]$. In this case, 60% of the experts believe that the area of possible values of x is the interval $[-a, a]$ of length $2a$, and the remaining 40% believe that x is in the interval $[-2a, 2a]$ (of length $4a$). If the experts from this 60% majority are right, then we need $\sim \log_2(2a) - \log_2 \varepsilon$ binary questions. If the minority experts are right, then we need $\sim \log_2(4a) - \log_2 \varepsilon$ questions. Since all the experts are considered equally good, it is reasonable to assume that in general, in 60% of the cases the majority is right, and in 40% of cases, the minority is right. Therefore, the average number of binary questions that we have to ask in order to locate x in an interval of length ε, is $\sim 0.6(\log_2(2a) - \log_2 \varepsilon) + 0.4(\log_2(4a) - \log_2 \varepsilon) = (0.6 \log_2(2a) + 0.4 \log_2(4a)) - \log_2 \varepsilon$.

In the general case, the h_1-th part of all the experts believe that x belongs to the set $\{x : \pi(x) \geq h_1\}$, $(h_2 - h_1)$ of them believe that $x \in \{x : \pi(x) \geq h_2\}$, $(h_3 - h_2)$ of them believe that $x \in \{x : \pi(x) \geq h_3\}$, ..., and $h_k - h_{k-1}$ of them believe that $x \in \{x : \pi(x) \geq h_k\}$. If $x \in \{x : \pi(x) \geq h_1\}$, then we need $Q(\{x : \pi(x) \geq h_1\}, \varepsilon)$ questions to determine x with the precision ε. If $x \in \{x : \pi(x) \geq h_2\}$, then we need $Q(\{x : \pi(x) \geq h_2\}, \varepsilon)$ questions, etc.

Therefore, according to the opinion of h_1 of experts, we need $Q(\{x : \pi(x) \geq h_1\}, \varepsilon)$ questions. Next, according to the opinion of $(h_2 - h_1)$ of the experts, we need to ask $Q(\{x : \pi(x) \geq h_2\}, \varepsilon)$ questions, and similarly for the remaining data.

It becomes natural to define the expected number of questions as

$$Q(\pi, \varepsilon) = \sum_{i=0}^{k-1} (h_{i+1} - h_i) Q(\pi_{\{x : \pi(x) \geq h_{i+1}\}}, \varepsilon).$$

The case of non-normalized membership function can be handled along the same lines. Suppose, for example, that some term from the natural language is represented by a (fuzzy) membership function $\pi(x)$ which is equal to 0.6 for all x from $[a,b]$, and to 0 for all other x. In view of the above interpretation of $\pi(x)$, it means that only 60% of all the experts had any opinion about what the initial natural language term means, and the others simply gave no answers. In this case, we can take into consideration only the opinions of those who made an actual statement.

As we have already argued, it is reasonable to assume that in 60% of cases the majority is right. Therefore, the only thing that we know about the average number of questions is that it is $\geq 0.6 Q(\pi_{[a,b]}, \varepsilon) \sim 0.6(\log_2(b-a) - \log_2(2\varepsilon))$. When the remaining 40% of the experts make their decisions, there may be more questions, but right now, when we are given the above-described membership function $\pi(x)$, this number $0.6Q(\pi_{[a,b]}, \varepsilon)$ is the only estimate that we can get. Again, it is reasonable to take it as the description of uncertainty of this membership function $\pi(x)$.

It justifies the following definition:

DEFINITION 7 *Let $\pi(x)$ be a piecewise-constant membership function, that takes only the values $h_0 = 0 < h_1 < \cdots < h_k$. Its ε-uncertainty $Q(\pi, \varepsilon)$ is*

$$Q(\pi, \varepsilon) = \sum_{i=0}^{k-1} (h_{i+1} - h_i) Q(\pi_{\{x : \pi(x) \geq h_{i+1}\}}, \varepsilon).$$

PROPOSITION 6 *For an arbitrary piecewise constant normalized membership function $\pi(x)$, uncertainty $(U(\pi), m(\pi))$ is dfined, $m(\pi) = 1$, and*

$$U(\pi) = \sum_{i=0}^{k-1} (h_{i+1} - h_i) \log_2(\mu\{x : \pi(x) \geq h_{i+1}\}).$$

It brings us to the general case of an arbitrary membership function $\pi(x)$ that can appear in control problems. In our earlier work we postulated some regularity conditions to assure existence and useful asymptotic properties of uncertainty expressions. All such restrictions can be removed through a judicial application of *function rearrangements*. It is a technique originated in the theory of inequalities [4], esp. in applied mathematics, and used recently for continuous fuzzy uncertainty [15].

Here, however, we aim at application to typical fuzzy control problems. And membership functions that are usually considered in fuzzy control [12] are all rather regular. Namely, their domain can be divided into finitely many intervals of monotonicity motivating the following definition.

DEFINITION 8 *Membership function $\pi(x)$ is regular if $\pi(x) \to 0$ as $|x| \to \infty$, and if there exist values $x_1 < x_2 < ... < x_n$ such that on each of the intervals $(-\infty, x_1)$, (x_i, x_{i+1}), $1 \leq i \leq n-1$, and (x_n, ∞), the function $\pi(x)$ is either monotonic increasing or monotonic decreasing, and for each i, the value $\pi(x_i)$ coincide either with the left-side limit $\lim_{\varepsilon > 0, \varepsilon \to 0} \pi(x - \varepsilon)$, or with the right-side limit $\lim_{\varepsilon > 0, \varepsilon \to 0} \pi(x + \varepsilon)$.*

Although not needed in the future, we remark that $\pi(x) \to 0$ for $x \to -\infty$ and $\pi(x) \geq 0$ for all x. Therefore, a regular function $\pi(x)$ can only be increasing on $(-\infty, x_1)$ nad only decreasing on $[x_n, \infty)$.

When we define a membership function in mathematical terms, we say that its values $\pi(x)$ are real numbers. To describe a real number with infinite precision, one needs infinitely many bits. Therefore, a procedure that allows to describe the experts' degree of belief with ever increasing precision would, after any finite number of questions, only approximate the value of $\pi(x)$.

There is a similarity between this argument and an earlier one. There we knew x with some precision, while here we can know $\pi(x)$ only with some precision.

At each stage of determining $\pi(x)$, it is known with some precision δ. At this stage there are only finitely many distinguishable degrees of belief $h_0 = 0 < h_1 < h_2 < ... < h_k = 1$ such that $|h_{i+1} - h_i| \leq \delta$. In other words, all the values of $\pi(x)$ from $h_0 = 0$ to h_1 are indistinguishable from 0, all the values from h_1 to h_2 are indistinguishable from h_1, etc. Changing the values $\pi(x)$ to the corresponding values h_i resultsin a piecewise constant function $\bar{\pi}(x)$ that is (at this stage) quite possible. Therefore, its uncertainty $U(\bar{m})$ is a possible value of uncertainty $U(\pi)$.

We get an entire interval of the possible values of $U(\bar{m})$. Each additional measurement diminishes the set of possible functions $\bar{m}$, making the resulting interval smaller. For $\delta \to 0$ this interval shrinks to a point, making natural to consider the limit of uncertainties of membership function π. Let us now state formal definitions.

DEFINITION 9 *A monotonic sequence $h_0 = 0 < h_1 < h_2 < \cdots < h_k = 1$ is δ-precise, $\delta > 0$, if $h_{i+1} - h_i \leq \delta$ for all i.*

DEFINITION 10 *For arbitrary monotonic $\vec{h}$ its projection $pr_{\vec{h}} : [0,1] \to [0,1]$ is $pr_{\vec{h}}(y) = 0$ for $y < h_1$ and $pr_{\vec{h}}(y) = h_i$ for $h_i \leq y < h_{i+1}$.*

They have an immediate conclusion.

PROPOSITION 7 *If $\pi(x)$ is a membership function, and $\vec{h}$ is any sequence, then $\bar{\pi}(x) = pr_{\vec{v}}(\pi(x))$ is a piecewise constant membership function.*

It follows that for each of such projected membership functions $\bar{\pi}(x) = pr_{\vec{v}}(\pi(x))$, the uncertainty $U(\bar{\pi})$ is well defined. We need two more definitions.

DEFINITION 11 *Let $\delta > 0$, and $\pi(x)$ be a regular membership function. We say that a pair of real numbers (u, m) is a δ-possible value of uncertainty for π, if $u = U(pr_{\vec{h}}(\pi(x)))$ and $m = m(pr_{\vec{h}}(\pi(x)))$ for some δ-precise sequence $\vec{h}$.*

DEFINITION 12 *Let $\pi(x)$ be a regular membership function. We say that its uncertainty is defined and equal to (u, m) if whenever (u_n, m_n) is a δ_n-possible value of uncertainty for π, then $(u_n, m_n) \to (u, m)$. These values u and m will be denoted by $U(\pi)$ and $m(\pi)$.*

We can now state the main results.

THEOREM 1 *For an arbitrary regular membership function $\pi(x)$ its uncertainty is defined, $m(\pi) = \sup_x \pi(x)$, and*

$$U(\pi) = \int_0^{\sup_x \pi(x)} \log_2(\mu(\{x : \pi(x) \geq h\})) \, dh.$$

An immediate consequence is that for a normalized regular function $\pi(x)$, $m(\pi) = 1$ and

$$U(\pi) = \int_0^1 \log_2(\mu(\{x : \pi(x) \geq h\})) \, dh.$$

For a monotonic normalized function $\pi(x)$, this expression can be simplified further.

THEOREM 2 *Let $x_0 > 0$ be a positive real number. If $\pi(x) = 0$ for $x < 0$, $\pi(0) = 1$, $\pi(x) = 0$ for $x \geq x_0$, and for $x \in [0, x_0)$, $\pi(x)$ is continuous and decreasing, then*

$$U(\pi) = \log_2(x_0) - \frac{1}{\ln(2)} \int_0^{x_0} \frac{1 - \pi(x)}{x} \, dx.$$

The first term in this formula is the uncertainty contributed just just belonging to $[0, x_0]$. It implies that if we know only that $x \in [0, x_0]$, than we estimate the uncertainty in the value of x as $\log_2(x_0)$. If we include the additional information that x belongs to a fuzzy set described by a membership function $\pi(x)$, then we lower the certainty to the value $U(\pi) = \log_2(x_0) - (1/\ln(2)) \int_0^{x_0} (1 - \pi(x))/x \, dx$. This decrease in uncertainty $U(\pi_{[0,x_0]}) - U(\pi)$ measures the information that is brought on by this additional knowledge. For numerical representation of this information we can take this difference in uncertainties

$$I(\pi) = (1/\ln(2)) \int_0^{x_0} (1 - \pi(x))/x \, dx.$$

It is interesting to mention that the resulting expression for the information practically coincides with the expression $I(\pi) = \int_0^{x_0} (1 - \pi(x))/x \, dx$ that was introduced (from different assumptions) in [15, 17]. The factor $1/\ln(2)$ that makes them different is simply due to the fact that we use $\log_2$ for entropy, while some other authors use natural logarithms. This correspondence can be demonstrated formally by a suitable rearrangement of the membership functions.

4 COMPARISON OF UNCERTAINTIES

We will look now for the operations that lead, respectively, to the greatest and the least uncertainty. To formalize that, we must learn to compare uncertainties of different membership functions.

When we use entropy $S(p)$ as a definition of uncertainty of a probability distribution p, we just say that the uncertainty of a distribution p_1 is greater than the uncertainty of p_2 if $S(p_1) \geq S(p_2)$.

To be able to define uncertainty as a pair of numbers we introduce a relevant ordering.

DEFINITION 13 *Uncertainty* (u_1, m_1) *is greater than* (u_2, m_2), *denoted* $(u_1, m_1) \succ (u_2, m_2)$ *if there exists an* ε_0 *such that for all* $\varepsilon < \varepsilon_0$,

$$u_1 - m_1 \log_2(2\varepsilon) > u_2 - m_2 \log_2(2\varepsilon).$$

The resulting order structure is lexicographic.

PROPOSITION 8 $(u_1, m_1) \succ (u_2, m_2)$ *if and only if either* $m_1 > m_2$, *or* $m_1 = m_2$ *and* $u_1 > u_2$.

4.1 Maximum uncertainty operations

Let us consider a typical fuzzy control setting. Assume given a set $S \subset R^n$, whose elements $\vec{x} = (x_1, \ldots, x_n) \in S$ ar the states of the system. Informally, the values $x_1, \ldots, x_n$ describe everything that we need to know to make a control decision. For example, if we control a heater/cooler, then $n = 1$, and the only variable we need to know is the difference $x_1 = t - t_0$ between the actual and the desired temperature. If we are controlling a spaceship, then we need to know its coordinates x_1, x_2, x_3, its current velocity vector (three more variables $x_4 = \dot{x}_1, x_5 = \dot{x}_2, x_6 = \dot{x}_3$), and two angles that describe the orientation. So, for a spaceship, $n = 8$.

Now let us fix a finite set $\mathcal{P}$ of continuous membership functions and view its elements as *fuzzy properties*. Examples are offered by such concepts as *big, medium* etc.

DEFINITION 14 *Elementary formula E is an expression of the form* $P_i(x_i)$, *where* P_i *is a fuzzy property. A rule is an expression of the form* $E_1, \ldots, E_m \rightarrow P(u)$, *where* E_i *are elementary formulae, P is a fuzzy property, and u is a special variable reserved for control. Formulae* E_i *are called conditions, while* $P(u)$ *is the conclusion of the rule. By a knowledge base we understand a finite set of rules.*

As an example one can consider rule $N(x_1) \rightarrow N(u)$, stating that if the difference $t - t_0$ between the actual and the desired temperatures is negligible, then the control should be negligible. Another possible rule is $SP(x_1) \rightarrow SN(u)$, meaning that if the difference $t - t_0$ is small positive, then we need to apply a small negative control (switch on the cooler for a little bit.) A similar rule $SN(x_1) \to SP(u)$ tells that if it becomes a little bit cold, it is necessary to switch on the heater for a while.

If we have a set of rules, then we can say that a control u is appropriate if and only if one the rules is applicable, and u appropriate according to this rule. Let us denote the statement *control u is appropriate* by $C(u)$. Then, for the three rules that describe the cooler/heater, we have the informal *formula* that describes when a control u is appropriate:

$$C(u) \equiv (N(x)\&N(u)) \vee (SP(x)\&SN(u)) \vee (SN(x)\&SP(u)).$$

Since $N(x)$, $N(u)$, ..., are fuzzy statements, we can get only fuzzy conclusions about the control, whereby $C(u)$ also becomes a fuzzy statement. To extract the precise values, we need to choose some operations that would describe & and ∨ for fuzzy values. We can make a strong case for selecting control that corresponds to the biggest uncertainty.

When we make a choice of &− and ∨−operations, we restrict the set of possible control strategies. Since a wrong choice can lead to a low quality control, it sounds reasonable to try to loose as few possibilities as possible. In other words, it sounds reasonable to choose &− and ∨−operations in such a way that the uncertainty corresponding to $\pi_C(u)$ is the biggest possible.

This methodology is well known in the case when the uncertainty is probabilistic; it is called a *maximum entropy approach*, and it is widely applied to various problems ranging from processing physical data to processing uncertainties in expert systems [5, 8, 2, 9].

Initially, Zadeh [21] proposed using min and max, though stressing that these operations "are not the only operations in terms of which the union and intersection can be defined," and "which of these ...definitions is more appropriate depends on the context." Accordingly, we propose using the maximally general operations. Choice of operations that correspond to & and ∨, permits using the above formula to describe, for each u, what is the reasonable degree of belief that this value u is an appropriate control. In other words, we will be able to generate a membership function $\pi_C(u)$ that corresponds to a control rule. Following that we need some *defuzzification* procedure to transform this membership function into a single recommended control value.

To establish precise definitions, we need to consider what &− and ∨− operations could be appropriate. For the &−operation (we will denote it by

$f_\&(a,b)$) let us suppose that we have two statements A and B. Our degree of belief in A is equal to a, and belief in B is equal to b. If we have no other information about A and B, what must the reasonable degree of belief in $A\&B$ equal to? This reasonable degree of belief will be denoted by $f_\&(a,b)$. In the same situation, a reasonable degree of believe in $A \vee B$ will be denoted by $f_\vee(a,b)$, and $f_\vee$ will be called an $\vee$−operation.

In describing uncertainty of a membership function, we used the interpretation of membership values $\pi(x)$ as frequencies. Namely, we assumed that as a truth value $t(A)$ of an uncertain statement A, we take the ratio $t(A) = N(A)/N$, where $N(A)$ is the number of experts who believe in A, and N is the total number of experts that were questioned. In this interpretation, the following inequalities are true

$$N(A \vee B) \leq N(A) + N(B)$$
$$N(A \vee B) \leq N$$
$$N(A \vee B) \geq N(A)$$
$$N(A \vee B) \geq N(B)$$

If we divide both sides of these inequalities by N, and combine them into one, we get

$$\max(t(A), t(B)) \leq (A \vee B) \leq \min(t(A) + t(B), 1),$$

hence

$$\max(a,b) \leq f_\vee(a,b) \leq \min(a+b, 1).$$

Likewise, from $N(A\&B) \leq N(A)$ and $N(A\&B) \leq N(B)$ we conclude that

$$t(A\&B) \leq \min(t(A), t(B)),$$

and thus $f_\&(a,b) \leq \min(a,b)$.

If beliefs in A and in B were independent, then we would have $t(A\&B) = t(A)t(B)$. In real life situations beliefs are not independent—if an expert has strong beliefs in several statements that later turn out to be true, then this means that he is really a good expert, and therefore it is reasonable to expect that his degree of belief in other statements that are true is greater. If A and B are complicated statements, then most of those experts who believe in A are presumably quite accomplished, and therefore they believe in B as well (and hence in $A\&B$). Therefore, the total number $N(A\&B)$ of experts who believe in $A\&B$ must be greater than the same number in the case when beliefs in A and B were uncorrelated random events. We conclude that the following inequality is reasonable to postulate

$$t(A\&B) \geq t(A)t(B),$$

hence, $f_\&(a,b) \geq ab$. In statistical terms, we can interpret it by saying that A and B are non-negatively correlated. We arrive at the following definitions

DEFINITION 15 *The and-or pair of continuous functions are $f_\&, f_\vee : [0,1] \times [0,1] \to [0,1]$, which are non-decreasing in both variables, and satisfy*

- $\max(a,b) \leq f_\&(a,b) \leq \min(a+b,1)$
- $f_\vee(a,b) \leq \min(a,b)$
- $f_\&(0,a) = 0, f_\&(1,a) = a, f_\vee(0,a) = a, f_\vee(1,a) = 1$
- $f_\vee(a,b) = f_\vee(b,a), f_\&(a,b) = f_\&(b,a)$

An *and-or* pair is called *correlated* if $f_\&(a,b) \geq ab$ for all a and b. The considerations behind these definition are quite simple:.

- If A is false, then $A\&B$ is also false, and $f_\&(0,a) = 0$ for all a.

- If A is true, then $A\&B$ is true if and only if B is true. In this case $t(A\&B) = t(B)$, hence $f_\&(a,1) = a$ for all a.

- The statements that A and B are both true or that B and A are both true are equivalent. Therefore, $t(A\&B)$ must be always equal to $t(B\&A)$, or $f_\&(a,b) = f_\&(b,a)$ for all a,b.

- If our degree of belief in A increases, then our degree of belief in $A\&B$ cannot become less. The function $f_\&$ must be non-decreasing in both variables.

- If our degrees of belief in A and B change a slightly, then our degree of belief in $A\&B$ cannot change substantially. The smaller is the change in $t(A)$, $t(B)$, the smaller must be the change in $t(A\&B)$. In other words, the function $f_\&$ must be continuous.

Similar arguments justify the conditions on $f_\vee$.

It is convenient to extend our notation to more than two arguments. For three numbers a, b, c, we write $f_\&(a, b, c) = f_\&(f_\&(a, b), c)$, and in general, for $a, b, \ldots, c$

$$f_\&(a, b, \ldots, c) = f_\&(\ldots(f_\&(f_\&(a, b), \ldots)c).$$

Likewise

$$f_\vee(a, b, \ldots, c) = f_\vee(\ldots(f_\vee(f_\vee(a, b), \ldots)c).$$

The notation reflects the common understanding that $A\&B\&C = (A\&B)\&C$, and $A \vee B \vee C = (A \vee B) \vee C$. Now, we can define fuzzy control.

DEFINITION 16 *Assume given a knowledge base* $K = \{R_1, R_2, \ldots\}$, *an and-or pair* $(f_\&, f_\vee)$, *and a state* $\vec{x} \in S$. *The membership function corresponding to a rule* $P_1(x_{i_1}), \ldots, P_m(x_{i_m}) \rightarrow P(u)$, *is given by a function* $\pi_R = f_\&(P_1(x_{i_1}), \ldots, P_m(x_{i_m}), P(u))$. *Control membership function which corresponds to the knowledge base and the state* $\vec{x}$ *is*

$$\pi_C(u) = f_\vee(\pi_{R_1}, \pi_{R_2}, \ldots),$$

where $R_1, R_2, \ldots$ *are all the rules from* K.

We are now in a position to establish the operators that offer maximum uncertainty for the ensuing control rules.

THEOREM 3 *Suppose that* K *is a knowledge base, and* $\vec{x}$ *is a state. Let us denote by* $\pi_C(u)$ *the control membership function, that corresponds to an arbitrary and-or pair* $(f_\&(a, b), f_\vee(a, b))$, *and by* $\tilde{\pi}_C(u)$ *the control membership function that corresponds to the and-or pair* $(\min(a, b), \min(a + b, 1))$. *Then,* $(U(\tilde{\pi}_C), m(\tilde{\pi}_C)) \succeq (U(\pi_C), m(\pi_C))$.

In other words, the greatest uncertainty is attained when we use $\min(a, b)$ for $\&$, and $\min(a + b, 1)$ for $\vee$. The proof is an immediate consequence of the definitions and the earlier remarks about inequalities among the uncertainty values. An interesting interpretation is provided by [11] that these operations lead to maximally stable controls. This agrees with a common-sense perception—we minimized the lost opportunities, and therefore ended up with the best possible control.

4.2 Minimization of uncertainty

Some of the earlier arguments are reasonable only if we are ready to apply various defuzzification techniques to extract the best control from $\pi_C(u)$. However, in industrial applications, a defuzzification rule is usually fixed. Since this rule may not necessarily be the most appropriate [20, 11], it is reasonable to try to rely on it only to a smallest extent feasible. In other words, in such cases, it is reasonable to choose & and ∨− operations on the premise that the uncertainty related to $\pi_C(u)$ is the *least* possible. The following result is again immediate.

THEOREM 4 *Let $\pi_C(u)$ be the control membership function corresponding to an arbitrary correlated and-or pair $(f_\&(a,b), f_\vee(a,b))$, and let $\tilde{\pi}_C(u)$ be the control membership function for the pair $(ab, \max(a,b))$. Then*

$$(U(\pi_C), m(\pi_C)) \succeq (U(\tilde{\pi}_C), m(\tilde{\pi}_C)).$$

In other words, uncertainty is the least when we use ab for &, and $\max(a,b)$ for ∨. It has been shown in [11] that these very and-or operations lead to maximally smooth controls. This result also has a natural interpretation: since we are extremely cautious, we end up with a very smooth control.

5 CLOSING REMARKS

We already noted some analogies between our information and probabilistic entropy. It is, of course, the case of general complexity based informations. In a recent article [14] this is analyzed in considerable detail.

Similar correspondence can be established with fuzzy *continuous* information measures. We remarked on how the use of rearrangements helps to consolidate the results. Rearrangements other than decreasing are also of interest; in particular, those symmetric about the midpoint of the domain are likely to be most useful [6]. Further analysis is a subject of the on-going work.

ACKNOWLEDGEMENTS

The second author was supported by NSF Grant No. CDA-9015006, NASA Research Grant No. 9-482 and the Institute for Manufacturing and Materials Management grant. The authors are thankful to Bob Lea, Ron Yager, and John Yen for the stimulating discussions.

References

[1] S. S. L. Chang and L. A. Zadeh. *On fuzzy mapping and control*, IEEE Transactions on Systems, Man and Cybernetics, 1972, Vol. SMC-2, pp. 30–34.

[2] P. Cheeseman. *In defense of probability* in Proceedings of the 8-th International Joint Conference on AI, Los Angeles, 1972, pp. 1002–1009.

[3] D. Dubois and H. Prade. *Possibility Theory.* Plenum Press, New York 1988.

[4] G. H. Hardy, J. E. Littlewood, and G. Polya. *Inequalities.* Cambridge University Press, Cambridge 1934.

[5] E. T. Jaynes. *Where do we stand on maximum entropy?* in R. D. Levine and M. Tribus (Eds.) *The maximum entropy formalism*, MIT Press, Cambridge, MA, 1979.

[6] B. Kawohl. *Rearrangements and Convexity of Level Sets in PDE.* Lecture Notes in Mathematics 1150, Springer-Verlag 1985.

[7] G. J. Klir and T. A. Folger. *Fuzzy sets, uncertainty and information.* Prentice Hall, Englewood Cliffs, NJ, 1988.

[8] O. M. Kosheleva, V. Kreinovich. *A letter on maximum entropy method.* Nature, 1979, Vol. 281, pp. 708–709.

[9] V. Kreinovich. *Entropy approach for the description of uncertainty in knowledge bases*, Tech. Rep., Center for the New Informational Technology "Informatika", Leningrad, 1989 (in Russian).

[10] V. Kreinovich, C. Quintana, and R. Lea. *What procedure to choose while designing a fuzzy control? Towards mathematical foundations of fuzzy control*, Working Notes, 1st Int. Workshop on Industrial Appl. Fuzzy Control and Intelligent Systems, College Station, TX, 1991, pp. 123–130.

[11] V. Kreinovich, C. Quintana, R. Lea, O. Fuentes, A. Lokshin, S. Kumar, I. Boricheva, and L. Reznik. *What non-linearity to choose? Mathematical foundations of fuzzy control.* Proc. 1992 Int. Fuzzy Systems and Intelligent Control Conference, Louisville, KY, 1992, pp. 349–412.

[12] C. C. Lee. *Fuzzy logic in control systems: fuzzy logic controller.* IEEE Trans. Systems, Man and Cybernetics, 1990, Vol. 20, No. 2, pp. 404–435.

[13] E. H. Mamdani. *Application of fuzzy algorithms for control of simple dynamic plant*, Proceedings of the IEE, 1974, Vol. 121, No. 12, pp. 1585–1588.

[14] E. W. Packel, J. F. Traub, H. Wozniakowski. *Measures of uncertainty and information in computation*, Information Sciences, Vol. 65, pp. 253–273.

[15] A. Ramer. *Concepts of fuzzy information measures on continuous domains*, Int. J. General Systems, 1990, Vol. 17, pp. 241–248.

[16] A. Ramer, J. Hiller, C. Padet. *Principle of Maximizing Fuzzy Uncertainty*, Int. J. General Systems, 1993 (to appear.)

[17] A. Ramer, R. Yager. *Analysis of specificity of fuzzy sets*, Proceedings of the IEEE International Conference on Fuzzy Systems, San Diego, CA, 1992, pp. 1097–1104.

[18] J. F. Traub, H. Wozniakowski. *A General Theory of Optimal Algorithms.* Academic Press, New York 1980.

[19] J. F. Traub, G. W. Wasilkowski, H. Wozniakowski. *Information-Based Complexity.* Academic Press, New York 1988.

[20] J. Yen, N. Pfluger, and R. Langari. *A defuzzification strategy for a fuzzy logic controller employing prohibitive information in command formulation*, Proc. IEEE Int. Conference on Fuzzy Systems, San Diego, CA, March 1992.

[21] L. A. Zadeh. *The concept of a linguistic variable and its application to approximate reasoning*, Part 1, Information Sciences, 1975, Vol. 8, pp. 199–249.

Chapter 5
Alternative Structures for Knowledge Representation in Fuzzy Logic Controllers

Introduction, *100*
Basic Structures of Fuzzy Logic Controllers, *100*
Certainty Qualified Antecedents, *107*
Alternative Formulations for Rule Outputs, *109*
Chaining of Fuzzy Rules, *114*
Decoupled Inputs, *121*
Hierarchical Representation of Rules, *123*
Conclusion, *134*
References, *134*

This chapter looks at the fuzzy logic controller and describes the basic assumptions inherent in the Mamdani model. A distinction is made between rule firing based upon possibility and upon certainty qualification. Rules are looked at as a partitioning of the input space, and different representations of the rule consequent are discussed. It is shown how to reason when having a chaining of fuzzy rules and when having a decoupling of rules. A new structure for the representation of rules in fuzzy systems is introduced in this chapter. Called the *hierarchical prioritized structure* (HPS), this new structure allows for a natural framework for learning rules in addition for providing a useful structure for representing knowledge.

ALTERNATIVE STRUCTURES FOR KNOWLEDGE REPRESENTATION IN FUZZY LOGIC CONTROLLERS

Ronald R. Yager

Machine Intelligence Institute

Iona College

1. INTRODUCTION

Recent interest has developed in the use of fuzzy set theory for the modeling of complex systems. We call such a representation a **Fuzzy Model of a System (FMS)**. Fuzzy Logic Control (FLC) which has recently found numerous applications.[1-6] can be seen as a special case of this more general class of systems. The construction of these systems involves modeling a, usually non-linear, relationship between system input and output. In this work we look at structures that are useful for the development of these types of fuzzy models.

One significant feature of this work is the introduction of a new structure for the representation of knowledge fuzzy systems modeling. We call these structures Hierarchical Prioritized Structures and denote them as HPS.

2. BASIC STRUCTURES OF FUZZY LOGIC CONTROLLER

The fuzzy logic controller is generally made up of two components as shown in Figure 1.

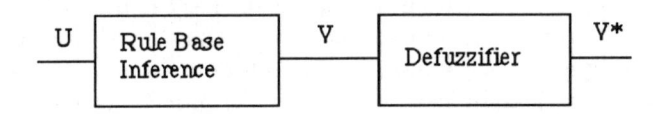

Figure 1. Fuzzy Logic Controller

The rule base[1] of the fuzzy logic controller usually consists of a collection of

---

[1] In this work we shall use the currently popular terminology of refering to the model of the input/output relationship as a rule or knowledge base.

rules involving U and V. The rule base is a model of the input/output relationship. This portion of the system provides as an output a fuzzy subset F over the base set of V. In the framework of FLC F can be seen as the fuzzy recommended action. The defuzzifier unit takes as its input the fuzzy subset F and based upon this subset selects one element y* as the crisp (defuzzified) controller output. The defuzzification process has been most often accomplished by one of two methods, the Center of Area (COA) or Mean of Maxima (MOM)[5, 6]. Recently Filev and Yager [7] have provided for a generalization of this defuzzification operation based upon a parametrization procedure which admits the COA and MOM as special cases. The advantage of this parametrization is that it allows for an adaptive learning of the appropriate defuzzification process.

In this work we shall focus on the representation used in the rule base portion of the fuzzy controller. Conceptually a fuzzy logic controller or more generally any FMS can be seen as a function f mapping input to output (see Figure 2).

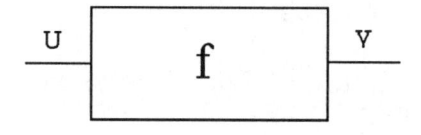

Figure 2.

Ideally our knowledge of the relationship between input and output, the function f, is a function as shown in Figure 3.

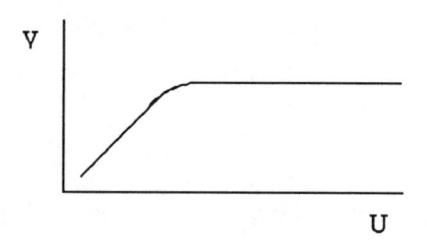

Figure 3. Ideal Knowledge of System Controller

Essentially f can be then seen as a mapping **f:** $X \rightarrow Y$, $y = f(x)$, where X and Y are the base sets of U and V respectively. Formally we can represent this mapping as a relationship F on $X \times Y$ such that $(x, y) \in F$ if $f(x) = y$. In this framework we see that

$$F = \{(x_1, y_1), (x_2, y_2), \ldots, (x_n, y_n)\}$$

where each pair (x_i, y_i) is a solution of f, $f(x_i) = y_i$. If we denote $S_i = \{(x_i, y_i)\}$ then we can write

$$F = \bigcup_i S_i.$$

Each module S_i can be seen as being semantically interpreted as a proposition of the form

When U is X_i it is the case that V is Y_i

where $X_i = \{x_i\}$ and $Y_i = \{y_i\}$.

The above statement can be interpreted using the theory of approximate reasoning [8-11] as the proposition

$$(U, V) \text{ is } S_i$$

where $S_i = X_i \times Y_i = (X_i \cap Y_i)$. Thus S_i is the cartesian product of X_i and Y_i. Using this representation we get that our function can be expressed as

$$F = \bigcup_i (X_i \cap Y_i).$$

An input to the system consists of a proposition

$$U \text{ is } D$$

where $D = \{x^*\}$, a crisp input value.

Conjuncting these two pieces of information we get

$$(U, V) \text{ is } H$$

where $H = S \cap D$, therefore

$$H = \bigcup_i S_i \cap D = \bigcup_i X_i \cap Y_i \cap D$$

Furthermore we note that

$$X_i \cap D = x^* \quad \text{for } x_i = x^*$$
$$X_i \cap D = \phi \quad \text{for } x_i \neq x^*.$$

Thus $H = X^* \cap Y^*$. Finally taking the projection [8, 10] onto V we get

$$V \text{ is } Y^*$$

thus

$$V = y^*.$$

In most real world control problems the complexity of the situation precludes our knowing the function f in such a specific crisp manner as that shown in Figure 3. Figure 4 shows a more realistic situation in regards to our knowledge of the function

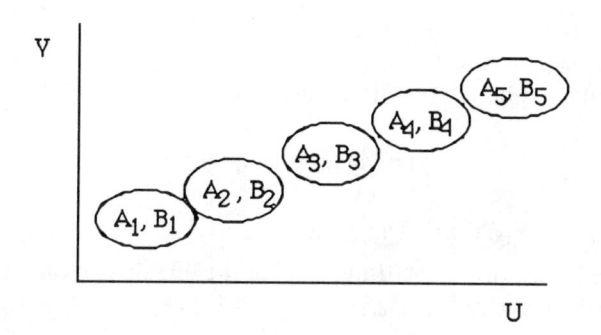

Figure 4. Granular Knowledge of System Input/Output Relationship

The above Figure 4 is meant to indicate that our knowledge about the controller(system) comes in the form of a collection granules rules of the type:

when U is A_i it is the case the V is B_i

where A_i and B_i are fuzzy subsets of X and Y respectively. The intent of the above is to indicate that when U lies in the set A_i it is the case that V lies in the set B_i. The important idea here is that we have partitioned the input space into ranges each of which has associated with it its own particular output value. We call the construction of such a collection of rules a FMS. The use of a FMS can be viewed as being obtained from either of two directions. In the first case we have a very complex nonlinear system which we simplify by using such a model. In the second case because of our lack of knowledge about f we approximate it with such a system.

The procedure for calculating the output of this FMS for a given input follows directly the procedure of the specific case. First we get that $S_i = A_i \times B_i$ ($A_i \cap B_i$) where S_i is a relationship on $X \times Y$ such that

$$S_i(x, y) = \text{Min}(A_i(x), B_i(y)).$$

Furthermore our complete relationship F becomes $F = \bigcup_i S_i = \bigcup_i (A_i \cap B_i)$. If our input is a proposition *U is D* where D is also fuzzy subset of X then we get

$$(U, V) \text{ is } H$$

where

$$H = \bigcup_i (A_i \cap B_i) \cap D,$$

hence $H(x, y) = \text{Max}_i[A_i(x) \wedge B_i(y) \wedge D(x)]$.

In order to obtain the value of V we must take the projection on V, thus the output is *V is G* where $G(y) = \text{Max}_x[H(x, y)]$. In this case

$$G(y) = \text{Max}_i[\text{Poss}[A_i|D] \wedge B_i(y)]$$

where $\text{Poss}[A_i|D] = \text{Max}_x[A_i(x) \wedge D(x)]$.[12]

The value $\text{Poss}[A_i|D]$ is the firing level of the ith rule and can be denoted $\mathfrak{I}_i$. We note in the special case where D is a singleton, $D = \{x^*\}$ then $\text{Poss}(A_i|D) = A_i(x^*)$ and therefore

$$G(y) = \text{Max}_i[A_i(x^*) \wedge B_i(y)].$$

Figure 5 shows a systematic view of the fuzzy logic controller.

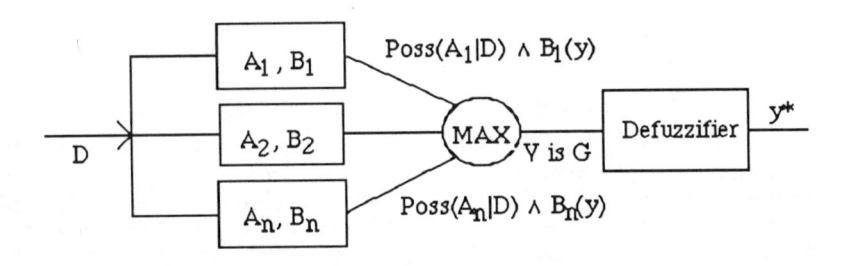

Figure 5. Structure of Fuzzy Logic Controller

In the preceding we have assumed that the antecedent consists of a single variable U. For the issues of interest of us in the following the consideration of multiple antecedents doesn't significantly change the situation and throughout this paper for the most part we shall continue to consider only single valued antecedents.

The model used to represent the function f will be called the **Basic Fuzzy Logic Controller Model** or **Mamdani Model**. The salient features of this model are the following:

1. Antecedents of the form U is A_i, where A_i is a normal fuzzy subset.
2. Consequents of the form V is B_i, where B_i is a normal fuzzy subset.
3. The individual rule representation is a conjunction of the antecedent and consequent: $\boxed{A_i, B_i}$ as (V, U) is S_i where $S_i = A_i \cap B_i$. We call this the **Mamdani type representation/ translation**.

4. Output of each individual rule in the form $\underline{V \text{ is } G_i}$ where $G_i(y) = \Im_i \wedge B_i(y)$ and $\Im_i$ the firing level equal to $\text{Poss}(A_i|D)$. ($\text{Poss}(A_i|D)$ reduces to $A_i(x^*)$ when the input is a singleton).

5. Aggregation of the individual rule outputs to obtain the overall system output via the max operation: $G(y) = \text{Max}_i[G_i(y)]$.

A number of interesting properties of this model can be pointed out.

Theorem: If D and $\hat{D}$ are two inputs such that $D \subset \hat{D}$ then $G \subset \hat{G}$.

Proof: For any rule the firing level $\Im_i = \text{Poss}[A_i|D]$ and $\hat{\Im}_i = \text{Poss}[A_i|\hat{D}]$. If $D \subset \hat{D}$ then $\Im_i \leq \hat{\Im}_i$ for all i and therefore $G_i \subset \hat{G}$.

The implication of this theorem is that as the input becomes less specific, more imprecise, the output gets larger. We see that this kind of system deals with uncertainty in the input by increasing the output. We note that in most imprecise situation $D = X$, $\Im_i = 1$ for all i and hence $G = \bigcup_i B_i = B^\circ$. It is obvious then for any D^* the corresponding $G^* \subset B^\circ$. It should be clear that in the FLC then if we have a completely unknown input, X, then the crisp output is the COA of B°.

Theorem: Let K be our knowledge base in a FLC, it is the collection of rules. Assume that under K if the input is D then the output is G:

1. If $K^+ = K \cup R$ where R is some additional rules then with input D the output G^* is such that

$$G \subset G^+$$

2. If $K^- = K - R$, where R is some subset of K then with input D the output G^- is such that

$$G^- \subset G.$$

Increasing the rules increases the size of the fuzzy output of the system., while decreasing the decreases the size of the output of the system. It should be emphasized that an increase in rules doesn't necessary clarify the selection, they could be contradictory..

Theorem: Assume the input to the fuzzy logic controller is the normal set D and let G be the output of the FLC. Then

$$\text{Max}_y G(y) = \text{Max}_i \Im_i.$$

Proof: The output of a FLC is $G = \bigcup_i \hat{B}_i$ where $B_i = \Im_i \wedge B_i$, and $\Im_i = \text{Poss}[A_i|D]$.

From the above $\text{Max}_y\, G(y) = \text{Max}_y[\text{Max}_i[\Im_i \wedge B_i(y)]]$. Let $\text{Max}_i(\Im_i)$. Assume $\Im_i$ attain its maximum value in the kth rule and let y^* be such that $B_k(y^*) = 1$. For any y, $G(y) = \text{Max}_i[\Im_i \wedge B_i(y)]$ but for any i and y $\Im_i \wedge B_i(y) \leq \Im^* \wedge B_k(y^*)$, thus for any

$$G(y) \leq \text{Max}_i[\Im^* \wedge B_i(y)] = \Im^* \wedge B_i(y) \leq \Im^*.$$

For y^* $G(y^*) = \text{Max}_i[\Im_i \wedge B_i(y)] = \Im^*$. Thus $\text{Max}_y\, G(y) = \Im^* = \text{Max}_i[\Im_i]$.

An implication of this theorem is that the maximum value of the fuzzy output of the controller can be seen as a measure of the degree to which the rule base has a rule relevant for the current input. The maximum value of the fuzzy output of the controller can be used as a measure of relevancy of the rule base for a given input.

We should also note that for any rule i

$$\text{Max}_y\, \hat{B}_i(y) = \Im_i.$$

The final step in the operation of the FLC is to use the fuzzy output G to select a singleton $y^* \in Y$ as the output of the controller. This process is called the defuzzification process. The most commonly used defuzzification procedure is the Center of Area (COA) method [1, 6]. In the case where the case set Y is finite, $Y = \{y_1, \ldots, y_m\}$ the COA method obtains y^* using

$$y^* = \frac{\sum_{i=1}^{m} G(y_i) \times y_i}{\sum_{i=1}^{m} G(y_i)}$$

As described in [7, 13] the COA defuzzification procedure involves a normalization of the fuzzy output set G to give us a probability distribution over the base set G such that for each y_i the probability p_i of y_i is given by

$$p_i = \frac{G(y_i)}{\sum_{i=1}^{m} G(y_i)}.$$

This normalization is then followed by the taking of an expected value over the set Y with respect to the probability distribution,

$$y^* = \sum_i p_i y_i.$$

As described by Filev and Yager [7, 13] alternative forms of defuzzification can be had by first transforming the fuzzy subset G into a new fuzzy subset on Y, $\hat{G}$, before the normalization process is implemented. The transformation from G to $\hat{G}$ is generally a kind of dilation or concentration operation [13, 14]. As discussed in [7, 13, 14] this transformation can be parameterized where the parameter can be interpreted as a measure of our confidence in the model.

3. CERTAINTY QUALIFIED ANTECEDENTS

In the Mamdani model of the FLC having antecedent condition V is A_i, A_i assumed normal, the firing level or relevancy of the ith rule for the normal input V is D, is

$$\Im_i = \text{Poss}[A_i|D] = \text{Max}_x[A_i(x) \wedge D(x)].$$

As we noted in the special case where $A_i = \{x^*\}$, is a singleton, $\Im_i$ reduces to $A_i(x^*)$. Essentially this form of firing level implies that the firing of the rule is determined by the possibility of A_i given the input D, this means that the firing is related to the *degree of intersection* between A_i and the input D. We can more specifically denote the antecedent condition in this basic case as

if V is A_i **is possible** then . . . .

It is noted that as D increases $\Im_i$ will not decrease. In particular a decrease in specificity of the input causes an increase in firing level or relevancy of the rules.

There exists an alternative, and as we shall show, more stringent requirement we can make on the antecedent condition. In particular we suggest that an alternative antecedent requirement is

"if V is A_i is **certain** then . . . "

We mean to indicate that the ith rule fires if we can certain the value for V lies in A_i. Again we assume the input to be V is D, where D is normal. Using the results found in [15] we can implement this kind of certainty requirement by defining the firing level γ_i as

$$\gamma_i = \text{Cert}[A_i|D] = 1 - \text{Poss}[\overline{A_i}|D].$$

In this case the degree of firing is related to *the degree of inclusion* of D in A_i. Using this formulation for firing level or relevancy of a rule we get as of output of each rule, G_i, where

$$G_i(y) = \gamma_i \wedge B_i(y)$$

and that the overall output is

$$G = \cup_i G_i$$

In the next theorem we justify our statement that the above approach is more stringent.

Theorem: Assume A and D are two normal fuzzy subsets of X. Letting $\Im = \text{Poss}[A|D]$ and $\gamma = 1 - \text{Poss}[\overline{A}_i|D]$ it is always the case that

$$\Im \geq \gamma.$$

Proof: $\Im = A(x_1) \vee D(x_1) \vee A(x_2) \vee D(x_2) \vee \ldots,$
With D normal, and $D(x^*) = 1$ then

$$\Im \geq A(x^*).$$

$$\gamma = (A(x_1) \vee \overline{D}(x_1)) \wedge (A(x_2) \vee \overline{D}(x_2)), \ldots$$
$$\gamma = (A(x^*) \vee 0) \leq A(x^*)$$

hence

$$\Im \geq \gamma.$$

Thus we see that the use of certainty qualification always leads to a smaller relevancy or rules.

We now show that using this certainty qualified antecedent as the specificity of input set increases the level of firing decreases.

Theorem: Assume D and $\hat{D}$ are two normal inputs such that $D \subset \hat{D}$. Let γ and $\hat{\gamma}$ be the firing level for antecedent A under these conditions, then $\gamma \geq \hat{\gamma}$.

Proof: $\gamma = 1 - \text{Poss}[\overline{A}|D]$ and $\hat{\gamma} = 1 - \text{Poss}[\overline{A}|\hat{D}]$. Since $D \subset \hat{D}$, then $\text{Poss}[\overline{A}|\hat{D}] \geq \text{Poss}[\overline{A}|D]$ hence $\gamma \geq \hat{\gamma}$.

This theorem shows that going to a less specific input can't increase the firing level.

Corollary: For $D \subset \hat{D}$ it is the case that for any rule $\hat{G}_i \subset G_i$.

Proof: Since $G(y) = \gamma_i \wedge B_i(y)$ and $\hat{G} = \hat{\gamma}_i \wedge B$.

Corollary: For $D \subset \hat{D}$ it is the case that the overall outputs are related by $\hat{G} \subset G$.
Thus a decrease specifically doesn't imply an increase of output.

We also see that since $G_i(y) = \gamma_i \wedge B_i(y)$ and the B_i's are assumed normal

$$\text{Max}_y G(y) = \text{Max}_i[\gamma_i]$$

Thus we see here also that the maximum firing value of the output is a measure of the degree to which the rule base has been relevant for the current input.

A question of particular interest is the determination of the conditions required of the input, D, to cause the maximum firing of a rule with antecedent A_i. The following theorem provides this information.

Theorem: For a rule with normal antecedent A_i the maximal possible firing level, $\gamma_i = 1$, is attained if D is such that

$$\overline{A_i} \cap D = \Phi,$$

that is $D(x) > 0$ only for those values of x for which $A_i(x) = 1$.

Proof: γ_i attain its maximum value when $\gamma_i = 1 = 1 - \text{Poss}[\overline{A_i}|D]$, that is $\text{Poss}[\overline{A_i}|D] = 0$, this requires $\overline{A_i} \cap D = \Phi$.

Since we have assumed D to be normal in the FLC, we see that maximum firing occurs if $D^* = \{x^*\}$ where $A(x^*) = 1$. Furthermore any $\hat{D}$ such that $D^* \subset \hat{D}$ also fires at this level only if $\hat{D}$ has non zero membership grades for only those elements which have membership grade of one in A.

4. ALTERNATIVE FORMULATIONS FOR RULE OUTPUTS

As we indicated in the previous section the knowledge base of the fuzzy logic controller (FLC) involves a relationship between the input and output variables. We can of course consider that we have multiple inputs, $u_1, u_2, \ldots, u_r$ each over the base set X_i. Thus formally the FLC knowledge base is a representation of a function

$$v = f(u_1, u_2, \ldots, u_r).$$

The function f is in general a complex ill defined non-linear relationship between the input variables and the output. The spirit taken in the development of FLC is to partition the input space, $X = X_1 \times X_2 \times X_r$ into regions in which we can assume a simple relationship exists between the inputs and output. Essentially each rule describes a region of the input space, R_j. Thus if the jth rule in our FLC says

when U_1 is A_{j_1} and U_2 is A_{j_2} and U_r is A_{j_r} then V is B_j

where A_{j_i} is a fuzzy subset of X_i we are saying in the subspace of X defined by

$A_{j_1} \times A_{j_2}, \ldots, \times A_{j_r}$, which we denote as R_j, the effective value for V is B_j. The fuzziness associated with the R_j's imply there is some overlap between the regions of the rules. Ideal design of these controllers requires that the partitioning by the R_j's covers X.

In the approach suggested in [1, 16, 17] which we shall call the <u>Mamdani model</u> it is assumed in each region of the partition, each R_j, the output *is a constant* all though fuzzy value, B_j.

In [18, 19] an alternative more general form of the output of the rules was suggested by Takagi, Sugeno and Kang. We shall call this the TSK model. In the TSK model the form for the value of V in each region was expressed as a linear function of the input variables. Thus in the TSK model the right hand side of the jth rule was of the form

$$V \text{ is } b_{j0} + b_{j1}u_1 + b_{j2}u_2, \ldots, b_{jn}u_r$$

where the b_{ji}'s are crisp numbers. As noted by Filev [20] this can be seen as a quasi-linear model.

In the TSK model it was also suggested to use a different methodology for implementing the rule aggregation and defuzzification. In this model these two steps are combined into one step. To describe this difference we shall use the following notation. Let the inputs to the system be the crisp values a_i, $u_i = a_i$. Let us denote the level of firing for each rule as $\mathfrak{I}_j$, where

$$\mathfrak{I}_j = \text{Min}_i[A_{ji}(a_i)].$$

We note this is the same level of firing as the Mamdani model. Finally let

$$z_j = b_{j0} + b_{j1}a_1 + b_{j2}a_2, \ldots, + b_{jr}a_j$$

be the output of the jth rule for this input.

In the TSK model the output of the controller is

$$z^* = \frac{\sum_j \mathfrak{I}_j \times z_j}{\sum_j \mathfrak{I}_j}$$

This TSK aggregation/defuzzification procedure can be seen to be closely related to the standard procedure as used in Mamdani's approach. Representing the TSK model in a Mamdani type model each rule would generate as its output a fuzzy subset

$$V \text{ is } G_j$$

where with z_j as described above $G_j = \left\{\frac{\mathfrak{I}_j}{z_j}\right\}$.

We particularly note that in the TSK model each G_j just has one element in its support set, z_j. Using this standard approach we would then calculate the overall output of the system, G, as the union of the outputs of the individual rules

$$G = \cup_j G_j,$$

$$G(y) = \text{Max}_j G_j(y).$$

The defuzzified value in this situation using COA is

$$y^* = \frac{\Sigma_i G(y_i) \times y_i}{\Sigma_i G(y_i)}$$

It should be noted that if all the z_j's are different then the application of the standard aggregation/defuzzification procedure would lead to the same result as the TSK approach to aggregation/defuzzification. However in the case when a value z appears as the output in more than one rule the results are different.

One way of characterizing the TSK process is by seeing that the defuzzification process as the standard one, COA, but the aggregation of the individual rules as different. In particular we say that the resulting aggregated fuzzy subset in TSK is $\hat{G}$ where

$$\hat{G} = \uplus G_j$$

where $\uplus$ is defined such that

$$\hat{G}(y) = \Sigma_j G_j(y).$$

Thus the TSK aggregation replaces the max with a summation. One advantage of this aggregation is that it allows for reinforcement of repeated suggestions, z_j's.

Formally we should note that the object $\hat{G}$ is not truly a fuzzy subset, this is so because it is possible for $\hat{G}(y)$ to be greater than zero.

We should note that the defuzzification process in the TSK model is essentially normalization followed by a taking of an expected value

$$z^* = \frac{\Sigma_i \hat{G}(y_i) \times y_i}{\Sigma_i \hat{G}(y_i)}$$

Alternative aggregation procedures are possible which can to some degree capture this reinforcement. In particular one can use another t-conorm [21-23] then the max to aggregate the G_j's. Thus one can let

$$G^S = \underset{S}{\cup} G_j$$

where S is some t-conorm. In this case

$$G^s(y) = \mathbf{S}_j(G_j(y))$$

Examples of t-conorms [21-23], other than min, are

$$S(a, b) = a + b - ab$$

$$S(a, b) = (a + b) \wedge 1.$$

However these t-conorms do saturate at one. Another alternative approach to aggregate is to use the OWA operators developed in [24].

There exists an alternative view to looking at the type of aggregation suggested by the TSK method, this view is based upon the concept of a fuzzy bag [25, 26]. We recall that a bag differs from a set in that it allows multiple copies of a element. In a fuzzy bag we also allow multiple copies.

Assume Y is a set. A bag M associates with each $y \in Y$ a membership set,

$$M(y) = \bigcup_\alpha \{(M_{y,\alpha}, \alpha)\}.$$

In this representation $M_{y,\alpha}$ are the number of times the element y appears in the bag with level α.

Assume A and B are two bags of Y the bag union of these two bags is defined as follows

$$A \uplus B = D$$

where

$$D_{y,\alpha} = A_{\alpha,y} + B_{y,\alpha}.$$

Thus the count of y at level α in D is simply the sum of y at the level α in both A and B.

We next can define the cardinality of an element y in the bag A, we denote this as Card(y|A) and define it as

$$\text{Card}(y|A) = \sum \alpha_i A_{\alpha_i, y}$$

Finally we can define the cardinality of the bag as

$$\text{Card}(A) = \sum_j \text{Card}(y_j|A).$$

With this brief introduction to bags we see that the TSK aggregation is simply a bag aggregation of the G_j's. Thus

$$\hat{G} = \biguplus G_j.$$

The defuzzification associated with the bag $\hat{G}$ is

$$z^* = \frac{\sum_i \text{Card}(y_i/\hat{G}) * y_i}{\text{Card}(\hat{G})}$$

This is exactly the result as in the TSK approach.

In the TSK model it is assumed that the right hand side is a linear non-fuzzy function of the input values. Sugeno and his co-workers restricted the right hand side to be non-fuzzy and linear because they were also interested in developing techniques for system and parameter identification from data. The assumption of crispness and linearity significantly simplifies this problem. One natural extension of this model is to drop the requirement of linearity and allow the outputs of each rule to be a crisp non-linear function of the inputs. Thus dropping the linearity condition we would have for each rule

$$y \text{ is } f_i(u_1, u_2, \ldots, u_r)$$

where f_i is a non-linear function of the inputs. Filev has called these types of models Quasi-nonlinear fuzzy models [27].

More generally we also can drop the restriction of crispness of the output of each rule and look at the situation in which the right hand side is a general fuzzy relationship between the input and the output. Consider a rule base consisting of n rules. Each rule is of the form

When u_1 is A_{1i} **and** u_2 is A_{2i} **and** u_r is A_{ri} **then** $(u_1, \ldots, u_r, v)$ is R_i.

In the above A_{ji} is a fuzzy subset of the base set of the variable u_i, X_i, Y is the base set of V and R_i is a fuzzy relationship on the cartesian product space $Z = X_1 \times X_2 \times X_r \times Y$

We shall assume that the input to the system are the crisp collection of values

$$u_j = a_j \qquad j = 1, 2, \ldots, r.$$

Using the Mamdani interpretation for the rule translation we get for each rule

$$(u_1, u_2, \ldots, u_r, V) \text{ is } A_{1i} \times A_{2i} \times A_{3i}, \ldots, \times A_{ri} \times R_1 \times B_1 \times B_2, \ldots, \times B_r$$

where $B_j = \left\{\frac{1}{a_j}\right\}$. From this model, using the projection operation, we get as the output provided by the ith rule <u>V is Q_i</u> where Q_i is a fuzzy subset of the output space Y such that for each $y \in Y$

$$Q_i(y) = (\text{Min}_j[A_{ji}(a_j)]) \wedge R_i(a_1, \ldots, a_r, y)$$

If we denote $\Im_i = \text{Min}_j[A_{ji}(a_j)]$ as the firing level or *relevancy* of the ith rule then

$$Q_i(y) = \Im_i \wedge R_i(a_i, \ldots, a_r, y).$$

Having obtained Q_i for each rule these individual output values must be aggregated to get the overall output value Q. In order to accomplish this we can use any of the techniques we discussed for aggregation. For example

$$Q = \bigcup_i Q_i.$$

In [28] Yager suggested a softer aggregation technique based upon the use of the OWA [24] operators. In particular he suggested providing a set of weights w_k, $\sum w_k = 1$ and then calculating

$$Q^+(y) = \sum w_k b_k(y)$$

where $b_k(y)$ is the kth largest of the $Q_i(y)$'s. The weights selected, the w_k's, should reflect an *orlike* aggregation [28].

Two special cases of the relationship R_i are worth noting. If

$$R(a_1, a_2, \ldots, a_r, y) = R_i(\hat{a}_1, \hat{a}_2, \ldots, a_r, y) \text{ for all } (a_1, \ldots, a_r) \text{ and } (\hat{a}_1, \ldots, \hat{a}_r)$$

then it is essentially the case that $R_i(a_1, \ldots, a_r, y) = R_i(y)$. In this case we have the Mamdani model, constant fuzzy outputs for each rule.

A second special case occurs if $R_i(a_1, \ldots, a_r, y) \in \{0, 1\}$, R_i is just a crisp relationship. Furthermore if we put the additional requirement that for any tuple $(a_1, \ldots, a_r)$ there is only one $y \in Y$ which has non-zero membership we have obtained the quasi-nonlinear version of the TSK model.

5. CHAINING OF FUZZY RULES

The typical fuzzy controller is of the form of a function of the type $V = f(U)$ where V and U may be joint variables, consist of more than single variables. However the situation is essentially a one step input-output environment.

In some cases we may have situations in which a chaining of variables takes place.

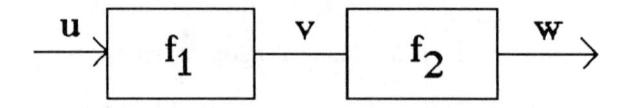

$$\xrightarrow{u} \boxed{f_1} \xrightarrow{v} \boxed{f_2} \xrightarrow{w}$$

Figure 6. Chained System

This situation is typified by model shown in Figure 6.

In this figure we see that the input collection of variables U are first transformed via function f_1 into a collection of intermediate variables V which are in turn transformed via function f_2 into the output collection of variables W. Thus we have a two step operation.

One case is which we can get this kind of model is that in which the controller complexity is of such a nature as to require this kind of representation. A second situation which we induce this kind of situation is shown in Figure 7.

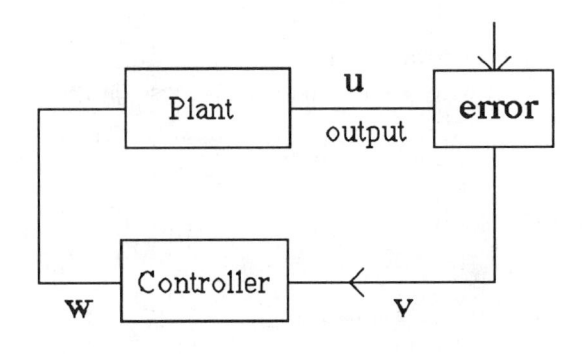

Figure 7.

In this case U is the output of the plant, V is the input to the controller and W the output of the controller used to correct the plant output. In this case we consider that the process used to generate the error, V, may be a complex of fuzzy process. For example we may specify the ideal as a fuzzy set and we desire that the output U be close to the idea or some function of the ideal.

In the following we shall let X, Y and Z be the base sets of the variables U, V, and W respectively. Furthermore we shall assume that the representation of the function f_1 is of the Mamdani type. Thus f_1 is represented by a collection of rules of the form

When U is A_i then V is B_i

where $i = 1, \ldots, n$. In addition we shall also let f_2 be represented by a Mamdani type representation. Thus f_2 is represented by a collection of rules of the form

When V is C_j then W is D_j

where $j = 1, \ldots, m$.

In the above the A_i's are fuzzy subsets of X, the B_i's and C_j's are fuzzy subsets of Y and the D_j's are fuzzy subsets of Z.

Using the theory of approximate reasoning we can represent the function f_1 is a proposition

$$(U, V) \text{ is } F_1$$

where F_1 is a fuzzy relation of $X \times Y$. As discussed in the previous section the form of F_1 is

$$F_1 = \bigcup_{i=1}^{n}(A_i \cap B_i).$$

Thus F_1 is the <u>union</u> of the individual rules.

In a similar manner we can represent f_2 as a proposition

$$(V, W) \text{ is } F_2$$

In this case F_2 is a fuzzy relationship on $Y \times Z$ such that

$$F_2 = \bigcup_{j=1}^{m}(C_j \cap D_j).$$

We shall assume that the input to our controller is the knowledge that

$$U \text{ is } A$$

where A is a fuzzy subset of X.

The first step in the process of getting the system output is to put this input into the function f_1 to get a value for V. This step is accomplished by conjuncting F_1 with A to give us

$$(U, V) \text{ is } H$$

where $H = F_1 \cap A = \bigcup_{i=1}^{n}(A_i \cap B_i) \cap A$.

Applying the projection principle [29] to this to get a value for V we get <u>V is G</u> where

$$G = \bigcup_{i=1}^{n} \widehat{B}_i.$$

In the above $\widehat{B}_i$ is defined via the projection operation as

$$\widehat{B}_i(y) = \text{Max}_x[A_i(x) \wedge A(x) \wedge B_i(y)]$$

$$\widehat{B}_i(y) = \text{Poss}[A_i|A] \wedge B_i(y)$$

The value of V, G, now is used as the input to the second function f_2 to get the output of the system W. This second step of the process is accomplished by conjuncting F_2 with G to give us

$$(V, W) \text{ is } M,$$

where M is a fuzzy relationship on $Y \times Z$ such that

$$M = F_2 \cap G = \bigcup_{j=1}^{m}(C_j \cap D_j) \cap G = \bigcup_{j=1}^{m}(C_j \cap D_j \cap G) = \bigcup_{j=1}^{m}(C_j \cap D_j \cap \bigcup_{i=1}^{n} \widehat{B}_i)$$

It is important to note that in the formulation of M the relationship between F_2 and F_1 is one of conjunction.

The final step in the process is to take the projections of M onto Z. Thus we get

$$W \text{ is } E$$

where

$$E(z) = \text{Max}_y M(y, z).$$

Performing this operation we get $E(z) = \text{Max}_j[\text{Poss}[C_j|G] \wedge D_j(z)]$. In the preceding

$$\text{Poss}[C_j|G] = \text{Max}_y[C_j(y) \wedge \text{Max}_i(\widehat{B}_i(y))]$$

$$\text{Poss}[C_j|G] = \text{Max}_i[\text{Max}_y(C_j(y) \wedge \widehat{B}_i(y))]$$

$$\text{Poss}[C_j|G] = \text{Max}_i \, \text{Max}_y(C_j(y) \wedge B_i(y) \wedge \text{Poss}[A_i|A])$$
$$\text{Poss}[C_j|G] = \text{Max}_i[\text{Poss}(A_i|A) \wedge \text{Poss}(C_j|B_i)]$$

Thus the final form for E is

$$E(z) = \text{Max}_j[\text{Max}_i[\text{Poss}(A_i|A) \wedge \text{Poss}(C_j|B_i) \wedge D_j(z)]]$$

To get an intuitive feel for the form we shall consider the case when $n = m = 2$. Here we get

$$E(z) = \text{Poss}(A_1|A) \wedge \text{Poss}(C_1|B_1) \wedge D_1(z)$$
$$\vee \text{Poss}(A_2|A) \wedge \text{Poss}(C_1|B_2) \wedge D_1(z)$$

$$\vee \ Poss(A_1|A) \wedge Poss(C_2|B_1) \wedge D_2(z)$$
$$\vee \ Poss(A_2|A) \wedge Poss(C_2|B_2) \wedge D_2(z)$$

From a semantic point of view we can see the above process as generating a new input–output function between V and W. This new function, which we denote as f, is a composition of the two original function f_1 and f_2. There are two different ways to view this new function:

I. Let R_1 be defined such that

$$R_1(x) = A_1(x) \wedge Poss(C_1|B_1) \vee A_2(x) \wedge Poss(C_1|B_2)$$

and let R_2 be defined such that

$$R_2(x) = A_1 \wedge Poss(C_2|B_2) \vee A_2(x) \wedge Poss(C_2|B_2)$$

then we see this system as consisting of two rules relating U and V

When U is R_1 then W is D_1
When U is R_2 then W is D_2.

More generally if we have i = n and j = m. We can see that f is represented as a collection of rules of the form

When U is R_j then W is D_j

where j = 1, ..., m and $R_j(x) = Max_i[A_i(x) \wedge Poss(C_j|B_i)]$.

II. Let T_1 and T_2 be defined such that

$$T_1(z) = D_1(z) \wedge Poss(C_1|B_1) \vee D_2(z) \wedge Poss(C_2|B_1)$$
$$T_2(z) = D_1(z) \wedge Poss(C_1|B_2) \vee D_2(z) \wedge Poss(C_2|B_2)$$

then we see the system is

When U is A_1 then W is T_1
When U is A_2 then W is T_2

In the more general setting when i = n and j = m we get a representation of f as

When U is A_i then W is T_j

where i = 1, ..., n and $T_i(z) = Max_j[D_j(z) \wedge Poss(C_j|B_i)]$.

A system of equations of the above type can be considered as normal if

(1) For each output D_j there exists some input that fires it
(2) for each input A_i there is some complete firing

To satisfy (1) we need that for each R_j there exists at least one X such that $R_j(x) = 1$. Assume $A_i(x)$ is normal for each i, what is required is the for each j

$$Max_j Poss[C_j|B_i] = 1.$$

Thus for each j

$$\bigcup_{i=1} C_j \cap B_i$$

is normal, has at least one element with membership grade one.
Furthermore the second condition requires

$$Max_j Poss[C_j|B_i] = 1$$

Thus for each i,

$$\bigcup_{j=1}^{m} C_j \cap B_i$$

is normal.

In the typical setting of the fuzzy logic controller the input to the system is usually a crisp singleton, $U = x^*$. In this situation we can replace $Poss[A_i|A]$ throughout by $A_i(x^*)$.

In the following Figure 8 we see a systematic representation of the above process.

In the above the aggregations performed by the Max operator can be replaced by any other t–conorm. We can also use an aggregation in the spirit of the TSK model. In order to provide this we must use a bag type aggregation. This will greatly complicate the process.

If we replace the Mamdani representation of the output by a quasi-linear TSK representation we essentially replace each by B_i by $b_{i0} + b_{i1}U$ and each D_j by $d_{j0} + d_{j1}V$. In this case if the input is $u = x^*$ then each rule has as its output a singleton

$$\widehat{B}_i = \left\{ \frac{A_i(x^*)}{b_{i0}+b_{i1} x^*} \right\} = \left\{ \frac{A_i(x^*)}{\widehat{b}_i} \right\}$$

where $\widehat{b}_i = b_{i0} + b_{i1} x^*$. Using a max type aggregation we get as our output for the first phase

$$V \text{ is } G$$

where

$$G = \cup_i \left\{ \frac{A_i(x^*)}{b_{i0}+b_{i1} x^*} \right\} = \cup_i \left\{ \frac{A_i(x^*)}{\widehat{b}_i} \right\}$$

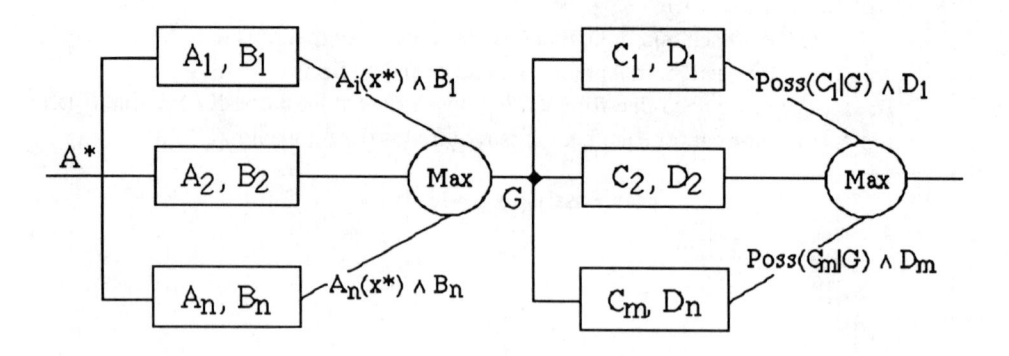

Figure 8. Chained System

For each rule j of the second portion we get

$$V \text{ is } \widehat{D}_j$$

where

$$\widehat{D}_j = \text{Poss}[C_j|G] \wedge \left\{\frac{1}{d_{j0}+d_{j1}\ G}\right\}$$

We first note that Poss $C_j|G = \text{Max}_i[A_i(x^*) \wedge C_j(\widehat{b}_i)]$. Next the set

$$\left\{\frac{1}{d_{j0}+d_{j1}\ G}\right\} = \cup_i \left\{\frac{A_i(x^*)}{d_{j0}+d_{j1}\times(b_{i0}+b_{i1}\ x^*)}\right\}$$

Thus

$$\widehat{D}_j = \cup_i \left\{\frac{\text{Poss}[C_j|G] \wedge A_i(x^*)}{d_{j0}+d_{j1}\times b_{i0}+d_{j1}b_{ij}\ x^*}\right\}$$

Finally the output is

$$W \text{ is } \widehat{D}$$

where

$$\widehat{D} = \cup_j \widehat{D}_j.$$

6. DECOUPLED INPUTS

In the previous section we reconsidered the situation in which we had a chaining of functions to give us the desired output. In this section we consider the case in which there is a decoupling of input variables. Figure 9 shows a systematic representation of this situation.

In the above each U_i corresponds to a collection of input variables. One way to conjecture the occurrence of this type of situation is the case in which we have multi-independent measures of performance. For each of these performance measures a different suggested action is recommended.

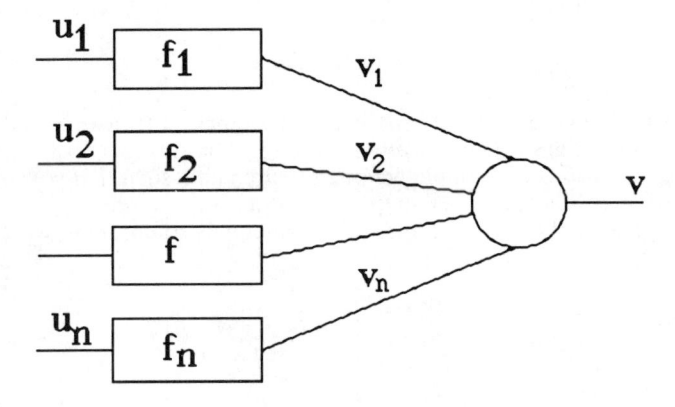

Figure 9. Decoupled System

For each function f_k we have a representation as a collection of rules with antecedents A_{ki} and consequent B_{ki}. We assume that the kth function has i_k rules. If the inputs to the structure are $u_k = a_k$ then the firing level of the ith rule in the function f_k is $A_{ki}(a_k)$ which we denote $\Im_{ki}$. The output of the ith rule in the function f_k is G_{ki} where G_{ki} is a fuzzy subset of Y such that

$$G_{ki}(y) = \Im_{ki} \wedge B_{ki}(y).$$

The output of the function f_k is the knowledge that

$$V \text{ is } G_k$$

where $G_k = \bigcup_i G_{ki}$ and thus $G_k(y) = \text{Max}_i[G_{ki}(y)]$.

The next step is to combine the individual functional outputs, the G_k, to obtain an overall output V is G where

$$G = \text{Comb}[G_k].$$

The central question becomes how do we combine these individual outputs, what is the form of the function Comb ? At one extreme is to take the position that if one function indicates that y is a good solution then y is an overall good solution in the case

$$G(y) = \text{Max}_k\, G_k(y).$$

At the other extreme is to say that if one function says that y is a bad solution then y is an overall bad solution. In this case

$$G(y) = \text{Min}_k\, G_k\, C(y).$$

One way to avoid either of these two extremes is to use the OWA operators suggested by Yager. These operators provide for aggregations lying between the extremes of *and* and *or*. They allow for implementing requirements such that y is a good solution if "most" or "some" of the individual outputs say its a good solution.

We recall that if we have n individual outcomes to combine we need an OWA weighting vector

$$W = \begin{bmatrix} w_1 \\ \vdots \\ w_n \end{bmatrix}$$

such that

(1) $w_i \in [0, 1]$

(2) $\sum_i w_i = 1$

Using this approach for each $y \in Y$, $G(y) = f(G_1(y), G_2(y), \ldots, G_n(y))$ where

$$G(y) = \sum_j w_j b_j$$

with b_j being the jth largest of the $G_i(y)$.

We see that if W is such that $w_1 = 1$ we simply get the maximum. If W is such that $w_n = 1$ we get the minimum. If W is such that $w_i = 1/n$ for all i we get an averaging,

$$G(y) = \frac{1}{n}\sum_k G_k(y).$$

More detail about OWA aggregation can be found in [24].

7. HIERARCHICAL REPRESENTATION OF RULES

In [30] Yager provides an example in which the Mamdani type representation and aggregation leads to unsatisfactory results. As a motivation for the introduction of a new structure for the representation of rules in fuzzy modeling we repeat this example.

Example: Consider a rule base consisting to two rules

$$\text{if U is 12 then V is 29} \quad \text{I.}$$

$$\text{If U is A then V is B} \quad \text{II.}$$

where $A = [10\text{-}15]$ and B is $[25\text{-}30]$. We see that if $U = 12$ then the first rule specifically tells us to set $V = 29$. If we use the mechanism of the fuzzy logic control reasoning system in this situation both rules would fire and we would get $\underline{V \text{ is } G}$ where $G = G_1 \cup G_2$. Since

$$G_1 = \{29\} \text{ and } G_2 = [25\text{-}30]$$

we get that $G = [25 - 30]$. The application of the defuzzification process leads to a selection of $V = 27.5$. Thus we see that the very specific instruction was not followed.

The problem with the technique used is that the most specific information was swamped by the less specific information. In this section we shall provide for a new form for the representation of the relationship $V = f(U)$. The representational form introduced in this section is called a **H**ierarchical **P**rioritized **S**tructure (HPS) representation. In addition to overcoming the problem illustrated in the previous example this HPS representation has an inherent capability to emulate the learning of general rules. In the following figure #10 shows in a systematic view the form of this new HPS representation.

In Figure 10 the overall function f, relating the input U to the output V, is comprised of the whole collection of subboxes. In the HPS each subbox, denoted f_i, is a collection of rules relating the system input, U, and the current iteration of the output, V_{i-1}, to a new iteration of the output. We note that within this HPS V_i is used to indicate the value for V provided as an output of the ith set of rules. The output of the nth subsystem, V_n, becomes the overall output of the system, V. As we envision this system *for $i < j$ we say that f_i has a higher priority than f_j.* We anticipate that the higher priority boxes would have less general information, consist of rules with more specific antecedents then those of lower priority. We recall that specifically is closely related to the cardinality of the set and inclusion [31, 32]. Thus we see the boxes of higher priority consisting of more specific antecedent information, which generally reflects into more specific consequent information. As we envision this system working an input value for U is provided, if it matches one or more of the rules in the first (highest priority) level then it doesn't bother to fire any of the less specific rules in the lower priority levels.

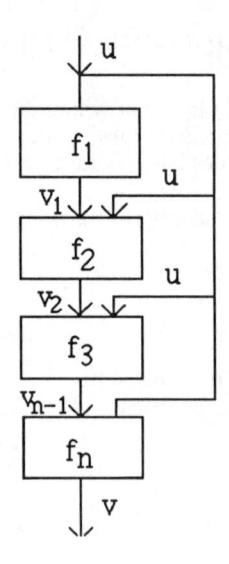

Figure 10. Hierarchical Prioritized Structure

In the following we describe the formal operation of this HPS. As we indicated V_j denotes the output of the jth level. We shall assume $V_0 = \Phi$. In the HPS we shall use the variable $\hat{V}_j$ to indicate the **maximum membership grade** associated with the output of the jth level, V_j. Thus if V_j is A and $A = \{\frac{.6}{x_1}, \frac{.2}{x_2}, \frac{.4}{x_3}, \frac{.8}{x_4}, \frac{.3}{x_5}\}$ then $\hat{V}_j$ is .8.

In the HPS representation each f_j is for $j = 1, \ldots, n-1$ (accept for the lowest level, $j = n$) a collection of rules m_j

I   *When U is $\underline{A_{ji}\ is\ certain}$ and $\hat{V}_{j-1}$ is low then V_j is B_{ji}*

plus the rule

II V_j is V_{j-1}

The representation and aggregation of rules at each level is of the standard Mamdani type, disjunction of the individual rules.

We see that in I the rule fires if we are certain that the input U lies in A_{ji} and $\hat{V}_{j-1}$ is low. Since $\hat{V}_{j-1}$ is the maximum membership grade of V_{j-1}, it can be seen as a measure of how much matching we have up to this point. Essentially this term is saying that if the higher priority rules are relevant, $\hat{V}_{j-1}$ is not low, then don't bother using this information. On the other hand if the higher priority rules are not relevant, not to much matching $\hat{V}_{j-1}$ is low, then try using this information.

The rule II just assures us that the output of sub–box j is the union of the output of the previous boxes plus whatever is contributed by box j. We can replace this rule by requiring that

$$V_j = V_{j-1} \cup \widetilde{B_j}$$

where $\widetilde{B_j}$ is the value of V_j obtained by the aggregation of the outputs of the collection of rules in I

The representation of the box f_n is a collection of rules

When U **is** A_{ni} **and** $\widehat{V}_{n-1}$ **is lower then** V **is** B_{ni}

plus the rule

$$V \text{ is } V_{n-1}.$$

The notable difference between the lowest priority box and the other ones is that the antecedent regarding U is certainly quality in the higher boxes. The need for this becomes apparent when the input is not a singleton[2].

We shall use the Mamdani translation of the rules as the conjunction of the antecedents with the consequent.

In the HPS structure $\widehat{V}_{j-1}$ is the highest membership grade in V_{j-1} and as such the term $\widehat{V}_{j-1}$ *is low* is used to measure the degree to which the higher prioritized information have matched the input data. We note that low is a fuzzy subset on the unit interval. Any definition for low must satisfy the following conditions:

(1) low (0) = 1 **boundary condition/normality**
(2) low (1) = 0 **boundary condition**
(3) low (a) $\geq$ low (b) if b > a **monotonicity**

One simple definition for low is

$$\text{low}(x) = 1 - x.$$

Let us look at the formal functioning of this HPS representation. Assume that the input to the system is the crisp value $U = x^*$. Then since $V_0 = \Phi$ the output of the first, highest priority, subsystem is

$$V_1 \text{ is } G_1$$

where

---

[2] I would like to thank D. Filev for pointing out the necessity of making this distinction.

$$G_1 = \bigcup_{j=1}^{n_1} A_{1j}(x^*) \wedge B_{1j}$$

We shall denote $g_i = \text{Max}_y G_i(y)$.

The output of the second subsystem is

$$V_2 \text{ is } G_2$$

where

$$G_2 = \bigcup_{j=1}^{n_2} \text{low}(g_1) \wedge A_{2j}(x^*) \wedge B_{2j} \cup G_1$$

More generally after the i subsystem kicks in we get

$$G_i = \text{low}(g_{i-1}) \wedge \bigcup_{j=1}^{n_i} A_{ij}(x^*) \wedge B_{ij} \cup G_{i-1}.$$

The final output of the system is

$$V \text{ is } G$$

where

$$G = \text{low}(g_{n-1}) \wedge \bigcup_{j=1}^{n_n} A_{nj}(x^*) \wedge B_{nj} \cup G_{n-1}.$$

It should be clear that if $n = 1$, then

$$G = G_1 = \bigcup_{j=1}^{n_1} A_{1j}(x^*) \wedge B_{1j}$$

and the system reduces to that initially proposed by Mamdani.

We should point out that more sophisticated formulations for the output of the individual rules, such as in the TSK model, can be used. Thus we see that the Mamdani and TSK models are special cases of this more general formulation of fuzzy logic controllers.

We note that if the input to the system is a fuzzy subset, U is A, then we must replace $A_{ij}(x^*)$, to take into account the take of specificity of A. If we denote $\Im_{ij}$ as the degree of firing (or relevancy) of A_{ij} under the input U is A then

$$\Im_{ij} = \text{Cert}[A_{ij}|A] = 1 - \text{Poss}[\overline{A_{ij}}|A] \quad \text{for } i = 1, \ldots, n-1$$

$$\Im_{ij} = \text{Poss}[A_{nj}|A] \text{ for } i = n.$$

We note that in the special case where $A = \{x^*\}$ then $\Im_{ij} = A_{ij}(x^*)$ for all i. Using this more general notation we get

$$G_i = \bigcup_{j=1}^{n_i} \text{low}(g_{i-1}) \wedge \Im_{ij} \wedge B_{ij} \cup G_{i-1}.$$

Noting that low (g_{i-1}) is independent of j and taking advantage of the distributing property we can write this as

$$G_i = \text{low}(g_{i-1}) \wedge \bigcup_{j=1}^{n_i} \Im_{ij} \wedge B_{ij} \cup G_{i-1}.$$

We recall that g_{i-1} is the maximum membership grade in G_{i-1}, and hence low(g_{i-1}) indicates the degree to which the maximum membership grade satisfies the condition of being low.

We note that if $G_{i-1} = \Phi$ then $g_{i-1} = 0$ and low$(g_{i-1}) = 1$ and hence

$$G_i = \bigcup_{j=1}^{n_i} \Im_{ij} \wedge B_{ij}.$$

As a matter of fact this is the condition on the first level. We also note that if the system just consists of one level then $G_i = G$ and the HPS reduces to the classic FLC control system.

For intuitive simplicity, we shall in the following, define low(a) as $(1 - a)$. With this substitution we get as a general form for the output of each level

$$G_i = ((1 - g_{i-1}) \wedge \bigcup_{j=1}^{n_i} \Im_{ij} \wedge B_{ij}) \cup G_{i-1}.$$

Since

$$g_{i-1} = \text{Max}_y G_{i-1}(y) = \text{Poss}[G_{i-1}|Y] = \text{Poss}[G_{i-1}]$$

then

$$G_i = ((1 - \text{Poss}[G_{i-1}] \wedge \bigcup_{j=1}^{n_i} \Im_{ij} \wedge B_{ij}) \cup G_{i-1}.$$

We notice that the term $(1 - g_{i-1})$ bounds the allowable contribution of the ith subsystem to the overall output. We see that as we get at least one element y to be, a good answer (an element in G_{i-1}) we limit the contribution of the lower priority subsystems. It is this characteristic of a kind of saturation along with the prioritization that allows us to avoid the problem described earlier.

Let us look at the dynamics of the term g_i, we recall this is the largest membership grade in G_i. Consider the term

$$T_i = \bigcup_{j=1}^{n_i} \Im_{ij} \wedge B_{ij}$$

for a fixed i. It is easy to see, because of the assumed normality of the B_{ij}, that

$$\text{Max}_y[T_i(y)] = \text{Max}_j(\Im_{ij}).$$

Let us denote

$$M_i = \text{Max}_j[\Im_{ij}].$$

Thus M_i is the maximal firing level of any U antecedent in i^{th} sub-box.

Since G_0 is initiated as Φ, then $g_0 = 0$ and hence

$$G_1 = \bigcup_{j=1}^{n_1} \Im_{1j} \wedge B_{1j}.$$

From this we see that $g_1 = \text{Max}_j[\Im_{1j}] = M_1$. Thus g_1 is the maximal firing level of any rule in the first level of hierarchy.

We see next that since

$$G_2 = (1 - g_1) \wedge \bigcup_{j=1}^{n_2} \Im_{2j} \wedge B_{2j} \cup G_1$$

we have

$$g_2 = \text{Max}_y G_2(y) = (1 - g_1) \wedge \text{Max}[\bigcup_{j=1}^{n_2} \Im_{2j} \wedge B_{2j}] \vee \text{Max}_y G_1(y)$$

$$g_2 = 1 - g_1 \wedge \text{Max}_j \Im_{2j} \vee g_1$$

$$g_2 = (\bar{g}_1 \wedge M_2) \vee g_1$$

More generally we see that

$$g_i = (\bar{g}_{i-1} \wedge M_i) \vee g_{i-1}$$

Thus we see that $\bar{g}_i$ controls the allowable effect for the rule set i.

We note that we can more suggestively write

$$\text{Poss}[G_i] = ((1 - \text{Poss}[G_{i-1}]) \wedge M_i) \vee \text{Poss}[G_{i-1}]$$

however, since $\text{Cert}[A] = 1 - \text{Poss}[\bar{A}]$, we get $\text{Poss}[G_i] = (\text{Cert}[\bar{G}_{i-1}] \wedge M_i) \vee \text{Poss}[G_{i-1}]$.

The structure

$$G_i = \bar{g}_{i-1} \wedge \bigcup_{j=1}^{n_i} \Im_{ij} \wedge B_{ij} \cup G_{i-1} \qquad \textbf{(I)}$$

for aggregation of information of the different levels of the hierarchy devised from our HPS structure while appealing has one characteristic which we may not always desire. $\bar{g}_{i-1}$ limits the amount of contribution that we can get from the ith level. The term $\bar{g}_{i-1} \wedge \bigcup_{j=1}^{n_i} \Im_{ij} \wedge B_{ij}$ has $\bar{g}_{i-1}$ as its highest value. Since this term is connected to G_{i-1} via a union operation, once $\text{Max} G_{i-1}$ gets above .5, the largest terms can't grow and we get a kind of saturation.

In the following we suggest an alternative and perhaps more suitable approach to the aggregation between the levels of the HPS. Let G_i be the output of the ith level, let $\bar{g}_{i-1} = \text{Max}_y G_{i-1} = \text{Poss}[G_i|Y]$ and let

$$T_i = \bigcup_{j=1}^{n_i} \Im_{ij} B_{ij},$$

the aggregation of the rules in the ith level for the input U. Then

$$G_i(y) = T_i(y) * (1 - g_{i-1}) + G_{i-1}(y) \qquad \textbf{(II)}$$

$$G_i(y) = T_i(y) * (1 - \text{Poss}[G_{i-1}]) + G_{i-1}(y)$$

Notice that we have essentially replaced the union operation by an addition. While we have also replaced the min by a product this is not as fundamental a change as using the sum. We see that the multiplication of T_i by $(1 - g_{i-1})$ keeps the sum from going beyond one and thus $G_i(y)$ is still a fuzzy subset. What is happening in this structure is that as long as we have not found one y with membership grade 1, $\text{Poss}[G_{i-1}] \neq 1$, we add some of the output of the current subbox to what we already have. Each element y gets $1 - \text{Poss}[G_{i-1}]$ portion of the contribution at that level, T_i. The amount $\text{Poss}[G_{i-1}]$ is determined by the highest membership grade of any element in G_{i-1}.

We should point out that the aggregation performed in the hierarchical structure, whether we use I or II, is not a pointwise operation. This means that the value of $G_i(y)$ doesn't only depend on the membership grade of y in G_{i-1} and T_i but on membership grades at other points. In particular through the term $\bar{g}_{i-1} = 1 - \text{Max}_y G_i(y)$, it depends upon the membership grade of all elements from Y in G_{i-1}.

We should note that implicit in this structure is a new kind of aggregation. Assume A and B are two fuzzy subsets we define the combination of these sets as the fuzzy subset D, denoted $D = \gamma(A, B)$ where

$$D(x) = (1 - \text{Poss}(A)) * B(x) + A(x).$$

In the previous part we have described the formal mechanism used for the reasoning and aggregation process in the HPS. While the formal properties of the new aggregation structure are important and very useful a key to the usefulness of the HPS in fuzzy modeling is the semantics used in the representation of the information via this structure.

In constructing an HPS representation to model a system we envision that the knowledge of the relationship contained in the HPS structure be stored in the following manner. At the highest level of priority, $i = 1$, we would have the most specific precise knowledge. In particular we would have point to point relationships,

When U is 3 then V is 7
When U is 9 then V is 13

This would be information we know with the greatest certainty.

At the next level of priority the specificity of the antecedent linguistic variables, the A_{2j}'s, would decrease. Thus the second level would contain slightly more general knowledge. The support sets of the antecedents would be wide and will as the range of the consequent.

In Figure 11 we show a prototypical example of the widths of the support sets of the antecedents, A_{ij}, at different levels. In Figure 11 $\hat{A}_{ij}$ is the support sets of A_{ij}

$$\hat{A}_{ij} = \{x | A_{ij}(x) > 0\}.$$

What we see is that as we go to higher levels the cardinality of the support sets increases.

Generally this increase in size of the antecedent is accompanied by an increase in the support set of the consequent. We have less specific rules.

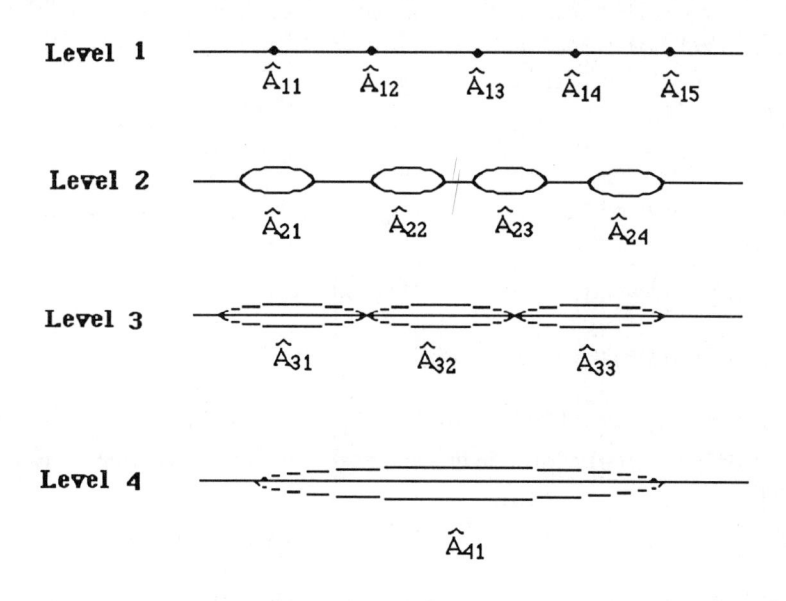

Figure 11.

Essentially what we envision is that at the highest level we have specific point information. The next level encompass these points and in addition provides a more general and perhaps fuzzy knowledge. We note that the lowest most level can be used to tell us what to do if we have no knowledge up to this point. In some sense the lowest level is a default value [33, 34].

Example: Assume we are using an HPS representation to model a function $V = f(U)$, where the base set for U is [0, 100]. A typical HPS representation could be as follows.

      LEVEL #1
R_{11} When U is 5 then V is 13
R_{12} When U is 75 then V is 180
R_{13} When U is 85 then V is 100
      LEVEL #2
R_{21} When U is "about 10" then V is "about 20"
R_{22} When U is "about 30" then V is "about 50"
R_{23} When U is "about 60" then V is "about 90"
R_{24} When U is "about 80" then V is "about 120"
R_{25} When U is "about 100" then V is "about 150"

(we assume triangular fuzzy subsets)

LEVEL #3

R_{31} When U is "low" then V is "about 40"
R_{32} When U is "medium" then V is "about 85"
R_{31} When U is "high" then V is "about 130"

LEVEL #4

R_{41} U is *anything* then V is 2u.

Having defined our knowledge base we now look at the performance of this system for various inputs;

<u>Case 1:</u> U = 75. At level one we get $T_1 = \{180\}$ hence since

$$G_1(y) = \bar{g}_0 * T_1(y) + G_0(y).$$

Since $G_0 = \Phi$ then $g_1 = 1$ which give us $G_1 = T_1 = \left\{\frac{1}{180}\right\}$. We now see that $\bar{g}_1 = 0$ and hence no other rules will fire lower in the hierarchy. This system provides as its output for U = 75 that

$$V \text{ is } 180.$$

<u>Case 2:</u> U = 80. At level one no rules fire, $\Im_{ij} = 0$ for all j. Thus $T_1 = \Phi$ hence

$$G_1 = \bar{g}_0 * T_1 + G_0 = \Phi$$

and therefore $\bar{g}_1 = 1$. At level two

$$G_2 = \bar{g}_1 * T_2 + \Phi = T_2.$$

For U = 80 we assume that R_{24} fires completely, $\Im_{24} = 1$ and that all other rules don't fire, $\Im_{2j} = 0$, for $j \neq 2$. Thus $T_2 = $ "about 120" and $G_2 = $ "about 120". Since $g_2 = 1$ then $\bar{g}_2 = 0$ and no rules at lower priority will fire thus G_2, "about 120", is the output of the system for U = 80.

<u>Case 3:</u> U = 20. No rule at level one will fire, hence $G_1 = G_0 = \Phi$. At level two we shall assume that R_{21} fires to degree .3 and R_{22} also fires to degree .3. Thus

$$T_2 = .3 \wedge B_1 \cup .3 \wedge B_2 = .3 \wedge (B_1 \cup B_2)$$
$$T_2(y) = .3 \wedge (B_1(y) \vee B_2(y)).$$

We note B_1 and B_2 are "about 20" and "about 30" respectively. Hence

$$G_2(y) = (1 - g_1) * T_2(y) + G_1(y) = T_2(y)$$

At level three R_{31} fires to degree 1 while R_{31} and R_{32} don't fire at all. Hence

$$T_3 = \text{"about 40"}$$

Since Max $[G_2] = .3$ thus $1 - g_2 = .7$ and therefore

$$G_3(y) = .7 * T_3(y) + G_2(y)$$

Since Max $T_3(y) = 1$ we see that the process stops here and G_3 is the output of the system.

What we see with this HPS representation is that we have our most general rule stored at the *lowest* level of priority and we store exceptions to this rule at higher levels of priority. In some cases the exceptions to general rules may themselves be rules, we would then store exceptions to these rules at still higher levels of priority. As the previous example illustrates in the HPS system for a given input we first look to see if the input is an exception, that is what we are essentially doing by looking at the high priority levels.

The HPS representation is a formulation that has an inherent structure for a natural human like learning mechanism. We shall briefly describe the type of learning that is associated with this structure.

Information comes into the system in terms of point by point knowledge, data pairs between input and output. We store these points at the highest level of priority. Each input/output pair corresponds to a rule at the highest level. If enough of these points cluster in a neighborhood in the input/output space we can replace these points by a general rule (see Figure 12).

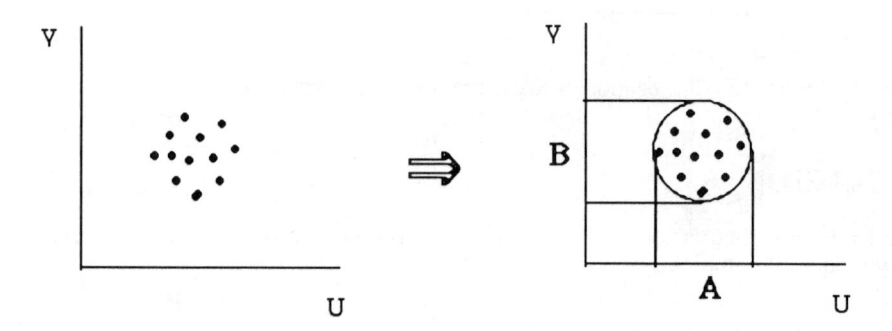

Figure 12. Formulation of Rules for Input/Output Pairs

Thus from the dots, input/output pairs, we get a relationship that says *if U is in A then V is B*. We can now forget about the dots and only save the new relationship. We save this at the next lowest priority in the system, in subbox 2.

We note that the introduction of the rule essentially extends the information contained in the dots by now providing information about spaces between the dots. We can also save storage because we have eliminated many dots and replaced them by one circle. One downside to this formulation is that in generalizing we have lost some of the specificity carried by the dots.

It may occur that there are some notable exceptions to this new general rule. We are able to capture this exception by storing them as high level points.

We further note that new information enters the system in terms of points. Thus we see that the points are either new information or exceptions to more general rules. Thus specific information enters as points it filters its way up the system in rules.

We see that next that it may be possible for a group of these second level rules to be clustered to form new rules at the third level.

In Figure 13 the large bold circle is seen as a rule which encompasses the higher level rules to provide a more general rule. The necessity to keep these more specific rules, thus in level 2, depends upon how good the less specific rule captures the situation.

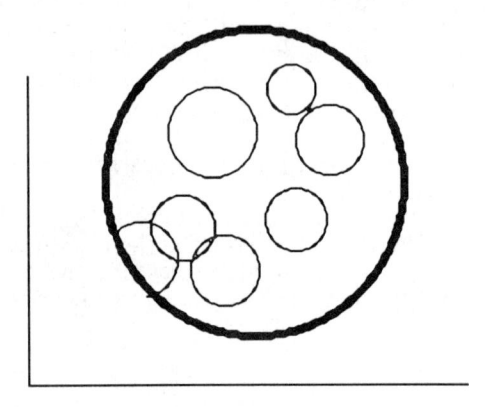

Figure 13. Aggregation of Rules into More General Rules

8. CONCLUSION

We have looked at various structure for the representation of knowledge in rule based fuzzy models of complex systems.

REFERENCES

[1]. Mamdani, E. H. and Assilian, S., "An experiment in linguistic synthesis with a fuzzy logic controller," *Int. J. of Man-Machine Studies*, 7, 1-13, 1975.

[2]. Tong, R. M., "An annotated bibliography of fuzzy control," in *Industrial Applications of Fuzzy Control*, Sugeno, M. (Ed.), North-Holland: Amsterdam, 249-269, 1985.

[3]. Sugeno, M., *Industrial Applications of Fuzzy Control,* North-Holland: Amsterdam,

1985.

[4]. Bernard, J. A., "Use of a rule-based system for process control," *IEEE Control Systems Magazine* 8, No. 5, 3-13, 1988.

[5]. Lee, C. C., "Fuzzy logic in control systems: fuzzy logic controller, Part 1," *IEEE Trans. on Systems, Man and Cybernetics*, 20, 404-418, 1990.

[6]. Lee, C. C., "Fuzzy logic in control systems: fuzzy logic controller, Part II," *IEEE Trans. on Systems, Man and Cybernetics*, 20, 419-435, 1990.

[7]. Filev, D. and Yager, R. R., "A generalized defuzzification method under BAD distributions," *International Journal of Intelligent Systems*, 6, 687-697, 1991.

[8]. Zadeh, L. A., "A theory of approximate reasoning," in *Machine Intelligence*, Vol. 9, Hayes, J., Michie, D., & Mikulich, L.I. (Eds.), New York: Halstead Press, 149-194, 1979.

[9]. Yager, R. R., "Approximate reasoning as a basis for rule based expert systems," *IEEE Trans. on Systems, Man and Cybernetics*, 14, 636-643, 1984.

[10]. Yager, R. R., "Deductive approximate reasoning systems," *IEEE Transactions on Knowledge and Data Engineering*, 3, 399-414, 1991.

[11]. Dubois, D. and Prade, H., "Fuzzy sets in approximate reasoning Part I: Inference with possibility distributions," *Fuzzy Sets and Systems*, 40, 143-202, 1991.

[12]. Zadeh, L. A., "Fuzzy sets as a basis for a theory of possibility," *Fuzzy Sets and Systems*, 1, 3-28, 1978.

[13]. Yager, R. R. and Filev, D. P., "On the issue of defuzzification and selection based on a fuzzy set," Technical Report# MII-1201, Machine Intelligence Institute, Iona College, 1991.

[14]. Yager, R. R. and Filev, D. P., "SLIDE: A simple adaptive defuzzification method," *IEEE Transactions on Fuzzy Systems*, (To Appear).

[15]. Zadeh, L. A., "Fuzzy sets and information granularity," in *Advances in Fuzzy Set Theory and Applications*, Gupta, M.M., Ragade, R.K. & Yager, R.R. (Eds.), Amsterdam: North-Holland, 3-18, 1979.

[16]. Mamdani, E. H., "Application of fuzzy algorithms for control of simple dynamic plant," *Proc. IEEE*, 121, 1585-1588, 1974.

[17]. Mamdani, E. H. and Sembi, B. S., "Process control using fuzzy logic," in *Fuzzy Sets*, Wang, P. P. and Chang, S. K., (Eds.), Plenum Press: New York, 249-266, 1980.

[18]. Takagi, T. and Sugeno, M., "Fuzzy identification of systems and its application to modeling and control," *IEEE Transactions on Systems, Man and Cybernetics*, 15,

116-132, 1985.

[19]. Sugeno, M. and Kang, G. T., "Structure identification of fuzzy model," *Fuzzy Sets and Systems*, 28, 15-33, 1988.

[20]. Filev, D., "Fuzzy modelling of complex systems," I*nt. J. of Approximate Reasoning*, 4, 281-290, 1991.

[21]. Alsina, C., Trillas, E. and Valverde, L., "On some logical connectives for fuzzy set theory," *J. Math Anal. & Appl.*, 93, 15-26, 1983.

[22]. Dubois, D. and Prade, H., "A review of fuzzy sets aggregation connectives," *Information Sciences*, 36, 85 - 121, 1985.

[23]. Bonissone, P. P., "Selecting uncertainty calculi and granularity: An experiment in trading off precision and complexity," in *Proceedings of the First Workshop on Uncertainty in Artificial Intelligence*, Los Angeles, 57-66, 1985.

[24]. Yager, R. R., "On ordered weighted averaging aggregation operators in multi-criteria decision making," *IEEE Transactions on Systems, Man and Cybernetics*, 18, 183-190, 1988.

[25]. Yager, R. R., "On the theory of bags," *Int. J. of General Systems*, 13, 23-37, 1986.

[26]. Yager, R. R., "Cardinality of fuzzy sets via bags," *Mathematical Modeling*, 9, 441-446, 1987.

[27]. Filev, D., "Towards the concept of quasilinear fuzzy systems," *Proc. of the Int. Conf. on Fuzzy Logic and Neural Nets*, Iizuka, Japan, 761-765, 1990.

[28]. Yager, R. R., "A general approach to rule aggregation in fuzzy logic control," *Applied Intelligence*, (To Appear).

[29]. Yager, R. R., "A note on projections of conditional possibility distributions in approximate reasoning," *Kybernetes*, 15, 185-187, 1986.

[30]. Yager, R. R., "An alternative procedure for the calculation of fuzzy logic controller values," *Journal of Japanese Society of Fuzzy Technology*, SOFT 3, 736-746, 1991.

[31]. Yager, R. R., "Measures of specificity for possibility distributions," in *Proc. of IEEE Workshop on Languages for Automation: Cognitive Aspects in Information Processing*, Palma de Mallorca, Spain, 209-214, 1985.

[32]. Yager, R. R., "Specificity measures of possibility distributions," *Proceedings of the Tenth NAFIPS Meeting*, U. of Missouri, Columbia, MO, 240-241, 1991.

[33]. Reiter, R., "A logic for default reasoning," *Artificial Intelligence*, 13, 81-132, 1980.

[34]. Reiter, R., "Nonmonotonic reasoning," in *Annual Reviews of Computer Science*, Vol. 2, Palo Alto: Annual Reviews, 147-186, 1987.

PART B

METHODOLOGIES AND ALGORITHMS

Chapter 6
Dynamic Analysis of Fuzzy Logic Control Structures

Introduction, *142*
Fuzzy Control Structures, *143*
Dynamic Analysis and Design, *146*
Fuzzy Controllers with Consequent Functions, *149*
Example, *152*
Conclusions, *158*
References, *158*

This chapter studies the stability and robustness of fuzzy logic control systems. The authors summarize previous results on the application of the qualitative theory of nonlinear dynamical systems to the analysis of two fuzzy control structures: direct fuzzy control, and fuzzy auto-tuning of conventional controllers. New results for the analysis of fuzzy control structures with a function of the inputs in the consequent of the fuzzy rules are presented. The main issues are illustrated by the analysis of fuzzy logic control of a nonlinear system.

DYNAMIC ANALYSIS OF FUZZY LOGIC CONTROL STRUCTURES

A. García-Cerezo
Dpto. Ingeniería Sistemas y Automática
Universidad de Málaga.
Plaza El Ejido, 29013 Málaga (Spain).

A. Ollero
Dpto. Ingeniería Sistemas y Automática
Universidad de Málaga
Plaza El Ejido, 29013 Málaga (Spain).

J. Aracil
Dpto. Ingeniería Sistemas y Automática
ETS Ingenieros Industriales
Avenida Reina Mercedes, 41012 Sevilla (Spain)

1. INTRODUCTION

Fuzzy control systems have been succesfully applied to a wide variety of practical problems [1], [2]. It has been shown that these controllers may perform better than conventional model-based controllers, specially when applied to processes difficult to model, with nonlinearities, and when there is a significative heuristic knowledge from human operators.

However, in spite of its practical success, there is not a standard procedure for tuning fuzzy controllers. The design of a fuzzy controller is a very time consuming activity involving knowledge acquisition, definition of the control structure, definition of rules, and tuning a variety of gains and other controller parameters.

The relation between these structures and design parameters, and the performance of the fuzzy control loops is also difficult due to the inherent nonlinear nature of systems with fuzzy logic components in the control loop. In fact, there are not many results that can be used to guarantee basic dynamics properties such as the stability and robustness of fuzzy control systems.

In this chapter we consider the analysis of fuzzy control structures regarding the dynamic behavior of the fuzzy control loops. The objective is to provide tools relating directly the fuzzy controller characteristics with the dynamic performance of the closed loop control system, and to use these tools for the design of stable and robust fuzzy logic controllers.

Some preliminary results on the analysis of fuzzy control systems are based on the study of the relation matrix defined by the rules [3]. There are also some techniques based on the concept of linguistic trajectories [4].

The application of the circle stability criteria was suggested by [5]. This criterion is also proposed in [6]. In [7] and [8] the input-output stability approach is also used. In these works the conicity criterion, a generalization of the circle criterion for multivariable systems, is applied for the analysis of fuzzy control systems.

Energy criteria for stability analysis was proposed in [9]. In this work the performance evaluation is based on the structure of the fuzzy relation defining the system. This approach has been adopted by several authors [10], [11], [12], [13]. However it has been pointed out that these structural techniques are not appropiate to analytically predict the dynamic evolution of the system [11] and require very conservative conditions to assure the stability [14], [15].

In [16] we proposed the application of geometrical state-space approaches for the stability analysis of fuzzy control systems. Based on the qualitative theory of non linear systems, we presented in a later work [17] indices for the global analysis of expert control system, and particularly for fuzzy control systems, and we have provided some guidelines for the design [18]. In this chapter summarize these contributions and extend the results for the study of fuzzy controllers structures with consequent functions in the rules. These fuzzy control structures have demostrated efficiency in many applications, and are related with fuzzy modelling and identification techniques which are specially useful when it is difficult for the expert to specify the conventional fuzzy rules [19]. Furthermore, these structures can be related with learning and neural-network approaches for control.

This chapter is organized as follows. Section 2 presents the fuzzy control structures to be considered. The third section includes the basic expressions for the analysis and design of these control structures. In Section 4 we present results for the analysis of the above mentioned control systems with consequent functions in the rules and particularly for linear weighted functions. The results are illustrated in Section 5 with the analysis of the fuzzy logic control of a system with nonlinear damping. Finally we present the conclusions and references.

2. FUZZY CONTROL STRUCTURES

Several fuzzy control structures have been used for the last 15 years. Direct fuzzy control systems generate the control action from a set of rules representing the heuristic knowledge about the process to be controlled [20]. This structure is shown in Fig. 1 where $x = (x_1, ..., x_n)^T$ are the variables of the process to be controlled and u represents the control variables generated by a conventional rule-based fuzzy logic controller defined by a set of rules with the form:

IF x_1 is lx_1, x_2 is lx_2, ..., x_n is lx_n THEN u is lu

where $lx_i \in \chi$ (i=1,.., n) and $lu \in \Lambda$ are linguistic labels defining respectively regions in the range of variation of x_i (i=1,.., n) and u, wherein the rule is valid. For example,

IF x_1 is positive_big, x_2 positive_small THEN u is positive

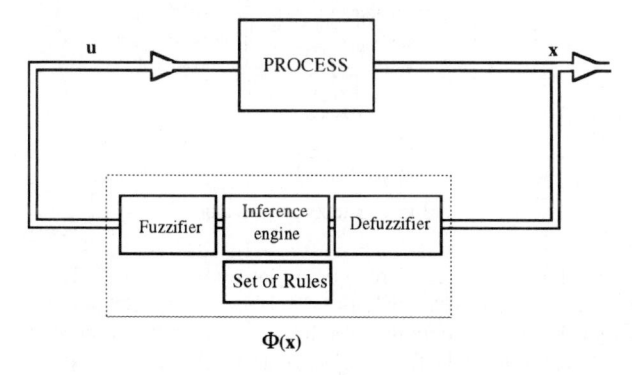

Fig.1. Fuzzy Controller

The fuzzy controller in Fig. 1 consists in a set of fuzzy control rules, the inference mechanism, and the interfaces to the process to be controlled (fuzzifier for the controller input, and defuzzifier for the controller output). It has been shown [21] that these controllers can be represented by means of the nonlinear function,

$$u = \Phi(x) \tag{1}$$

In the following we assume $\Phi(0) = 0$.

On the other hand, fuzzy logic techniques can also be applied for automatic tuning and supervision of conventional controllers such as PID controllers and state feedback controllers [22]. In these cases the set of fuzzy rules provides adaptation of the conventional controller for different working conditions. The structure is shown in Fig. 2. Consider a conventional non-fuzzy controller defined by:

$$u = \gamma(x, p), \tag{2}$$

where x represents again the process variables and p is a set of controller parameters. In this case the fuzzy component consists in a set of fuzzy rules to infer the appropriated value of each control parameter:

IF is lx_1, x_2 is $lx_2, ..., x_n$ is lx_n THEN p is lp

where lx_i is as above and $lp \in \Lambda P$ is a linguistic label defining a region in the interval of variation of the controller parameter p (i.e. *positive_smal* increment of the gain, or *positive_big* gain).

For example,

IF x_1 is positive_big, x_2 is positive_small THEN p is positive_big

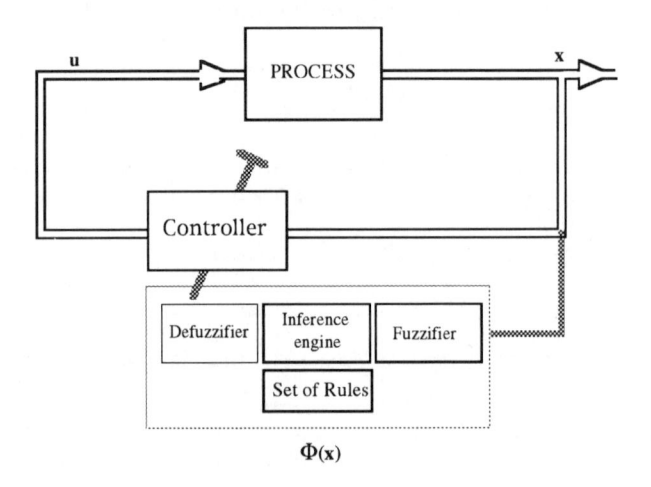

Fig.2. Fuzzy Supervisor

Consider again the basic direct fuzzy control structure of Fig. 1. There are also other alternative approaches to the conventional rules describing the actions in terms of the process variables. Thus, models with rule outputs given by a consequent function of inputs have been used [1]. Figure 3 represents this fuzzy logic control structure. The controllers have R_i *(i=1,...,nr)* rules such as

IF x_1 is lx_1, x_2 is $lx_2, ..., x_n$ is lx_n THEN $u^i = \Phi(x_1, ..., x_n, c_0, ..., c_r)$ (3)

where Φ is a consequent function usually defined as polynomials of the controller inputs: $x_1, ..., x_n$, and u^i is the numerical output of the rule R^i.

The coefficients $c_0, ..., c_r$ are parameters that can be modified to optimize the controller's performance. These parameters can also be obtained by means of modelling and learning methods from the human operator control actions.

The most widely used function is the linear combination:

$$u^i = c_0 + c_1 x_1 + ... + c_n x_n \qquad (4)$$

When the controller inputs are provided by the process interface, the truth value w_i of each rule (i=1, ..., nr) is computed and the output of the controller is given by the weighted average:

$$u = \frac{\sum_{i=1}^{nr} w^i u^i}{\sum_{i=1}^{nr} w^i} \qquad (5)$$

This structure has revealed interest for several applications in which is difficult or unrealistic to extract and quantify the rules from the human operators heuristic knowledge. Moreover, this technique can be also applied for the structure shown in Fig. 2.

Other modelling techniques based on the modification of the relation matrix defined by the rules have also been proposed in the fuzzy logic control literature. In these approaches the basic idea is the computation of an inverse model of the process to be controlled.

There is also predictive approaches based on these fuzzy modelling techniques [23]. These structures will be not considered in the next sections. However, they can be studied with straightforward extensions of the proposed methods.

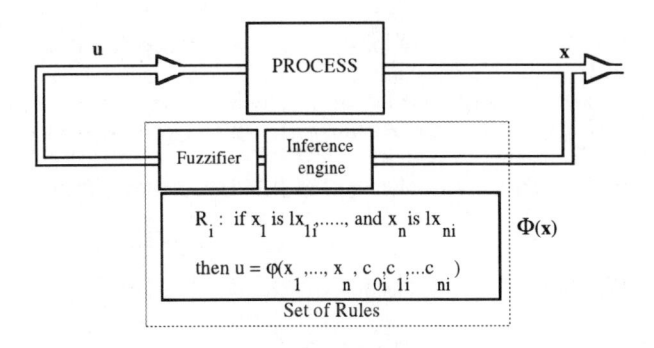

Fig.3. Fuzzy Controller with Weighted Output

3. DYNAMIC ANALYSIS AND DESIGN

All the above mentioned fuzzy control structures involve nonlinear control systems. Assume some knowledge of the process to be controlled can be obtained. In fact, approximate knowledge about the dynamic relations involved is normally available from process analysis and experimentation even in complex processes. Fuzzy [24] and neural-network based modelling techniques can also be used to obtain a nonlinear relation representing the process to be controlled through experimentation. Thus, let the approximate model of the closed loop system be represented by:

$$\frac{dx}{dt} = f(x) + Bu \qquad (6)$$

where $x \in X \subset \Re^n$ is a n-vector of process variables, $f(x)$ is a nonlinear function describing the open loop dynamic with $f(0)= 0$, and $u \in U \subset \Re^m$ represent the control variables.
We remark this model can be considered as an approximate one. Then, if we apply any analysis and design technique, the robustness of the resulting control strategy must be studied.

Direct fuzzy control

In [17] we study the stability of fuzzy control systems represented by (6) with u given by (1). By substituting (1) into (6) we have the closed loop representation:

$$\frac{dx}{dt} = f(x) + B(\Phi(x)) \qquad (7)$$

with $x(0)=0$ being the equilibrium of the system.
Stability and robustness indices are defined in the framework of the qualitative theory of dynamical systems. Two stability indices that provide a measure of the relative stability of the equilibrium point at the origin can be defined. These indices are related to the ways a system can loss its stability.
The first index copes with the stability loss when one real eigenvalue of the system crosses the imaginary axis. This index is given by the expresion:

$$I_1 = (-1)^n det\ J \qquad (8)$$

where J is the Jacobian matrix of the dynamical closed loop system at the equilibrium point, and det stands for the determinant.
The second index is related to the loss of stability as a consequence of the occurence of a Hopf bifurcation, and the appearance of a limit cicle. This happens when two complex imaginary eigenvalues cross the imaginary axis.
Let $P_n(s)$ be the characteristic polynomial of J at the equilibrium point:

$$P_n(s) = s^n + a_1 s^{n-1} + \ldots + a_{n-1}s + a_n \qquad (9)$$

This polynomial can also be expressed as:

$$P_n(s) = \Theta(s^2 + \omega^2) + \theta_1 s^{n-2} + \theta_2 s^{n-3} + \ldots + \theta_{n-3}s + \theta_{n-2} \qquad (10)$$

where Θ is an order n-2 polynomial. The condition for $P_n(s)$ to have two imaginary roots is:

$$\theta_1 = \theta_2 = \ldots = \theta_{n-2} \qquad (11)$$

For example, for $n=3$ the application of (10), (11) gives $I_2 = a_1 a_2 - a_3$. The index I_2 becames negative when the roots of $P_n(s)$ have positive real components, and positive for negative real components.

It can be shown [17] that, for $n=2$, the Hopf bifurcation condition can be expressed as:

$$a_1 = 0 \tag{12}$$

and the index is given by:

$$I_2 = -(trJ) \tag{13}$$

where tr stands for the trace. The above indices can be used to study the relative stability of the equilibrium point at the origin. The larger their positive values are, the larger the degree of relative stability is. The annulation of anyone of them, gives rise to the stability loss of the system; that is the system becomes unstable.

The third way to loss global stability is associated to the appearance of other equilibria, and then, the partition of the state space in many attractions bassins. Global stability analysis involve conditions to preclude the appearance of other equilibria. To introduce these conditions consider a dimension 1 system, and assume that $b=1$. In this case, it can be shown from geometrical considerations [17] that the global stabilty condition is

$$|\Phi(x)| < |f(x)| \tag{14}$$

If this condition is not fullfilled, a bifurcation is produced given rise to the appearance of a new attractor. Thus, a measure of the degree of global stability is given by the distance between $\Phi(x)$ and $f(x)$ excluding a region around the origin in which an equilibrium point is supposed to be. Then, the index can be expressed as

$$I_3 = min_{\bar{\beta}}(\Phi(x) + f(x)) \tag{15}$$

where min is over the full range of x except a region $\beta = [\beta_1, \beta_2]$ around the origin. The bound values β_1 and β_2 are defined as the values closest to the origin so that $\Phi'(\beta_1) = -f'(\beta_1)$ and $\Phi^T(\beta_2) = -f'(\beta_2)$ where the apostrophe stands for the derivatives with respect to x.

For $n=2$, $b = [b_1, b_2]^T$ and the condition $f(x) = b\Phi(x)$ for the appearance of a new new attractor only can occur in the region of the state space where $f(x)$ takes the direction (b_1, b_2) of the control component $b\Phi(x)$. This region is given by the auxiliary subspace defined by:

$$\frac{f_1(x)}{b_1} = \frac{f_2(x)}{b_2} \tag{16}$$

For dimension n systems and $m>1$ controller outputs, the auxiliary subspace is given [17] by:

$$\frac{f_1(x)}{\sum_{j=1}^{m} b_{1_j} \Phi^j} = \frac{f_2(x)}{\sum_{j=1}^{m} b_{2_j} \Phi^j} = \ldots = \frac{f_n(x)}{\sum_{j=1}^{m} b_{n_j} \Phi^j} \tag{17}$$

where Φ^j is the j-th component of the vector Φ.

Fuzzy autotuning

In this section we deal with the fuzzy adaptation of the parameters of a conventional controller. For example, if we consider the closed loop monovariable (one control variable) control system (6), and a linear state feedback controller is used, the control variable is given by

$$u = k^T x = k_1 x_1 + k_2 x_2 + \ldots + k_n x_n \tag{18}$$

In this case, it is intended to adapt the parameters k_i, $i=1..n$, by using a set of heuristc rules. Two different structures are considered:

<u>Additive structure</u>: In this case the control law is defined by:

$$u = -(k+s)^T x = (k_1 + s_1) x_1 + (k_2 + s_2) x_2 + \ldots + (k_n + s_n) x_n \tag{19}$$

where s_i, $i = 1..n$, can be obtained by using rules such as:

IF x_1 is *positive_big*, x_2 is *positive_medium*, ..., and x_n is *negative_small*
THEN s_i is *positive_big*

Thus, by using the same formalisms that in the above section, we can write:

$$s = \Phi(x) \tag{20}$$

$$s_i = \Phi^i(x) \tag{21}$$

where Φ^j is the j-th component of the vector Φ that correspond to parameter s_i.

Observe how $s^T(x)$ can be interpreted as additional feedbacks to compensate perturbations or changes in the system.

Consider the two-dimensional problem. The Jacobian at the equilibrium point (origin $x_1=0$, $x_2=0$) is given by :

$$J = \begin{bmatrix} f_{11} + b_1 (k_1 + \Phi^1(0)) & f_{12} + b_1 (k_2 + \Phi^2(0)) \\ f_{21} + b_2 (k_1 + \Phi^1(0)) & f_{22} + b_2 (k_2 + \Phi^2(0)) \end{bmatrix} \tag{22}$$

where

$$f_{ij} = \frac{\partial}{\partial x_j} f_i(x) \tag{23}$$

Thus, we can compute the indices I_1 and I_2 by using (8) and (13) respectively. Then, the stability at the origin and the robustness can be studied as in the direct control case. The index I_3 can also be computed in the auxiliary subspace (16).

<u>Multiplicative structure</u>: In this case the control law can be written as:

$$u = (k_1 c_1) x_1 + (k_2 c_2) x_2 + \ldots + (k_n c_n) x_n \tag{24}$$

where c_i ($i=1, ...n$) are obtained from a set of rules.

Thus, we have:

$$c_i = \Phi^i(x_i) \qquad (25)$$

In this structure, the Jacobian at the origin for the two-dimensional case is given by:

$$J = \begin{bmatrix} f_{11} + b_1(k_1\Phi^1(0)) & f_{12} + b_1(k_2\Phi^2(0)) \\ f_{21} + b_2(k_1\Phi^1(0)) & f_{22} + b_2(k_2\Phi^2(0)) \end{bmatrix} \qquad (26)$$

and we can apply the same techniques to compute the indices I_1, I_2, and I_3 for stability and robustness studies.

Finally, notice that the above concepts can also be applied for the automatic expert supervision of other controllers such as the conventional PID regulators. In this case the control action can be written as:

$$u = K(e(t) + (1/T_i)(\int e(t)dt) + T_d \frac{d}{dt}e(t)) \qquad (27)$$

where $e(t)$ is the error between a reference signal and the output of the process. The parameters K, T_i and T_d can be automatically tuned taking into account rules relating its values with some characteristic of the process output.

4. FUZZY CONTROLLERS WITH CONSEQUENT FUNCTIONS

Consider the fuzzy controller with output given by (5). By substituting (4) in (5) the controller output is given by:

$$u = \frac{\sum_{i=1}^{nr} w^i(c_0 + c_1 x_1 + \ldots + c_n x_n)}{\sum_{i=1}^{nr} w^i} = \Phi(x_1, x_2, \ldots, x_n) \qquad (28)$$

Now assume the following restrictions on the fuzzy sets and rules:

i) The fuzzy subsets $\mu_{Aj}(x_i)$ associated with the controller input variables x_i are triangular type as shown in Fig.4. The membership functions of these fuzzy subsets satisfy the following relations:

$$\mu_{Aj}(x_i) + \mu_{Aj+1}(x_i) = 1, \mu_{Ak}(x_i) = 0 \qquad (29)$$

$$k \neq j, j+1$$

where $\mu_{Ai_j}(x_i)$ and $\mu_{Ai_{j+1}}(x_i)$ are the membership functions of two consecutive fuzzy subsets.

ii) There is a rule $R^0 \in \{R_1, \ldots, R_{nr}\}$ centered at the origin. That is, the truth value of this rule is

$$w^0(0, \ldots, 0) = 1$$

iii) $\Phi(x) = -\Phi(-x)$ near the origin (the fuzzy rules R^i and R^{-i} adjoints to R^0 in the positive and negative directions of x_i are symmetric).

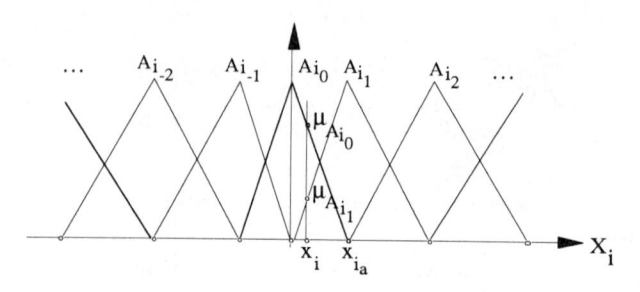

Fig. 4. Triangular Fuzzy Subsets

Under these conditions the linearization of the controller inference function at the origin is given by:

$$u = \sum_{i=1}^{n} \Phi_i x_i \tag{30}$$

where

$$\Phi_i = \left.\frac{\partial}{\partial x_i}\Phi(x_1, x_2, \ldots, x_n)\right|_{x=0}, i = 1, \ldots, n$$

A graphical interpretation of the linearized controller for a dimension 2 system is shown in Fig. 5.

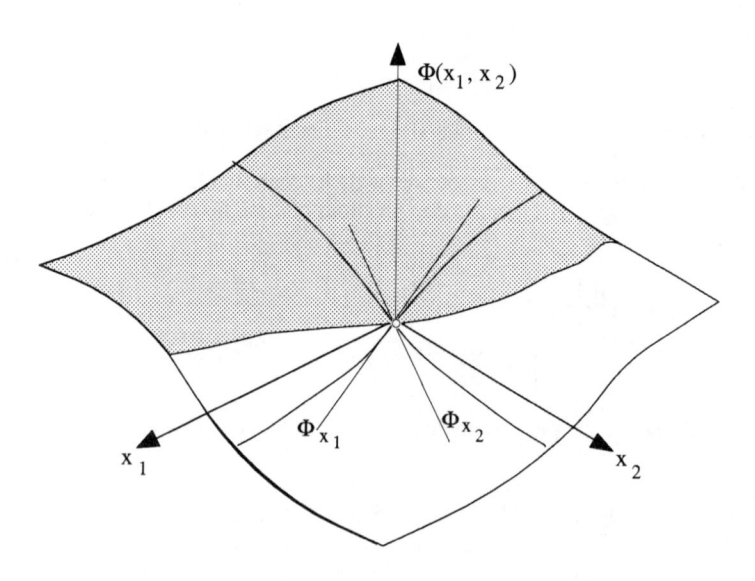

Fig. 5. Graphical Interpretation of Linearized Controller

From conditions i), ii) and iii) it can be seen that only two rules affect the computation of the derivatives at the origin. By using (29) the following expressions can be obtained:

$$\mu_{A_{i_j}}(x_i) = 1 - \frac{x_i}{x_{i_a}} \qquad (31)$$

$$\mu_{A_{i_{j+1}}}(x_i) = \frac{x_i}{x_{i_a}} \qquad (32)$$

where x_{i_a} is defined as in Fig. 4. Thus, the truth values of the rule R^0 are given by:

$$w^0(x_1, 0, \ldots, 0) = \left(1 - \frac{x_1}{x_{1a}}\right)$$
$$w^0(0, x_2, \ldots, 0) = \left(1 - \frac{x_2}{x_{2a}}\right)$$
$$\ldots$$
$$\ldots$$
$$w^0(0, 0, \ldots, x_n) = \left(1 - \frac{x_n}{x_{na}}\right) \qquad (33)$$

For the adjoint rules R^i we have:

$$w^1(x_1, 0, \ldots, 0) = \left(\frac{x_1}{x_{1a}}\right)$$
$$w^2(0, x_2, \ldots, 0) = \left(\frac{x_2}{x_{2a}}\right)$$
$$\ldots$$
$$\ldots$$
$$w^n(0, 0, \ldots, x_n) = \left(\frac{x_n}{x_{na}}\right) \qquad (34)$$

Then, the inference vector is given by:

$$\Phi(0, \ldots, x_i, \ldots, 0) = \left(1 - \frac{x_i}{x_{i_a}}\right)(c_0^0 + c_i^0 x_i) + \frac{x_i}{x_{i_a}}(c_0^i + c_i^i x_i)$$

$$\text{for } 0 < x_i < x_{i_a} \qquad (35)$$

Taking into account the assumption $\Phi(0) = 0$, we have $c_0^0 = 0$. Then

$$\Phi(0, \ldots, x_i, \ldots, 0) = ((c_i^i - c_i^0) x_i) \frac{x_i}{x_{i_a}} + (c_i^0 x_i) + \frac{x_i}{x_{i_a}} c_0^i \qquad (36)$$

Thus, the derivatives of (36) at $x=0$ are:

$$\frac{\partial}{\partial x_i}\Phi(x) = \frac{1}{x_{ia}}[2(c_i^i - c_i^0)x_i] + c_i^0 + \left.\frac{c_0^i}{x_{i_a}}\right|_{x=0} = c_i^0 + \frac{c_0^i}{x_{i_a}} = \tilde{c}_i \qquad (37)$$

From (37) the linearized controller output (30) at the origin can be written:

$$u = \Phi(x) = \sum_{i=1}^{n} \tilde{c}_i x_i \qquad (38)$$

These expressions can be used to study the stability of the closed loop fuzzy control system at the origin. Let A be the $n \times n$ Jacobian matrix:

$$a_{ij} = f_{ij} = \frac{\partial}{\partial x_j} f_i(x) \qquad (39)$$

of the free system

$$\frac{dx}{dt} = f(x)$$

The linearized system at $x=0$ can be represented as

$$\frac{dx}{dt} = A + b\,\tilde{c} \qquad (40)$$

where $b = \begin{bmatrix} b_1 & \ldots & b_n \end{bmatrix}^T$. For $n=2$ the Jacobian matrix of the closed loop system at the origin is given by:

$$J = \begin{bmatrix} f_{11} + b_1 \tilde{c}_1 & f_{12} + b_1 \tilde{c}_2 \\ f_{21} + b_2 \tilde{c}_1 & f_{22} + b_2 \tilde{c}_2 \end{bmatrix} \qquad (41)$$

Thus, we can compute the indices I_1 and I_2 by using (8) and (13) respectively. Furthermore, the auxiliar subspace (21) can also be defined and the global stability studied by means of the index I_3.

5. EXAMPLE

Consider the nonlinear dynamical system given by the diferential equation:

$$\frac{d^2y}{dt^2} + a_1(1-y^2) + a_2 y = b_2 u \qquad (42)$$

where $a_1 = 2.2165$; and $b_2 = a_2 = 12.7388$ corresponding to a second order system with nonlinear damping. The saturation in the actuators imposes $|u| \leq 15$.

In state space form, the model can be written:

$$\dot{x}_1 = x_2$$
$$\dot{x}_2 = -(a_2 x_1) - (a_1 x_2) + a_1 x_1 x_2 + b_2 u$$

The system (42) with $u=0$ has a stable zone around the origin. This area is delimited by an unstable limit cycle (see the phase portrait in Fig. 6).

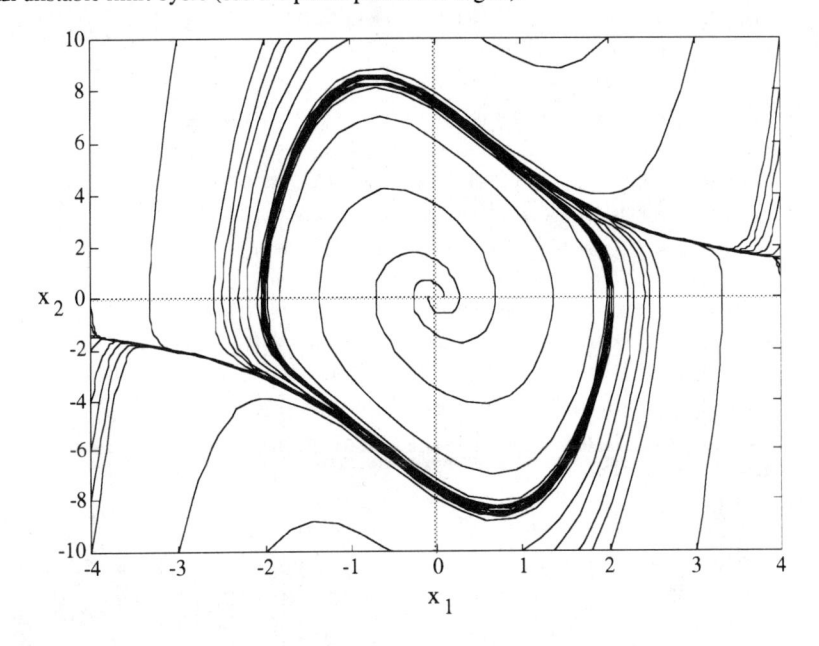

Fig. 6. Free System Response

To improve the dynamic behaviour, a fuzzy controller with rule weighted outputs has been designed. This controller has been designed according to the following criteria:

i) obtain a reasonable robust behaviour measured by indices I_1, I_2 and I_3; and
ii) increase the stable area of influence of the unstable limit cycle.

The fuzzy set of rules is shown in Table 1. The fuzzy linguistic labels associated with the input variables are defined in Fig. 7. In the following we analyze the resulting fuzzy logic control system to verify if conditions i) and ii) are fullfilled.

For $n=2$, the indices I_1 and I_2 are always linear functions of Φ_x. The values of Φ_x are obtained directly from the coefficients of the output as we shown in (37)-(38). For this example we have:

$$\Phi_x = \begin{bmatrix} 0.53 & -0.47 \end{bmatrix} \tag{43}$$

Then, the Jacobian matrix of the closed loop system is given by

$$J = \begin{bmatrix} 0 & 1 \\ -5.987 & -8.2 \end{bmatrix} \tag{44}$$

and the corresponding characteristic polinomial is:

$$s^2 + 8.2\,s + 5.987 = 0$$

Using (8) and (13) the values of the indices I_1 and I_2 are

$$I_1 = 8.2;\ I_2 = 5.987$$

Table 1: Fuzzy Set of Rules

Rule0: If x_1 is PZ and x_2 is VZ then $u = -0.25 x_1 - 0.33 x_2$
Rule1: If x_1 is PP and x_2 is VZ then $u = 1.56 - 1.28 x_1 - 2.6084 x_2$
Rule2: If x_1 is PN and x_2 is VZ then $u = -1.56 - 1.28 x_1 - 2.6084 x_2$
Rule3: If x_1 is PP and x_2 is VP then $u = 1.5 - 2.84 x_1 - 0.7192 x_2$
Rule4: If x_1 is PPB and x_2 is VZ then $u = 1.6250 - 0.8203 x_1 - 3.0156 x_2$
Rule5: If x_1 is PP and x_2 is VN then $u = -1.809 + 2.3 x_1 - 0.8840 x_2$
Rule6: If x_1 is PPB and x_2 is VN then $u = -1.588 + 0.8113 x_1 - 1.6679 x_2$
Rule7: If x_1 is PPB and x_2 is VP then $u = -5.0112 - 0.2156 x_1 - 1.1406 x_2$
Rule8: If x_1 is PZ and x_2 is VN then $u = 1.4002 - 0.1091 x_1 - 0.3778 x_2$
Rule9: If x_1 is PN and x_2 is VP then $u = 1.809 + 2.3 x_1 - 0.8840 x_2$
Rule 10: If x_1 is PZ and x_2 is VP then $u = -1.4002 - 0.1091 x_1 - 0.3778 x_2$
Rule11: If x_1 is PN and x_2 is VN then $u = -1.5 - 2.84 x_1 - 0.7192 x_2$
Rule12: If x_1 is PNB and x_2 is VP then $u = 1.588 + 0.8113 x_1 - 1.6679 x_2$
Rule13: If x_1 is PNB and x_2 is VZ then $u = -1.6250 - 0.8203 x_1 - 3.0156 x_2$
Rule14: If x_1 is PNB and x_2 is VN then $u = 5.0112 - 0.2156 x_1 - 1.1406 x_2$

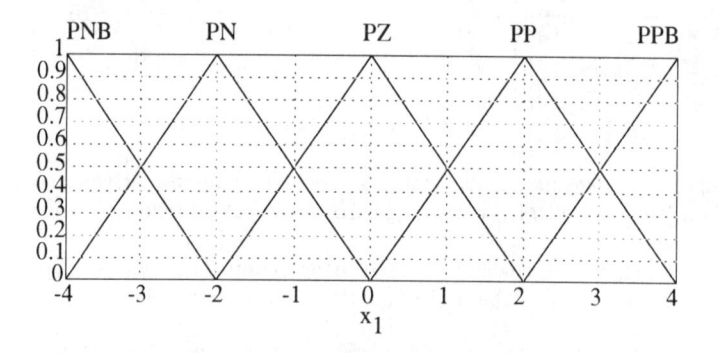

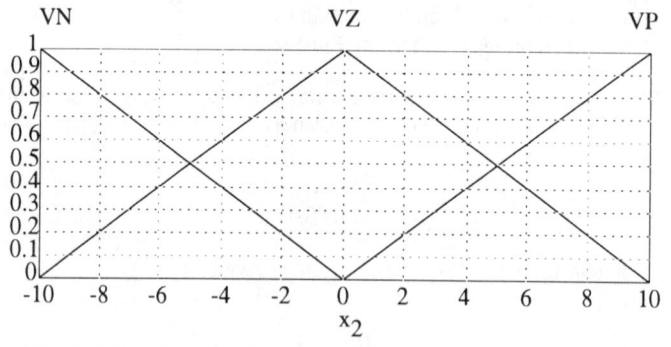

X1 and X2 represents Position and Velocity respectively.
PPB, PP, PZ, PN, PNB represents Position positive big, positive, near zero, negative and negative big
VP, VZ, VN represents Velocity positive, near zero and negative respectively.

Fig. 7. Fuzzy Subsets of Input Variables

Auxiliar subspace analysis.

The auxiliar subspace is given by the equation $x_2 = 0$. In Fig.8 the functions $|f(x)|$, $|b\Phi(x)|$ and $|f(x) + b\Phi(x)|$ are represented. It should be noted that there is not a local minimum in the function $|f(x) + b\Phi(x)|$. Then, index I_3 is undefined. This result can be interpreted as a good robustness with respect to the appearance of new equilibrium points.

In Fig.9, the phase plane of the closed loop system is shown. Note that the stable area is greater that the original one. However, the global stablility is not assured due to the effect of the saturation in the control action.

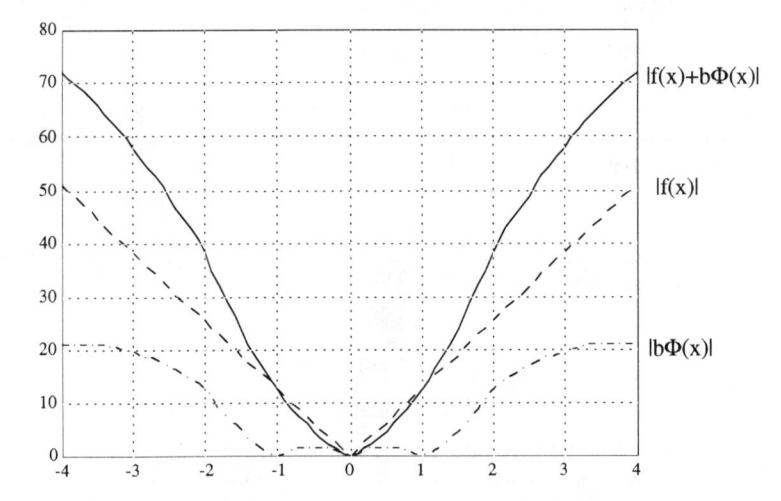

Fig. 8. Auxiliar Subespace Analysis (Table 1 Fuzzy Controller)

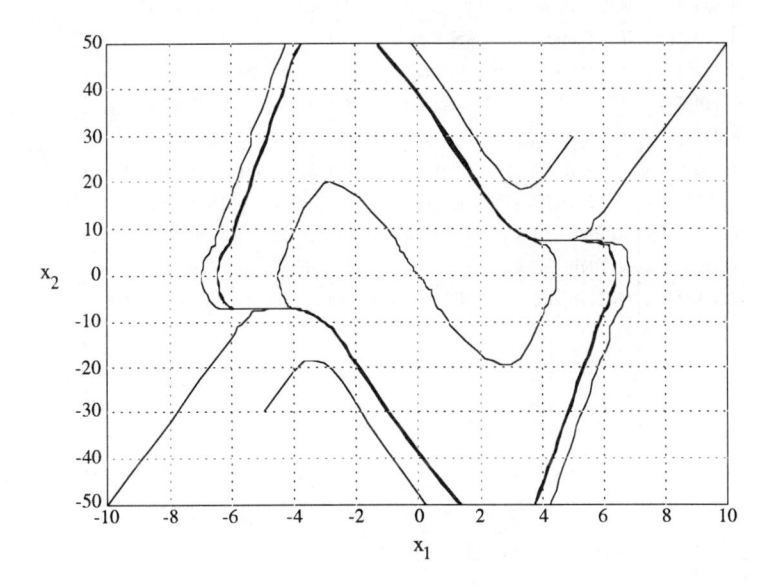

Fig. 9. Closed Loop Response (Table 1 Fuzzy Controller)

Finally, in Table 2, the rules of an instabilizing controller are shown. Note that only rules R_0, R_1 and R_2 are different from the controller in Table 1. The Jacobian matrix of the closed loop system is given by

$$J = \begin{bmatrix} 0 & 1 \\ -19.49 & 0.21 \end{bmatrix}$$

Then, the values of the indices I_1 and I_2 are:

$$I_1 = -0.21;\ I_2 = -19.49$$

which correspond to the unstable nature of the closed loop system. In Fig. 10 the functions $|f(x)|$, $|b\Phi(x)|$ and $|f(x) + b\Phi(x)|$ are represented. In this case, there is a local minimum of the function $|f(x) + b\Phi(x)|$. Then, index I_3 is given by

$$I_3 = 12.7$$

In Fig. 11, the phase portrait of the closed loop system with the table 2 controller is shown.

Table 2: Fuzzy Set of Rules

Rule0: If x_1 is PZ and x_2 is VZ then u = 0.25 x_1 + 0.33 x_2
Rule1: If x_1 is PP and x_2 is VZ then u = -1.56 + 1.28 x_1 + 2.6084 x_2
Rule 2: If x_1 is PN and x_2 is VZ then u = 1.56 + 1.28 x_1 + 2.6084 x_2
Rule3: If x_1 is PP and x_2 is VP then u = 1.5 - 2.84 x_1 - 0.7192 x_2
Rule4: If x_1 is PPB and x_2 is VZ then u = 1.6250 - 0.8203 x_1 - 3.0156 x_2
Rule5: If x_1 is PP and x_2 is VN then u = -1.809 + 2.3 x_1 - 0.8840 x_2
Rule6: If x_1 is PPB and x_2 is VN then u = -1.588 + 0.8113 x_1 - 1.6679 x_2
Rule7: If x_1 is PPB and x_2 is VP then u = -5.0112 - 0.2156 x_1 - 1.1406 x_2
Rule8: If x_1 is PZ and x_2 is VN then u = 1.4002 - 0.1091 x_1 - 0.3778 x_2
Rule9: If x_1 is PN and x_2 is VP then u = 1.809 + 2.3 x_1 - 0.8840 x_2
Rule10: If x_1 is PZ and x_2 is VP then u = -1.4002 - 0.1091 x_1 - 0.3778 x_2
Rule11: If x_1 is PN and x_2 is VN then u = -1.5 - 2.84 x_1 - 0.7192 x_2
Rule12: If x_1 is PNB and x_2 is VP then u = 1.588 + 0.8113 x_1 - 1.6679 x_2
Rule13: If x_1 is PNB and x_2 is VZ then u = -1.6250 - 0.8203 x_1 - 3.0156 x_2
Rule14: If x_1 is PNB and x_2 is VN then u = 5.0112 - 0.2156 x_1 - 1.1406 x_2

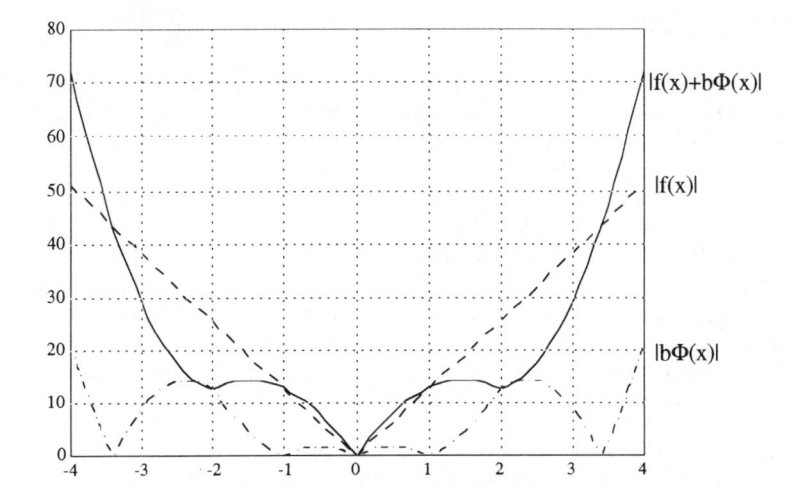

Fig. 10. Computation of I_3 on the Auxiliar Subespace
(Table 2 fuzzy controller)

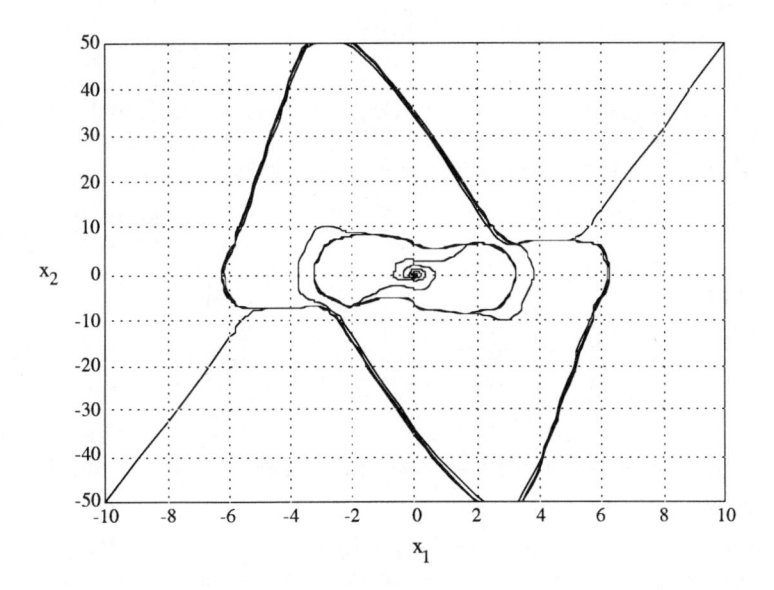

Fig. 11. Unstable Closed Loop Response
(Table 2 fuzzy controller)

6. CONCLUSIONS

Fuzzy logic control systems are inherently nonlinear dynamical systems. Then difficulties exists for the definition of general mathematical tools that could be used to analyze the dynamic behaviour of a fuzzy control system and to provide fuzzy controllers design guidelines. In this paper we have shown how concepts from the qualitative theory of nonlinear dynamical systems can be used to study the dynamic behaviour of fuzzy control structures. Particularly, three indices have been defined to study the stability at the origin and the global stability of the fuzzy logic control loops. These indices are functions of the fuzzy controller parameters and can be used to evaluate the robustness of fuzzy control structures. In the paper we have summarized previous results for the analysis of direct fuzzy controllers, and fuzzy automatic tuning of conventional state feedback controllers.Furthermore, we have presented new results for the analysis of fuzzy control systems with linear functions in the consequents of the rules. These results can be used to define design algorithms to increase the indices.

It can be argued that the application of the proposed methods requires a model of the process to be controlled. In fact any stability analysis needs knowledge about the dynamic behaviour. In many control problems this knowledge exists or can be obtained through experimentation. Note, for example, how the proposed stability analysis at the origin only require estimations of the derivatives at this point.

On the other hand, fuzzy control methods are not restricted for the complexity of the process model as control theory methods are, and fuzzy controllers with functions in the consequents of the rules can be obtained from human control operations.

Finally, in should be pointed out, that, in most cases, the above mentioned models can only be considered as approximated ones, and so the analysis of the robustness of the fuzzy control loops is essential. Remain that the presented indices provide a measure of the robustness and can be used to design stable fuzzy logic control systems working far from the stability loss.

REFERENCES

[1] Takagi, T. and Sugeno M. "Fuzzy Identification of Systems and its Applications to Modelling and Control". IEEE Trans on Sist. Man and Cib. Vol SMC-15, No 1, pp 116-132, 1985.

[2] Takashima S., "100 Examples of Fuzzy Theory Applications mostly in Japan", Trigger, July, 1989.

[3] Tong M.R., "Some Problems With the Design and Implementation of Fuzzy Controllers", British Steel Copr., Res. Associate. Control and management System Group. Univ. of Cambrigde, England, 1977.

[4] Braae M. and D.A. Rutherford, "Theoretical and Linguistic Aspects of the Fuzzy Logic Controller", Automatica, 15, pp. 553-577, 1979.

[5] Kickert W. and E.H. Mamdani, "Analysis of a Fuzzy Logic Controller", Fuzzy Sets and Systems, Vol.1, pp.29-44, 1978.

[6] Kumar S.R. and D.D. Majunder, "Application of Circle Criteria for Stability Analysis of

Linear SISO and MIMO System Associated with Fuzzy Logic Controllers", IEEE Trans. Sist. Man. Cyber., Vol.SMC-14, 2, pp.345-349, 1984.

[7] Aracil, J., A.Garcia-Cerezo, A. Barreiro and A. Ollero, "Fuzzy Control of Dynamical Systems. Stability Analysis Based on the Conicity Criterion", IFSA World Congress, Brussels, July 1991.

[8] Barreiro A, A. Garcia-Cerezo, A. Ollero. "Design of Robust Intelligent Control of Manipulators", IEEE International Conference on Systems Engeneering, pp. 225-228, Dayton (Ohio), August 1991.

[9] Kiszka J.B., M.M. Gupta, and P.N. Nikiforuk, "Energetic Stability of Fuzzy Dynamic Systems", IEEE Trans. Syst., Man, Cyber., Vol.15, no.6, pp.783-792, 1985.

[10] Tong R.M., "Some Properties of Fuzzy Feedback Systems", IEEE Trans. Pattern Analysis Mach.Intell.,Vol.10,no.6, pp.327-330, 1980.

[11] Kania A.A., J.B. Kiszka, M.B. Gorzalczany, J.R. Maj, and M.S. Stachowicz, "On stability of Formal Fuzzyness Systems", Inform. Sci. 22, pp.51-68, 1980.

[12] Cumani A., "On a Possibilistic Approach to the Analysis of Fuzzy Feedback System", IEEE Trans. Sist., Man, Cyber., Vol.12, no.3, pp.417-422, 1982.

[13] De Glas M., "Invariance and Stability of Fuzzy Systems", J. Math. Analisys Appl., Vol.199, pp.299-319, 1984.

[14] Chen Y.Y., and T.C. Tsao, "A Description of the Dynamical Behavior of Fuzzy Systems", IEEE Trans. Syst., Man, Cyber., Vol.19, no.4, pp.745-755, 1989.

[15] Langari G., and M. Tomizuka, "Stability of Fuzzy Linguistic Control Systems", Proc. 29th IEEE Conf. Decision and Control, pp. 2185-2190, 1990.

[16] Aracil, J., A. Garcia-Cerezo and A. Ollero, "Stability Analysis of Fuzzy control Systems. A Geometric Approach", in Kulikowski et al (Ed.): Artificial Intelligence, Expert Systems and Languages in Modelling and Simulation. North-Holland, pp. 323-330, 1988.

[17] Aracil, J., A. Ollero and A.Garcia-Cerezo, "Stability Indices for the Global Analysis of Expert Control Systems", IEEE Transactions on System, Man and Cybernetics", vol SMC 19, no. 5, pp. 988-1007, September, 1989.

[18] Ollero A, A.García-Cerezo and J.Aracil, "Design of Rule-based Expert Controllers", European Control Conference, Grenoble, July, 1991.

[19] Sugeno M. and Kang. G.T. "Structure identification of fuzzy models". Fuzzy Sets and Sistems. Vol. 28. pp 15-23. 1988

[20] Mandani, E.H. "Application of fuzzy algorithms for control of simple dynamic plant". Proc. IEEE, vol. 121, pp. 1858-1888. 1974.

[21] Garcia-Cerezo, A, A. Ollero and J. Aracil. "Stability of Fuzzy Control Systems by using Nonlinear System Theory". IFAC Simposium on Artificial Intelligence in Real-Time Control, pp. 171-176. 1992.

[22] Ollero A. and A. Garcia-Cerezo, "Direct Digital Control Auto-tunning and Supervision Using Fuzzy Logic", Fuzzy Sets And Systems, Intl. Journal, Vol.30, pp. 135-153, 1989.

[23] Nakamori, Suzuki and Yamamaka, "Model predictive control usingfuzzy dynamic models". V IFSA Congress Vol Engineering. pp. 135-138. 1991.

[24] Smith S.M., and D.J. Comer, "Automated Calibration of a Fuzzy Logic Controller Using a Cell Space Algorithm", IEEE Control Systems, August, pp. 18-28, 1991.

Chapter 7
Intelligent Fuzzy Controller for Event-Driven, Real Time Systems and its VLSI Implementation

Fuzzy Logic Finite State Machine, *162*
Algorithm of Creating a Multiple-Input Fuzzy Model, *164*
Hardware Accelerator, *168*
VLSI Implementation, *172*
Conclusions, *179*
References, *179*

Most linguistic models known are essentially static, that is, time is not a parameter in describing the behavior of the object's model. This chapter shows a model for synchronous finite state machines based on fuzzy logic. Such finite state machines can be used to build both event-driven time-varying rule-based systems and also the control unit section of a fuzzy logic computer. The architecture of a pipelined intelligent fuzzy controller is presented, and the linguistic model is represented by an overall fuzzy relation stored in a single rule memory. A VLSI integrated circuit implementation of the fuzzy controller is proposed. At a clock rate of 30 MHz, the controller can perform 3 MFLIPS on multi-dimensional fuzzy data.

INTELLIGENT FUZZY CONTROLLER FOR EVENT-DRIVEN, REAL TIME SYSTEMS AND ITS VLSI IMPLEMENTATION

Janos Grantner[1]
Department of Electrical Engineering
University of Minnesota, Minneapolis, MN 55455

Marek Patyra and Marian S. Stachowicz
Department of Computer Engineering
University of Minnesota, Duluth, MN 55812

1. FUZZY LOGIC FINITE STATE MACHINE

The general model of a finite state machine (FSM) is illustrated in Figure 1. Formally, a sequential circuit is specified by two sets of Boolean logic functions:

$$f_z(X, y) \rightarrow Z,$$
and
$$f_y(X, y) \rightarrow Y,$$

where X, Z, y, and Y stand for a finite set of input, output, the present and next states of the state variables, respectively. Function f_z maps the input and the present state of the state variables to the output, and function f_y maps the input and the present state of the state variables to the next state of the state variables.

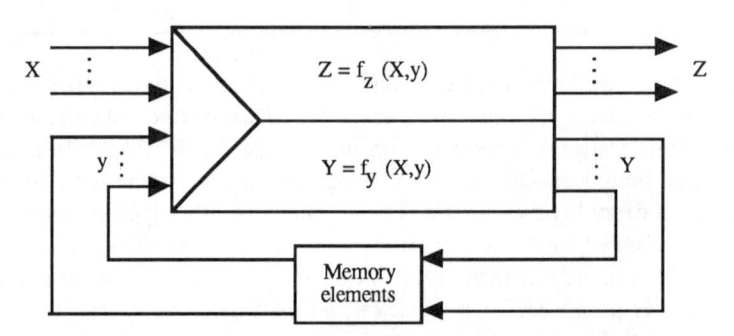

Fig. 1. General Model of a Finite State Machine (FSM).

The current states of the memory elements hold information on the past history of the circuit. The behavior of a synchronous sequential circuit can be defined from the knowledge of its signals at discrete instants of time. Those time instants are determined by

---

[1] Visiting Professor from Technical University of Budapest, Hungary.

a periodic train of clock pulses. The memory elements hold their output until the next clock pulse arrives.

We extend this model by introducing membership functions and fuzzy relations to map the changes which take place in fuzzy input data to both the fuzzy output and next state of the state variables.

With the model presented in this chapter, the definition of states will remain crisp- that is, the state of the system can be represented in one of the usual ways (i.e. by isolated flip-flops, registers or a microprogrammed control unit). The fuzzy outputs will be devised from a dynamically changing linguistic model since the response to a specific change at the fuzzy input will vary with different states of the FSM. We will refer to this model as Crisp-State-Fuzzy-Output FSM or CSFO FSM. A block diagram of the CSFO FSM is shown in Figure 2. X and Z stand for a finite set of fuzzy input and output, respectively.

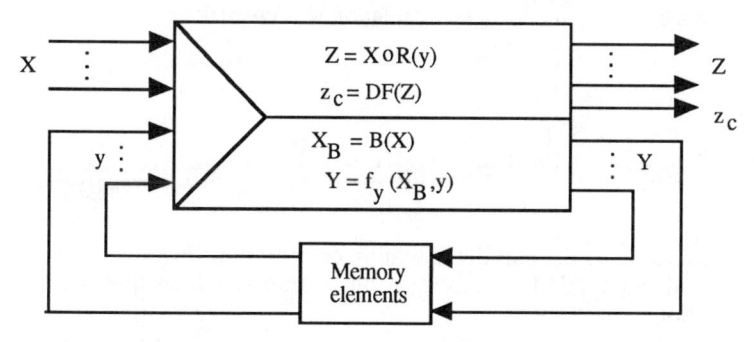

$$Z = X \circ R(y)$$
$$z_c = DF(Z)$$
$$X_B = B(X)$$
$$Y = f_y(X_B, y)$$

Fig.2. General Model of the Crisp-State-Fuzzy-Output Finite State Machine (CSFO FSM).

R stands for the object's model which is now function of the y present state of the state variables, and o is the operator of composition. The z_c crisp values of the fuzzy output are obtained by computing the DF defuzzification strategy. B stands for the transformation which maps the linguistic values of the X linguistic (fuzzy) variables to the X_B Boolean (two-valued) logic variables. Function f_y maps both the X_b Boolean logic variables and the y present state variables to the Y next state of the state variables.

To accelerate the mapping of the fuzzy input data X to both the new set of fuzzy output Z and crisp output z_c (i.e. to compute fuzzy inference and the DF defuzzification strategy), our pipelined fuzzy logic hardware accelerator model [5] will also be employed with the CSFO FSM. The next state of the state variables will be devised from both the present state variables and the X_B Boolean logic variables. For instance, a Boolean variable X1LOW is true if the position of the maximum in the membership function for linguistic variable X1 falls in the range 1 to 5. X1LOW is otherwise false.

The state transients will be completed simultaneously with the fuzzy pre-processing pipeline step. A new S_K state of the CSFO FSM will then select an overall fuzzy relation R_K which will in turn be used as the linguistic model in the fuzzy inference pipeline step while the system is in state S_K. With this model, the state variables will take their new values at the rate at which the pipeline steps proceed. The fuzzy outputs will be defuzzified in the last pipeline step. In the course of the learning process (eq. 3), an overall relation R_I is created for each state S_I (I = 1,..., N) of the CSFO FSM.

2. ALGORITHM OF CREATING A MULTIPLE-INPUT FUZZY MODEL

A linguistic model of a process can be built using a specialized software; fuzzy inference and defuzzification strategies can also be computed without using any dedicated hardware. However, in the case of real-time control applications, the pure software approach may not offer sufficient performance. We suggest a hardware accelerator for a multiple-input fuzzy logic controller. The accelerator is based upon the mathematical model as follows.

The process operation control strategy is created by analysis of input and output values, in which not only measurable quantities are taken into account but also parameters which cannot be measured, only observed [1]. On the basis of the verbal description, which is called a linguistic model, a fuzzy relation R is created:

$$R = \underset{I=1}{\overset{N}{*}} (XI \rightarrow YI) \quad (1)$$

In formula (1) $\rightarrow$ denotes the operation or operations by which fuzzy implications are defined, and the symbol $*$ represents an operation which interprets the sentence connective ALSO.

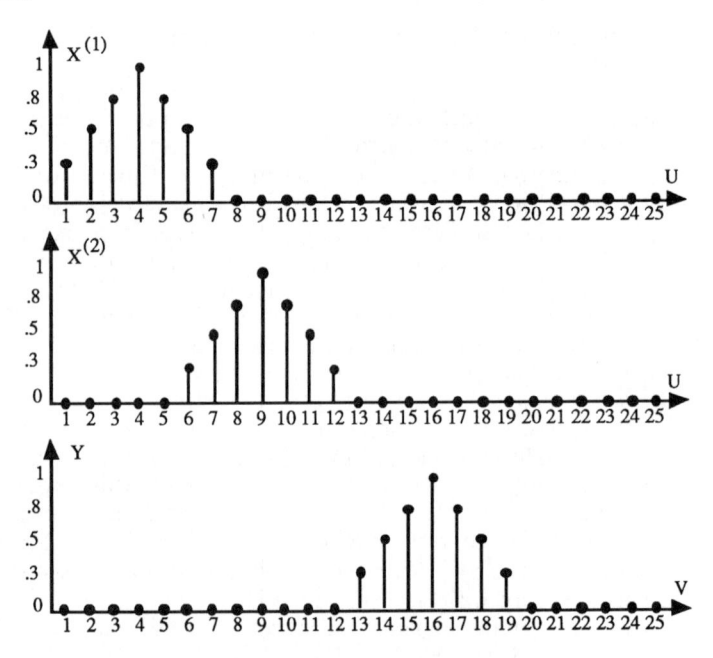

Fig. 3. Graphical Interpretation of Fuzzy Sets $X^{(1)}$, $X^{(2)}$, and Y.

We shall present the algorithm not only intended for creating a fuzzy model when a verbal description is given, but also for determining the model's answer to a given input [2].

The verbal description of the process performance contains N relations, and fuzzy sets describe the particular states which occur in the verbal description of inputs $X^{(1)}$ and $X^{(2)}$ and output Y be given in Formula 2. The graphic interpretations [4] of fuzzy sets $X^{(1)}$, $X^{(2)}$, and Y are illustrated in Figure 3.

R1:

IF $X^{(1)}$ is very small $(X^{(1)}{}_1)$ AND $X^{(2)}$ is medium $(X^{(2)}{}_1)$ THEN Y is medium (Y1)
    ALSO

R2:

IF $X^{(1)}$ is very small $(X^{(1)}{}_2)$ AND $X^{(2)}$ is medium $(X^{(2)}{}_2)$ THEN Y is medium (Y2)
    ALSO

        :

RN:

IF $X^{(1)}$ is very big $(X^{(1)}{}_N)$ AND $X^{(2)}$ is medium $(X^{(2)}{}_N)$ THEN Y is medium (YN)

(2)

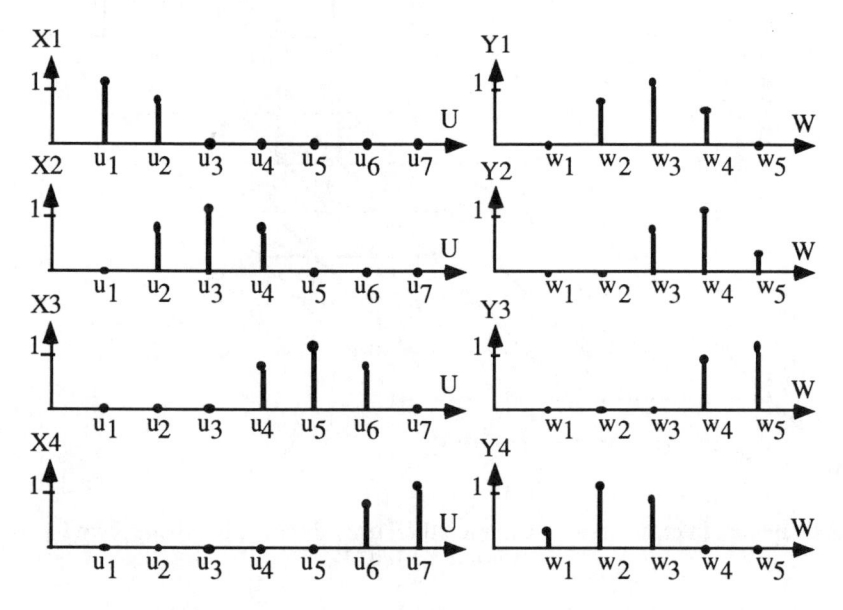

Fig. 4. Example of the Arbitrary Verbal Description Containing Four Fuzzy Relations:

        IF X=X1 THEN Y=Y1,   ALSO
        IF X=X2 THEN Y=Y2,   ALSO
        IF X=X3 THEN Y=Y3,   ALSO
        IF X=X4 THEN Y=Y4

In order to graphically illustrate introduced relations, an example of the verbal description of the object's performance which contains four fuzzy relations (the number four having been arbitrarily chosen) and fuzzy sets which describe the particular states which occur in verbal description of the input X and Y is presented in Figure 4.

The paragraphs below illustrate in turn:

Fuzzy Learning: A method of creating fuzzy relation R1 which represents the first fuzzy implication in the verbal description is interpreted as intersection. The remaining relations R2, R3,..., RN are created analogously by application of the same definition of fuzzy implication.

$$R1 = X1 \times Y1$$
$$\forall (u,w) \in U \times W \quad R1(u,w) = \min (X1(u), Y1(w)) \quad (3)$$
$$\forall u \in U \quad X1(u) = \min(X^{(1)}(u), X^{(2)}(u))$$

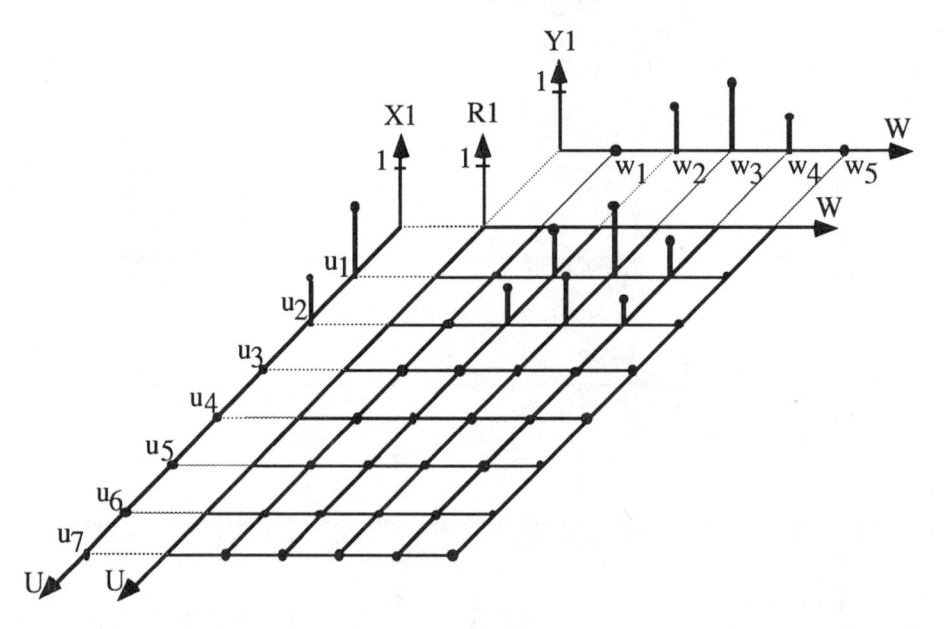

Fig. 5. Graphical Interpretation of Creation the Fuzzy Relation R1 Using Two Fuzzy Sets X1 and Y1 (eq. 3).

The final relation R (being the object's model) is obtained as the union of R1, R2,..., RN, since the sentence connective ALSO is defined as union.

$$R = R1 \cup R2 \cup ... \cup RN$$
$$\forall (u,w) \in U \times W \quad R(u,w) = \max[R1(u,w), R2(u,w),..., RN(u,w)] \quad (4)$$

Fuzzy Inference: The method of creating fuzzy answer Y to a fuzzy input X is to apply max-min composition.

$$\forall w \in W \quad Y(w) = \max_{u \in U} \, [\min(X(u), R(u,w))] \tag{5}$$

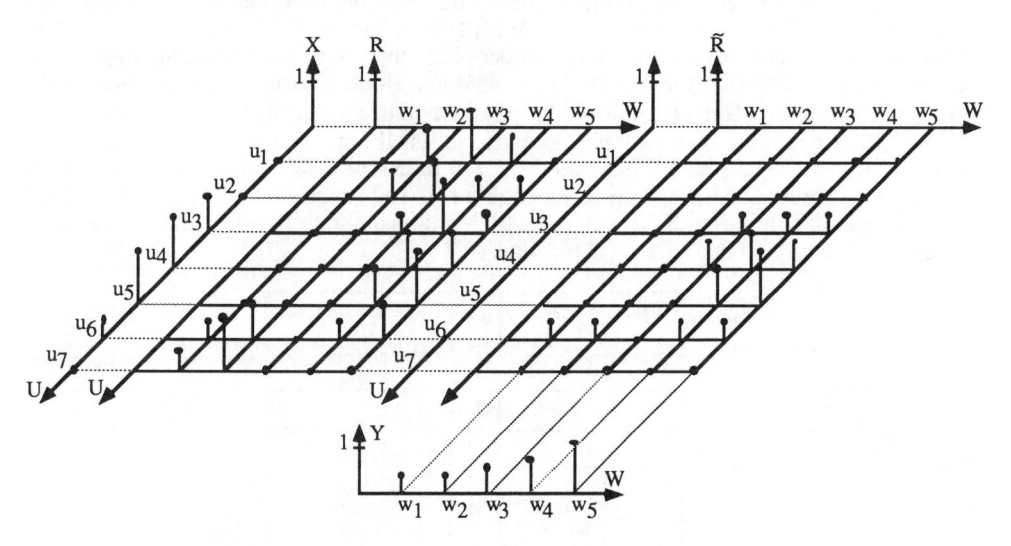

Fig. 6. Graphical Interpretation of Generation the Fuzzy Answer Y to the Fuzzy Input X Using the Max-Min Composition (eq. 5).

Defuzzification Strategy: The deterministic value of the answer (crisp value) is determined using the formula

$$y_C = \frac{1}{L} \sum_{J=1}^{L} w_J \tag{6}$$

where L is the number of points $w_J \in W$ in which output set Y reaches a maximum. The simple example shown in Fig. 7 illustrates the principle of defuzzification for five fuzzy numbers: 6, 7, 16, 17, and 19.

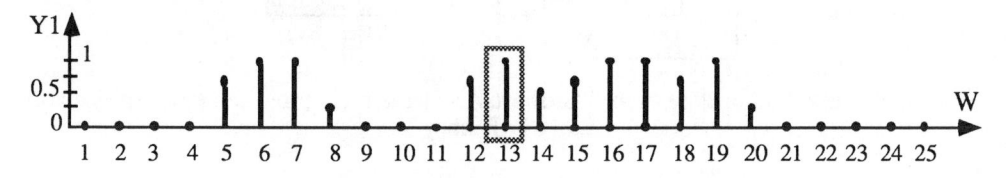

Fig. 7. Example of the Defuzzyfication Strategy.

3. HARDWARE ACCELERATOR

The hardware accelerator which performs the fuzzy learning, fuzzy inference, and defuzzification computation, that is, which maps the fuzzy inputs to fuzzy and/or crisp outputs, is summarized in this section.

Currently, in our research the degree of membership function is a discrete valued function with a 5-element domain set. With two-valued logic, three bits are used to represent each element of the set. The number of levels can be extended up to eight. The universe of discourse of a fuzzy subset is limited to a finite set with 25 elements ($u_{max} = w_{max} = 25$). Seventy five bits are used for digitization of the membership function.

The accelerator consists of four basic units: the host interface, the fuzzy pre-processing unit, the combined fuzzy model/fuzzy inference unit, and the defuzzifier unit. The last two are referred to as the fuzzy engine [3]. The functional block diagram of the accelerator is shown in Fig. 8. To achieve a high processing rate for real time applications, the units are connected in a four-level pipeline.

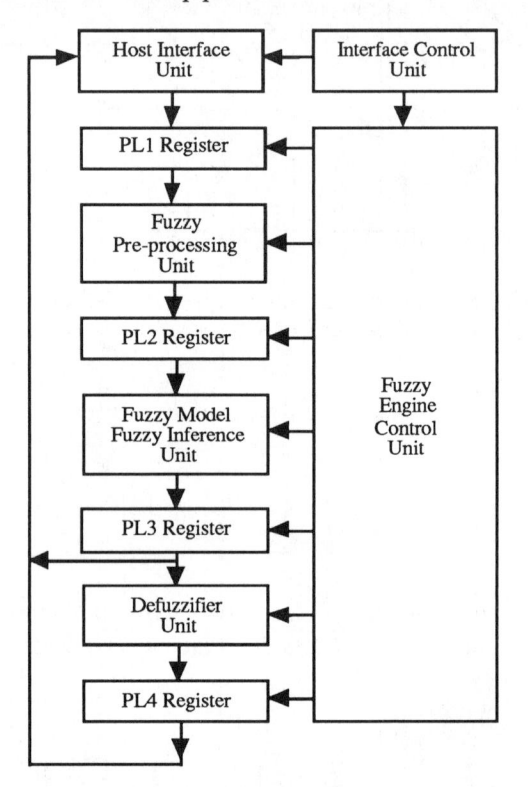

Fig. 8. Pipeline Architecture of the Hardware Accelerator. (Reproduced by permission of Larry L. Kinney.)

The core of the hardware accelerator is a fuzzy engine which implements the formuli (4) to (6). It is split into the fuzzy model/fuzzy inference unit and the defuzzifier unit. The functional block diagram of the fuzzy model/fuzzy inference unit (without increased parallelism) is shown in Figure 9.

After the XI and YI registers have been loaded, learning a multi-dimensional rule RK takes u_{max} clock periods. The MUX2 multiplexer at the input of the minimum unit selects the YI register. During the first clock step, u_I is paired with all w elements of YI and these pairs are fed to the inputs of the minimum unit. If the current rule is the first in a learning sequence, throughout the learning cycle 0 (nonmembership) elements will be paired with the outputs of the minimum unit and fed to the inputs of the maximum unit. The whole word of maximum values is stored at the first location of the R rule memory. During the jth clock step, u_j is compared to all w elements of YI simultaneously and the vector of the max elements is stored in the jth location of R.

If the current rule is not the first one in the learning process, the MUX3 multiplexer at the input of the maximum unit selects the ith row of R($1 \leq i \leq u_{max}$) during the ith clock step and the contents of this row in R will be updated from the outputs of the maximum unit.

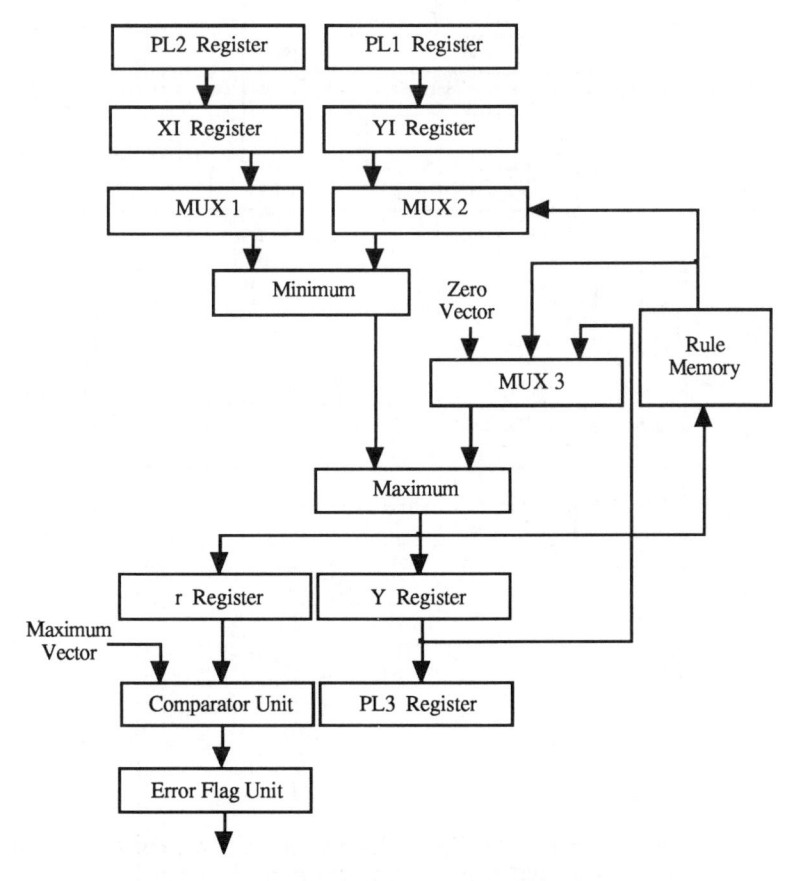

Fig. 9. Functional Block Diagram of the Fuzzy Model/Fuzzy Inference Unit.
(Reproduced by permission of Larry L. Kinney.)

Therefore the learning process of N rules takes $N \times u_{max}$ clock periods with the architecture shown in Figure 9. The clock steps needed to load registers XI and YI are ignored at this point. Computing the fuzzy inference (max-min composition) also takes u_{max} clock periods. This time the MUX2 multiplexer at the input of the minimum unit pairs the u_i element of XI with all r elements of the ith row in R. If i = 1 (first clock step), then the MUX3 multiplexer at the input of the maximum unit selects 0 as the other operand for each element at the output of the minimum unit. The outputs of the maximum unit are fed to the inputs of the Y register. From the second to the last clock steps, outputs of the Y register are fed back to the inputs of the maximum unit through the MUX3 multiplexer.

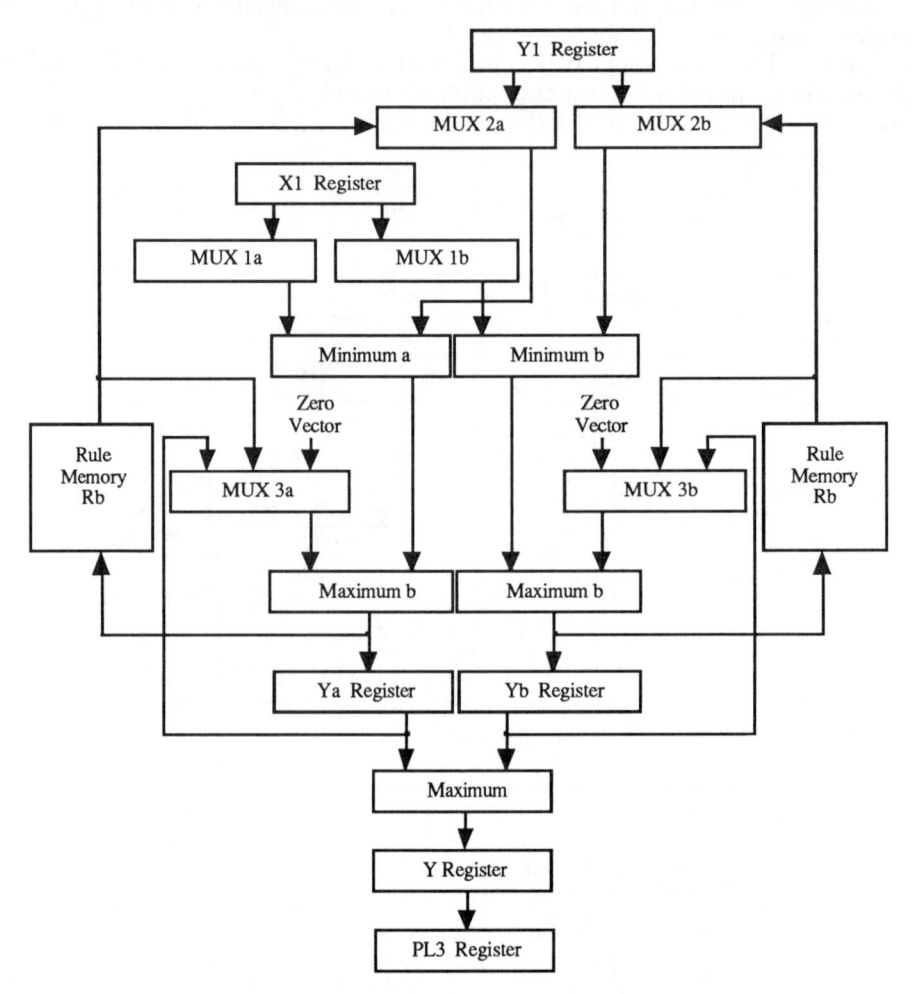

Fig. 10. Schematic Diagram of the Double Configuration of the Fuzzy Inference Unit. (Reproduced by permission of Larry L. Kinney.)

Contents of the R rule memory remain unchanged during the fuzzy inference process. After the last clock step, register Y holds the result of the XoR operation in the digitized fuzzy data format. To detect whether the condition: $\forall (u,w) \in U \times W$, $R(u,w) = 1$ is met, an error flag was added to the fuzzy engine. If the error flag is activated at the completion of the learning of a new rule, then all elements in the R rule memory become equal to 1 (full membership). This flag can be used to generate an interrupt request to the host machine. The system can then recover from this erroneous state by either downloading a "safe" model to the R memory or starting over the learning process with a modified model.

Due to the linear property of the max-min composition, by quadrupling the functional units of the basic architecture, the time required to complete the pipeline steps for either the fuzzy learning or the fuzzy inference process can be reduced to $[u_{max} \div 4] + 2$ clock periods. Fig. 10 shows the schematic diagram of the double fuzzy model/fuzzy inference unit. There are two such a blocks in actual design.

Since the precedence relation of the subtasks (I/O data transfer (T_1), the pre-processing of the multiple fuzzy inputs (T_2), the learning of a new rule or the performing a fuzzy inference operation (T_3), and the defuzzification (T_4)) are all linear operations, the four basic units of the hardware accelerator form a linear pipeline. The pipeline architecture allows the simultaneous operation of the four units. The space-time diagram in Figure 11 illustrates the overlapped operations of the pipeline units.

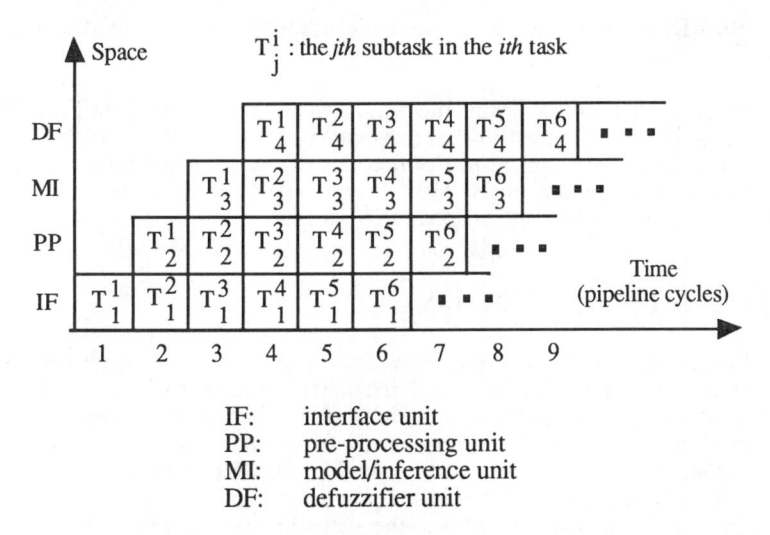

T^i_j : the *jth* subtask in the *ith* task

IF: interface unit
PP: pre-processing unit
MI: model/inference unit
DF: defuzzifier unit

Fig. 11. Overlapped Operations of the Pipeline Units. (Reproduced by permission of Larry L. Kinney.)

Assuming that the downloading of the fuzzy data from the host system to the accelerator and the reading of the fuzzy and/or crisp output data from there (subtask T_1) does not exceed $[u_{max} \div 4] + 2$ (9) clock periods, the accelerator produces new fuzzy and/or crisp output data every $[u_{max} \div 4] + 2$ clock periods once the pipeline is filled. Thus, at a clock rate of 30 MHz the fuzzy engine can perform over 3,000,000 fuzzy logical inferences per second with the current fuzzy data format.

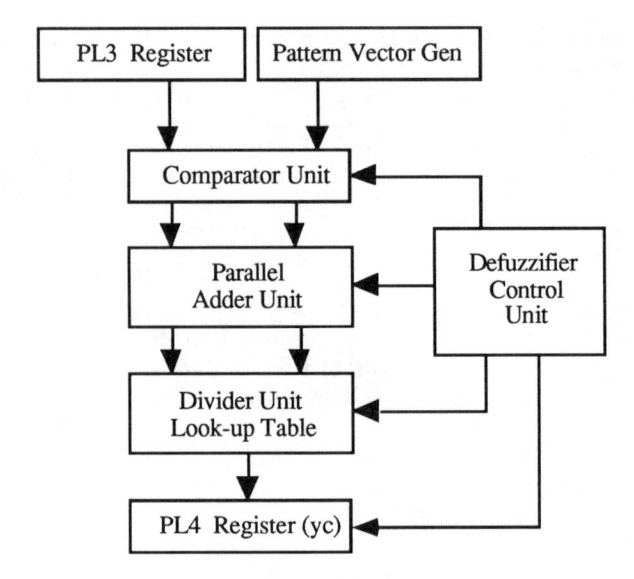

Fig. 12. Blok Diagram of the Defuzzifier Unit. (Reproduced by permission of Larry L. Kinney.)

The defuzzifier unit performs the defuzzification operation (see Fig. 12). Finding both the maximum and its position takes at most 4 clock periods. Two parallel adder networks are used to sum up the position codes of the maximum and obtain the total number of maximum simultaneously. Then, the crisp value is taken from the look-up table. If maximum value is zero then the crisp value is flagged.

4. VLSI IMPLEMENTATION

Although fuzzy logic has been successfully applied to control problems over the last 10 years, extensive integrated circuits implementations of fuzzy logic circuits have yet to be expected. Additionally, the booming market on fuzzy logic based consumer products in Japan has been created by the use of very simple fuzzy logic circuits, designed to be a new buzzword rather then based on the novel technology, and/or using a new approach. Profit oriented activity of major fuzzy logic based product players pushed the research from the area of innovative fuzzy technology into the second stage. Therefore, little progress has been observed during the last few years, especially in VLSI implementation of fuzzy logic circuits, where the results of our architectural research can be incorporated, thereby verifying our ideas.

This section is organized as follows. First, the fuzzy technology constraints are discussed. Second, the choice of the fuzzy information representation is presented. Then the design trade-offs are considered, followed by the VLSI implementation constraints. Finally, the scaled down, affordable version of the controller is discussed in detail.

One of the difficulties relating to issues of practicality is associated with the VLSI implementation. Therefore, the information provided in this section is based on both our estimates and previous experience with projects of a similar nature.

Due to our objective constraints, i.e. the MOSIS service presently available for chip fabrication, the full design version of the proposed controller will be designed, along with a scaled-down version, which will pass the constraints and then be fabricated.

There are two different versions of the fuzzy logic controller that could be useful in most practical implementations: a controller working stand alone (SA), or with an appropriate host computer (HC). These options will be taken into consideration.

Table 1. Preliminary Definitions of Signals for Full Scale Versions of HC and SA Fuzzy Logic Controller.

Full scale HC version	Full scale SA version
64b scaled address/data bus	64b digital address/data bus @
55 signal for data bus control %	32b XPROM interface
(Bus Parity)	4 XPROM control signals
(Command)	4 analog inputs %
(Status and CP)	3 signals for fuzzification control #
(Capability)	3 signal for inference control #
(Synchronization)	2 signals for defuzzification control #
(Arbitration)	4 signals for mode control
(Addresses)	2 or 3 analog outputs
3 signals for fuzzification control #	1 CLK global clock
3 signals for inference control #	1 STB strobe signal
3 signals for defuzzification control #	1 EN synchronization input
1 CLK global clock	1 RESET input
1 RESET input	1 CS Chip Select
1 CS Chip Select	8 power supply inputs
8 power supply inputs	

\# These options could be programmable.
% Preliminary number.
@ Can be used to substitute for a single analog input/output.

Let us discuss the VLSI implementation issues in more detail starting with the full scale design. According to our preliminary assumption, the descriptions of design signals devised are summarized in Table 1.

We assume that the proposed fuzzy controller will have three basic cycles of operation: fuzzy learning, fuzzy inference and stand-by. In case of the fuzzy learning and fuzzy inference operations the HC version will be supplied with fuzzy data through the host computer which performs the fuzzification of the analog inputs. It is obvious that HC version will be able to process only digital representation of the fuzzy data prepared by the host computer. In our first approach this version will not be cascadable.

The SA version of the chip will input the analog data and perform the fuzzification operation by itself. The stand-by mode will be common for both versions.

One can also see that the HC version will require a very detailed design of the interface to the bus system used by the host computer for data transmission while the SA version will need an A/D converter and a few D/A converter, which will be included in the chip design. It is assumed that the HC version will be communicating with the host computer through FUTUREBUS (Fig. 13).

```
                    ┌──────────┐
                    │   Host   │
                    │ Computer │
                    └────┬─────┘
                         ⇕
    ◄────────────────────┬────────────────────►
                  F U T U R E B U S
    ◄──────────┬─────────────────────┬─────────►
               ⇓                     ⇕
        ┌──────────┐          ┌────────────┐
        │ Address  │─────────►│FL Hardware │
        │ Decoder  │          │Accelerator │
        └──────────┘          └────────────┘
```

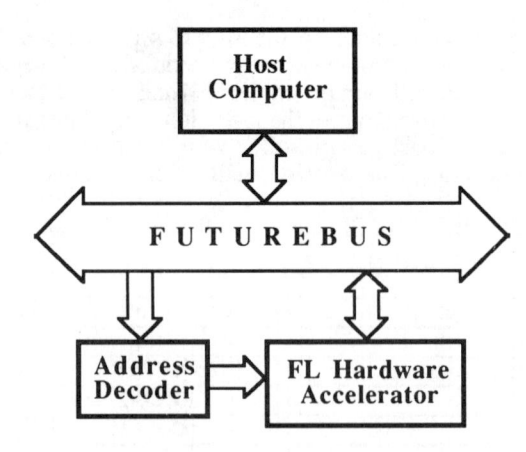

Fig. 13. Configuration for the Hardware Accelerator Working Under the Host Computer (HC version).

Each version has its own advantages and disadvantages basically due to communications issues, the number of pins, and the design efforts. It is important to point out that one can expect some instant differences in the performance of the four versions which will be further investigated.

Let us now focus on the scaled down implementations of the SA version of the proposed fuzzy logic controller. Two basic modes for chip operation will be designed: normal and programmed. The normal mode will include cascaded (parallel or serial) and noncascaded operation. The block diagrams illustrating these modes are shown in Fig. 14 and Fig. 15. In the programming mode, we assume that it will be possible to preprogram the fuzzifier, defuzzifier or inference engine, or any combination of these, in order to preserve a flexible operation tuned to the actual system under control. In order to achieve programmability, an EPROM/EEPROM type of memory block will be built-into the chip and will be controlled by the external source through both a memory I/O port and control signals. Our decision to fabricate this version is based on both the number of pins and also the number of signals needed to implement this version (no data interface is needed). The scaled down implementation of our design matching the objective constraints is presented as follows with respect to major design issues.

• Chip size and package

The reasonable MOSIS package has 132 pins and can contain a chip occupying 7.8mm*9.2mm of silicon area (max). The choice of CMOS technology leads us to the variety of available processes starting from *lambda*=1μm to *lambda*=0.6μm. Keeping in mind the maximum signal frequency for chip operation, which was originally set between 25MHz and 30MHz, as well as the maximum chip size, the n-well, double-metal CMOS technology with lambda=0.6μm will be adequate to achieve the design goals.

• Chip area and number of transistors

The maximum chip area of 72.68mm^2 (132 pins package) can contain about 600000 transistors for highly regular structure[2] with the standard CMOS technology (*lambda*=0.6μm). According to our estimations we will be able to put at most as four parallel fuzzy data processing paths into the chip. The single data processing path including the programming options (memoryless) is estimated to have about 50,000 transistors.

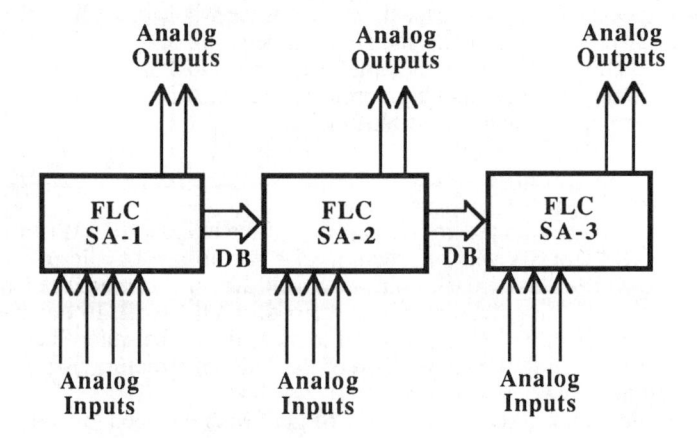

Fig. 14. Serial Configuration for Fuzzy Logic Controller (SA version).

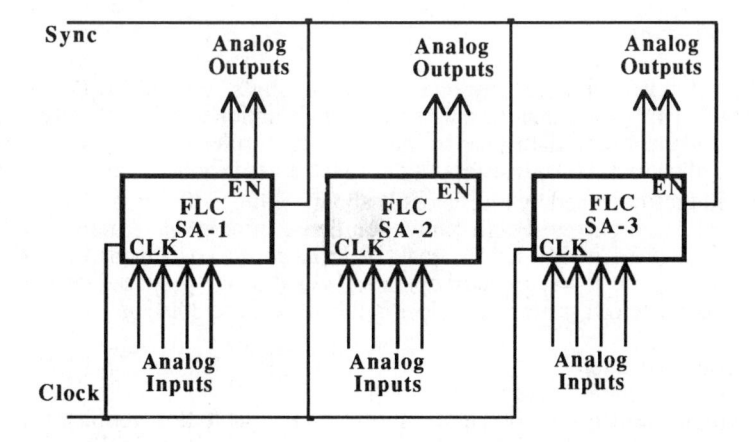

Fig. 15. Parallel Configuration for Fuzzy Logic Controller (SA version).

---

[2] Excluding the area occupied by the chip frame.

- Clock strategy and clock distribution

We decided to use single external clock signal (CLK) to generate an on-chip, two-phase nonoverlapping internal clock signals ($\Phi 1$ and $\Phi 2$), the two phase-clock system having the advantage of making hazard problems within the pipeline paths more easily identifiable. These phases will be distributed over the whole chip using a second metallization layer. Because the longest possible metal line is about 10mm, we chose a tree-like structure for phase distributions driven by high gain clock drivers. These drivers will be designated to drive an appropriate capacitive load of the whole clock line tree. According to the results of our previous research, the single processing path will have the ability to operate at 8 clock cycles/pipeline step. Setting the external clock rate at around 30MHz will enable us to operate the processor at a high processing rate. A future detailed investigation will help us to determine the highest possible clock rate.

- Rule memory

The major problem with the limited capacity of the internal SRAM (for HC version) of EPROM/EEPROM (for SA version) memory for the storage of rules of inferences in previous works [6-8] does not exist in our approach due to the strategy of building the global rule for the whole linguistic model described in Chapter 2. In our case only 1/4 Kbyte SRAM or EPROM/EEPROM is needed to store the global rule. Such an approach creates a luxury of increasing the parallelism of the internal structure by a factor of four, which is discussed in the next section.

It also should be noted that the idea of CSFO FSM is intended to be implemented in the SA version. Furthermore, the required extension of the rule memory (every FSM state will have assigned rule memory) will be evaluated. It is however unlikely that overall number of transistors for a single path will reach 100,000 transistors.

- Pipeline architecture

The estimated number of transistors for a single fuzzy data processing path is around 50,000. This means that the chip under consideration has the capacity containing at least four separate fuzzy data processing paths plus rule memory, which gives total estimation of about 300,000 transistors (look-up table used for defuzzification is included). The estimated area occupied by transistors is about 45 mm^2. The rest of the chip area will be used to provide high speed communication between processing units and the built-in memory (EPROM/EEPROM). It is expected that four parallel data paths will be designed in the chip increasing the actual speed of operation twice. In the proposed design, 3 MFLIPS performance is expected assuming the clock rate will be up to 30MHz.

- Fuzzifier implementation

The circuit implementation of the fuzzifier block has yet to be determined. The structure of the fuzzifier should evolve as a result of the research to be done in the first phase of our project. The fuzzifier is predicted to consist of a set of externally programmable membership function generators with a controlled number of discretization intervals as well as with the controlled number of membership function levels (see Fig. 16).

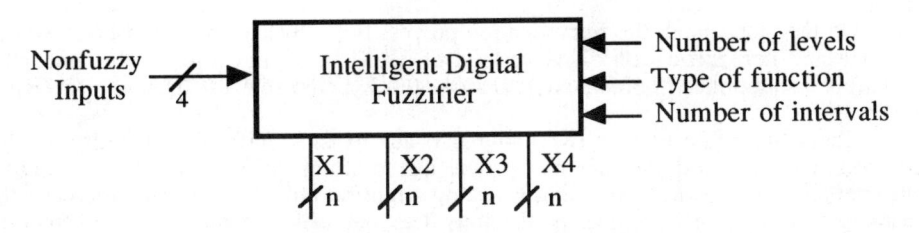

Fig. 16. The Idea of Intelligent Fuzzifier.

Based on the preliminary assumptions listed in Table1, two different pin configurations for the HC and SA versions on VLSI implementation of intelligent fuzzy logic controller have been proposed. The pin configurations are shown in Figure 17 and Figure 18 respectiively.

The HC version (Fig. 17) should be equipped with fast 64b address/data bus providing the communication with host processor. The Bus Parity, Command Status, Synchronization, Arbitration and Address signals should occupy about 55 pins.

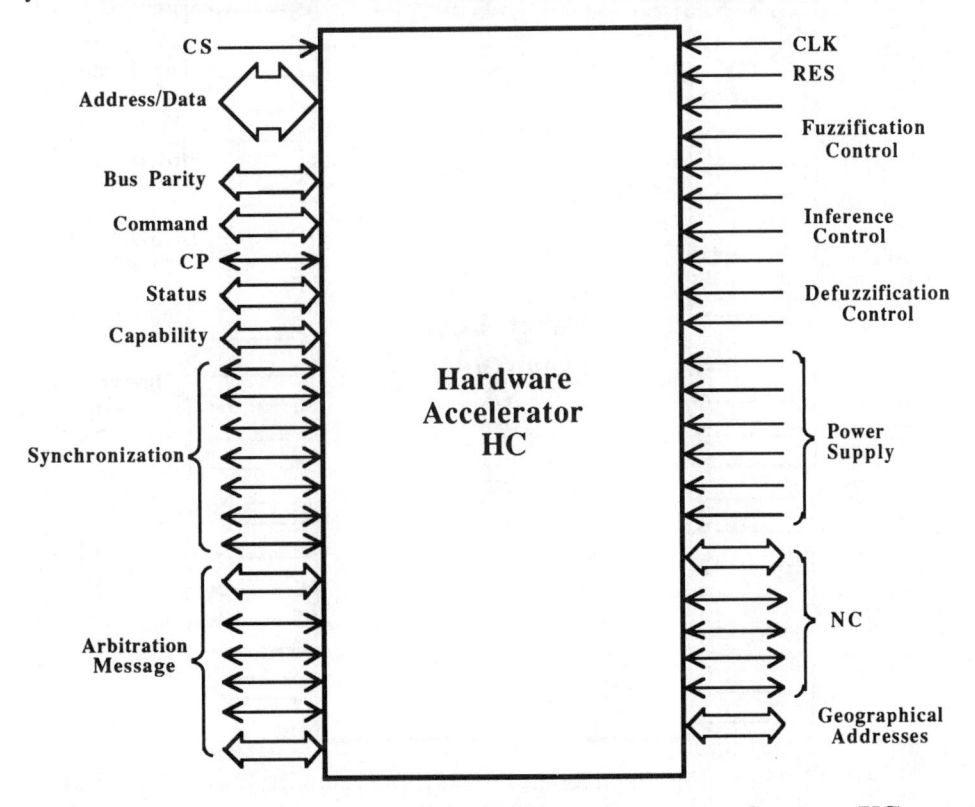

Fig. 17. Pin Configuration for Hardware Accelerator; Host Computer (HC) Configuration.

178   *Fuzzy Control Systems*

On the other hand, the fuzzyfication process is controlled by 3 signals as well as inference and defuzzification. This version of VLSI implementation is equiped with standard power supplies (8 pins), external clock (CLK), chip select (CS), and reset (RES) signals.

The stand alone version (SA) (Fig. 18) due to its specific functionality, is also equipped with fast address/data bus providing communication with external digital instruments as a substitute for single analog input/output. This feature increase the flexibility of the chip in a practical applications. The four analog inputs are used to enter the analog process data into the chip. Four analog outputs send the defuzzified results to the external instruments controlling the process. The fuzzy rules (the model of the process) are stored in the internal memory (EPROM OR EEPROM) which can be programmed externally using the appropriate signals (XPROM Control and XPROM I/O signals). The fuzzification, inference, and defuzzifivcation are performed analogously to the HC version. The programming and mode control is performed using a set of bidirectional signals which occupy separate pins. This version is also equiped with standard power supplies (8 pins), external clock (CLK), chip select (CS), and reset (RES) signals.

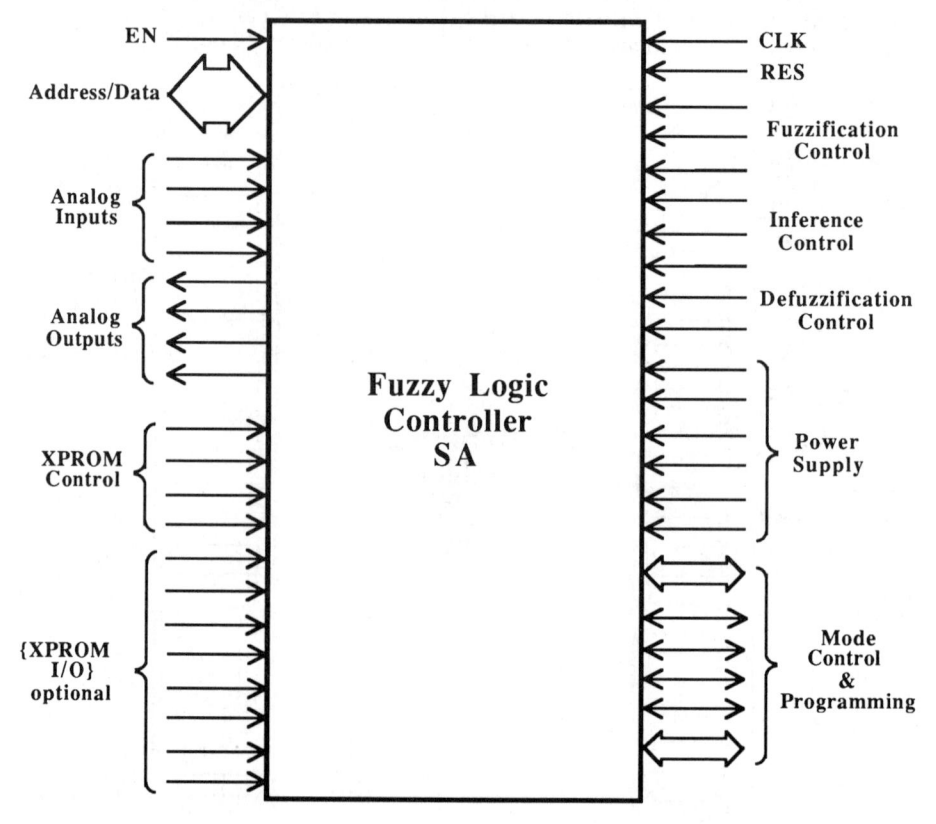

Fig. 18. Pin Configuration for the Fuzzy Logic Controller; Stand Alone (SA) Configuration.

5. CONCLUSIONS

The general model for fuzzy state machine (FSM) used to formulate fuzzy controllers for event-driven real-time systems has been described. An improved architecture for fuzzy logic controller has been defined based on the introduced Crisp-State-Fuzzy-Output Finite State Machine (CSFO FSM). The defined architechure provides a novel strategy for fuzzy model building enabling fuzzy inferences to be performed in a single stage of a hardware accelerator, an ability not common to previously published architechures. This proposed architecture for the fuzzy controller hardware accelerator, appropriately pipelined, reaches the speed of at least 3M fuzzy logical operations per second and has the ability to work in a real time environment. The VLSI Implementation issues of a proposed architecture, as well as the host computer (HC) version, the stand alone (SA) version, and the SA version's chip realization have been discussed.

REFERENCES

1. Zadeh L. A., "A Fuzzy Algorithmic Approach to the Definition of Complex or Imprecise Concepts", *International Journal Man Machine Studies*, 8, pp. 249-291, 1976.

2. Stachowicz M. S., "The Application of Fuzzy Modelling in Real-Time Expert Systems for Control", *Proc. 49th Ironmaking Conference*, Detroit, March 25-28 1990, pp. 503-512.

3. Stachowicz M. S., Grantner J., Kinney L.L., "Two-Valued Logic for Linguistic Data Acquisition", *NAFIPS Workshop '91*, University of Missouri-Columbia, pp. 168-172, May 14-17, 1991.

4. Stachowicz M. S., Kochanska M. E., "Graphic Interpretation of Fuzzy Sets and Fuzzy Relations", (Ballester A., Cardins D., Trillas E.) (eds): *Mathematics at the Service of Man*, West Berlin, Springer-Verlag, 1982, pp. 620-629.

5. Stachowicz M. S., Grantner J., Kinney L. L., "Pipeline Architecture Boosts Performance of Fuzzy Logic Controller", *IFSICC'92 International Fuzzy Systems and Intelligent Control Conference,* Louisville, Kentucky, March 15-18, 1992, pp. 190-198.

6. Togai M., Watanabe H., "Expert System on a Chip: An Engine for Real-Time Approximate Reasoning", *IEEE Expert*, pp. 55-62, Fall 1986.

7. Watanabe H., Dettloff W. D., Yount K. E., "A VLSI Fuzzy Logic Controller with Reconfigurable, Cascadable Architecture", *IEEE Journal of Solid-State Circuits*, Vol. 25, pp. 376-381, 1990.

8. FC 110 Digital Fuzzy Processor DFP^{TM}. Togai InfraLogic, Inc. 10/1991.

9. Patyra M. J., "VLSI Implementation of Fuzzy-Logic Circuits", *International Fuzzy Systems Association World Congress*, Brussels, Belgium, June 1991.

Chapter 8
Constraint-Oriented Fuzzy Control Schemes for Cart-Pole Systems by Goal Decoupling and Genetic Algorithms

Introduction, *182*
The Notion of Constraint-Oriented Fuzzy Control, *183*
Construction of Fuzzy Control Scheme by Decoupling the Goals on Cart and Pole, *185*
Construction of Fuzzy Control Systems by Genetic Algorithm Techniques, *188*
Construction of Control Scheme for Cart-Pole Systems by Genetic Algorithms, *191*
Conclusions, *193*
References, *194*

This chapter introduces a new method of fuzzy control based on the notion of constraint-oriented fuzzy inference. It shows that the goal of control can be naturally decomposed into independent subgoals that can be pursued independently by componential fuzzy control subrules. This modularized structure of constraint-oriented fuzzy control allows the construction of control systems using genetic algorithm techniques in such a manner that the genetic pool is gradually self-organized by trial-and-error experiences which results in efficient and adaptable control schemes. These approaches to the construction of fuzzy control schemes are applied to a cart-pole system to illustrate their effectiveness.

CONSTRAINT-ORIENTED FUZZY CONTROL SCHEMES FOR CART-POLE SYSTEMS BY GOAL DECOUPLING AND GENETIC ALGORITHMS

Osamu Katai[†], Masaaki Ida[†], Tetsuo Sawaragi[†],
Sosuke Iwai[†], Shinichi Kohno[*] and Tetsuji Kataoka[†]

[†]Dept. of Precision Mechanics, Faculty of Engineering, Kyoto University
Sakyo-ku, Kyoto 606-01, JAPAN

[*]Dept. of Aeronautic Engineering, Faculty of Engineering, The University of Tokyo
Minato-ku, Tokyo 153, JAPAN

1. INTRODUCTION

It is commonly acknowledged that *constraint-oriented* ways of problem solving are one of the most promising ways for dealing with large-scaled complex problems that are increasingly requiring our attention [1] - [3]. The reason for this is that the constraint-oriented ways of problem solving permit us to *decouple* the information (knowledge) on the problem itself from the way such knowledge is used to solve the problems, i.e., it provides us to use *declarative* knowledge representation for problem solving [4], [5]. This *modularity* of knowledge and the ways of problem solving also enable us to decompose the total goal into componential subgoals which can be pursued independently of each other.

In this study, we will construct fuzzy control systems for cart-pole systems in such a way that the total goal of control is decomposed into the goal on each componential subsystems, e.g., the cart subsystem and the pole subsystem. It is shown that this kind of modularized control can effectively be constructed by introducing the notion of *constraint-oriented fuzzy inference*, where fuzziness is regarded to be an *ordered collection of crisp constraints* thus enabling us to utilize the above structural modularity of constraint-oriented style of problem solving [6] - [9]. It will be shown that the pole and the cart can be controlled separately even though they actually interact with each other.

Also, in this study, *Genetic Algorithm* techniques [10], [11] are used to construct the constraint-oriented fuzzy control systems by utilizing the modular structure properties of the proposed control systems. Trial and error experiences are used to organize fragmental componential crisp if-then rules, the organization of which is expected to constitute the final constraint-oriented fuzzy control rules [12]. It is shown that this machine learning method of fuzzy control systems (production systems) can be effectively applied to yield smooth and stable controlled behavior of cart-pole systems.

2. THE NOTION OF CONSTRAINT-ORIENTED FUZZY CONTROL

In this study, we will regard a fuzzy inference rule as an *ordered collection* of crisp *constraint-interval rules* as follows:

$$\text{if } x \in X_i, \text{ then } y \in Y_i, \quad i = 1, 2, ..., n. \tag{1}$$

These intervals X_i's and Y_i's represent certain constraints on the if-part and the then-part variables x and y, respectively. The ordered collection (1) of these constraint-interval rules is called *constraint-interval fuzzy inference rule* where i stands for the level of membership grade, and we do not presume that the upper level constraint-intervals should be included in the lower level constraint-intervals [6] - [9].

In this new notion of fuzzy inference rule, we will have two different kinds of interpretations. The first is to interpret each crisp constraint rule as the representation of the constraint that should necessarily be satisfied without fail. The second, on the other hand, is to interpret this ordered collection of crisp constraints as a bundle of focal constraints that represent the desirable region to which the values of the pair of the if-part and the then-part variables should preferably belong. For instance, if we have a definite constraint C(x, y) on the observation variable (if-part variable) x and the control variable (then-part variable) y, we have to set intervals X_i on x and Y_i on y in such a way that C(x, y) is satisfied without fail no matter what values x and y will be provided such that they are in the intervals X_i and Y_i, respectively. In this case, the collection (1) of the constraint-interval rules gives us the former interpretation, and they constitute a *confining* fuzzy control rule. If $\{X_i\}_i$ is of "mountain-like" shape, then $\{Y_i\}_i$ is of "basin-like" shape, i.e., (the upper level) intervals Y_j's include (the lower level) intervals Y_k's provided that $j \geq k$ as shown in Fig. 1. On the other hand, as shown in Fig. 1, if the constraint regions $C_1, C_2, ... , C_n$ are set in such a way that $C_1 \supset C_2 \supset ... \supset C_n$ and these focal area represent the graded desirable area on the pair x and y, and if they are approximated by rectangular regions, then the sides of these rectangular regions constitute a constraint-interval fuzzy control rule of the second type. In this *goal-seeking* fuzzy rule,

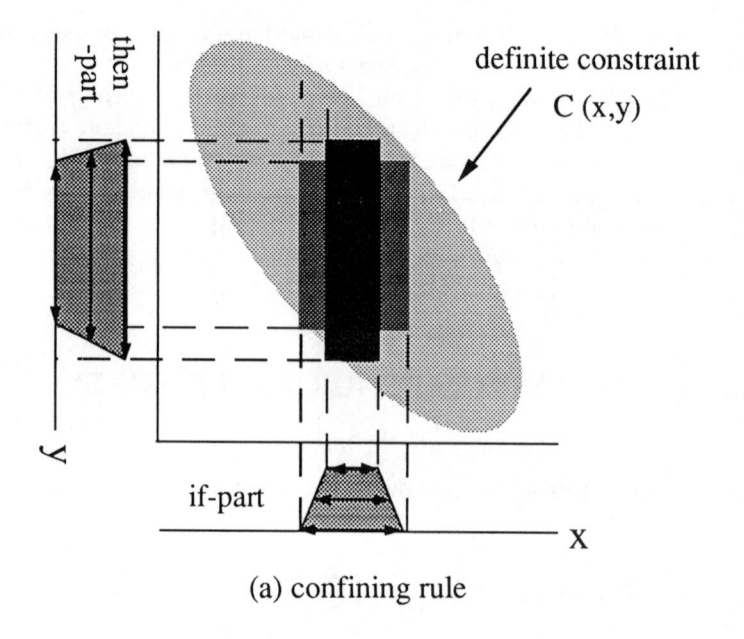

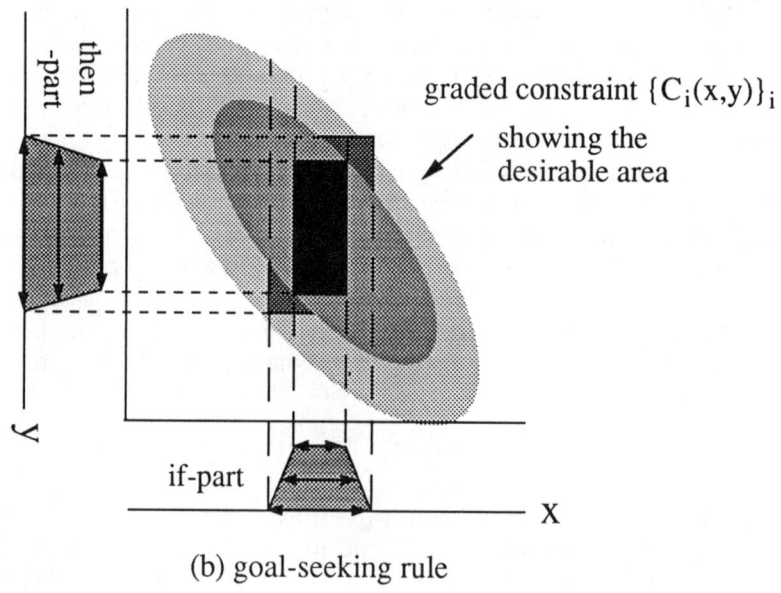

Fig. 1 Two Interpretations of Constraint-interval Fuzzy Rule

the collection of if-part intervals and that of then-part intervals are both "mountain-like" shaped as shown in Fig. 1.

3. CONSTRUCTION OF FUZZY CONTROL SCHEME BY DECOUPLING THE GOALS ON CART AND POLE

We will consider the control problem of cart-pole systems in which the rail of the cart is of bounded length. In this case, the goal of control G_{cp} of the total system is decomposed into two subgoals G_p and G_c that are on the pole and the cart, respectively. These goals are dependent on the observation $(\theta, \dot{\theta})$ and $(x, \dot{x})$, respectively. We will assume that these goals can be pursued separately, and hence the structure of control here is given as in Fig. 2.

Depending on the types of interpretation of the fuzzy control rules, i.e., whether they are interpreted as *confining* or *goal-seeking* rules, we will have two different kinds of G_c and G_p.

G_c: (C_g) "to set the cart position as close as possible to the center of the rail" (*goal-seeking* for the cart)
(C_c) "not to let the cart run out of the rail area" (*confining* the cart)
G_p: (P_g) "to set the pole as vertical as possible" (*goal-seeking* for the pole)
(P_c) "not to let the pole fall down" (*confining* the pole)

Hence, we will have four different kinds of control goals as follows:

$$\begin{aligned}
\text{case (1):} &\quad (C_g) \ \& \ (P_g) \\
\text{case (2):} &\quad (C_g) \ \& \ (P_c) \\
\text{case (3):} &\quad (C_c) \ \& \ (P_g) \\
\text{case (4):} &\quad (C_c) \ \& \ (P_c)
\end{aligned}$$

It should be noted that both of the constraint-intervals derived from the fuzzy inference on the cart and the pole are necessarily satisfied, and hence the "AND" composition of these then-parts (conclusive parts) are used to obtain the integrated conclusive part from which a modified version of the centroid method is used to yield the value of the control variable, in this case, the external force F to the cart.

Fig. 3 shows the result of the control of the cart-pole system under the initial condition

$$\theta = 0.25 \text{ (rad)}, \quad \dot{\theta} = 0 \text{ (rad/sec)} \tag{2}$$

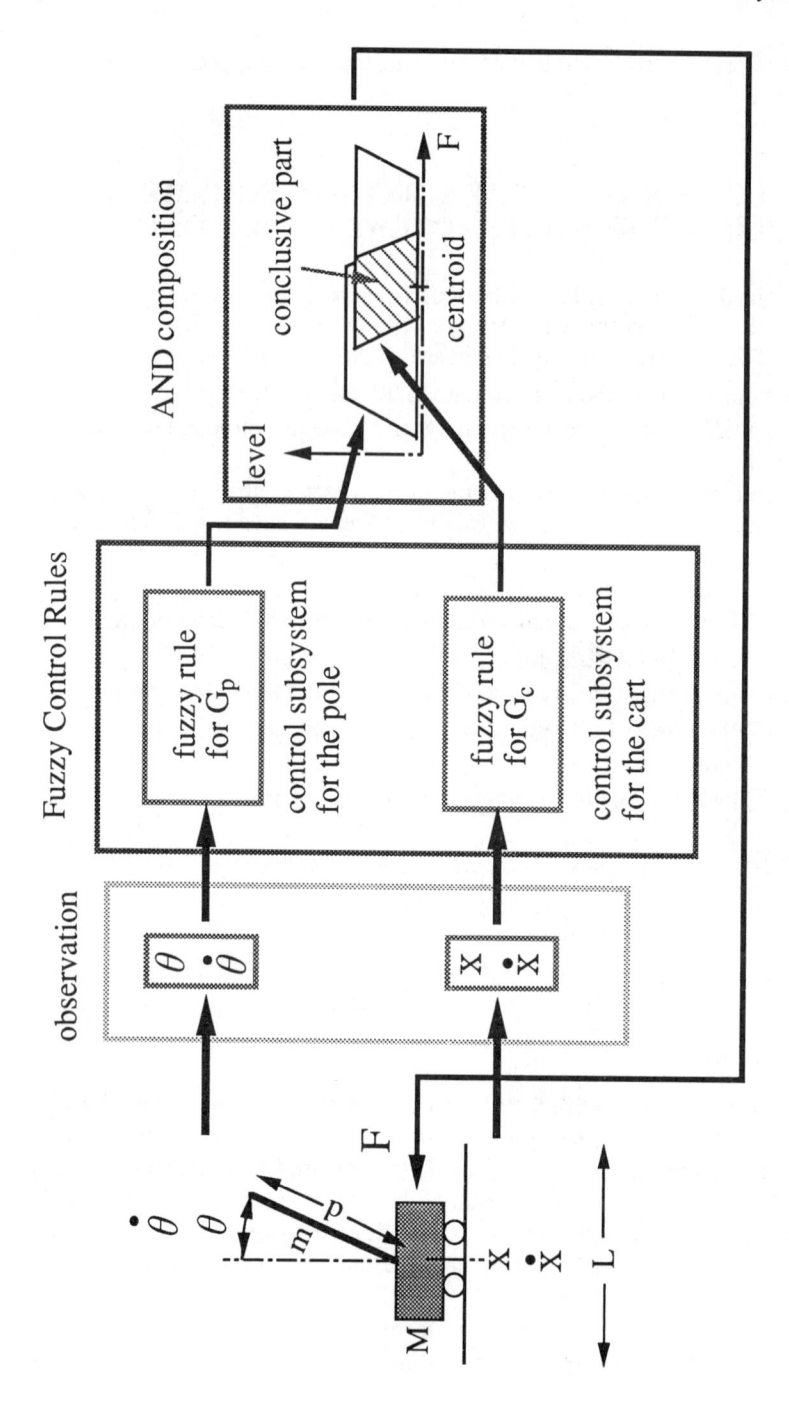

Fig. 2 Decoupled Fuzzy Control for Cart-Pole Systems

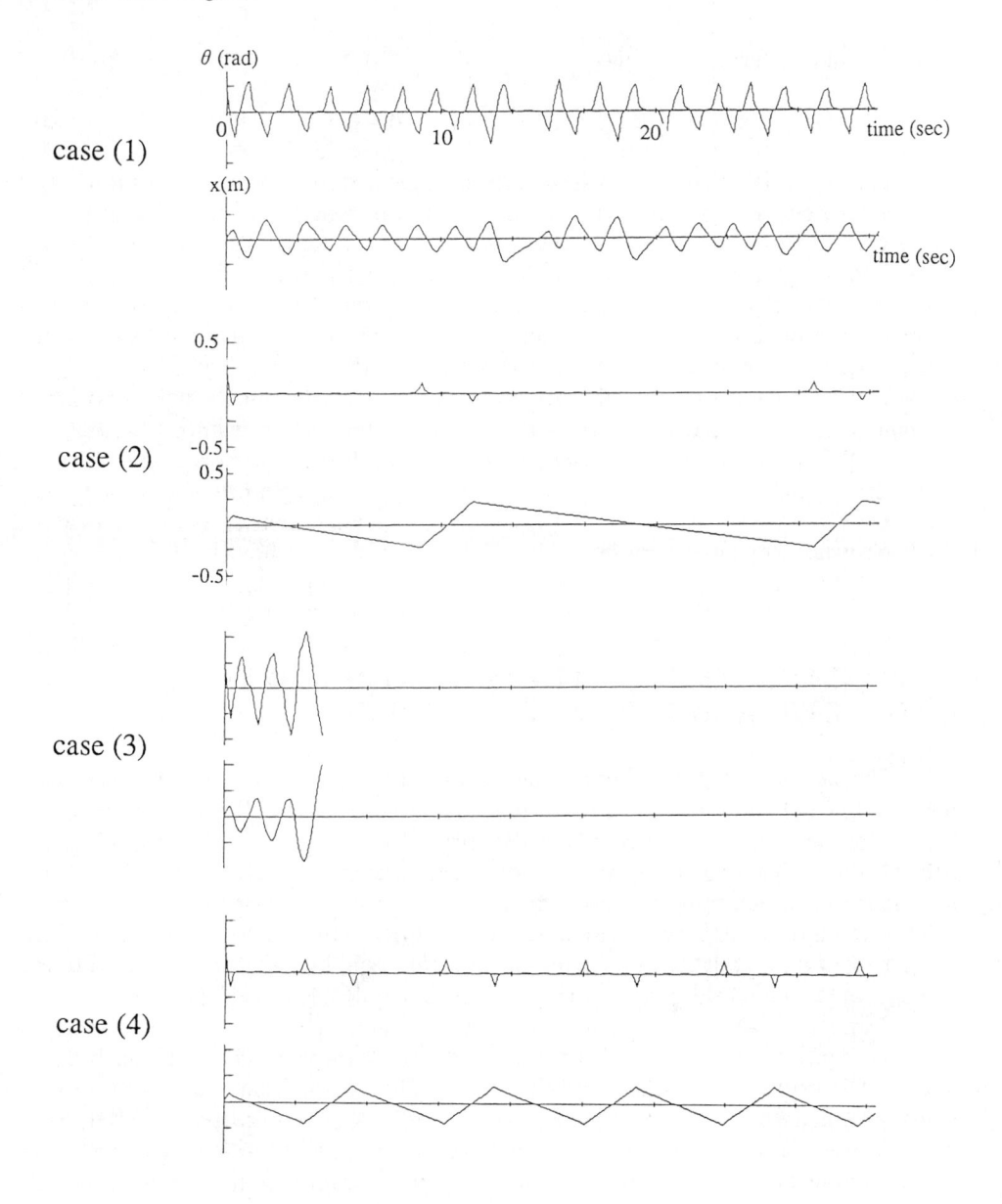

Fig. 3 Typical Behaviors of the Cart-Pole System by Decoupling the Goal of Control

and the set of parameters as follows:

$$M = 0.2 \text{ (kg)}, \quad m = 0.1 \text{ (kg)}, \quad L = 1.0 \text{ (m)}, \quad d = 0.5 \text{ (m)} \tag{3}$$

For case (1), the control subsystem for the pole and that for the cart respectively insist on their own goal-seeking activities, hence the compromise between them results in their swinging motions from left to right as shown in Fig. 3. In this case, the behavior of the system is rather stable; it is seldom that the cart runs out of the rail or the pole falls down unless the initial condition is too severely set. In case (2), the control subsystem for the pole insists on its own goal-seeking activity and the cart accommodates or adapts itself to the pole's behavior, hence the pole is held vertically and the cart swings smoothly right and left by using the full range of the rail. In case (3), on the contrary, the pole subsystem accommodates itself to the activity of the cart subsystem which results in so tight a condition that it is impossible to prevent the pole from falling down. Hence both of them move rather rapidly at first but are stopped suddenly by the falling down of the pole in the early stage. In case (4), both of the subsystems accommodate themselves to each other's behavior, hence they move very smoothly. The range of the cart movement is a bit small compared to that of case (3).

4. CONSTRUCTION OF FUZZY CONTROL SYSTEMS BY GENETIC ALGORITHM TECHNIQUES

The modularized structure of the constraint-interval fuzzy control provides us with a novel and flexible way of constructing a production system in a self-organizing manner. This is done by organizing the componental crisp constraint-interval rules to yield fuzzy control rules. One promising way of doing this is to use *Genetic Algorithm (GA)* techniques, in which trial and error experiences are used to refine the population of solution candidates which are coded into sequences of symbols called *chromosomes*. The refinement of the population is done through the process in Fig. 4 which is derived from analogy to the evolutionary genetic processes in creatures [10], [11], [13], [14].

Namely, in GA, each candidate for the optimal solution is coded into a sequence of symbols, e.g. binary coding, and these chromosomes are evaluated to calculate their *degree of fitness* to the environment (the problem). These chromosomes are then *selected* and *reproduced* by referring to their fitness values. A *crossover* between selected chromosomes subsequently takes place yielding the chromosomes, some of which are expected to be better than the original ones. Finally, *mutation* on the chromosomes is done to yield novel chromosomes. This operation is done by applying random numbers to each symbol in the chromosomes. The whole process of this evolution can be regarded as a multi-point search for the optimal solution. The crossover operation shifts the points of search in a global fashion, while the mutation operation does so in a local fashion.

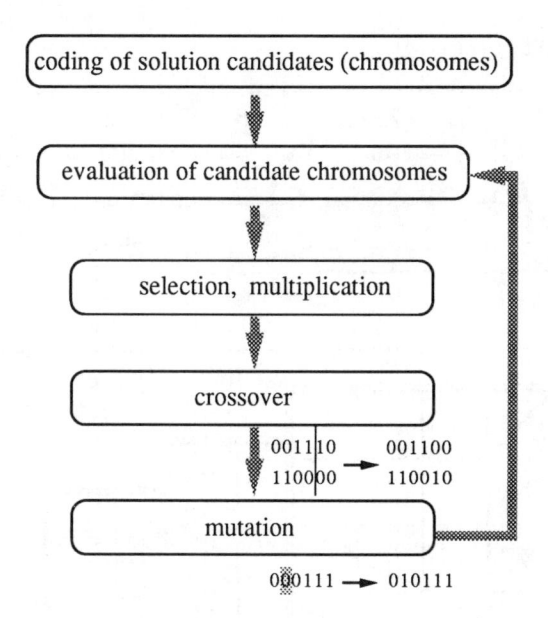

```
┌─────────────────────────────────────────┐
│ coding of solution candidates (chromosomes) │
└─────────────────────────────────────────┘
                    ↓
┌─────────────────────────────────────────┐
│ evaluation of candidate chromosomes     │←──┐
└─────────────────────────────────────────┘   │
                    ↓                          │
┌─────────────────────────────────────────┐   │
│        selection, multiplication         │   │
└─────────────────────────────────────────┘   │
                    ↓                          │
┌─────────────────────────────────────────┐   │
│               crossover                  │   │
└─────────────────────────────────────────┘   │
              0011|10     001100              │
              1100|00  →  110010              │
                    ↓                          │
┌─────────────────────────────────────────┐   │
│               mutation                   │───┘
└─────────────────────────────────────────┘
              000111  →  010111
```

Fig. 4 Evolutionary Process in Genetic Algorithm

In this study, we will regard each componential crisp constraint-interval rule in fuzzy rules as a chromosome; that is we will adopt the *Michigan approach* instead of the *Pittsburgh approach* for constructing production systems by GA [15], [16]. In the former approach, each componential production rule is coded as a chromosome, and in the latter approach, the whole production system itself is coded as a chromosome, thus the size of the chromosomes becomes huge compared to that in the former approach. The evolutional operations, on the contrary, becomes complex in the latter approach compared to the former one. The main reason for adopting the former approach, the Michigan approach, is that we are searching for a method that will yield *self-organizing* mechanisms in the constraint-interval fuzzy inference systems. Self-organization here is carried out by linking componential constraint-interval rules among different levels to yield a constraint-interval fuzzy rule as shown in Fig. 5.

In the Michigan approach, we have to evaluate the contribution of each fragmental componential rule making up the evolutional operations in the Michigan approach more difficult to be carried out than that of the Pittsburgh approach. Particularly, for the case of production systems for control, the evaluation on the whole control actions is usually done after a sequence of actions is applied, hence very complicated evaluation algorithms such as Bucket Brigade Algorithm [17] are used.

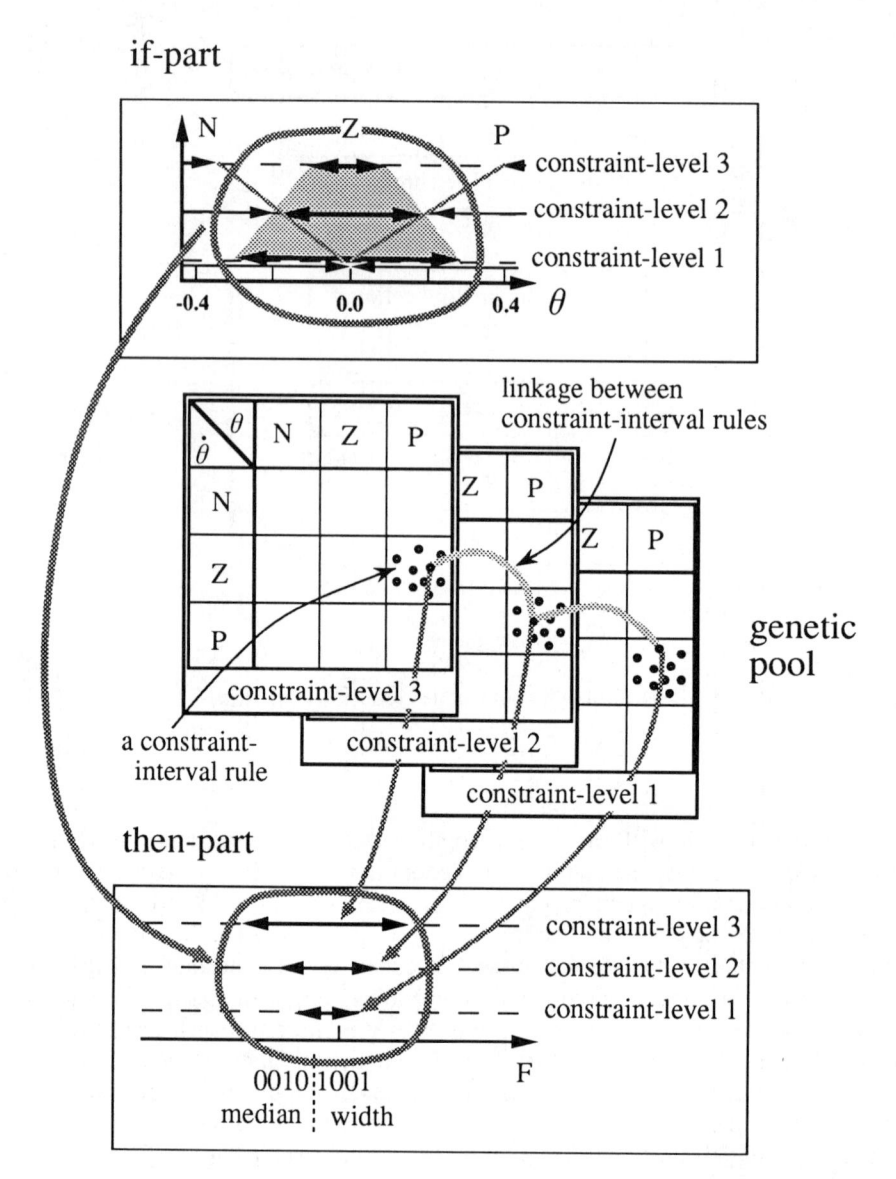

Fig. 5 Organization of a Fuzzy Control Rule by Linking Componential Constraint-interval Fuzzy Inference Rule

5. CONSTRUCTION OF CONTROL SCHEME FOR CART-POLE SYSTEMS BY GENETIC ALGORITHMS

In order to reduce the above mentioned difficulty in the evolutionary process of the Michigan approach, we first disregard the boundedness of the rail. Also, we set in advance the if-parts of the fuzzy rules, i.e., the constraint-intervals in the if-parts, and fix them as constant throughout the evolutionary process. Thus, the learning is done by searching for the best then-part intervals for each if-part interval. Namely, each chromosome corresponds to a then-part interval; it is coded into an eight bit binary sequence, the first half of which stands for its median and the latter half the width. Moreover, instead of using Bucket Brigade Algorithm, we make instantaneous evaluation of each control action. This is done by confining the firing rules at each time and also by referring to the ideal state of the system; the evaluation is done by calculating the amount of increase or decrease of the difference between the ideal and the current states. Hence, no bucket brigade-type credit assignment mechanisms are necessary. According to this evaluation, the selective reproduction operation is done, and also crossover and mutation operations are carried out, yielding better chromosomes. This evolutional operation is done per 100 times of the control actions. If the degree of the pole from the vertical axis is greater than 1rad, the pole is regarded as falling down, and another trial will be started by setting the randomly selected initial condition from the following region:

$$- 0.4 \leq \theta \leq 0.4, \quad - 0.4 \leq \dot{\theta} \leq 0.4. \tag{4}$$

Also, if the pole is within the region of $-1\text{rad} \leq \theta \leq 1\text{rad}$ during 2000 control actions, the trial is terminated and a new trial with randomly selected initial condition starts. The parameters of the evolution process are set as follows:

$$\text{Pcross} = 0.6, \ \text{Pmutation} = 0.01, \ \text{No. of chromosomes} = 90. \tag{5}$$

Fig. 6 shows the typical behavior of the pole during the learning process. In the early stage, it moves rather randomly and then suddenly stops by overranging the condition: $-1\text{rad} \leq \theta \leq 1\text{rad}$. In the intermediate stage, it moves more smoothly but still after a while it will fall down. In the final stage, it is held almost vertically and is expected to behave quite stably.

Fig. 7 shows the resultant constraint-intervals of the rule corresponding to the case NN (θ: Negative, $\dot{\theta}$: Negative). In this case, a "basin-like" then-part is obtained for "mountain-like" if-parts such as in *confining* rules. It is observed that in the final configuration of the *genetic (chromosome) pool*, there still remains a large amount of *diversity* of chromosomes for some sort of rules (if-parts). This means that the trial and error experience is still insufficient for converging on the optimal solution (control rules). It should be noted, however, that this diversity may be beneficial, i.e., it might be the origin of the flexibility of this learning fuzzy control scheme; the system gradually alters

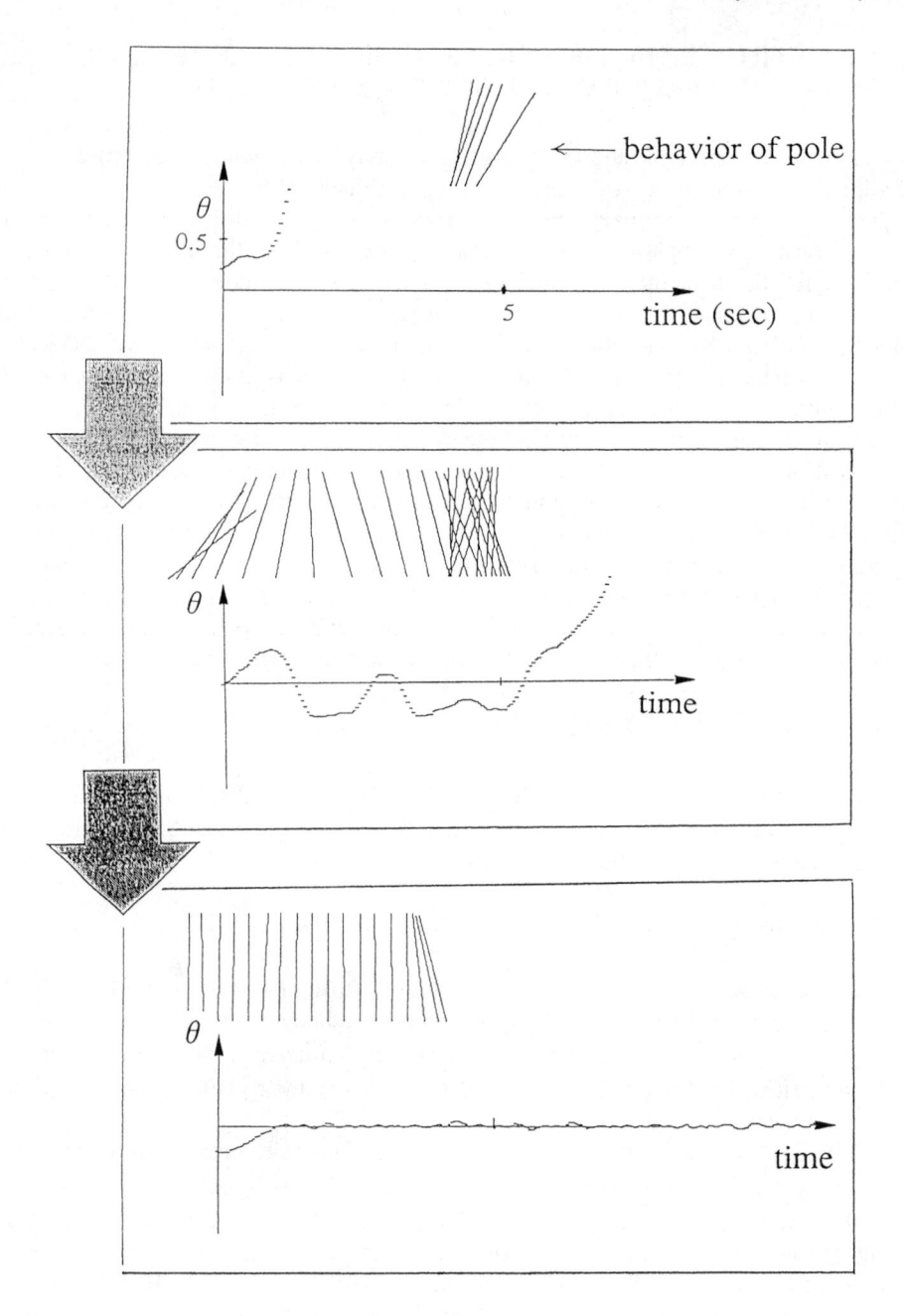

Fig. 6 Learning Behavior of the Cart-pole System by GA

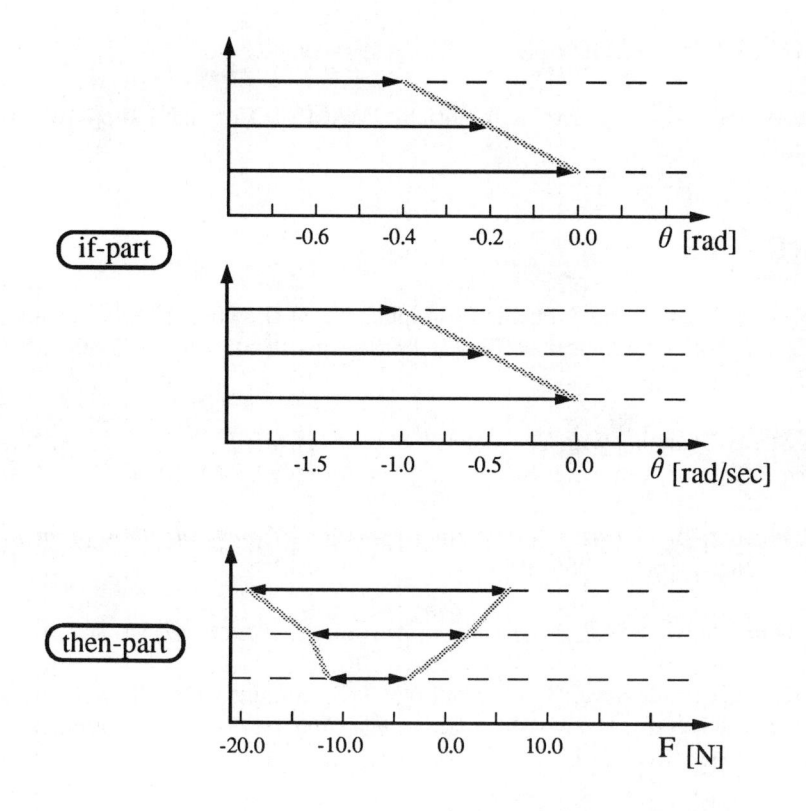

Fig. 7 The Resultant Fuzzy Rule for NN

the distribution of fitness on this divergent chromosome pool, thus shifting its control action in accordance with the changes in the environment such as the change in the values of the parameters of the poles and the carts.

6. CONCLUSIONS

We have shown that a novel way of treating fuzziness from *constraint-oriented* perspectives permits us to construct fuzzy control systems in a *decentralized* and *self-organizing* manner that is expected to reduce the complexity of problem solving and learning for the construction of fuzzy control systems. This new approach was then applied to construct control schemes of cart-pole systems through which we have shown the practical effectiveness of our approach.

ACKNOWLEDGEMENTS

The authors are grateful to Mr. Kiminori SHIMAMOTO of our Laboratory for his collaboration.

REFERENCES

[1] D. L. Waltz, "Generating Semantic Descriptions from Drawings of Scenes with Shadows," *Technical Report AI-TR-271*, Massachusettes Institute of Technology, 1972

[2] M. J. Stefic, "Planning with Constraints," *Stanford Heuristic Programming Project Memo HPP-80-2,* Computer Science Dept. *Report No. STAN-CS-80-784*, 1980

[3] E. C. Freuder, "Synthesizing Constraint Expression," *Communications of the ACM*, Vol.21, No.11, pp.958-966, 1978

[4] H. W. Gusgen, *CONSAT: A System for Constraint Satisfaction*, Pitman, 1989

[5] A. Aiba and R. Hasegawa, "Constraint Logic Programming - CAL, GDCC and Their Constraint Solvers -," *Proc. of the International Conf. on the Fifth Generation Computer Systems*, Vol.1, pp.113-131, 1992

[6] O. Katai, M. Ida, T. Sawaragi and S. Iwai, "Treatment of Fuzzy Concepts by Order Relations and Constraint-Oriented Fuzzy Inference," *Proc. of NAFIPS'90*, pp.300-303, 1990

[7] O. Katai, M. Ida, T. Sawaragi and S. Iwai, "Dynamic and Context-Dependent Treatment of Fuzziness from Constraint-Oriented Perspectives," *Proc. of IFSA'91*, Vol. on Artficial Intelligence, pp.101-104, 1991

[8] O. Katai, M. Ida, T. Sawaragi and S. Iwai, "Formalization of Fuzzy Set Concept and Fuzzy Inference Based on Order-Constraint Structures," *Trans. of the Society and Instrument and Control Engineers*, Vol.28, No.1, pp.30-39, 1992 (in Japanese)

[9] O. Katai, M. Ida, T. Sawaragi and S. Iwai, "Control of Inverted Pendulum and Vehicle Path Planning by Constraint-Interval Fuzzy Inference," *Trans. of the Society and Instrument and Control Engineers*, Vol.29, No.4, 1993 (in Japanese)

[10] D. E. Goldberg, *Genetic Algorithms in Search, Optimization, and Machine Learning*, Addison-Wesley, 1989

[11] C. L. Karr, "Genetic Algorithms for Fuzzy Controllers," *AI Expert*, Feb. pp.26-33, 1991

[12] O. Katai, M. Ida, T. Sawaragi and S. Iwai, "A Self-Organizing Adaptable Control Scheme Based on Constraint-Oriented Fuzzy Inference," *Proc. of the IMACS/SICE International Symposium on Robotics, Mechatronics and Manufacturing System '92*," Vol.1, pp 161-166, 1992

[13] C. L. Karr, "Design of an Adaptive Fuzzy Logic Controller Using a Genetic Algorithm," *Proc. of the 4th International Conference on Genetic Algorithm*, pp.450-457, 1991

[14] P. Thrift, "Fuzzy Logic Synthesis with Genetic Algorithms," *Proc. of the 4th International Conference on Genetic Algorithm*, pp.509-513, 1991

[15] J. H. Holland and J. S. Reitman, "Cognitive Systems Based on Adaptive Algorithms in D. A. Waterman and F. Hayes-Roth (eds.), *Pattern-Directed Inference Systems*, pp.313-329, Academic Press, 1978

[16] L. B. Booker, D. E. Goldberg and J. H. Holland, "Classifier Systems and Genetic Algorithms," *Technical Report No.8*, Cognitive Science and Machine Intelligence Laboratory, University of Michigan, 1987

[17] J. H. Holland, "Escaping Brittleness: The Possibilities of General-Purpose Learning Algorithms Applied to Parallel Rule-Based Systems," in R. S. Michalski, J. G. Carbonell and T. M. Mitchell (eds.), *Machine Learning*, Vol.2, pp.593-623, Morgan Kaufmann, 1986

Chapter 9
A Self Generating and Tuning Method for Fuzzy Modeling using Interior Penalty Method and its Application to Knowledge Acquisition of Fuzzy Controller

Introduction, *198*
Self Tuning Method using Interior Penalty Method, *199*
Hybrid Algorithm for Rule Generation and Parameter Tuning, *207*
Application to Knowledge Acquisition of Fuzzy Controller, *218*
Conclusion, *222*
References, *223*

This chapter proposes a new method for fuzzy modeling. It combines a self-tuning method for the fuzzy model using the interior penalty method, and a self-generating algorithm for new fuzzy rules which is invoked when the effect of parameter tuning is diminished. Given the number of fuzzy rules and the fuzzy partition, the self-tuning problem is formulated considering the constraints on the shape parameters of the membership functions. Then, the self-tuning method is proposed using the interior penalty method and gradient methods. Following that, a hybrid algorithm is proposed which consists of the self-tuning method and a procedure to generate new fuzzy rules in the regions with maximum inference error. Numerical examples demonstrate that the new hybrid algorithm can achieve specified model accuracy with less number of fuzzy rules than other tuning methods proposed in the literature. The hybrid algorithm is also applied to the knowledge acquisition and refinement of the fuzzy controller. It is shown that the algorithm is effective in reducing the number of rules for the feedback fuzzy controller, which are obtained from a skilled human operator or by other self-tuning methods.

A SELF GENERATING AND TUNING METHOD FOR FUZZY MODELING USING INTERIOR PENALTY METHOD AND ITS APPLICATION TO KNOWLEDGE ACQUISITION OF FUZZY CONTROLLER

Ryu Katayama, Yuji Kajitani, and Yukiteru Nishida
Information & Communication Systems Research Center, Sanyo Electric Co., Ltd.
1-18-13, Hashiridani, Hirakata, Osaka 573, Japan
E-mail: katayama@rd.sanyo.co.jp

1. INTRODUCTION

In recent years, intelligent consumer electronic products or industrial systems incorporating fuzzy logic have been widely and intensively developed. Furthermore, as the systems we must design become more complicated, the need for automatic knowledge acquisition or self-tuning methods for fuzzy rules grows more and more important. So far, various methods of fuzzy modeling such as the complex method [1,2], neural network [3], and the gradient method [4,5] have been proposed. Among them, the gradient method is considered to be most efficient. Fuzzy modeling for simplified fuzzy reasoning by the gradient method is proposed by Ichihashi[4], where only the conclusion parts (singletons) of the fuzzy rules are learned. Nomura et al. applied the steepest descent method to determine both shape parameters of the membership functions and the conclusion parts of the fuzzy rules [5]. Maeda et al. proposed the self-tuning algorithm for fuzzy membership functions using computational flow network [14], which essentially belongs to the gradient type methods. Moreover, a fuzzy modeling algorithm using fuzzy neural networks is reported by Horikawa et al. [15], where the membership functions are composed by the plural sigmoid functions. All of these fuzzy modeling methods [5,14,15] are regarded as gradient type algorithm for tuning both the membership functions of the premise parts and the conclusion parts of the fuzzy rules.

But the problem inherent in all of these methods [5,14,15] is that the constraints on the shape parameters of the membership functions are not considered explicitly. Therefore, this formulation has the following disadvantages during the tuning process

or when optimal parameters are obtained:
(1) There is a possibility that there exist regions in the input space where membership functions are not defined.
(2) There is a possibility that the order of the membership functions defined as initial knowledge base is not preserved.

In order to solve these problems, we propose an efficient fuzzy modeling method by gradient methods considering constraints on the membership functions [9,11]. In our formulation, inequalities on shape parameters of the membership functions (the vertex value of triangles) are treated explicitly so that the following requirements are satisfied during the tuning process or after an optimal solution is obtained:
(1) Membership functions are always guaranteed to be defined elsewhere in the input space.
(2) The order of vertex values of triangles defined as initial membership functions is always preserved.

In our formulation, the original constrained optimization problem is transformed to an unconstrained optimization problem using interior penalty method [6,7], and we apply gradient methods such as steepest descent method or conjugate gradient method to the obtained unconstrained problem. The explicit expression of the gradient of the augmented objective function and learning algorithm are derived.

Furthermore, we propose a hybrid algorithm [11] which consists of above self-tuning method and a self-generating algorithm [10] for new fuzzy rules in the regions with the maximum inference error. This rule generation procedure is invoked when the effect of parameter tuning is diminished. Numerical examples are also given which demonstrate that the new hybrid algorithm can achieve the specified model accuracy with less number of fuzzy rules than similar tuning methods [4,10] proposed so far.

The proposed hybrid algorithm is also applied to the knowledge acquisition and the refinement of the fuzzy controller. It is shown that the proposed algorithm is effective in reducing the number of the rules for feedback fuzzy controller, which are obtained by the skilled human operator or other self-tuning methods such as the adaptive learning gain algorithm with the hierarchical fuzzy inference mechanisms [12,13] for fuzzy control systems.

2. SELF TUNING METHOD USING INTERIOR PENALTY METHOD

2.1 Problem Formulation

Let $\underline{x} = (x_1, \cdots, x_i, \cdots, x_m) \in R^m$ be inputs and $\underline{y} = (y_1, \cdots, y_t, \cdots, y_v) \in R^v$ be outputs. The problem is to identify the nonlinear function $\underline{y} = \underline{f}(\underline{x}): R^m \rightarrow R^v$. Let input x_i be normalized such that

$$x_i \in X_i = \{x_i \mid a_i^0 \leq x_i \leq a_i^{Li+1}\}, \quad i=1,\cdots,m \qquad (1)$$

are satisfied, where a_i^0, $a_i^{L_i+1}$, $i=1,\cdots,m$, are given constants. Also, if $v \geq 2$, output y_t, $t=1,\cdots,v$, are assumed to be normalized such that

$$y_t \in Y_t = \{y_t \mid -1 \leq y_t \leq 1\}, \quad t=1,\cdots,v \qquad (2)$$

Select L_i points in the interval X_i satisfying (L_i+1) inequalities (3):

$$a_i^0 \leq a_i^1 \leq a_i^2 \leq \cdots \leq a_i^{j(i)} \leq \cdots \leq a_i^{L_i} \leq a_i^{L_i+1}, \quad i=1,\cdots,m \qquad (3)$$

The membership functions $A_i^{j(i)}(x_i)$, $j(i)=0,\cdots,L_i+1$ are defined by (4), where the form is expressed as triangle as shown in Fig.1, and whose vertex is $a_i^{j(i)}$ and width is ($a_i^{j(i)+1} - a_i^{j(i)}$), which intersects with adjacent ones with their grade value = 0.5.

$$A_i^{j(i)}(x_i) = \begin{cases} 1- (a_i^{j(i)} - x_i)/(a_i^{j(i)} - a_i^{j(i)-1}), & a_i^{j(i)-1} \leq x_i \leq a_i^{j(i)} \\ 1- (x_i - a_i^{j(i)})/(a_i^{j(i)+1} - a_i^{j(i)}), & a_i^{j(i)} \leq x_i \leq a_i^{j(i)+1} \\ 0, & x_i \leq a_i^{j(i)-1}, a_i^{j(i)+1} \leq x_i \end{cases} \qquad (4)$$

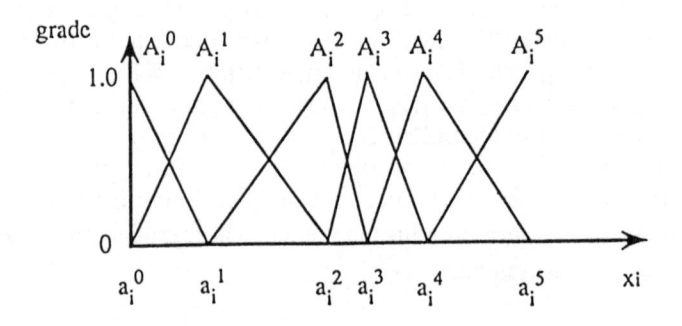

Fig.1 Membership Functions of Antecedent Part ($L_i=4$)

Then, $n \equiv \prod_{i=1}^{m} (L_i+2)$ rules for simplified fuzzy reasoning are given by (5):

Rule s:
  If x_1 is $A_1^{j(1)}(x_1)$ and $\cdots$ and x_i is $A_i^{j(i)}(x_i)$ $\cdots$ and x_m is $A_m^{j(m)}(x_m)$
  Then y_1 is $w_1^s, \cdots, y_t$ is $w_t^s, \cdots, y_v$ is w_v^s,

$$0 \leq j(i) \leq L_i+1, \quad i=1,\cdots,m, \quad s=1,\cdots,n \qquad (5)$$

Let μ_s be the membership value of antecedent part of rule s. Then, the output of the simplified fuzzy reasoning y_t, t=1,···,v, are computed by (6) and (7),

$$\mu_s = A_1^{j(1)}(x_1) \cdot A_2^{j(2)}(x_2) \cdot \cdots \cdot A_m^{j(m)}(x_m) \tag{6}$$

$$y_t = \sum_{s=1}^{n} \mu_s w_t^s / \sum_{s=1}^{n} \mu_s = \sum_{s=1}^{n} \mu_s w_t^s \qquad t=1,\cdots,v \tag{7}$$

since $\sum_{s=1}^{n} \mu_s = 1$ always holds.

Assume $\underline{a}_i \equiv (a_i^1,\cdots,a_i^{L_i}) \in R^{L_i}$, i=1,···,m, and $M \equiv \sum_{i=1}^{m} L_i$, then we introduce the shape parameter vector $\underline{a} \in R^M$ given by (8),

$$\underline{a} \equiv (\underline{a}_1,\underline{a}_2,\cdots,\underline{a}_i,\cdots,\underline{a}_m) \tag{8}$$

Inequalities (3) with respect to vector $\underline{a}$ are simply represented by (12) under the conditions (9)~(11).

$$g_i^{j(i)}(\underline{a}) \equiv a_i^{j(i)} - a_i^{j(i)+1},$$
$$j(i)=0,1,\cdots,L_i, \ i=1,\cdots,m \tag{9}$$

$$\underline{g}_i \equiv (g_i^0, g_i^1, \cdots, g_i^{L_i}) \in R^{L_i+1}, \ i=1,\cdots,m \tag{10}$$

$$\underline{g}_a \equiv (\underline{g}_1, \underline{g}_2, \cdots, \underline{g}_i, \cdots, \underline{g}_m) \in R^{M+m} \tag{11}$$

$$\underline{g}_a(\underline{a}) \leq \underline{0} \tag{12}$$

Similarly, we introduce $\underline{w} \in R^{nv}$ defined by (13),

$$\underline{w}_s \equiv (\underline{w}_1^s,\cdots,\underline{w}_v^s) \in R^v, \qquad s=1,\cdots,n$$
$$\underline{w} \equiv (\underline{w}_1, \underline{w}_2, \cdots, \underline{w}_s, \cdots, \underline{w}_n) \in R^{nv} \tag{13}$$

Let us introduce $\underline{z}=(\underline{a},\underline{w}) \in R^{M+nv}$, then (12) is equivalently expressed as (14),

$$\underline{g}(\underline{z}) \equiv \underline{g}_a(\underline{a}) \leq \underline{0}, \ \underline{g} \in R^{M+m} \tag{14}$$

Suppose desirable N input/output data obtained from experts are

$$(\underline{x}^1,\underline{y}^1),\cdots,(\underline{x}^p,\underline{y}^p),\cdots,(\underline{x}^N,\underline{y}^N)$$

and $*\underline{y}^p$ is a output vector of fuzzy reasoning using p-th input $\underline{x}^p$, p = 1,$\cdots$,N. Then, the fuzzy modeling problem is formulated as (15):

$$\min_{\underline{z}=(\underline{a},\underline{w})} E \equiv \frac{1}{2} \sum_{p=1}^{N} \sum_{t=1}^{v} (y_t^p - {}^*y_t^p)^2$$

subj. to

$$(4),(5),(6),(7),(14) \tag{15}$$

2.2 Self Tuning Method Using Interior Penalty Method

Since problem (15) is a constrained optimization problem with respect to $\underline{z}$, we apply an interior penalty method[6]. Let us introduce the following augmented objective function defined on int Z,

$$P(\underline{z};r) \equiv E + r\ \phi\ (g(\underline{z})) \tag{16}$$

where $Z = \{\ \underline{z}\,|\,g(\underline{z}) \leq \underline{0}\}$, $r > 0$, and the interior penalty function ϕ is continuous and defined as

$$\phi\ (g(\underline{z})) > 0 \quad \text{as } z \in \text{int Z} \tag{17a}$$

$$\phi\ (g(\underline{z})) \to \infty \text{ as } z \to \partial Z \text{ (the boundary of Z)} \tag{17b}$$

In order to apply the theory of interior penalty method, we assume the following assumption:

【Assumption 1】 The interior set of Z which is denoted by int $Z = \{\underline{z}|\ g(\underline{z}) < \underline{0}\}$ is not empty and its closure is equal to Z. ■

Let us consider the minimization problem of the augumented objective function

$$\min_{\underline{z}=(\underline{a},\underline{w})} P(\underline{z};r)$$

subj. to (4),(5),(6),(7) \hfill (18)

Let $\{\underline{z}_k\}$ be a sequence of optimal solution of (18) for positive penalty parameter $\{r^k\}$ which strictly decreases and converges to zero. Then, the following theorem for convergence holds [7,8].

【Theorem 1】 Assume that there exists an optimal solution $\underline{z}^*$ for the problem (15) and there exists an optimal solution $\underline{z}_k$ for the problem (18) with any positive penalty parameter r^k. Then, under Assumption 1 and the continuity of E with respect to z, the following properties hold:

(1) $\lim_{k\to\infty} P(\underline{z}_k ; r^k) = E(\underline{z}^*)$

(2) $\lim_{k\to\infty} r^k \cdot \phi(\underline{g}(\underline{z}_k)) = 0$

(3) $\lim_{k\to\infty} E(\underline{z}_k) = E(\underline{z}^*)$ ■

Therefore, we can obtain an optimal solution of the original problem (15) by solving (18) by appropriate gradient methods such as steepest descent method or conjugate gradient method for positive penalty parameter sequence $\{r^k\}$ which strictly decreases and converges to zero. The gradient of the augmented objective function $P(\underline{z}_k ; r^k)$ with respect to z is given by the following Theorem 2.

[Theorem 2] Assume that the penalty function $\phi(\underline{g}(\underline{z}))$ is given by (19)

$$\phi(\underline{g}(\underline{z})) \equiv \sum_{i=1}^{m} \sum_{j(i)=0}^{Li} (-1/g_i^{j(i)}(\underline{a}))$$

$$= \sum_{i=1}^{m} \sum_{j(i)=0}^{Li} (-1/(a_i^{j(i)} - a_i^{j(i)+1})) \quad (19)$$

Then, the gradients of the augmented objective function $P(\underline{z};r)$ with respect to w_j^s, $j=1,\cdots,v$, $s=1,\cdots,n$, and $a_i^{j(i)}$, $1\leq j(i)\leq Li$, $i=1,2,\cdots,m$, are given by (20) and (21) respectively.

$$\partial P(\underline{z};r)/\partial w_t^s = -\sum_{p=1}^{N} \{(y_t^p - {}^*y_t^p) \cdot \mu_s\},$$

$$t=1,\cdots,v, \ s=1,\cdots,n \quad (20)$$

$$\partial P(z;r)/\partial a_i^{j(i)} = -\sum_{p=1}^{N}\sum_{t=1}^{v} \{(y_j^p - {}^*y_j^p)\sum_{\nu=1}^{n}(w_t^\nu \cdot (\partial \mu_\nu/\partial a_i^{j(i)}))\}$$

$$+r\{-1/(a_i^{j(i)-1} - a_i^{j(i)})^2 + 1/(a_i^{j(i)} - a_i^{j(i)+1})^2\},$$

$$1 \leq j(i) \leq L_i, \ i=1,\cdots,m \quad (21)$$

where $\mu_\nu = \prod_{i=1}^{m} A_i^{h(i)}(x_i)$, and $\partial \mu_\nu / \partial a_i^{j(i)}$ of (21) is given by (22a~e) according to the following five cases (a)~(e) respectively.

(a): $h(i)=j(i)-1$ and $a_i^{j(i)-1} \leq x_i^p \leq a_i^{j(i)}$

(b): $h(i)=j(i)$ and $a_i^{j(i)-1} \leq x_i^p \leq a_i^{j(i)}$

(c): $h(i)=j(i)$ and $a_i^{j(i)} \leq x_i^p \leq a_i^{j(i)+1}$

(d): $h(i)=j(i)+1$ and $a_i^{j(i)} \leq x_i^p \leq a_i^{j(i)+1}$

(e): except (a)~(d)

$$\frac{\partial \mu_\nu}{\partial a_i^{j(i)}} = \prod_{\substack{u=1 \\ u \neq i}}^{m} A_u^{h(u)}(x_u^p) \frac{(x_i^p - a_i^{j(i)-1})}{(a_i^{j(i)} - a_i^{j(i)-1})^2} \tag{22a}$$

$$\frac{\partial \mu_\nu}{\partial a_i^{j(i)}} = \frac{-1}{a_i^{j(i)} - a_i^{j(i)-1}} \mu_\nu \tag{22b}$$

$$\frac{\partial \mu_\nu}{\partial a_i^{j(i)}} = \frac{-1}{a_i^{j(i)+1} - a_i^{j(i)}} \mu_\nu \tag{22c}$$

$$\frac{\partial \mu_\nu}{\partial a_i^{j(i)}} = - \prod_{\substack{u=1 \\ u \neq i}}^{m} A_u^{h(u)}(x_u^p) \frac{a_i^{j(i)+1} - x_i^p}{(a_i^{j(i)+1} - a_i^{j(i)})^2} \tag{22d}$$

$$\frac{\partial \mu_\nu}{\partial a_i^{j(i)}} = 0 \tag{22e}$$

∎

By use of Theorem 2, the learning algorithm for w_t^s and $a_i^{j(i)}$ is given by (23) and (24), where q denotes iteration number for the problem (18), and α is an optimal step size at the q-th iteration which is obtained by solving a linear search problem.

$$w_t^s(q+1) = w_t^s(q) - \alpha \cdot \partial P(\underline{z}^q; r)/\partial w_t^s,$$
$$t=1,\cdots,v \quad s=1,\cdots,n \qquad (23)$$

$$a_i^{j(i)}(q+1) = a_i^{j(i)}(q) - \alpha \cdot \partial P(\underline{z}^q; r)/\partial a_i^{j(i)},$$
$$1 \leq j(i) \leq L_i, \quad i=1,\cdots,m \qquad (24)$$

2.3 Numerical Example

[Example 1] In order to show the effectiveness of considering the constraints on the shape parameters of the membership functions, let us consider the following function approximation problem for a fixed number of fuzzy rules:

$$y = \begin{cases} 0.1/x^2, & (0.1 \leq x \leq 1.0) \\ 10.0 \sin(2\pi x), & (1.0 < x \leq 1.5) \end{cases} \qquad (25)$$

The total number of learning data (x,y) is N=29, where input x is given from 0.1 to 1.5.

Fig.2 represents the desired input/output relation given by (25). We set 7 rules and membership functions which are equally allocated in the input region as shown in Fig.3a.

Stop condition is given as

$$\varepsilon_2 = |P(\underline{z}^k, r^k) - P(\underline{z}^{k-1}, r^{k-1})| < \delta_2 = 10^{-3} \qquad (26)$$

and we use penalty parameter updating coefficient $C = 0.5$.

By applying the penalty method, the stop condition (26) is satisfied when k=21, $r^{21}=5.0\times 10^{-6}$, and we have $P(\underline{z}^{21}, r^{21}) = 12.71$. The learning curve is shown in Fig.4 and the membership functions after tuning are shown in Fig.3b. Thus, the order of vertex values of each membership function is preserved.

On the other hand, we solve the same function approximation problem (25) with the same conditions except that without considering the constraints on the shape parameters of the membership functions.

Fig.5 represents the shape of the acquired membership functions. This result indicates that the order of vertex a_1^4 and a_1^5 is reversed, since the constraints (3) are not satisified, and this yields the illegal shape of the membership functions as shown in Fig.5. Thus, the explicit treatment of constraints on the shape parameters of membership functions is inevitable in self-tuning methods for fuzzy rules.

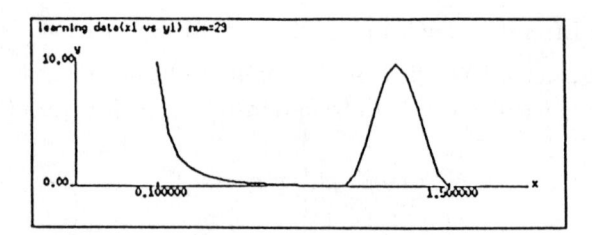

Fig.2 Desired Input/Output Relation

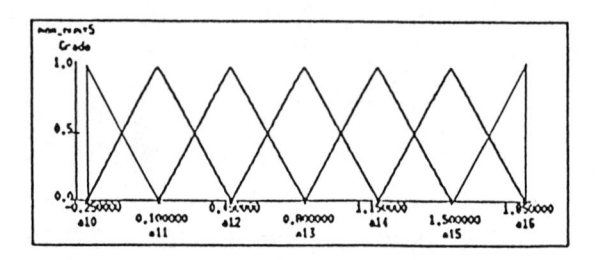

Fig.3a Initial Membership Functions

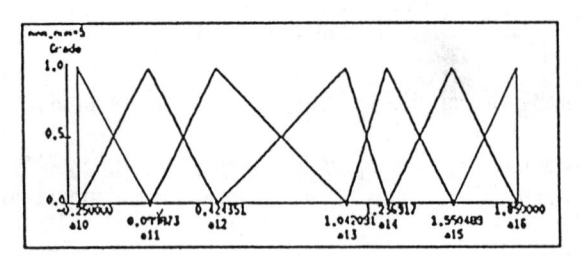

Fig.3b Membership Functions Obtained by Penalty Method

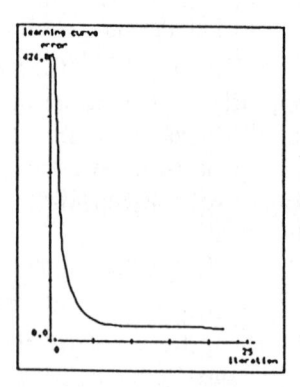

Fig.4 Learning Curve by Penalty Method

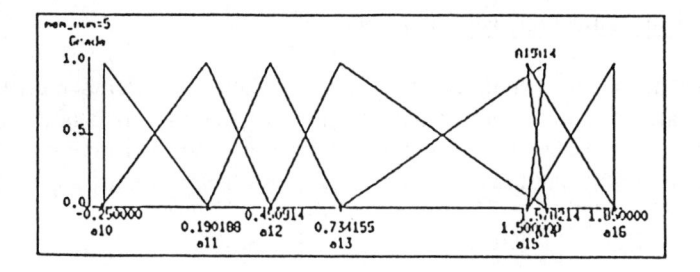

Fig.5 Membership Functions Obtained Without Considering Constraints

3. HYBRID ALGORITHM FOR RULE GENERATION AND PARAMETER TUNING

In this section, we propose a new hybrid algorithm for automatic fuzzy modeling which is composed of the parameter tuning method described in Section 2 and the rule generation algorithm proposed by Araki et al.[10]. The entire procedure of the hybrid algorithm consists of following three sub-procedures.

(1) Determination of the rule generation region.
(2) Generation of new membership functions and fuzzy rules.
(3) Parameter tuning for vertex values of membership function and conclusion parts (singletons) of fuzzy rules.

Sub-procedures (1) and (2) are almost the same as those which is proposed in Ref.[10], but are extended to the multiple output model in our formulation. Therefore, we describe the extended rule generation method in 3.1. The parameter tuning method presented in Section 2 is used for sub-procedure.

3.1 Rule Generation Method

Araki et al.[10] proposed the self rule generation method for single output fuzzy model, by which input space is iteratively fuzzy partitioned by the ratio of 1/2 in the sub regions where the value of output inference error takes the maximum. Therefore, we hereafter call this method as $1/2^n$ Fixed Grid Method. We summarize their method by extending the algorithm to the multiple output fuzzy model. The method consists of two major procedures: determination of the rule generation region, and generation of new membership functions and fuzzy rules.

(1) Determination of the rule generation region

Rules are newly generated in the region where the inference error takes the maximal value among errors computed in every region which is divided by the neighboring two membership functions of the premise part. For multiple output fuzzy model, we define the output inference error $e(D_h)$ for the sub-region D_h by (27),

$$e(D_h) = \frac{1}{N_h} \max_t \{ \sum_{\underline{x}^p \in D_h} |y_t^p - {}^*y_t^p| \} \qquad (27)$$

where N_h is a number of data $(\underline{x}^p, \underline{y}^p)$ which exist in the sub-region D_h.

An example of the region is shown in Fig.6. In this case, three membership functions are defined for both input x_1 and x_2. Therefore, the total number of fuzzy rules is nine, the grids denoted by ○ in Fig.6. Inference errors defined by (27) are computed in the four regions such as D_{10}, D_{11} for x_1, and D_{20}, D_{21} for x_2. In this example, let us assume that the selected region where inference error takes the maximal value is D_{21}.

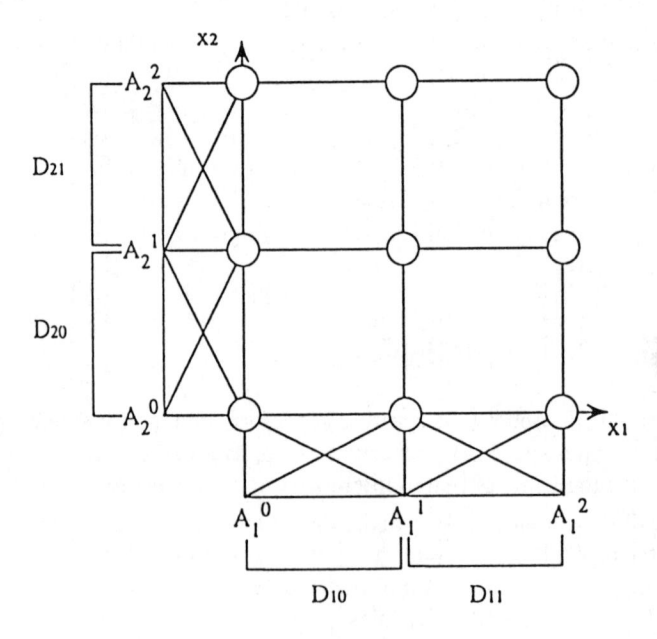

Fig.6 Example of Rule Generating Regions

(2) Generation of new membership functions and fuzzy rules

A membership function is generated dividing the region which is selected by the above procedure into two equal regions by its center vertex value. Then, the number of membership function j(i) is renumbered in the smallest order. Fig. 7 illustrates an example of the generation of the membership function in the region D_{21} shown in Fig.6. In this example, three rules denoted by ● are newly generated.

After generating the membership function, the real numbers of conclusion parts of the newly generated rules are defined by the output values of fuzzy inference by inputting the center values of the premise part of the generated rules using the old fuzzy rules.

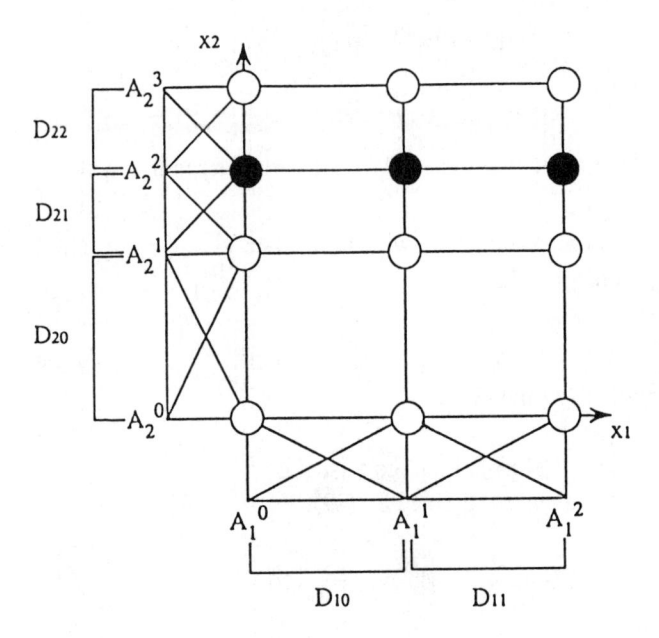

Fig.7 Example of Generated Rules

3.2 Hybrid Algorithm for Rule Generation and Parameter Tuning

We now describe the entire hybrid algorithm in detail.

【Hybrid Algorithm】
【Step 1】
Set the value of updating coefficient C < 1.0 for the penalty parameter r^k, stop criteria value δ_1 for tuning parameters with fixed penalty parameter, stop criteria value

δ_2 for tuning with the fixed number of fuzzy rules, and desired accuracy value δ_3 of the fuzzy model.

【Step 2】

Set the initial fuzzy rules with at least two membership functions of the antecedent part for every inputs.

【Step 3】

Read the N input/output data $(\underline{x}^1, \underline{y}^1), \cdots, (\underline{x}^p, \underline{y}^p), \cdots, (\underline{x}^N, \underline{y}^N)$.

【Step 4】

Set iteration number k = 0 for updating penalty parameter r^k. Compute initial penalty parameter value r^0 by (28), where $\underline{z}_k^{q(k)} = \underline{z}_0^0$ is initial parameter values.

$$r^0 = E(\underline{z}_0^0) / \phi(g(\underline{z}_0^0)) \qquad (28)$$

【Step 5】

Set the iteration number q(k) = 0 for the parameter tuning with fixed r^k.

【Step 6】

Compute output vector $*\underline{y}^p$ with input data $\underline{x}^p$, p=1,$\cdots$,N, with the fuzzy inference equations (4)~(7).

【Step 7】

Update w_t^s, t=1,$\cdots$,v, s=1,$\cdots$,n, $a_i^{j(i)}$, $1 \leq j(i) \leq L_i$, i=1,$\cdots$,m, using parameter learning equations (23) and (24).

【Step 8】

If q(k)=0, let q(k)=q(k)+1 and go to step 6.

If q(k)>0, compute ε_1 defined by (29),

$$\varepsilon_1 = |P(\underline{z}_k^{q(k)}, r^k) - P(\underline{z}_k^{q(k)-1}, r^k)| \qquad (29)$$

【Step 9】

If $\varepsilon_1 < \delta_1$, go to step 10.

Otherwise, let q(k)=q(k)+1 and go to step 6.

【Step 10】

If k=0, let k = k+1, $r^k = r^{k-1} \cdot C$, and go back to step 5.

If k>0, compute ε_2 defined by (30).

$$\varepsilon_2 = |P(\underline{z}_k^{q(k)}, r^k) - P(\underline{z}_{k-1}^{q(k-1)}, r^{k-1})| \qquad (30)$$

【Step 11】

If $\varepsilon_2 < \delta_2$, go to step 12.

Otherwise, let k=k+1, $r^k = r^{k-1} \cdot C$, and go back to step 5.

【Step 12】
Compute inference error ε_3 by (31),

$$\varepsilon_3 = \frac{1}{N} \sum_{p=1}^{N} \| \underline{y}^p - {}^*\underline{y}^p \|^P \tag{31}$$

where $\|\underline{y}\|^P$ is l_p-norm of vector $\underline{y} \in R^v$, which is defined as

$$\| y \|^P = \left(\sum_{t=1}^{v} |y_t|^P \right)^{1/P} \tag{32}$$

Usually we use P=1,or 2.
If $\varepsilon_3 < \delta_3$, finish the entire identification procedure.
Otherwise, go to step 13.

【Step 13】
Compute the inference errors for every region partitioned by the neighboring two membership functions. The inference error is calculated using max-norm of output error vector. Determine the region where new rules are to be generated as described in Section 3.1.

【Step 14】
Compute the conclusion parts of the generated new rules by the un-updated rules. Go back to step 4. ∎

We show the computational flow chart of the hybrid algorithm in Fig.8.

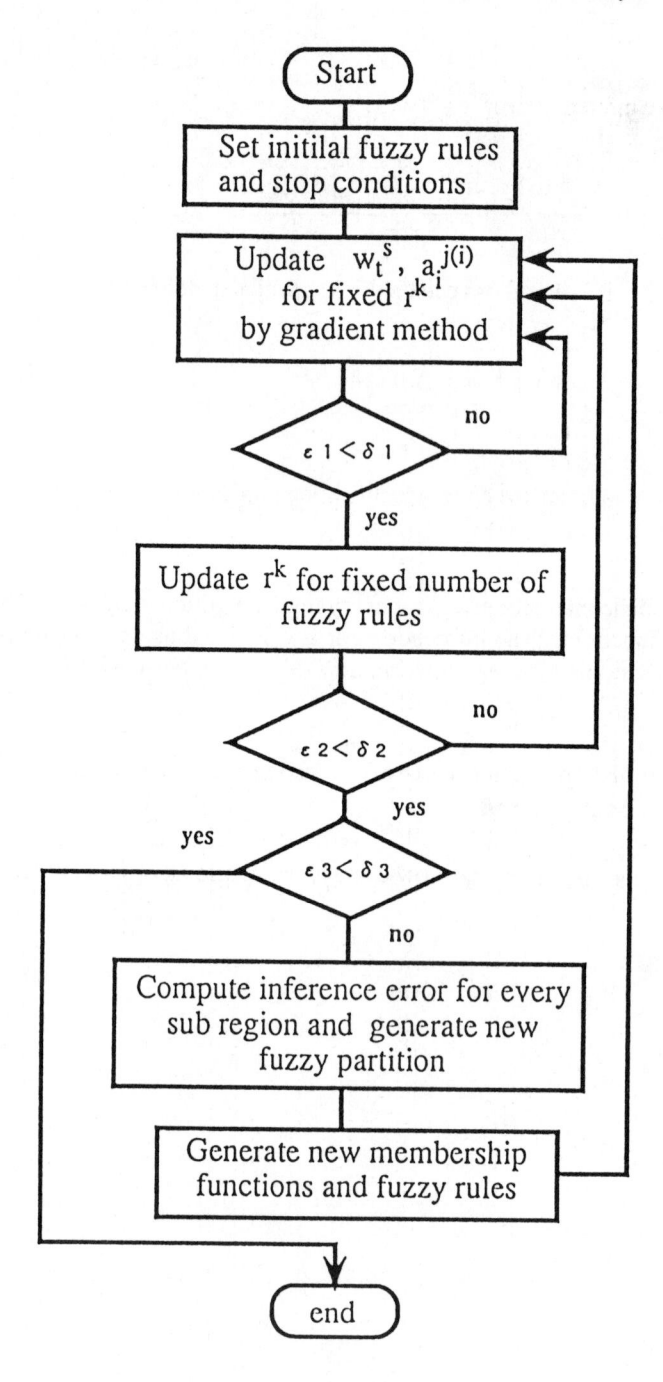

```
                    Start
                      │
                      ▼
        ┌─────────────────────────┐
        │ Set initilal fuzzy rules│
        │   and stop conditions   │
        └─────────────────────────┘
                      │
                      ▼
        ┌─────────────────────────┐
        │  Update  $w_t^s$, $a_i^{j(i)}$  │◄────┐
        │    for fixed $r_i^k$        │◄────┤
        │    by gradient method    │◄────┤
        └─────────────────────────┘     │
                      │                  │
                      ▼                  │
                  ╱ $\epsilon 1 < \delta 1$ ╲ ── no ──┤
                   ╲        ╱              │
                      │ yes               │
                      ▼                    │
        ┌─────────────────────────┐       │
        │ Update $r^k$ for fixed number│       │
        │       of fuzzy rules    │       │
        └─────────────────────────┘       │
                      │                    │
                      ▼                    │
                  ╱ $\epsilon 2 < \delta 2$ ╲ ── no ──┘
                   ╲        ╱
                      │ yes
                      ▼
        yes     ╱ $\epsilon 3 < \delta 3$ ╲
        ┌──────╲        ╱
        │             │ no
        │             ▼
        │  ┌─────────────────────────┐
        │  │ Compute inference error │
        │  │  for every sub region   │
        │  │ and generate new fuzzy  │
        │  │        partition        │
        │  └─────────────────────────┘
        │             │
        │             ▼
        │  ┌─────────────────────────┐
        │  │ Generate new membership │
        │  │ functions and fuzzy rules│
        │  └─────────────────────────┘
        │             │
        └─────────────┤
                      ▼
                    end
```

Fig.8 Computational Flow Chart of Hybrid Algorithm

3.3 Advantage of the Hybrid Method Over Similar Methods

In this section, let us discuss the relationship among the similar gradient type self-tuning methods [4,10,11] so far proposed. Those are as follows.
(1) Ichihashi's iterative fuzzy modeling method[4] (Equal Fixed Grid Method)
(2) Araki's self-generating method[10] ($1/2^n$ Fixed Grid Method)
(3) Proposed hybrid algorithm[11] (Free Grid Method)

In the iterative fuzzy modeling method, a method to increase the number of fuzzy partitions of the premise part, and the rule generation procedure is proposed. However, in their method, the number of fuzzy partitions is so increased to divide each input range into equally partitioned regions at any number of membership functions. Since the premise parts are fixed during the tuning process of conclusion parts, we call this method the Equal Fixed Grid Method.

In case of the Araki's self-generating method, the number of fuzzy partitions is automatically determined, since the rules are iteratively generated in the region with the largest inference error among errors computed in every regions divided by the neighboring two membership functions. Since the vertex values of generated membership functions are located at the $1/2^n$ partitioned grids of the input variable range, and the premise parts are fixed during the tuning process of conclusion parts similarly to the iterative fuzzy modeling method, we call this method the $1/2^n$ Fixed Grid Method.

On the other hand, in the proposed hybrid method, the vertex position of the premise of the membership functions can move freely satisfying the constraints (3) as well as the conclusion parts during the tuning process. Therefore, we call this method the Free Grid Method.

Now, let us denote the feasible model parameter space as $Z_1(n)$, $Z_2(n)$, $Z_3(n)$, according to the Equal Fixed Grid Method, $1/2^n$ Fixed Grid Method, Free Grid Method respectively, where n is the number of fuzzy rules. Also let us denote the optimal solutions of the original identification problem (15) as $\underline{z}_1^*$, $\underline{z}_2^*$, $\underline{z}_3^*$ for the feasible parameter space $Z_1(n)$, $Z_2(n)$, $Z_3(n)$, respectively.

Since the relation

$$Z_1(n) \subset Z_3(n), \quad Z_2(n) \subset Z_3(n) \text{ for } \forall n \tag{33}$$

always holds as shown in Fig.9, we have

$$E(\underline{z}_3^*) \leq E(\underline{z}_1^*), \quad E(\underline{z}_3^*) \leq E(\underline{z}_2^*) \text{ for } \forall n \tag{34}$$

Therefore, we expect the best solution by the Free Grid Method compared to the other two methods. We show this result by the numerical examples in the next section.

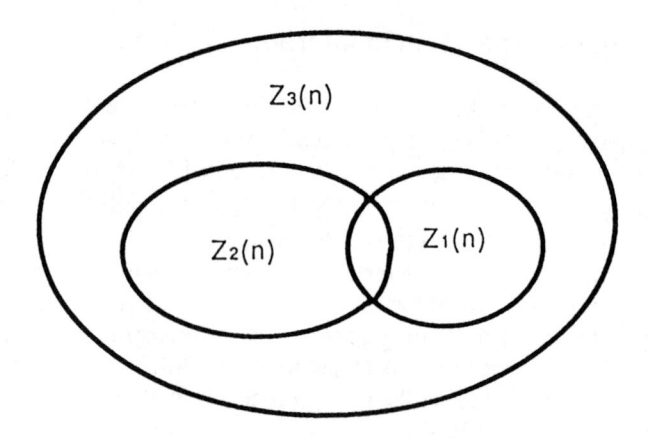

Fig.9 Comparison of Gradient Type Methods Concerning Feasible Parameter Space

3.4. Numerical Examples

[Example 2] In order to evaluate the proposed algorithm, let us consider the following problem to identify the one input/one output function represented by (35),

$$y = \begin{cases} |x| &, (-1.0 \leq x \leq 1.0) \\ (x-2.0)^2 &, (1.0 < x \leq 2.0) \end{cases} \quad (35)$$

The total number of learning data (x,y) is N=31, where input x is given from -1.0 to 2.0.

Fig.10 represents the desired input/output relation given by (35). We use penalty parameter updating coefficient $C = 0.5$, and positive constants for stop criteria as

$$\delta_1 = 10^{-6}, \ \delta_2 = 10^{-7}, \ \delta_3 = 10^{-2}.$$

For the identification problem (35), we applied the following three methods.
(1) Equal Fixed Grid Method
(2) $1/2^n$ Fixed Grid Method
(3) Free Grid Method

The membership functions obtained by the three method are shown in Fig.11a ~Fig.11c. Also, the relation of the output inference error ε_3 with respect to the number of fuzzy rules for the three methods is illustrated in Fig.12.

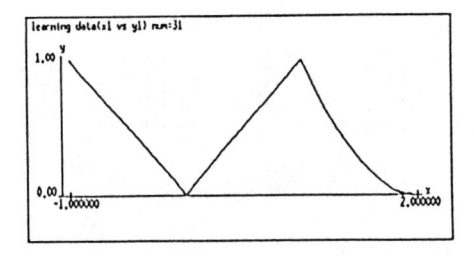

Fig.10 Desired Input/Output Relation

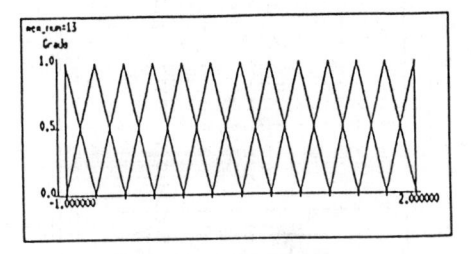

Fig.11a Membership Functions Obtained by Equal Fixed Grid Method

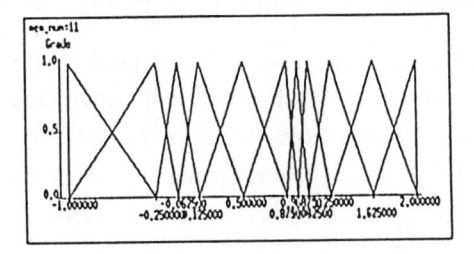

Fig.11b Membership Functions Obtained by $1/2^n$ Fixed Grid Method

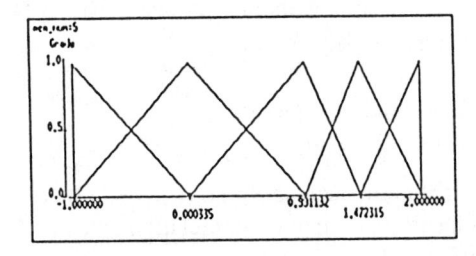

Fig.11c Membership Functions Obtained by Free Grid Method

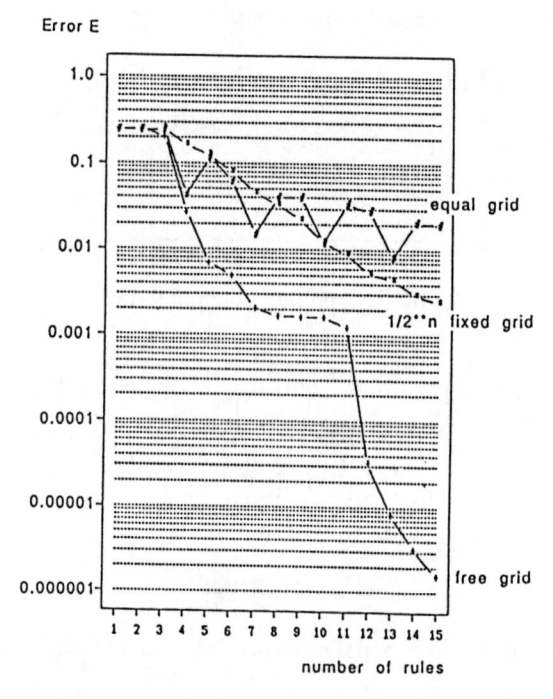

Fig.12 Model Error vs Number of Fuzzy Rules for Three Methods

The result indicates that the proposed Free Grid Method (hybrid method) requires only 5 rules to satisfy the specified model inference accuracy δ_3, while the Equal Fixed Grid Method (iterative fuzzy modeling method [4]) requires 13 rules, and the $1/2^n$ Fixed Grid Method (self-generating method [10]) requires 11 rules. Also, from Fig.12, the proposed hybrid method achieves minimal output error value ε_3 for any rule number n.

【Example 3】 Now, let us examine another problem. This is two input/one output identification problem defined by (36):

$$y = \sin(\pi x_1) + \cos(\pi x_2),$$
$$(-1.5 \leq x_1 \leq 1.5, -1.5 \leq x_2 \leq 1.5) \quad (36)$$

The total number of learning data is N=64.

Fig.13 represents the desired input/output relation given by (36). We use penalty parameter updating coefficient C = 0.5, and positive constants for stop criteria as

$$\delta_1 = 10^{-6}, \delta_2 = 10^{-7}, \delta_3 = 10^{-2}.$$

The membership functions obtained by the Free Grid Method are shown in Fig.14. Also, the relation of the output inference error ε_3 with respect to the number of fuzzy rules for the three methods are shown in Fig.15.

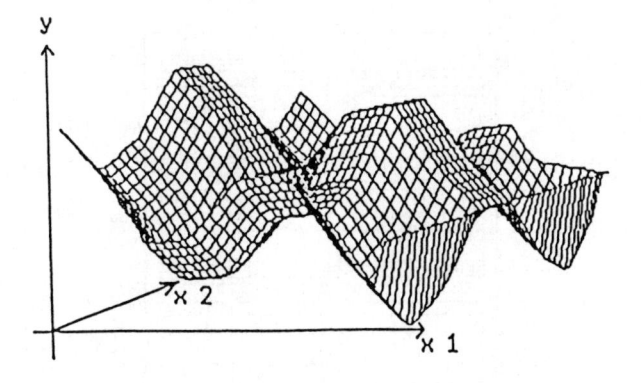

Fig.13 Desired Input/Output Relation

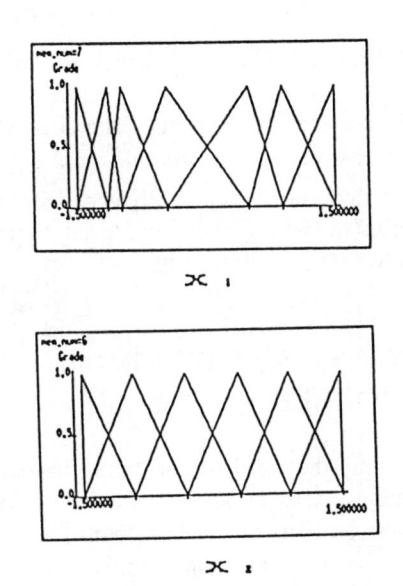

Fig.14 Membership Functions Obtained by Free Grid Method

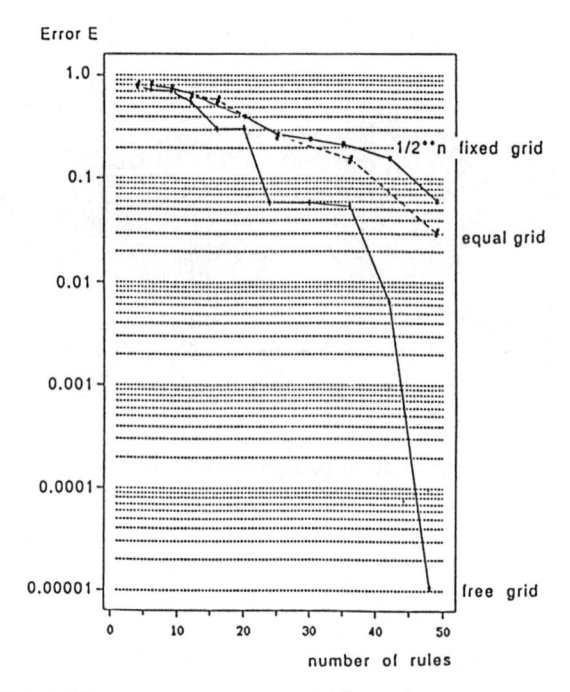

Fig.15 Model Error vs Number of Fuzzy Rules for Three Methods

The result also indicates that the proposed Free Grid Method (hybrid method) achieves minimal output error value ϵ_3 for any rule number n.

Therefore, the hybrid algorithm automatically generates the least number of fuzzy rules to achieve the specified model accuracy, while satisfying the two requirements imposed on the shape of the membership functions.

4. APPLICATION TO KNOWLEDGE ACQUISITION OF FUZZY CONTROLLER

In this section, we apply the hybrid algorithm to the design of a fuzzy control system. One of the effective means to design a fuzzy controller is to linguistically describe the skilled operator's experience and/or control engineer's knowledge using fuzzy rules. However, if the operator cannot express linguistically what kind of appropriate action he takes in a particular situation, it is very useful to model his control actions using numerical data [1,2]. In this case, the design problem of the fuzzy controller is reduced to the fuzzy modeling problem. Therefore, the hybrid algorithm described in Section 3 is directly applied to this kind of knowledge acquisition problems of the fuzzy controller.

Another potential application of the hybrid algorithm is the knowledge refinement of the fuzzy controller, although this is essentially in the realm of the knowledge acquisition problems. Since the hybrid algorithm is superior in realizing specified model error with relatively small number of the fuzzy rules, the method is expected to be effective in reducing the number of the optimal fuzzy rules for feedback fuzzy controller, which are obtained by the skilled human operator or other self-tuning methods such as the adaptive learning gain algorithm [12,13], which is a kind of the on-line tuning method for fuzzy control system. This feature is very useful especially for the microcomputer based applications with limited memory resources for storing fuzzy rules. We show this kind of knowledge refinement problem of the fuzzy controller.

[Example 4] Let us consider the model reference adaptive fuzzy control system [12,13] shown in Fig.16. We assume that the plant is a 2nd order dumping system, whose transfer function G(s) is given by (37):

$$G(s) = C/(s^2 + As + B), \quad A = 0.638, B = 0.034, C = 1.228 \qquad (37)$$

The transfer function H(s) of the reference model is given by (38):

$$H(s) = 1/(1 + Ts)^5, \quad T = 1.8 \qquad (38)$$

The desired reaching time TR = 25 sec, and we assume the velocity type fuzzy PI controller. Since only the conclusion parts of the fuzzy rules are tuned in the adaptive learning gain algorithm [12,13], we assume 7×7 membership functions of input e and de/dt shown in Fig.17 for the controller. Therefore, the number of fuzzy rules is 49. The goal is to design such a fuzzy controller that the output transition of the plant for the step change of the setting value r gets closer to that of the reference model as much as possible.

We applied the adaptive learning gain algorithm. The ideal output y^o of reference model, and the convergence of the plant output y and manipulated value m is shown in Fig.18. It takes about five learning cycles for convergence. The optimal input/output relation of the fuzzy controller thus obtained is shown in Fig.19.

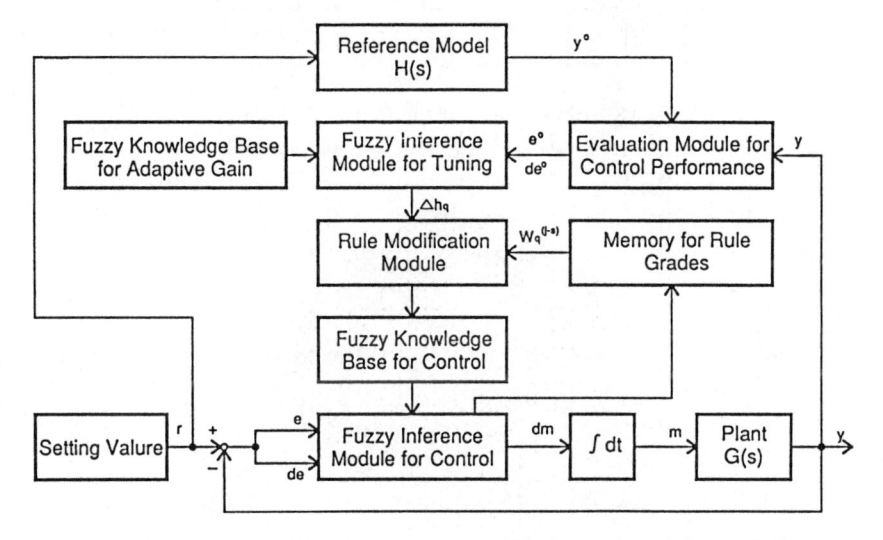

Fig.16 Diagram of Model Reference Adaptive Fuzzy Control System

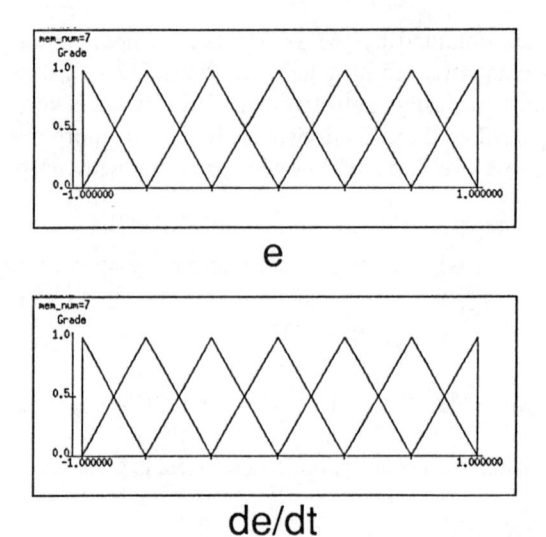

Fig.17 Membership Functions of the Fuzzy Controller for On-Line Tuning

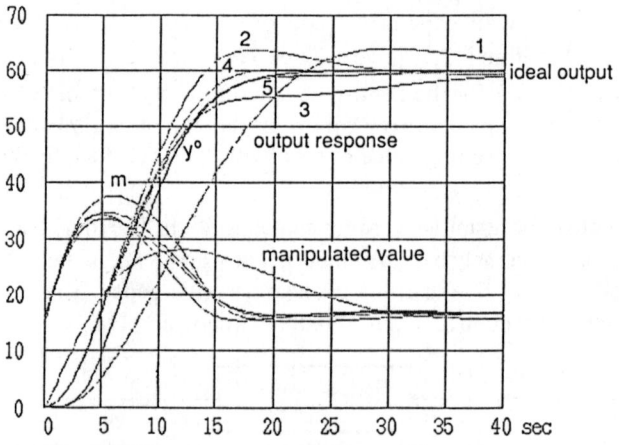

Fig.18 Convergence of the Output Response
Using the Adaptive Learning Gain Method

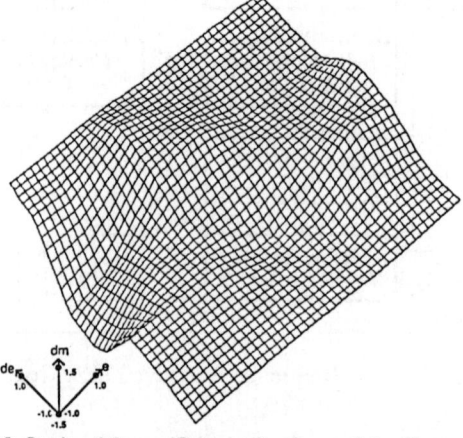

Fig.19 Optimal Input/Output Surface of the Fuzzy Controller
(Rule Number=49)

Now, we applied the hybrid algorithm to reduce the number of fuzzy rules. Fig.20 represents the relation of the output inference error ϵ_3 with respect to the number of fuzzy rules. The membership functions and approximated input/output surface of the fuzzy controller with $\epsilon_3 = 0.032$ are shown in Fig.21 and Fig.22, respectively. In this case, the total number of rules is reduced to 36 from 49. Also, the output response of the plant using the fuzzy controller with these reduced rules is illustrated in Fig.23. This means that the controller with these reduced 36 rules has almost the same performance as the optimal controller with 49 rules. Thus, the hybrid method is proved to be useful for the knowledge acquisition and the refinement of the fuzzy controller.

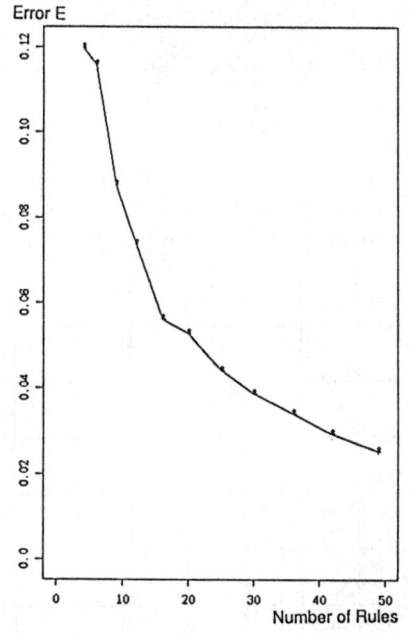

Fig.20 Model Error vs Number of Fuzzy Rules for Free Grid Method

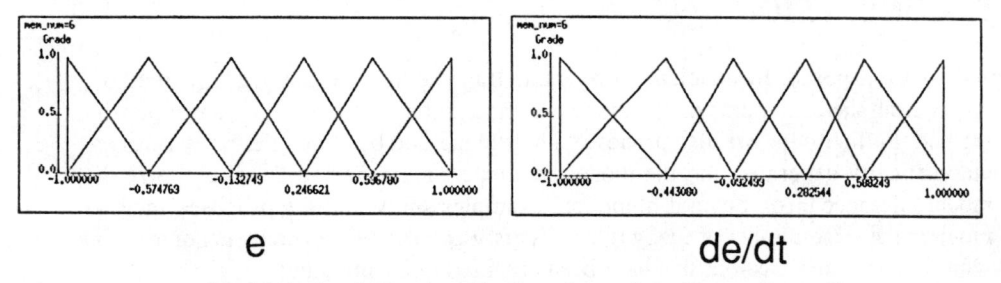

Fig.21 Refined Membership Functions of the Fuzzy Controller

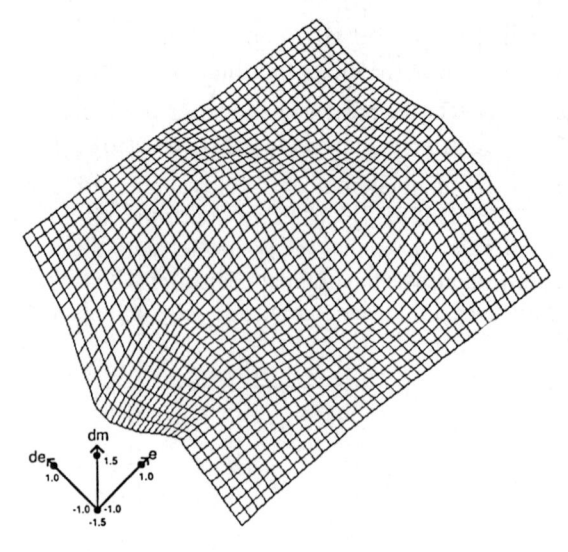

Fig.22 Approximated Input/Output Surface of the Fuzzy Controller with Reduced Rules (Rule Number=36)

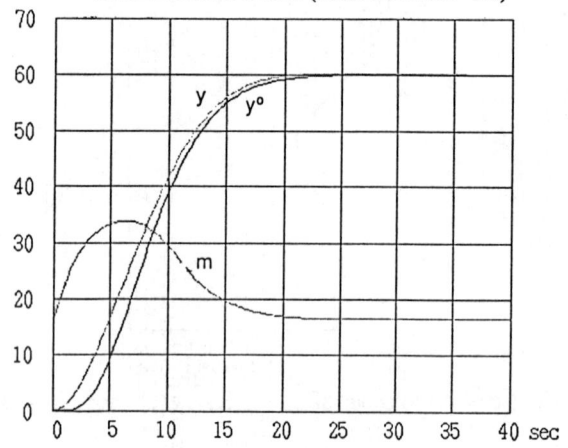

Fig.23 Output Response Using Fuzzy Controller with Reduced Rules

5. CONCLUSION

We proposed an efficient fuzzy modeling method by the gradient method, considering the constraints on the membership functions and using an interior penalty method. Furthermore, we also proposed a hybrid method by which the fuzzy rules are automatically generated and the model parameters are tuned so as to minimize the model inference error. Several numerical examples show that the proposed method requires the least number of fuzzy rules to satisfy any desired accuracy in comparison with similar tuning methods that have been proposed in the literature.

Finally, we applied the hybrid method to the knowledge acquisition and the refinement of the fuzzy controller. We would like to stress that the knowledge refinement process using the hybrid algorithm is very useful especially for the microcomputer based applications with limited memory resources for storing fuzzy rules.

Although the shape of the membership functions treated here is restricted to the triangle type, the proposed algorithm can be naturally extended to the fuzzy model with more smooth C^n or C^∞ class membership functions. This extended method will be reported in the near future.

REFERENCES

[1] T.Takagi, M.Sugeno: "Fuzzy Identification of Systems and its Applications to Modeling and Control", IEEE Trans. on Systems, Man, and Cybernetics, vol.SMC-15, no.1, pp.116-132,1985.

[2] M.Sugeno, G.T.Kang: "Structure Identification of Fuzzy Model", Fuzzy Sets and Systems, vol.28, pp.15-33, 1988.

[3] H.Takagi,I.Hayashi: "NN-Driven Fuzzy Reasoning", International Journal of Approximate Reasoning, 5, pp.191-212, 1991.

[4] H.Ichihashi, T.Watanabe: "Learning Control System by a Simplified Fuzzy Reasoning Model", IPMU'90, Paris-France, pp.417-419, 1990.

[5] H.Nomura, I.Hayashi, N.Wakami: "A Self-Tuning Method of Fuzzy Reasoning Control by Descent Method", Proc. of 4th IFSA Congress, Brussels,vol.Eng., pp.155-158, 1991.

[6] M.Avriel: Nonlinear Programming, Prentice Hall, 1976.

[7] A.V.Fiacco, G.P.McCormick: Nonlinear Programming: Sequential Unconstrained Minimization Techniques, John Wiley & Sons, New York, 1968.

[8] K.Kuwata, Y.Kajitani, R.Katayama, Y.Nishida: "A Model Determination Support System for Neuro and Fuzzy Model by Plural Performance Indices", Proc. of the 43th Congress of Information Processing Society of Japan, pp.281-282,1991. (in Japanese)

[9] Y.Kajitani, K.Kuwata, R.Katayama, Y.Nishida: "An Automatic Fuzzy Modeling with Constraints of Membership Functions and a Model Determination for Neuro and Fuzzy Model by Plural Performance Indices", Proc. of the International Fuzzy Engineering Symposium '91 (IFES'91, Yokohama, Japan, Nov. 13-15,1991), pp.586-597.

[10] S.Araki, H.Nomura, I.Hayashi: "A Self-Generating Method of Fuzzy Inference Rules", Proc. of the International Fuzzy Engineering Symposium'91 (IFES'91, Yokohama, Japan,Nov. 13-15,1991), pp.1047-1058.

[11] R.Katayama,Y.Kajitani, Y.Nishida: "A Self Generating and Tuning Method for Fuzzy Modeling using Interior Penalty Method", Proc. of the 2nd International Conference on Fuzzy Logic and Neural Networks (IIZUKA'92, Iizuka, Japan, July 17-22, 1992), pp.357-360

[12] R.Katayama, Y.Kajitani, K.Matsumoto, M.Watanabe, Y.Nishida: "An Automatic Knowledge Acquisition and Fast Self Tuning Method for Fuzzy Controller Based on Hierarchical Fuzzy Inference Mechanisms", Proc. of 4th IFSA Congress, Brussels, vol. Artificial Intelligence, pp.105-108, 1991.

[13] R.Katayama, Y.Kajitani, K.Kuwata, Y.Nishida: "A Fast Self Tuning Method for Fuzzy Controller Based on Adaptive Learning Gain Algorithm", Proc. of 7th Fuzzy Symposium (Nagoya, Japan, Jun. 12-14,1991), pp.21-24, 1991. (in Japanese)

[14] A.Maeda, R.Someya, M.Funabashi: "A Self-Tuning Algorithm for Fuzzy Membership Functions using Computational Flow Network", Proc. of 4th IFSA Congress, Brussels, vol. Artificial Intelligence, pp.129-132, 1991.

[15] S.Horikawa, T.Furuhashi, Y.Uchikawa, T.Tagawa: "A Study on Fuzzy Modeling Using Fuzzy Neural Networks", Proc. of the International Fuzzy Engineering Symposium'91 (IFES'91, Yokohama, Japan,Nov. 13-15,1991), pp.562-573, 1991.

Chapter 10
Fuzzy Control of VSS Type and its Robustness

Introduction, *226*
Fuzzy Control of VSS Type, *226*
Rule Generation of Fuzzy Controller, *232*
Parameter Adjustment of Fuzzy Controller, *234*
Robustness of Fuzzy Controller, *238*
Conclusions, *238*
References, *238*

Recently there have been many applications of fuzzy control based on the intuition and experience of human operators. A fuzzy controller is regarded as a set of heuristic decision rules and can be easily implemented on a computer. However, a design method for fuzzy controllers, which guarantees stability and robustness, has not been established yet. In this chapter, the basic idea of designing fuzzy controllers to guarantee stability is presented based on the notion of *variable structure systems*. The robustness of the system is discussed, and the detailed distribution of the linguistic variables required for a robust fuzzy controller is shown.

FUZZY CONTROL OF VSS TYPE AND ITS ROBUSTNESS

Shigeyasu Kawaji and Nobutomo Matsunaga
Department of Electrical Engineering and Computer Science
Kumamoto University
Kurokami 2-39-1, Kumamoto 860, Japan

1 INTRODUCTION

Recently, fuzzy control [1] has found significant applications in many engineering fields. Fuzzy controller is regarded as a set of heuristic decision rules and is implemented easily in a computer. While the robustness of fuzzy controllers has been reported in several applications, the design method of fuzzy control systems which guarantees stability and robustness has not been established yet.

In this chapter, the basic idea of a design method of fuzzy control systems to guarantee stability is presented based on the Variable Structure Systems (VSS) approach [2]. The key is the following interpretation of fuzzy control rules: the rule is generally a fuzzy relation R_i of the form [3]:

$$R_i : \text{If}(A_i, B_i) \text{ Then } C_i$$

where A_i and B_i are fuzzy quantities representing process measurements and C_i is a fuzzy quantity representing a control signal. We remark that the rule R_i constitutes a "structure" of fuzzy control system and is switched according to the states of process. Thus, the separation of states based on "structure" is regarded as a VSS controller. It follows that we can treat a fuzzy control system as a kind of VSS and utilize its useful results on stability and robustness.

In the following section, some results of VSS theory are briefly reviewed and a fuzzy controller based on basic VSS theory is introduced. In Sections 3 and 4, a method for generating fuzzy rules and membership functions is suggested. Finally, in Section 5 the robustness of the proposed fuzzy control system is discussed.

2 FUZZY CONTROL OF VSS TYPE

The problem in designing fuzzy control systems is how to determine the free parameters of the fuzzy controller, i.e., rules and membership functions. Therefore, in order to design the fuzzy control system, we must first understand the concept of

the laws, the output is calculated using the grade of fuzziness. In this sense, we can fuzzy control as a kind of VSS control as shown in Fig. 1.

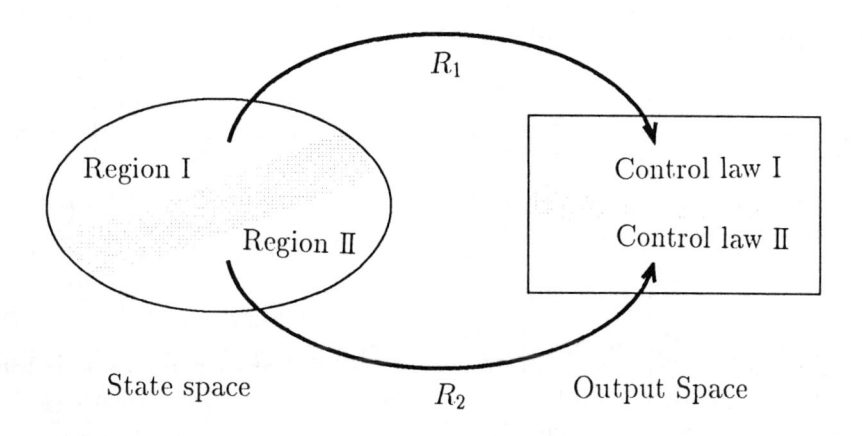

Fig. 1: The Concept of Fuzzy Control

2.1 VSS Controller

VSS theory has found many applications in the control of a wide range of processes. The central idea of VSS control is to switch a different control structure at each side of a given switching surface. Its salient feature is that a sliding mode occurs on the switching surface, and in this mode, the system remains insensitive to parameter variations and disturbances. Let us first briefly summarize the basic idea of a simple VSS control[2].

The basic law of VSS control is of the form:

$$u = -W sgn(s) \qquad (1)$$

where W is a constant parameter, $sgn(\cdot)$ is the sign function, and s is a switching variable given by:

$$s = f^T x \qquad (2)$$

The surface defined by $s = 0$ is called the switching surface, and represents the desired dynamics.

The new system properties are obtained by composing a desired trajectory from parts of the trajectories of different structures. An even more fundamental aspect of VSS is the possibility to obtain trajectory not inherent in any of the structures. These trajectories describe a new type of motion, the sliding mode. For example, if the s is determined as:

$$s = \dot{x} + ax = 0 \qquad (3)$$

and the sliding mode occurs, the state x is expressed as:

$$x = exp(-at)\, x(0) \qquad (4)$$

Thus, the state x depends on a and $x(0)$, i.e., the switching surface and the initial state.

The problem in designing a sliding mode controller is to determine the parameters $\{f^T, W\}$ which satisfy:

$$\begin{aligned}(i) & \quad s(t)\dot{s}(t) < 0 \\ (ii) & \quad x(t) \to 0 \text{ on } s(t) = 0\end{aligned} \qquad (5)$$

Condition (i) implies that the trajectories are constrained to point towards the surface $s = 0$, as illustrated in Fig.2. The condition refers to the existence of the sliding mode. Condition (ii) implies that the tracking error tends exponentially to zero once on the surface.

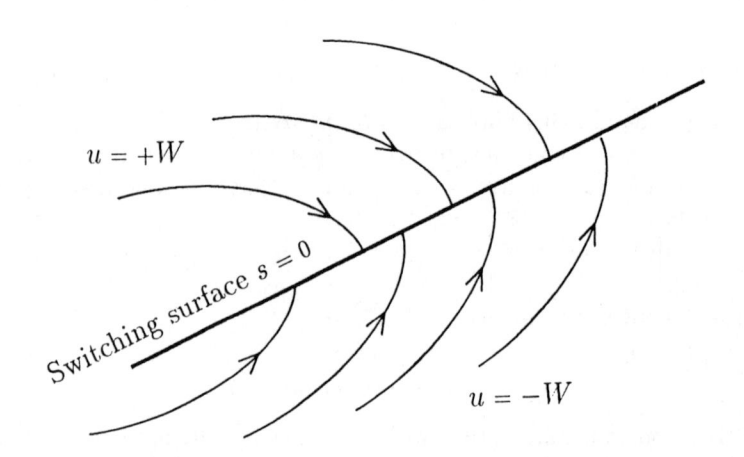

Fig.2: The Concept of Sliding Mode

Various design methods of sliding mode controllers have been reported in the literatures. Here we will give the following approach [4]: Let us consider a single-input system represented by:

$$\dot{x} = Ax + bu \qquad (6)$$

where $x \in R^n$, u is a scalar input, and A and b are real constant matrices.

Using a cheap control technique, we let

$$f^T = \frac{1}{\epsilon} B^T P \qquad (7)$$

where P is the positive definite solution of the Ricatti equation

$$PA + A^T P - \frac{1}{\epsilon^2} PBB^T P + Q = 0 \qquad (8)$$

At first, condition (i) will be checked. If $s > 0$, the closed loop system with the control law (1) is described as:

$$\dot{x} = Ax - bW \qquad (9)$$

Hence,

$$\dot{s} = f^T Ax - \frac{1}{\epsilon} b^T PbW \qquad (10)$$

When $|x|$ and ϵ are sufficiently small and W is sufficiently large, the positive-definiteness of P implies

$$f^T Ax << \frac{1}{\epsilon} b^T PbW \qquad (11)$$

or

$$s\dot{s} < 0 \qquad (12)$$

In case $s < 0$, the same discussion holds. It follows that, if x(0) is in a certain region, s tends to the surface $s = 0$ and is constrained on it. The magnitude parameter W should then be determined to satisfy condition (ii).

Next, after the state x has reached the switching surface, it might drift on the surface. Hence, the stability of the drifting behavior, or the sliding motion, has to be checked. This constrained motion is approximated as:

$$\begin{cases} \dot{x} = Ax + bu \\ s = f^T x = 0 \end{cases} \qquad (13)$$

The poles of the system (13) are identical to the zeroes of the open loop system with output

$$y = f^T x \qquad (14)$$

Since we select only stable eigenvalues, the resulting system is stable. Thus, the sliding motion is stable and condition (ii) is satisfied.

We should remark that the theoretical sliding mode is idealized. Due to the switching delay and negligible small time constant, it is rare in real system that ideal sliding mode occurs. In particular, the trajectories chatter along the switching surface. This situation is remedied by smoothing out the control discontinuity in the thin boundary layer neighboring the switching surface. A boundary layer is illustrated in Fig.3 for $n = 2$.

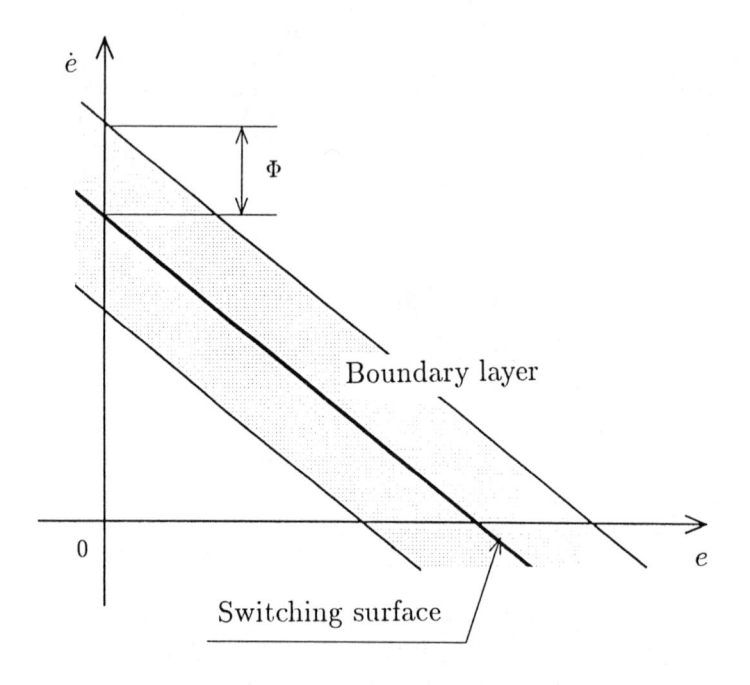

Fig.3: The Construction of Boundary Layer

By introducing the boundary layer, the switching law (1) can be modified as [5]:

$$u = \begin{cases} -W & , \quad \Phi < s \\ -W\dfrac{s}{\Phi} & , \quad -\Phi \le s \le \Phi \\ W & , \quad -\Phi > s \end{cases} \quad (15)$$

where Φ is the thickness of the boundary layer.

2.2 Fuzzy VSS Controller

In this section, we consider a fuzzy controller which is equivalent to a VSS controller. We consider simple fuzzy rules, e.g., for motor control, expressed as

R_1 : If s is Positive Then u is Negative
R_2 : If s is Negative Then u is Positive

It is noted that in this inference only two fuzzy quantities, Positive or Negative, are used and their membership functions are shown in Fig.4. Then, the output of proposed fuzzy controller can be described in terms of s, Φ and W as :

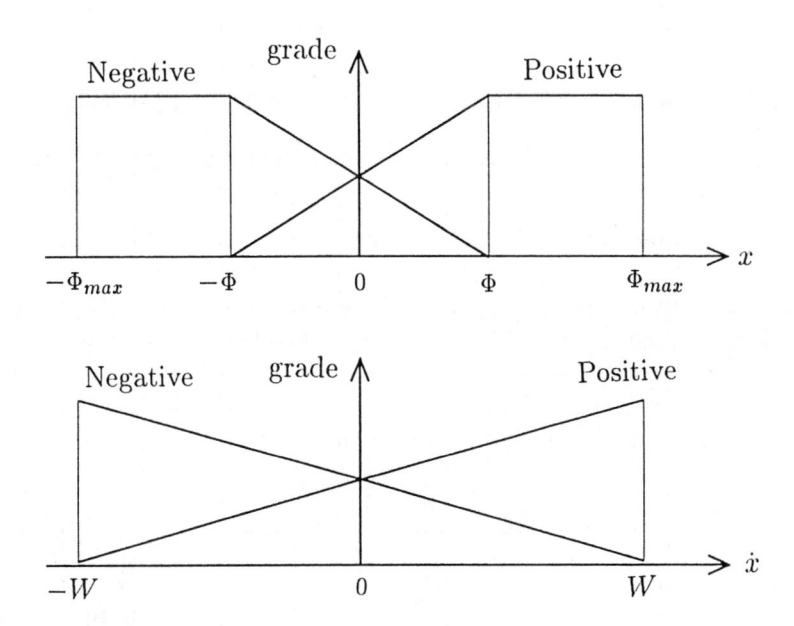

$$u = \begin{cases} -W , & \Phi < s < \Phi_{max} \\ -W\dfrac{s}{\Phi} , & -\Phi \leq s \leq \Phi \\ W , & -\Phi > s > -\Phi_{max} \end{cases} \quad (16)$$

Fig.4: Fuzzy Membership Functions

Comparing (16) with (15), we see that the fuzzy control laws (16) are equivalent to the VSS controller with the boundary layer (15) in the region $[-\Phi_{max}, \Phi_{max}]$, and that the potential field (dotted area in Fig.3) in the boundary layer depends on the shape of the membership functions.

From the facts described above, we may conclude that:
1. The fuzzy controller (16) is identical to the VSS controller with boundary layer.

2. The antecedent and consequent membership functions specify the potential in the boundary layer.

3 RULE GENERATION OF FUZZY CONTROLLER

In the preceding chapter, it was shown that the simple fuzzy controller is identical to the VSS controller with boundary layer. But, the distance from the switching surface is used as the input for the fuzzy controller of the VSS-type. So, it is difficult to use this fuzzy controller for intelligent control systems because of the following reasons:

(i) It is difficult to design control laws using human intuition or experience because the distance from the switching surface is used whereas the human cannot use it.

(ii) In case human intelligence is required, many linguistic values for rich signals must be used.

In order to overcome these difficulties, we use directly sensed quantities instead of s.

It was reported that the potential field of the fuzzy controller is very complicated when directly sensed quantities are used [7]. Therefore, in order to discuss the stability problem, we must consider a concrete plant. To this end, we use the DC servomotor which is widely used in engineering fields. A typical scheme of a fuzzy control system for the DC servomotor is shown in Fig.5.

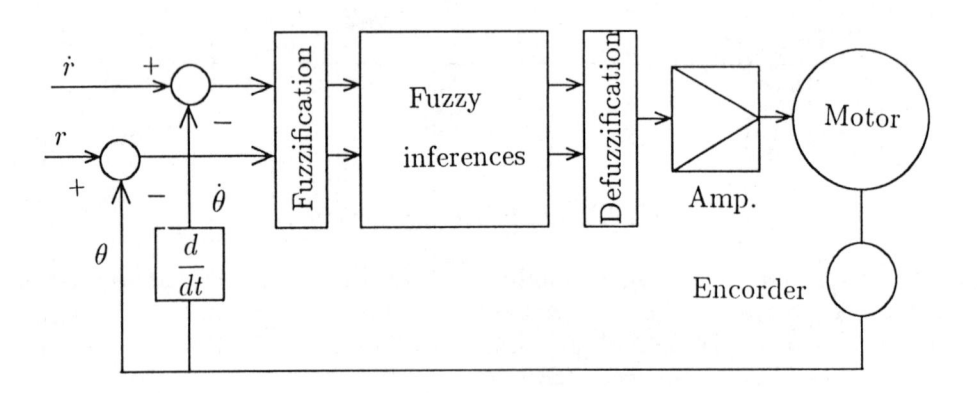

Fig.5: Fuzzy Control System for DC Servomotor

Here we assume that the basic structure and rough values of the parameters are known. This assumption is not too severe in realistic situations. The dynamic equation of the DC servomotor is given by the following equation:

$$\dot{x} = \begin{bmatrix} 0 & 1 \\ 0 & -\dfrac{D}{M} \end{bmatrix} x + \begin{bmatrix} 0 \\ \dfrac{K}{M} \end{bmatrix} u \qquad (17)$$

where $x = [\theta, \dot{\theta}]$, θ is the notational angle of the motor, M and D are the inertia and the damping of the motor, and K is the gain of the amplifier.

The fuzzy closed loop system whose stability is guaranteed can be constructed as follows. Let V be a semipositive-definite function such as:

$$V = (\dot{e} + \alpha e)^2 \qquad (18)$$

where $e = r - \theta$ is the error and r is the reference angle. The function (18) is a Lyapunov function along a line perpendicularly intersecting the following stable switching surface:

$$s = \dot{e} + \alpha e = 0 \qquad (19)$$

Consequently, from Lyapunov theorem, this fuzzy controller makes the closed-loop system stable if $\dot{V} < 0$, from which the following inequality equations are satisfied:

$$\beta \dot{e} < u, \quad \text{if } \dot{e} > -\alpha e \quad \text{(Region I)} \qquad (20)$$
$$\beta \dot{e} > u, \quad \text{if } \dot{e} < -\alpha e \quad \text{(Region II)} \qquad (21)$$

where $\beta = -M(\alpha - M/D)/K$.

Then, the rules for the fuzzy controller can be determined by using the phase plane in Fig.6 under the assumption that the antecedent and consequent membership functions are given a-priori. For example, a rule close to a point P in the figure is determined as follows: At the neighborhood of point P, the antecedent linguistic part is defined as "e is NS and $\dot{e}$ is NS", whereas the consequent linguistic part is defined from the above discussion as "u is NM (or NB)". Therefore, a rule close to the point P is decided as:

IF e is NS and $\dot{e}$ is NS Then u is NM (or NB)

4 PARAMETER ADJUSTMENT OF FUZZY CONTROLLER

4.1 Rule Generation of Fuzzy Controller

Since the stability condition is expressed as inequality equations, there may exist many choices of consequent linguistic values. To choose the best consequent linguistic values, we order them as shown in Table 1. In this case, full rules are used so that unexpected local minimum of potential is not generated. Let us characterize the rules by the following two parameters:

LV: Constraint force to the switching surface
α : The slope of the switching surface

and refer to the rule as Rule (LV, α).

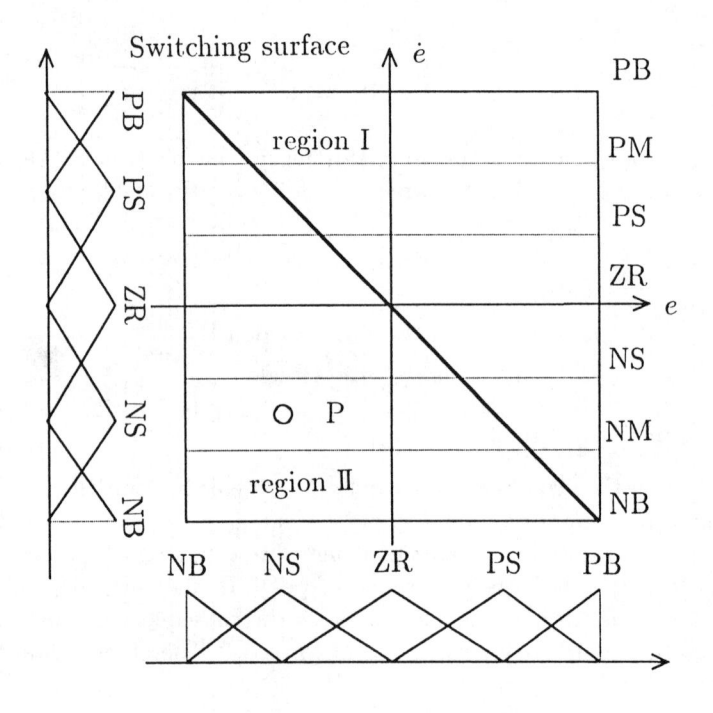

Fig.6: Phase Plane for Deciding a Rule

Fig.7 shows the step responses according to the rules in Table 1. From the figure, the overshoot and the settling time for Rule (B, ∗) are the smallest. But, if the load of the DC servomotor is increased, an oscillation occurs after all while Rule (B, B) is used. Therefore, the rules can be chosen as Rule (B,S).

Table 1: Table for Deciding a Rule

LV \ α	S	B
S	PM / PS / ZR / NS / NM	PM / PS / ZR / NS / NM
B	PB / PB / ZR / NB / NB	PB / PB / ZR / NB / NB

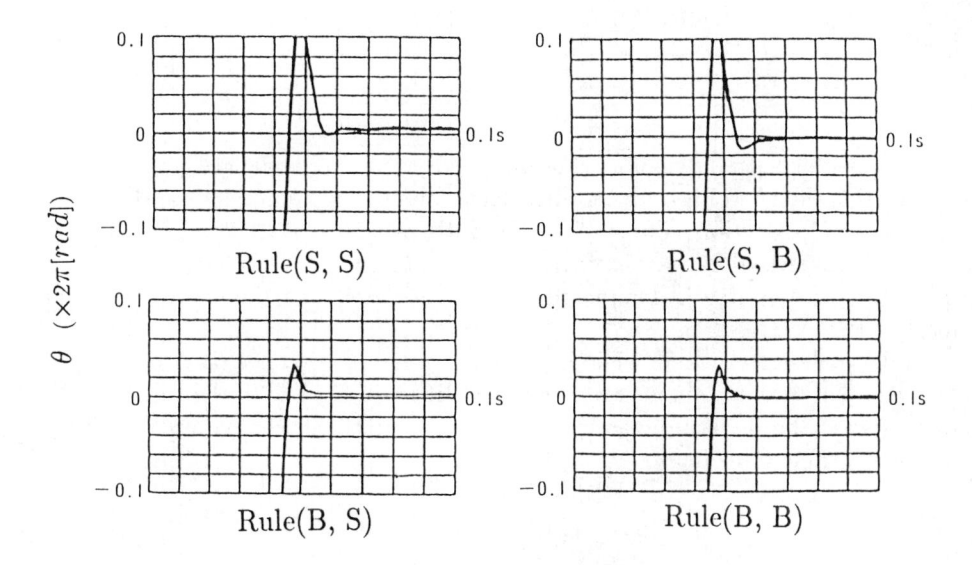

Fig.7: Experimental Response of DC Servomotor

4.2 Adjustment of Membership Functions

The membership functions are used for setting the switching surface. To simplify the calculation, we use membership functions of the triangle and singleton type for the antecedent and consequent membership functions, respectively. Here we consider the typical rules shown in Table.2. The fuzzy rules indicate that the switching surface exists in the 2nd and 4th quadrants, but that the shape of the switching surface is not clear. Therefore, we consider in this section the design of membership functions to realize the desired dynamics shown in Fig.8.

In Fig.8, the switching surface implies that the system has fast dynamics $(-m_r)$ in the rising area and slow dynamics $(-m_s)$ in the settling area.

Table 2: Typical Rules of Fuzzy VSS Controller

$\dot{e} \backslash e$	NB	NS	ZR	PS	PB
PB	ZR	NB	NB	NB	NB
PS	PB	ZR	NB	NB	NB
ZR	PB	PB	ZR	NB	NB
NS	PB	PB	PB	ZR	NB
NB	PB	PB	PB	PB	ZR

To realize the desired dynamics shown in Fig.8, we use the triangle-type membership function shown in Fig.9 for the antecedent membership functions. In the figure, the membership functions consist of five linguistic variables $\{NB, NS, ZR, PS, PB\}$, and are characterized by the parameters $\{L_e, m_e, L_{\dot{e}}, m_{\dot{e}}\}$. The behavior of the system in the settling area is dominated by $\{NS, ZR, PS\}$ and the one in the rising area by $\{PB, NB\}$.

In order to design the membership functions, we assume that the range of e, $\{L_e, m_e\}$ are given. The assumption is not severe in realistic situations. In the case that the parameters of the desired dynamics $\{m_r, m_s\}$ are specified, the parameters of the membership function $\{L_{\dot{e}}, m_{\dot{e}}\}$ are given by:

$$\begin{cases} L_{\dot{e}} = m_r L_e - (m_r - m_s)\dfrac{L_e}{m_e} \\ m_{\dot{e}} = \dfrac{m_e L_{\dot{e}}}{L_e m_s} \end{cases} \quad (22)$$

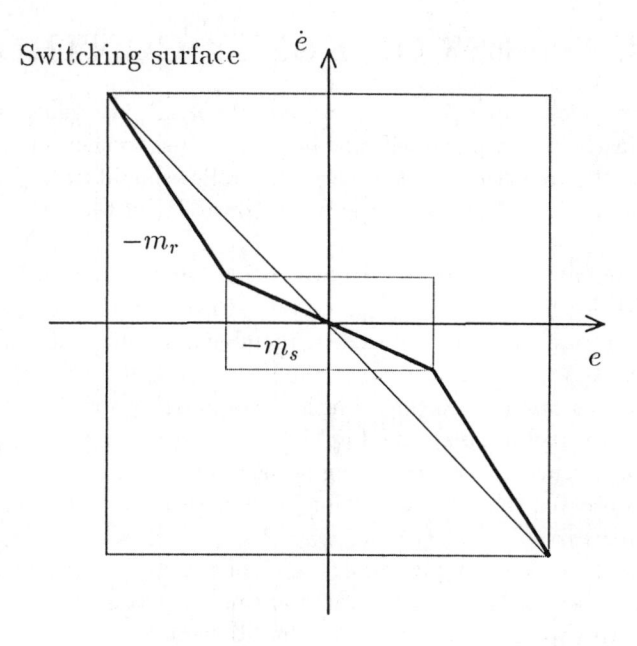

Fig.8: The desired dynamics of the system

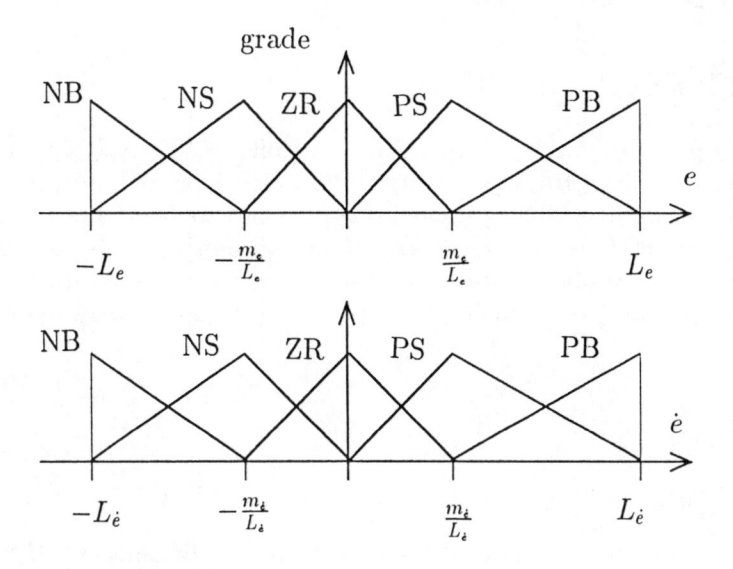

Fig.9: Triangle-type Membership function

5 ROBUSTNESS OF FUZZY CONTROLLER

Since the fuzzy controller operates as an interpolator of output signals, it is expected tha the states in the phase plane will not be constrained to the switching surface. For this reason, the robustness of the fuzzy controller should be checked. In this section, we discuss the robustness of the fuzzy controller of the VSS-type through simulations.

Let the parameters of the DC servomotor be $M = 0.1, D = 0.1$ and $K = 2$, and those of the desired dynamics $\{m_r, m_s\} = \{10, 5\}$. In the cases that 5 and 7 linguistic variables are used, the antecedent membership functions are obtained as shown in Fig.10.

The phase planes of the resulting transient responses using 5 and 7 linguistic variables are shown in Fig.11(a) and Fig.11(b), respectively. From Fig.11(a), the trajectory is constrained on the switching surface only in the settling area. From Fig.11(b), it is observed that the constraint force increases as the separation of the state space increases.

For comparison, the output potentials, $u^T u$, of the fuzzy controller are shown in Fig.12. It can been seen that the states are not constrained on the switching surface because the boundary layer is wide due to the fuzziness of the states. The robustness in the sense of VSS theory is guaranteed when the states remain on the switching surface. We may conclude that the robustness is guaranteed whenever many linguistic values are used.

6 CONCLUSIONS

We have proposed in this chapter a fuzzy controller of VSS type based on the essential features of fuzzy control. Furthermore, under the condition that the structure and the rough values of the parameters of the plant are know, the parameterization of the fuzzy controller, i.e., the rules and membership functions, has been clarified.

From the view point of robustness, it has been shown that a detailed distribution of linguistic variables is required in order for the proposed fuzzy control system to be robust.

References

[1] W. Pedrycz: *Fuzzy Control and Fuzzy Systems*, John Wiley, 1989

[2] V. I. Utkin: "Variable structure systems with sliding modes", *IEEE Trans. on Automatic Control*, Vol. AC-22, pp. 212-222, 1977

[3] M. Sugeno: "An introductory survey of fuzzy control", *Information Science*, 36, pp. 59-83, 1985

(a) 5 linguistic variables

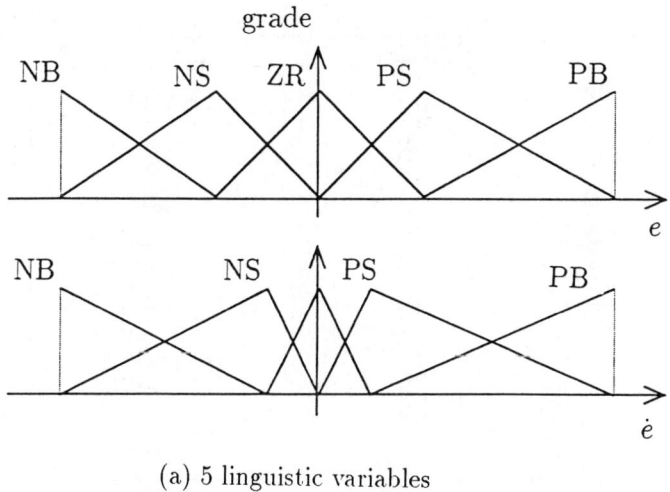

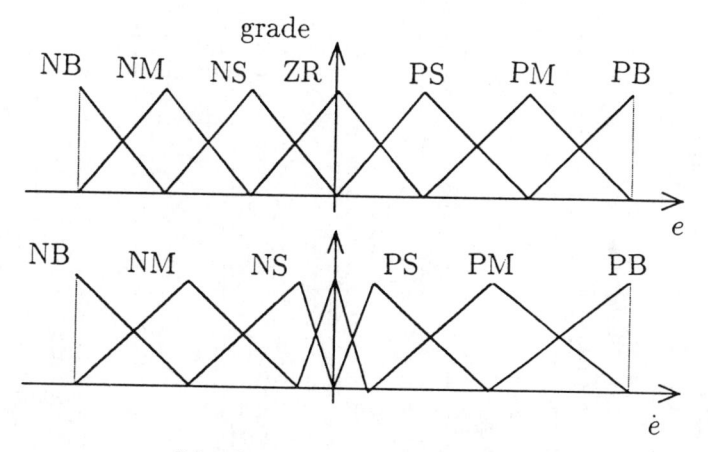

(b) 7 linguistic variables
Fig.10: Antecedent membership functions

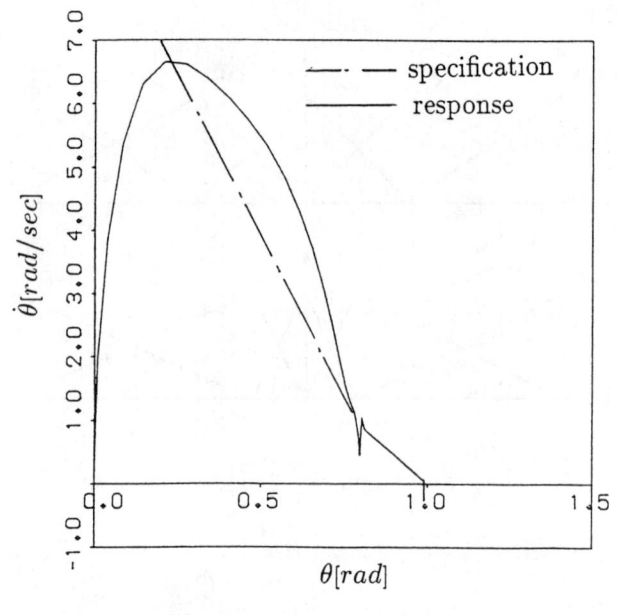

(a) 5 linguistic variables

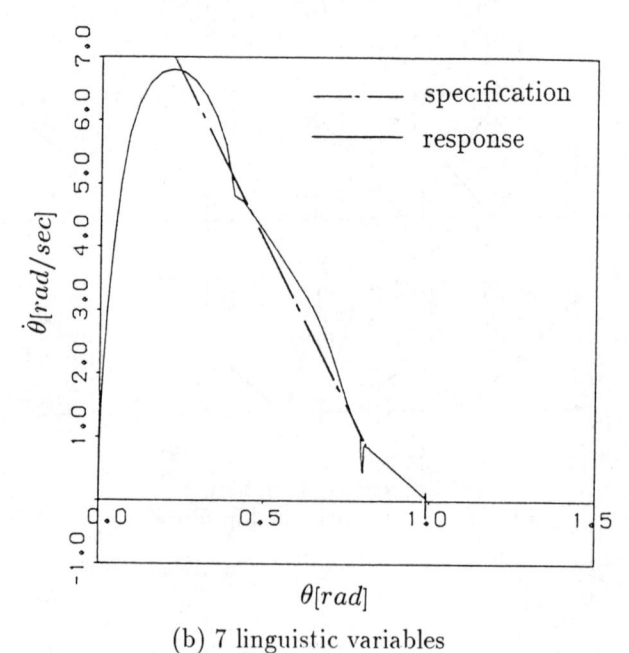

(b) 7 linguistic variables
Fig.11: Phase plane of the fuzzy control sysytem

(a) 5 variables

(b) 7 variables
Fig.12: The outputs of the fuzzy controller

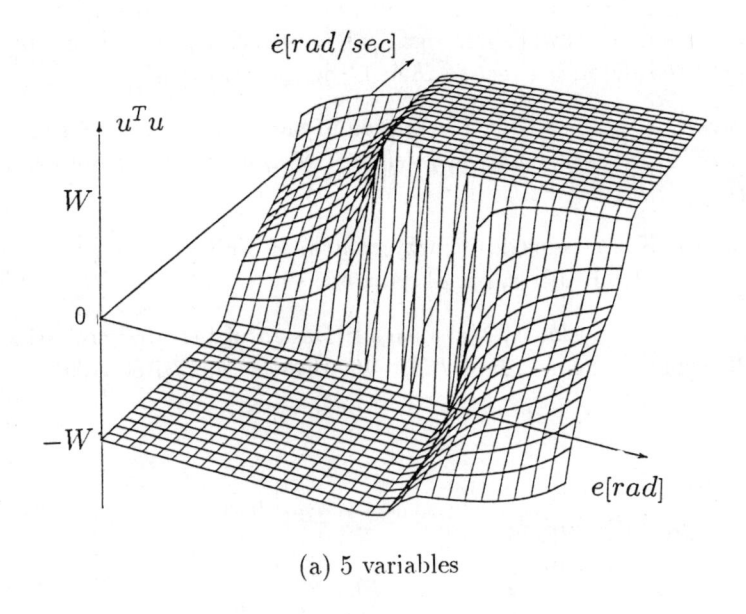

[4] T. Shiotsuki, S. Kawaji: "Design and implementation of VSS controller on personal computers", *Proc. 88' KACC*, Seoul, pp. 848-851, 1988

[5] S. Kawaji, N. Matsunaga: "Generation of fuzzy control rules for servomotor", *Proc. IEEE Int. Workshop on Intelligent Motion Control*, Istanbul, pp. 77-82, 1990

[6] S. Kawaji, N. Matsunaga: "Fuzzy control of VSS type and its robustness", *Proc. IFSA '91*, Brussels, pp. 81-84, 1991

[7] S. Kawaji, R. Maeda, N. Matsunaga: "Design of fuzzy control system based on PD control scheme", *Proc IFSA '91*, Brussels, pp. 77-80, 1991

Chapter 11
The Composition of Heterogeneous Control Laws

Introduction, *244*
Qualitative Descriptions of Incomplete Knowledge, *245*
A Heterogeneous Controller for the Water Tank, *248*
Guarantees, *250*
Simulation Results, *255*
Integral Action, *256*
Relationship to Fuzzy Logic Control, *259*
References, *260*

To design a control system to operate over a wide range of conditions, it may be necessary to combine control laws which are appropriate to the different operating regions of the system. The fuzzy control literature, and industrial practice, provide certain non-linear methods for combining heterogeneous control laws, but these methods have been very difficult to analyze theoretically. This chapter describes an alternate formulation and extension of this approach that has several practical and theoretical benefits. First, the elements to be combined are classical control laws, which provide high-resolution control and can be analyzed by classical methods. Second, operating regions are characterized by fuzzy set membership functions. The global heterogeneous control law is defined as the weighted average of the local control laws, where the weights are the values returned by the membership functions, thereby providing smooth transitions between regions. Third, the heterogeneous control system may be described by a qualitative differential equation, which allows it to be analyzed by qualitative simulation even in the face of incomplete knowledge of the underlying system or the operating region membership functions. An example of heterogeneous control is given for level control of a water tank, and two alternate analysis methods are presented.

THE COMPOSITION OF HETEROGENEOUS CONTROL LAWS

Benjamin Kuipers
Department of Computer Sciences
University of Texas at Austin
Austin, Texas USA

Karl Åström
Department of Automatic Control
Lund Institute of Technology
Lund, Sweden

1 INTRODUCTION

Much control theory is based on linear models. This works very well for steady state regulation at a fixed operating point. To make a control system that can operate over wide regions it is however necessary to introduce nonlinearities. There are several ways to do this. Linear feedback control can be combined with logic for switching between several linear feedback laws. Selectors that choose between different control laws depending on signal levels can be introduced. Systems of these types are common in industry. Due to their complexity they are however poorly understood theoretically. Their design is based on engineering experience combined with extensive simulation.

Fuzzy logic control [12, 6] is another approach to obtain nonlinear control systems. In this approach the measured variables are represented as fuzzy variables. A representation of the control signal as a fuzzy variable is computed from the measurements using fuzzy logic. The fuzzy variable is converted to a real variable using some type of "defuzzification."

In this paper we take an alternate view of the problem of switching between differennt control strategies. A heterogeneous control problem is decomposed into multiple, possibly overlapping, operating regions. The domain of each operating region is characterized by a fuzzy set membership function. This makes it possible to express smooth transitions between adjacent regions. Each operating region is associated with a qualitative description of the system state, e.g. the low, normal, or high level of water in a tank. The fuzzy set membership functions may be regarded as a measure of the *appropriateness* of applying a given qualitative description to the system state. It is assumed that, for any given system state, the operating region membership functions sum to 1.0.

Each region is associated with a control law, and the control signal applied to the plant is a weighted average of the control signals for each region, where the weights are provided by the membership functions of each region. This approach to fuzzy control was pioneered by Takagi and Sugeno [9] and Sugeno and Kang [8].

This approach makes it possible to decompose the design of a heterogenous controller into two relatively independent decisions: (1) the specification of natural, qualitatively distinct operating regions, and (2) the specification of a control law for each region. The weighted sum combination method provides smooth transitions from one region to another, and facilitates local and global analysis. The idea of combining simple linear feedback units with operations such as average, min, max, etc, is widely used industrially. The intent of this paper is to provide a mathematical basis for the design of such systems, and for local and global analysis of their properties.

Heterogeneous control is also related to gain scheduling. There are however some differences. In gain scheduling a specific control law is selected for a given operating region and the parameters of the controller are changed with the region. In heterogeneous control the values of the control signal for different regions are computed and averaged.

Classical control theory [2] provides a rich set of methods for local analysis of the individual control laws and for describing their behavior. A global analysis of the behavior of a heterogenous control system is expressed as a transition graph, where the nodes correspond to operating regions and the directed edges correspond to possible transitions between regions. We provide a methodology for deriving this global analysis from the individual control laws and the membership functions of the operating regions, even on the basis of an incomplete, qualitative description of the structure of the system and its controller.

The global analysis methodology makes it possible to derive the assumptions under which a discrete transition-graph abstraction captures the essential properties of a continuous, heterogenous dynamic control system. It also identifies situations where a proposed abstraction may fail, and helps identify the additional constraints required to guarantee the desired global behavior.

The basic concepts of heterogeneous control will be introduced with a simple level controller for a water tank.

2 QUALITATIVE DESCRIPTIONS OF INCOMPLETE KNOWLEDGE

There are at least two fundamentally different types of qualitative descriptions of incomplete knowledge of scalar quantities.

- **"Fuzzy" Values** are qualitative descriptions without precise boundaries.

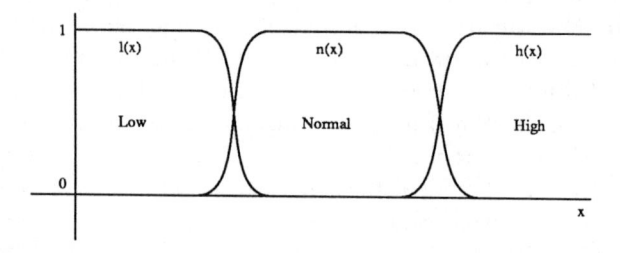

Figure 1: Appropriateness Measures (i.e. Fuzzy Set Membership Functions) for Qualitative Terms Describing Levels.

- **Landmark Values** are precise "natural joints" that break a continuum into qualitatively distinct regions.

2.1 The Fuzzy-Set Representation

Fuzzy sets were originally developed by Zadeh [12,11] to formalize qualitative concepts without precise boundaries. For example, the level of water in a tank might be characterized by the qualitative descriptive terms, *low*, *normal*, and *high*. There are no meaningful landmark values representing the boundaries between *low* and *normal*, or *normal* and *high*.

Zadeh [11] formalizes linguistic terms such as these as referring to *fuzzy sets* of numbers, in this case, levels of water in the tank. A fuzzy set, S, within a domain, D, is represented by a *membership function*, $s : D \rightarrow [0,1]$. For our purposes, we will interpret the value $s(x)$, for an element $x \in D$, as a measure of the *appropriateness* of describing x with the descriptor S. Figure 1 includes three membership functions defining the appropriateness of applying the qualitative descriptors $\{low, medium, high\}$ to quantitatively-defined levels.[1]

A *fuzzy logic controller* consists of a collection of simple control laws whose inputs and outputs are both fuzzy values [12, 6]. For example,

If water level is *high*, then set drain opening to *wide*;

where *high* and *wide* are qualitative terms described by fuzzy sets over their quantitative domains.

All controller rules are fired in parallel, and the recommended actions are combined according to fuzzy value combination rules, weighted by the degree of satisfac-

---

[1] *Appropriateness measure* is technically synonymous with the terms *membership function* and *possibility measure* as used in the fuzzy research community. However, the English connotation of *appropriateness measure* seems better to capture the relationship between a linguistic term and a quantitative measure.

tion of the antecedent. Some process of "defuzzification" is required to convert the resulting fuzzy set description of an action into a specific value for a control variable.

2.2 Landmark-Based Representation

Frequently, qualitative categories are defined by *landmark values*: precise boundary points separating qualitatively distinct regions of a continuum. For example, water temperature can be described qualitatively with respect to the landmarks

$$\cdots\cdots Freezing \cdots\cdots Boiling \cdots\cdots$$

and angles in a triangle can be described in terms of the landmarks

$$Zero \cdots\cdots Right \cdots\cdots Straight.$$

A value can be described qualitatively either as equal to a landmark value or in the open interval between two landmark values, even when numerical information is unavailable. It is often easier to obtain or justify the qualitative description of a quantity than its numerical value, particularly when knowledge is incomplete. Fortunately, landmark-based descriptions support *qualitative simulation*, to derive qualitative descriptions of the possible behaviors of a system from a qualitative description of its structure [3].

An ordinary differential equation describes a system in terms of a set of variables which vary continuously over time, along with constraints such as addition, multiplication, and differentiation, on the relationships among those variables. A *qualitative differential equation* (QDE) describes a system in much the same terms, except that (1) the values of variables are described qualitatively, and (2) certain functional relationships between variables may be incompletely known and qualitatively described. For example, air resistance on a moving body increases monotonically with velocity, and flow of water through an orifice increases monotonically with pressure. Both of these relations are non-linear, but useful qualitative conclusions can be drawn purely from monotonicity. It is useful to define the class M^+ of monotonic functions, and the class S^+ of monotonic functions with saturation.

- M^+ is the set of continuously differentiable functions $f : \Re \to \Re$ such that $f'(x) > 0$ for all $x \in \Re$. In a QDE, we may write $M^+(pressure, outflow)$ or $outflow = M^+(pressure)$ to mean that there is some $f \in M^+$ such that $outflow = f(pressure)$. M_0^+ is the subset of M^+ such that $f(0) = 0$, and M^- is the set of f such that $-f \in M^+$.

- S^+ is the set of continuously differentiable functions $f : \Re \to \Re$ such that, for specified pairs of landmark values (x_1, y_1) and (x_2, y_2),

    - $f(x) = y_1$ for all $x \leq x_1$,

- $f(x) = y_2$ for all $x \geq x_2$,
- $f'(x) > 0$ for all $x_1 < x < x_2$.

The turning points (x_1, y_1) and (x_2, y_2) must be specified as landmark values whenever the S^+ constraint is used. Notice, in figure 1, that the fuzzy set membership functions $h(x) \in S^+$ and $l(x) \in S^-$. We can also treat $n(x)$ as belonging to a composite set $S^+ \cdot S^-$. This qualitative description expresses a state of incomplete knowledge where we know only that the membership functions behave monotonically when they are between the landmarks 0 and 1.

Qualitative simulation using the QSIM algorithm [3, 5] takes a QDE and a qualitative description of an initial state, and derives a tree of qualitative state descriptions, where the paths from the root to the leaves of the tree represent the possible behaviors of the system. This set of behaviors is guaranteed to include every solution of every ordinary differential equation consistent with the given QDE and initial state. Thus, a result which follows from a qualitative description of a system, must apply to every fully-specified instance of that description.

The tree of possible behaviors of a qualitatively described system can be a powerful analytical tool. In particular, if a qualitative property (e.g. stability or zero-offset) holds on every branch of the tree, it must hold for every behavior of the system. The importance of the *qualitative* level of description is that the tree of behaviors for a given QDE may be finite, whereas the corresponding set of ordinary differential equations and their solutions is typically uncountably infinite.

In particular, we will use fuzzy set membership functions to define a non-linear, heterogeneous controller with smooth transitions between operating regions. The system, the control laws, and the operating region definitions can all be represented as a qualitative differential equation, supporting analysis both by classical means and by qualitative simulation.

3 A HETROGENEOUS CONTROLLER FOR THE WATER TANK

3.1 The Water Tank

Consider control of the amount x of water in a tank, where the inflow rate q may vary, and the area u of the drain opening is the control variable. The function $p(x)$ is a monotonically increasing function of x; for a cylindrical tank, $p(x)$ is proportional to the square root of the pressure. The dynamic behavior of the system is described by:

$$\dot{x} = f(x, u) = q - u \cdot p(x).$$

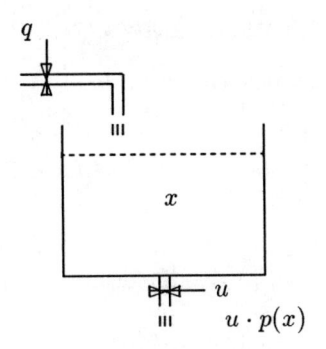

$$\dot{x} = f(x, u) = q - u \cdot p(x).$$

Figure 2: The Water Tank

3.2 Overlapping Operating Regions

The system has separate control laws in three operating regions, *Low*, *Normal*, and *High*, with overlapping membership functions, as shown below.

Note that there is a "pure" region over the intervals $[0, a]$, $[b, c]$, and $[d, \infty)$, and overlapping regions on (a, b) and (c, d). We assume that the setpoint x_s is in (b, c).

The membership functions $l(x)$, $n(x)$, and $h(x)$, for the three operating regions are not known completely. All that is known is that they rise or fall smoothly and monotonically between their plateaus, where the boundaries of the plateaus are characterized by the landmark values, a, b, c, and d. They are normalized, so that $l(x) + n(x) + h(x) = 1$. This state of knowledge can be expressed in terms of the QSIM S^+ constraint by introducing the two functions,

$$s_1(x) \in S^+_{(a,0),(b,1)} \qquad s_2(x) \in S^+_{(c,0),(d,1)}$$

such that

$$\begin{aligned} l(x) &= 1 - s_1(x) \\ n(x) &= s_1(x)(1 - s_2(x)) \\ h(x) &= s_2(x) \end{aligned}$$

Because we specify the membership functions qualitatively, and depend only on properties of the qualitative class, the properties we derive apply to every member of the class.

3.3 Heterogeneous Control Laws

The control laws[2] for the three regions are:

$$u(x) = \begin{cases} 0 & \text{if } x \in Low \\ k(x - x_s) + u_s & \text{if } x \in Normal \\ MAX & \text{if } x \in High \end{cases}$$

where the bias term u_s is adjusted to give the desired set point x_s for a nominal inflow q_s.

The global heterogenous control law is the average of the individual control laws, weighted by the membership functions of their regions; hence

$$u(x) = l(x) \cdot 0 + n(x) \cdot [k(x - x_s) + u_s] + h(x) \cdot MAX.$$

Figure 3 summarizes the heterogeneous controller for the water tank.

4 GUARANTEES

We want to prove that the heterogeneous controller brings the system back to the *Normal* operating region under some range of disturbances, and that an equilibrium in the region is obtained for constant disturbances. More importantly, we want to determine any quantitative constraints on the design of the controller (e.g. the value for MAX), and the range of possible disturbances on q, that the controller can handle.

There are two methods for doing this (fig. 4), which are elaborated on below.

1. (a) Determine the qualitative behavior of the system within each operating region.
   (b) Combine the qualitative descriptions.

2. (a) Combine the local laws into a single global law using the weighted average combination rule.
   (b) Determine the qualitative behavior of the global system.

4.1 Qualitative Combination of Local Properties

The direction of motion of the system as specified by each control law individually is determined first. The properties of the membership functions are not required for this analysis. Then, in the regions of overlap, if the directions of change agree, the global

---

[2] We are assuming here that the state variable x is directly observable, rather than separating out measurements, $y = g(x, u)$. In that case, a measurable variable such as level is linearly, or at least monotonically, related to x, so omitting it simplifies the presentation without reducing generality.

The water tank:

x : the level in the tank (sensed)
u : the drain opening (controlled)

The operating regions:

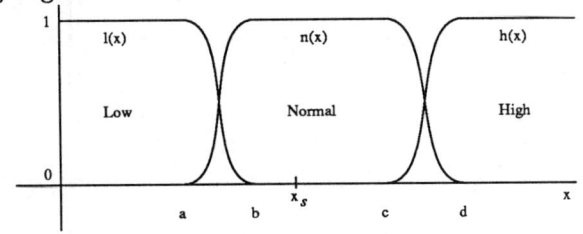

The local control laws:

$$Low \Rightarrow u_l(x) = 0$$
$$Normal \Rightarrow u_n(x) = k(x - x_s) + u_s$$
$$High \Rightarrow u_h(x) = MAX$$

The global control law:

$$\bar{u}(x) = l(x)u_l(x) + n(x)u_n(x) + h(x)u_h(x).$$

The discrete abstraction:

$$\boxed{Low} \longrightarrow \boxed{\boxed{Normal}} \longleftarrow \boxed{High}.$$

Figure 3: A Heterogenous Controller for the Water Tank.

```
                    Local         Global
                        2(a)
    Control Law  •─────────────→•
                 │               │
                 │1(a)           │2(b)
                 ↓               ↓
    Behavior     •─────────────→•
                        1(b)
```

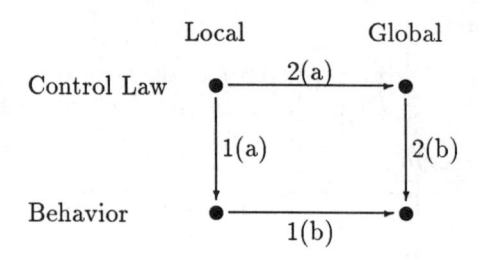

Figure 4: Two Approaches to Analyzing a Heterogeneous Controller.

law for the heterogenous controller must give motion in the same direction. If the different control laws give motion in opposite directions, qualitative simulation provides the possible behaviors. Alternatively, order-of-magnitude or semi-quantitative analysis may be able to clarify the system's behavior in the overlap regions.

In order to guarantee that the system always ends up within the "pure" operating region, (b, c) of the *Normal* controller, we need to impose constraints on (1) the range of inflow perturbations to be handled, and (2) the magnitude of the *High* response.

1. From the *Normal* model:
$$q_b \leq q \leq q_c$$
where q_b (resp. q_c) is the value of q that results in steady state at $x = b$ (resp. $x = c$).

2. From the *High* model:
$$q/MAX \leq p(c).$$

These conditions simply require that the drain area can be made sufficiently large so that the outflow at the desired level can be made to match the disturbance inflow. This guarantee does *not* depend on other constraints, in particular on the shapes of the membership functions.[3]

Once we have established these qualitative properties of the system and its heterogeneous controller, they can be expressed as a finite transition graph in which the nodes correspond to the operating regions.

```
    ┌─────┐      ┌────────┐      ┌──────┐
    │ Low │ ──→  │ Normal │  ←── │ High │
    └─────┘      └────────┘      └──────┘
```

---

[3]The individual steps of this analysis can be established automatically by QSIM simulation of the individual controllers.

where the double box signifies that the *Normal* region includes a steady state, and so can persist indefinitely, while the other regions can persist only for a finite time.

The abstraction relation is defined as follows:

- The state of the system corresponds to a node of the transition graph if it is in the *interior* of the corresponding "pure" operating region, where its membership function is equal to 1.

- The links between nodes correspond to the overlap between operating regions. The system must move from one pure region to another in finite time.

Figure 5 summarizes this analysis.

4.2 Qualitative Analysis of the Global Control Law

A global analysis of the heterogeneous system is possible when we can establish suitable relations among the individual control laws.

Suppose we can establish that the global control law $u(x)$ is a monotonic function of x. Then the closed-loop system can be described as

$$\dot{x} = q - u(x)\,p(x) = q - f(x), \text{ for some } f \in M^+.$$

Since this is a first-order system, the analysis is straight-forward. An equilibrium exists if q is in the range of f. The solution is unique since f is monotone. The solution is stable because $f' > 0$, since $f \in M^+$.

It is necessary to introduce some compatibility conditions in order to avoid pathological behavior of the system. To see this, consider the case where only two controllers are combined (e.g., the *Normal* and *High* controllers over the range (b, ∞) in the water-tank example). The control signal is then

$$u(x) = n(x)\,u_n(x) + h(x)\,u_h(x).$$

It is natural to have controllers such that

$$\frac{du_n}{dx} \geq 0 \text{ and } \frac{du_h}{dx} \geq 0.$$

Unfortunately, these conditions do not guarantee that u is monotone. To obtain this, some auxiliary conditions are required.

Consider

$$u' = n\,u'_n + n'\,u_n + h\,u'_h + h'\,u_h$$

$$n + h = 1$$
$$n' + h' = 0$$

- Overlapping operating regions for the local laws.

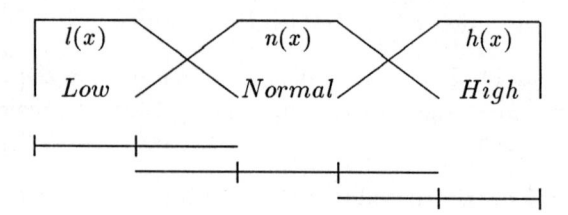

- Require qualitative agreement where local laws overlap.

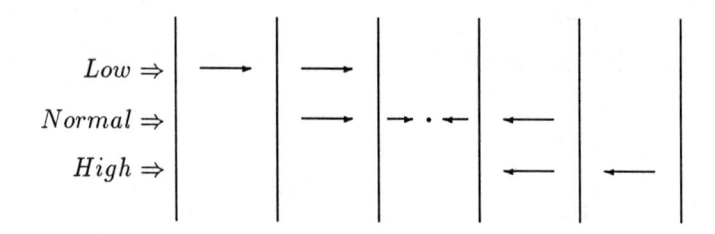

- Abstract the control law to a finite transition diagram.

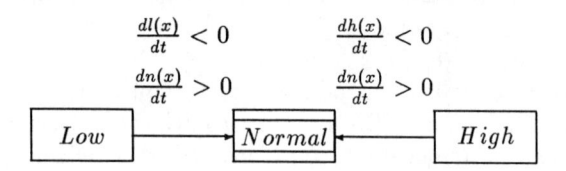

$$\frac{dl(x)}{dt} < 0 \qquad \frac{dh(x)}{dt} < 0$$
$$\frac{dn(x)}{dt} > 0 \qquad \frac{dn(x)}{dt} > 0$$

$$\boxed{Low} \longrightarrow \boxed{Normal} \longleftarrow \boxed{High}$$

Figure 5: Qualitative Combination of Properties of Local Laws.

The problem is that n' is negative. However, we can conclude:

$$u' = n\, u'_n + h\, u'_h + h'(u_h - u_n)$$

This assures us that $u' > 0$, and hence that $f(x) = u(x)\, p(x)$ is in M^+, if we impose the natural condition

$$u_n(x) \le u_h(x).$$

This condition needs to hold only for x where the two regions overlap. The argument obviously extends to more complex heterogeneous controllers, such as the water tank, where no more than two regions overlap at any point.

Consider the case where three regions overlap.

$$l + n + h = 1$$
$$l' + n' + h' = 0$$

The analysis produces a similar result:

$$\begin{aligned} u' &= l\, u'_l + l'\, u_l + n\, u'_n + n'\, u_n + h\, u'_h + h'\, u_h \\ &= l\, u'_l + n\, u'_n + h\, u'_h + n'(u_n - u_l) + h'(u_h - u_l) \end{aligned}$$

Thus the constraint,

$$u_l(x) \le u_n(x) \le u_h(x),$$

guarantees that $u' \ge 0$ even when all regions overlap.

Notice that this constraint does not require that the local control laws be linear. Furthermore, a local control law needs to satisfy this constraint only where its membership function is non-zero.

5 SIMULATION RESULTS

We can illustrate the performance of a heterogeneous controller on a water tank, in comparison with a proportional controller.

The capacity of the tank is 1000 liters of water. The nominal inflow rate is 100 liters/minute. The setpoint, x_s, is 700 liters. The offset u_s in the *Normal* control law u_n is set so that the steady state is at the setpoint when the inflow is nominal. The gain k is set so that $u_n(0) = 0$. The proportional controller simply uses u_n as the global control law.[4]

The operating regions for the HC controller are specified as in figure 3, with $a = 600$, $b = 650$, $c = 750$, $d = 800$, and $MAX = 50$.

---

[4]This comparison is for illustration only, since the proportional controller has an unrealistically low gain. With a higher gain, however, the physical limits on the valve make the proportional controller behave like a heterogeneous controller, but without smooth transitions or explicit design and validation.

Figure 6 shows the two control laws, and contrasts the behavior of the two controllers at constant nominal inflow, starting from initial states with the tank full and empty. Figure 7 shows the response of the two controllers to random variation in inflow.

6 INTEGRAL ACTION

The bias term in the proportional controller was introduced to make it possible for the controller to keep the level at the set point. Integral action may be viewed as an automatic adjustment of the bias term (See Figure 2.2 in [1]). For a simple PI controller the bias is generated as

$$T\frac{du_s}{dt} + u_s = ke + u_s$$

or

$$T\frac{du_s}{dt} = u_p = ke \qquad (1)$$

where u_p is the output of the PI controller, e the error $x - x_s$, k the proportional gain and T is the integration time. For a composite controller like the one used in heterogeneous control u_p should be replaced by the output of the heterogeneous controller.

Analysis of a controller with integral action is more complicated because the closed loop system is described by a second order differential equation and a simple monotonicity argument like the one used previously does not apply directly.

Just as there were two alternative approaches to the qualitative analysis of the heterogeneous "proportional" controller, there appear to be three basic approaches to analyzing the "integral" component of a heterogeneous controller.

1. The bias term u_s is adjusted, at a slower time-scale, by a heterogeneous P-controller as a function of the steady-state error, $x_s - x(\infty)$ as discussed below.

2. Local control laws, even with integral action, can be analyzed qualitatively, and associated with overlapping operating regions in the phase plane. If the directions of flow in the overlap regions are compatible, the qualitative descriptions can be combined into a discrete transition-graph representing behavior in the phase plane [7].

3. The local laws may be combined into a global control law using the weighted average combination rule, which may then be analyzed qualitatively.

One possibility is to exploit the fact integral action is a slow process. The idea of time scale separation introduced in [4] can then be applied. The full details will be given elsewhere. Let us just outline the ideas of the reasoning. Provided that the

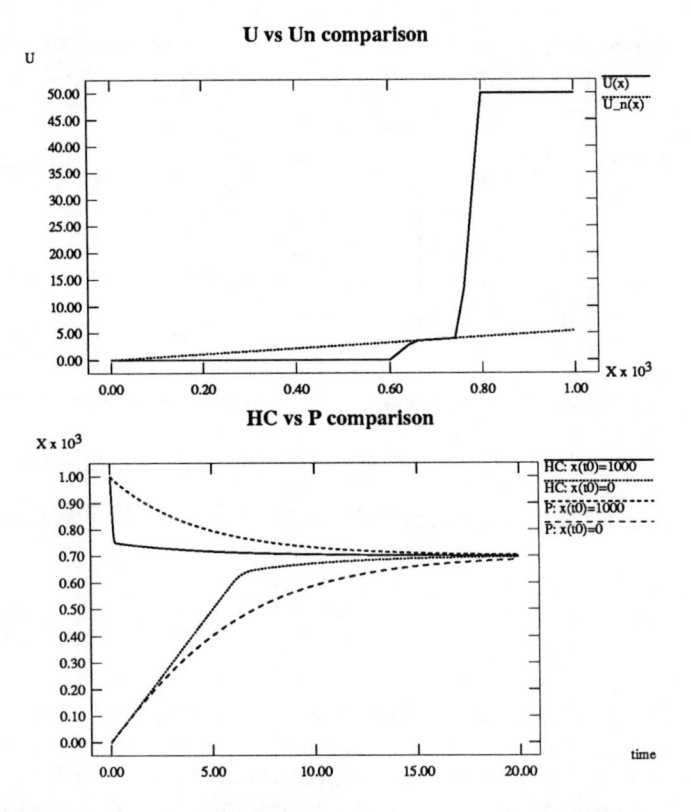

Figure 6: Comparison Between P and HC Controllers.

(a) The heterogeneous control law $u(x)$, and the proportional controller $u_n(x)$ are identical in the Normal region.

(b) The behaviors, $x(t)$, of the P- and HC-controllers, starting with the tank empty or full, with constant q at the nominal rate, so that steady state is at the setpoint.

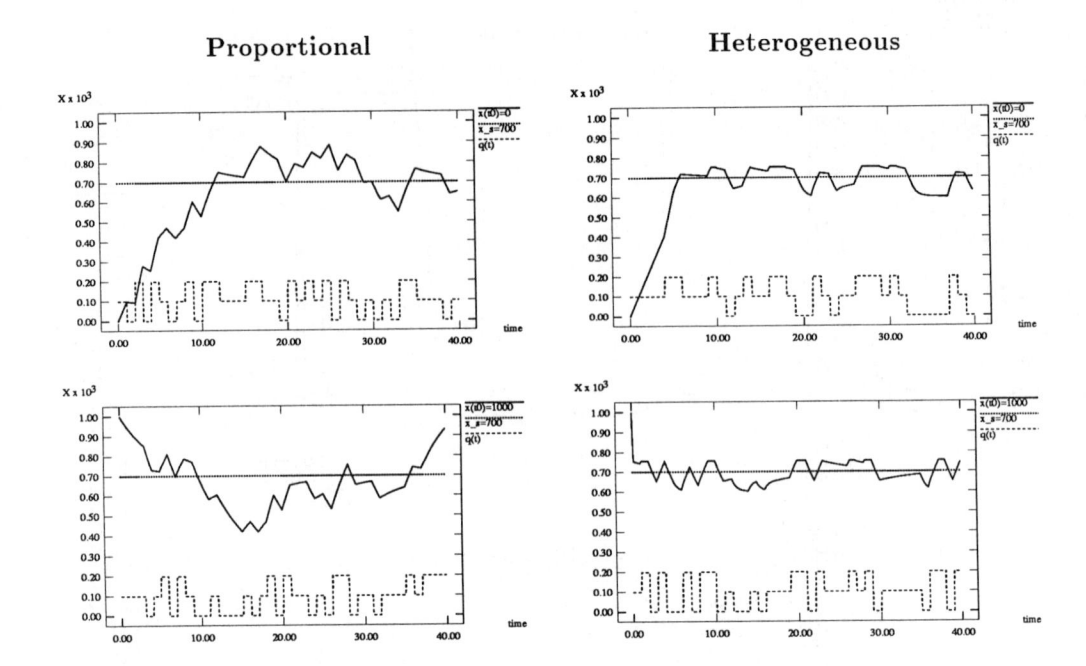

Figure 7: The Effect of Random Inflow Variation on P and HC Controllers. Inflow, q, plotted at the bottom of each graph, varies randomly between zero, nominal (100 liters/minute), and twice nominal. This figure shows the proportional controller (left column) and the heterogeneous controller (right column), with the tank initially empty (top row) or full (bottom row).

integration time T is sufficiently small the closed loop system can be decomposed into a fast system, where the bias term is considered constant, and a slow system, where the fast system is considered as a static system. The previous analysis then applies to the fast system. It follows from this analysis that the level goes to an equilibrium which may be different from the set point. At equilibrium the fast system can be described by

$$u_p = -f(u_s) \qquad (2)$$

where the function f belongs to M^+. The slow system is described by (2) and (1), i.e.

$$T\frac{du_s}{dt} = u_p = -f(u_s)$$

Since f is monotone this equation has a unique stable equilibrium

$$u_p = ke = 0$$

which implies that the error e must be zero when the slow system reaches equilibrium.

7 RELATIONSHIP TO FUZZY LOGIC CONTROL

Our approach to heterogeneous control shares many goals with, and draws much inspiration from fuzzy logic control [12, 6]. First, both approaches provide the ability to express and use incomplete knowledge of the system being controlled and the control law itself. Second, both approaches allow one to specify a complex control law as the composition of simple components. Third, both use fuzzy set membership functions to provide smooth transitions from region to region.

However, there are important differences between our approach and fuzzy logic control. Within the framework of fuzzy logic control, it is difficult to exploit, or even relate to, the methods or results of traditional control theory. Our approach uses landmark-based qualitative reasoning to combine the benefits of fuzzy logic control with the analysis methods of traditional control theory.

Granularity. A fuzzy logic controller is typically specified as a relatively fine-grained set of (fuzzy) regions, with a constant (fuzzy) action associated with each region. Within our framework, the design for a controller specifies a smaller set of qualitatively distinct operating regions, but with a classical control law associated with each region.

The net result of these two differences is that an HC controller requires a simpler specification, while providing the higher-precision control characteristic of classical control laws.

Ontology. We do not treat "linguistic values" or "linguistic variables" as objects in either the domain or range of our functions. Rather, the fundamental objects in heterogeneous control are real-valued, continuously differentiable functions, and sets of such functions defined by qualitative constraints.

Linguistic terms are treated simply as names for the operating regions of the mechanism. The specifications for the operating regions are tested for soundness by the qualitative analysis methods.

"Defuzzification." The output of a fuzzy logic control law is typically a constant action with a fuzzy magnitude. The fuzzy magnitude must then be mapped to a real value for output. Since control laws in our framework are classical control laws, they provide real, not fuzzy, outputs, and "defuzzification" is not necessary.

Compatability. The concepts underlying fuzzy logic control are relatively difficult to map into the classical framework, making it difficult to exploit existing methods for providing guarantees for the properties of fuzzy logic controllers.

In HC control, the individual regions can be analyzed using classical methods, and we have demonstrated qualitative methods for combining analyses of the individual regions into a global analysis.

Specificity. Although both approaches have the goal of representing incomplete knowledge, fuzzy set membership functions must be represented as specific real-valued functions.

While an appropriateness measure in an HC controller must also be fully specified, the *analysis* of the controller relies only on a qualitative description (e.g. S^+, S^-, or $S^+ \cdot S^-$) of the measure. This makes explicit the fact that a single guarantee applies to a whole class of appropriateness measures, allowing additional degrees of freedom for implementation decisions. The goal of qualitative analysis is to define the least restrictive description of the controller which provides a given performance guarantee.

REFERENCES

1. K. T. Åström and T. Hägglund. 1988. *Automatic Tuning of PID Controllers*. Research Triangle Park, NC: Instrument Society of America.

2. G.F. Franklin, J.D. Powell and A. Emami-Naeini. 1986. *Feedback Control of Dynamics Systems*. Reading, MA: Addison-Wesley.

3. B. J. Kuipers. 1986. Qualitative simulation. *Artificial Intelligence* **29**: 289 - 338.

4. B. Kuipers. 1987. Abstraction by time-scale in qualitative simulation. *Proceedings of the National Conference on Artificial Intelligence (AAAI-87)*. Los Altos, CA: Morgan Kaufman.

5. B. Kuipers. 1989. Qualitative reasoning: modeling and simulation with incomplete knowledge. *Automatica* **25**: 571-585.

6. E. H. Mamdani. 1974. Applications of fuzzy algorithms for control of a simple dynamic plant. *Proc. IEEE* **121**: 1585-1588.

7. E. Sacks. 1990. Automatic qualitative analysis of dynamic systems using piecewise linear approximations. *Artificial Intelligence* **41**: 313-364.

8. M. Sugeno and G. T. Kang. 1986. Fuzzy modelling and control of multilayer incinerator. *Fuzzy Sets and Systems* **18**: 329-346.

9. T. Takagi and M. Sugeno. 1985. Fuzzy identification of systems and its applications to modeling and control. *IEEE Transactions on Systems, Man, and Cybernetics* **15**: 116-132.

10. R. R. Yager, S. Ovchinnikov, R. M. Tong, and H. T. Nguyen (Eds.). 1987. *Fuzzy Sets and Applications: Selected Papers by L. A. Zadeh*. NY: John Wiley & Sons.

11. L. Zadeh. 1965. Fuzzy sets. *Information and Control* **8**: 338-353.

12. L. Zadeh. 1973. Outline of a new approach to the analysis of complex systems and decision processes. *IEEE Trans. Systems, Man, and Cybernetics* **SMC-3**: 28-44.

Chapter 12
Synthesis of Nonlinear Controllers via Fuzzy Logic

Introduction, *264*
Fuzzy Control Systems, *264*
Problem Statement, *267*
Discussion, *271*
Conclusion, *273*
References, *273*

This chapter presents an analytic formulation of a class of fuzzy logic control algorithms. This formulation makes use of the idea that characteristic functions of fuzzy sets may be considered to be algebraic functions of their arguments. Placing common constraints such as continuity and boundedness on these functions, and making use of a variation of the so-called centroid of area defuzzification rule, results in an analytic formulation of the control law that can be used to explain how fuzzy control algorithms function as they do, and makes formal analysis of fuzzy control systems, at least in principle, possible.

SYNTHESIS OF NONLINEAR CONTROLLERS VIA FUZZY LOGIC

Reza Langari
Department of Mechanical Engineering
Texas A&M University
College Station, TX 77843-3123

1 INTRODUCTION

Fuzzy Logic Control(FLC) is fast emerging as an alternative to conventional control techniques in situations where it may be infeasible to formulate an analytic model of the process or, alternatively, when performance measures are not readily transformed into standard control theoretic design objectives– as steady state or transient characteristics of the closed loop system. Industrial processes such as Gas Metal Arc Welding(GMAW), whose complexity exceeds the limitations of analytic modeling tools are prime candidate for application of FLC[1].

In spite of the success that FLC continues to demonstrate, however, precisely what function fuzzy logic control helps accomplish in a given situation remains to be fully understood. Attempts to deal with this issue have taken the form of empirically based comparative studies, such as comparison with PID control[2], or analytic studies, as for instance by Kickert and Mamdani[3], Buckley and Ying[4], and Langari[5]. The present article extends the results presented in the latter work and develops a more general analytic description that is applicable to a broader class of fuzzy logic control algorithms.

The remainder of the article is as follows. First an overview of fuzzy control, which is primarily meant to establish the notation for the presentation of our main results, is presented. Next, we derive the aforementioned analytic description and discuss its implications. We conclude the article by a summary of its major points.

2 FUZZY CONTROL SYSTEMS

In this article we consider fuzzy linguistic control system, taking the form shown in Figure 1, where the knowledge of operation of the plant is explicitly represented as *condition* → *action* rules of the form:

$R_{j,l}$: if $e(t)$ is $\tilde{A}_j$ and $de(t)$ is $\tilde{B}_l$ then $u(t)$ is $\tilde{C}_{j,l}$,

where $e(t)$ denotes the instantaneous value of the process error at time t and $de(t)$ is short for $\mathcal{D}(e;t)$, which stands for de/dt or $\int^t e d\tau$. Further, $\tilde{A}_j$, $\tilde{B}_l$ and $\tilde{C}_{j,l}$ belong to collections

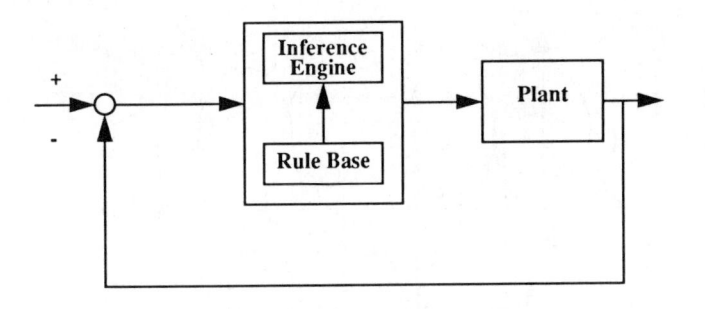

Figure 1 Architecture for Fuzzy Control

$\tilde{\mathcal{A}}$, $\tilde{\mathcal{B}}$, and $\tilde{\mathcal{C}}$ of fuzzy subsets defined over the domains of definition of the relevant variables(Figure 2.) Further, in above definitions, $R_{j,l}$ denotes the j,l^{th} rule in the rule set $\mathcal{R}$. In particular, $R_{j,l}$ may be viewed as associating elements, $\tilde{A}_j$ of $\tilde{\mathcal{A}}$ and $\tilde{B}_l$ of $\tilde{\mathcal{B}}$ with element $\tilde{C}_{j,l}$ of $\tilde{\mathcal{C}}$.

2.1 Control Computation

Suppose, at some instance t, the error $e(t)$ 'has positive grades of membership, $\mu_{\tilde{A}_j}(e(t))$ and $\mu_{\tilde{A}_{j+1}}(e(t))$ to some pair $\tilde{A}_j$ and $\tilde{A}_{j+1}$ of $\tilde{\mathcal{A}}$. Similarly, suppose $de(t)$ belongs to some pair $\tilde{B}_l$ and $\tilde{B}_{l+1}$ of $\tilde{\mathcal{B}}$. Accordingly, at this instant, the following control rules apply[1]:

$R_{j,l}$: if $e(t)$ is $\tilde{A}_j$ and $de(t)$ is $\tilde{B}_l$ then $u(t)$ is $\tilde{C}_{j,l}$,

$R_{j+1,l}$: if $e(t)$ is $\tilde{A}_{j+1}$ and $de(t)$ is $\tilde{B}_l$ then $u(t)$ is $\tilde{C}_{j+1,l}$,

$R_{j+1,l+1}$: if $e(t)$ is $\tilde{A}_{j+1}$ and $de(t)$ is then $u(t)$ is $\tilde{C}_{j+1,l+1}$,

$R_{j,l+1}$: if $e(t)$ is $\tilde{A}_j$ and $de(t)$ is $\tilde{B}_{l+1}$ then $u(t)$ is $\tilde{C}_{j,l+1}$,

with each rule satisfied to some degree. The corresponding *truth value* is defined, for instance for the first rule, by

$$\mu_{j,l} = min(\mu_{\tilde{A}_j}(e(t)), \mu_{\tilde{B}_l}(de(t))), \tag{1}$$

or, by

---

[1]. In principle it is possible for more than four rules to be applicable at any given time. The treatment as presented in this section, however, remains valid with the appropriate changes to the number of applicable rules where necessary.

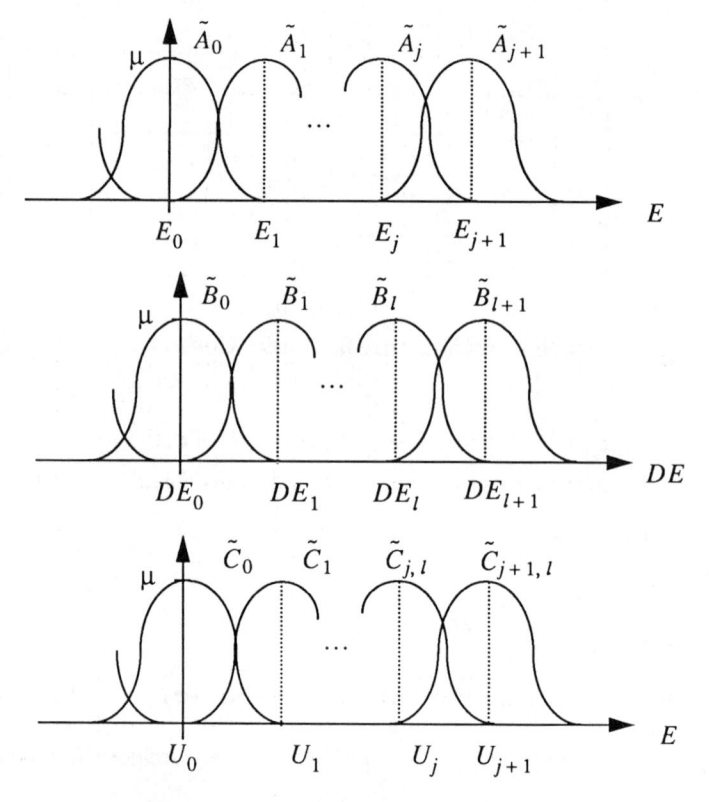

Figure 2 Collections $\mathcal{A}$, $\mathcal{B}$, and C of fuzzy subsets partitioning E, DE, and U.

$$\mu_{j,l} = \mu_{\tilde{A}_j}(e(t))\mu_{\tilde{B}_l}(de(t)). \qquad (2)$$

The truth values of other rules in the above set are similarly defined. Note that the *product* instead of *min* results in interactivity between the truth values of the components of the antecedent clause. This fact is essential to our analytic treatment, and will be made use of later in the article.

Now, representing the consequent clause of each $R_{j,l}$ rule, that is $\tilde{C}_{j,l}$, by its single representative, or *defuzzified*, value that is $U_{j,l}$, defined as

$$U_{j,l} = \frac{\int u\mu_{\tilde{C}_{j,l}}(u)}{\int \mu_{\tilde{C}_{j,l}}(u)}, \tag{3}$$

the control action, $u(t)$, is computed as

$$u(t) = \frac{\sum_{j,l} \mu_{j,l} U_{j,l}}{\sum_{j,l} \mu_{j,l}}, \tag{4}$$

where j and l range over the indices of all applicable rules.

Note that this approach is based on a variation of the Centroid of Area(COA) defuzzification rule[6], but as we shall see later has better analytical properties. For instance $U_{j,l}$ is the COA of $\tilde{C}_{j,l}$ and *would be* exactly equal to $u(t)$ if the ith rule were the *only* applicable rule. In effect the above approach superimposes the action of various rules and is exactly equivalent to the COA rule if triangular shaped membership functions were used. In other instances, the equivalence is approximate, but as we shall see shortly this distinction is irrelevant in so far as the presented results are concerned.

It is interesting to observe the fuzzy control action as a function of controller inputs. In particular, when the e and de are exactly at the center values of the relevant fuzzy subsets, for instance $\tilde{A}_j$ and $\tilde{B}_l$, or $\tilde{A}_{j+1}$ and $\tilde{B}_{l+1}$, then the control action is precisely that which is given by the defuzzified value of the rule, that is $U_{j,l}$. In other instances the fuzzy inference engine computes the control action, $u(t)$ by interpolation, that $u(t)$ is a weighted average of $U_{j,l}, U_{j+1,l}, \ldots$ and the associated weights are the truth values of the rules. This is the point that is rigorously established later on.

3 PROBLEM STATEMENT

We consider the single input, single output fuzzy logic control systems where the control rules take the form described in Section 2. The objective here is to develop an analytic representation of the control law, $u = FLC(e, de)$, which (*i*) remains faithful to the spirit in which the original rules were stated and (*ii*) offers a description of what the control law does and also what it does not do.

3.1 Assumptions and Definitions

Let us denote the domains of definition of e, de and u by E, DE, and U respectively. Then as shown in Figure 2, collections $\tilde{\mathcal{A}} = \{\tilde{A}_j\}$ of *unimodal, convex,* and *normal* fuzzy subsets[6] effectively partition E, DE, and U, respectively, as follows. Each element $\tilde{A}_j$, a fuzzy subset, is centered at some $E_j \in E$ and is characterized by some $\mu_{\tilde{A}_j}$. Similarly, each element $\tilde{B}_l \in \tilde{\mathcal{B}}$ is centered at some $DE_l \in DE$, and is characterized by $\mu_{\tilde{B}_l}$. In addition, each element $\tilde{C}_{j,\,l} \in \tilde{\mathcal{C}}$ is effectively represented by its *defuzzified value*, $U_{j,\,l}$. As discussed in Section 2.1, $U_{j,\,l}$ is the Centroid of Area(COA) representation of $\tilde{C}_{j,\,l}$. We shall further place the following constraint on $U_{j,\,l}$s.

Assumption 1 For each $U_{j,\,l}$, there exist a pair K_j and K'_l of real constants such that the following relationship holds.
$$U_{j,\,l} = K_j E_j + K'_l DE_l.$$

We then define, for each j and likewise for each l, ΔK_j and $\Delta K'_l$ as follows:
$$\Delta K_j = K_{j+1} - K_j, \tag{5}$$
$$\Delta K'_l = K'_{l+1} - K'_l. \tag{6}$$

Next we place some constraints on $\tilde{\mathcal{A}}$ and $\tilde{\mathcal{B}}$ as follows. First, we require that $\tilde{\mathcal{A}}$ and $\tilde{\mathcal{B}}$ form *approximate fuzzy partitions* of E and DE, respectively.

Assumption 2 Let $\tilde{\mathcal{A}} = \{\tilde{A}_j\}$ (and $\tilde{\mathcal{B}} = \{\tilde{B}_l\}$) be collection(s) of fuzzy subsets define over E (and DE). Then, for each element $e \in E$,
$$\sum_j \mu_{\tilde{A}_j}(e) \cong 1.$$
(A similar condition holds for $\tilde{\mathcal{B}}$.)

In other words, the sum total of membership of each element $e \in E$, for instance, to the elements of $\tilde{\mathcal{A}}$, should be nearly unity.

The interpretation of Assumption 2 is that, *externally, fuzzy classification must be compatible with feature based classification in terms of classical sets, where each element is categorized under one and only one class*. Note that in view of the statement of Assumption 2, compatibility is not strict. Indeed, if we were to interpret this definition in a strict sense, we would have to require that $\sum_j \mu_{\tilde{A}_j}(e) = 1$, as its is done in Langari[5]. We do, however, generally require that the deviation of $\sum_j \mu_{\tilde{A}_j}(e)$ from its nominal value of one be small. We may formalize this as
$$\delta\mu = \sup_e \left| \sum_j \mu_{\tilde{A}_j}(e) - 1 \right| \ll 1 \tag{7}$$

(A similar condition holds for $\tilde{\mathcal{B}}$.)

We shall also establish some notation as follows. Let $\bar{\mu}_{\tilde{A}_j}(e)$ denote the value of $\mu_{\tilde{A}_j}(e)$ that would satisfy the *ideal* true fuzzy partitioning condition. We then denote the difference between these two values by $\delta\mu_{\tilde{A}_j}(e)$:

$$\delta\mu_{\tilde{A}_j}(e) = \mu_{\tilde{A}_j}(e) - \bar{\mu}_{\tilde{A}_j}(e). \tag{8}$$

(A similar definition holds for $\tilde{\mathcal{B}}$. With this in mind, it is easy to see that $\delta\mu$ given by (7) can be written as

$$\delta\mu = \sup_e \left| \sum_j \delta\mu_{\tilde{A}_j}(e) \right|, \tag{9}$$

which of course as before must be small compared to one.

3.2 Formulation of the control law

Let us recall that any given instance the control action, as described in Section 2.1, can be written as

$$u(t) = \frac{\mu_{j,l} U_{j,l} + \mu_{j+1,l} U_{j+1,l} + \mu_{j+1,l+1} U_{j+1,l+1} + \mu_{j,l+1} U_{j,l+1}}{\mu_{j,l} + \mu_{j+1,l} + \mu_{j,l+1} + \mu_{j,l+1}}. \tag{10}$$

Now considering the fact that, following Section 2.1,

$$\mu_{j,l} = \mu_{\tilde{A}_j}(e) \mu_{\tilde{B}_l}(de), \tag{11}$$

which can be approximated as,

$$\mu_{j,l} \cong \bar{\mu}_{j,l} + \delta\mu_{j,l}, \tag{12}$$

with

$$\bar{\mu}_{j,l} = \bar{\mu}_{\tilde{A}_j}(e) \bar{\mu}_{\tilde{B}_l}(de), \tag{13}$$

and

$$\delta\mu_{j,l} = \bar{\mu}_{\tilde{A}_j}(e) \delta\mu_{\tilde{B}_l}(de) + \delta\mu_{\tilde{A}_j}(e) \bar{\mu}_{\tilde{B}_l}(de). \tag{14}$$

Accordingly, the expression on the right side of (10) can be written as

$$u(t) = \frac{\sum_{j,l} \bar{\mu}_{j,l} U_{j,l} + \sum_{j,l} \delta\mu_{j,l} U_{j,l}}{\sum_{j,l} \bar{\mu}_{j,l} + \sum_{j,l} \delta\mu_{j,l}}. \tag{15}$$

Further, the sum,

$$\sum_{j,l} \bar{\mu}_{j,l} = \bar{\mu}_{j,l} + \bar{\mu}_{j+1,l} + \bar{\mu}_{j+1,l+1} + \bar{\mu}_{j,l+1},$$

is 1, as shown in Langari[5], and $\sum_{j,l} \bar{\mu}_{j,l} U_{j,l}$ is the control action if the true fuzzy partitioning assumption discussed above were to hold. With this in mind, the expression for $u(t)$ can be written as

$$u(t) = \bar{u}(t) + \delta u(t), \qquad (16)$$

where

$$\bar{u}(t) = \sum_{j,l} \bar{\mu}_{j,l} U_{j,l}, \qquad (17)$$

$$\delta u(t) = \sum_{j,l} \delta\mu_{j,l} U_{j,l} - (\sum_{j,l} \bar{\mu}_{j,l}) \bar{u}(t). \qquad (18)$$

Now substituting for $U_{j,l}, U_{j+1,l}, \ldots$ their equivalent values from Assumption 1, we have:

$$U_{j+1,l} = U_{j,l} + \Delta, \qquad (19)$$

$$U_{j+1,l+1} = U_{j,l} + \Delta + \Delta', \qquad (20)$$

$$U_{j,l+1} = U_{j,l} + \Delta', \qquad (21)$$

where, denoting $E_{j+1} - E_j$ by ΔE_j and $DE_{l+1} - DE_l$ by ΔDE_l,

$$\Delta = \Delta K_j E_j + K_j \Delta E_j + \Delta K_j \Delta E_j, \qquad (22)$$

$$\Delta' = \Delta K'_l DE_l + K'_l \Delta DE_l + \Delta K'_l \Delta DE_l. \qquad (23)$$

Further, note that we can write $\bar{\mu}_{\tilde{A}_{j+1}}(e)$ as

$$\bar{\mu}_{\tilde{A}_{j+1}}(e(t)) = \left[\frac{1}{\Delta E_j} + F_j(e(t) - E_j)\right](e(t) - E_j), \qquad (24)$$

which, as shown in Figure 3, reflects the use of F_j to denote the deviation of $\bar{\mu}_{\tilde{A}_{j+1}}$ from its *extreme* linear form. (Similar approach is taken regarding $\bar{\mu}_{\tilde{B}_{l+1}}$.)

Substituting from above in the expression for $\bar{u}(t)$, given by (17), we have[1]:

$$\bar{u} = \bar{u}_1 + \bar{u}_2, \qquad (25)$$

where,

---

1. For simplicity of expression, and where no ambiguity may arise, we shall drop t from the algebraic expressions involving functions of time.

$$\bar{u}_1 = K_j e + \left[\frac{\Delta K_j}{\Delta E_j} E_{j+1} + F_j (e - E_j) \Delta\right] (e - E_j), \quad (26)$$

and,

$$\bar{u}_2 = K'_l de + \left[\frac{\Delta K'_l}{\Delta DE_l} DE_{l+1} + F'_l (de - DE_l) \Delta'\right] (de - DE_l). \quad (27)$$

Likewise, after some algebraic manipulations, we can show that the expression for δu, giv-

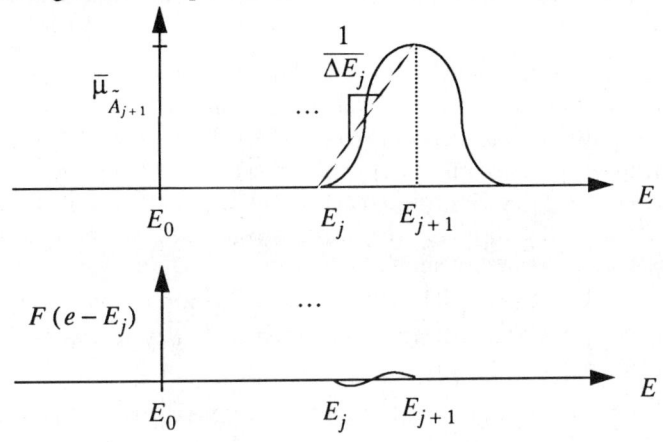

Figure 3 Deviation of $\bar{\mu}_{A_{j+1}}$ from its extreme linear form.

en by (18) yields

$$\delta u = \delta \mu_j \bar{u}_1 + \delta \mu'_l \bar{u}_2, \quad (28)$$

where

$$\delta \mu_j = \delta \mu_{A_j}(e) + \delta \mu_{A_{j+1}}(e), \quad (29)$$

$$\delta \mu'_l = \delta \mu_{B_l}(de) + \delta \mu_{B_{l+1}}(de). \quad (30)$$

4 DISCUSSION

The expression for the control law derived above can be interpreted as follows. First, u can be thought of as the sum of the contributions from $\bar{u}$ and δu. Now $\bar{u}$ is itself sum of the linear terms $K_j e$ and $K'_l de$ and the two additional terms,

$$\left[\frac{\Delta K_j}{\Delta E_j} E_{j+1} + F_j (e - E_j) \Delta\right] (e - E_j),$$

and,

$$\left[\frac{\Delta K'_l}{\Delta DE_l}DE_{l+1} + F'_l(de - DE_l)\Delta'\right](de - DE_l).$$

These terms approach, $\frac{\Delta K_j}{\Delta E_j}E_{j+1}(e - E_j)$, and $\frac{\Delta K'_l}{\Delta DE_l}DE_{l+1}(de - DE_l)$, respectively, in the limit[1], and thus $\bar{u}$ approaches its limiting form, which is *piecewise linear* in each of e and de. The terms δu is a perturbation on $\bar{u}$ and approaches zero under the assumption of true fuzzy partitioning, which is the limiting case of the constraint stated under Assumption 2 earlier.

The key idea here is that u is in the limit a piecewise linear function of each of its arguments. Such function may be thought of as being made up of quadrangular tiles forming a surface in the three dimensional space of the Cartesian coordinate frame, $e \times de \times u$. Such a function, can approximate *any* continuous function of its arguments over a finite interval to an arbitrary degree of accuracy[7]. The implication of this statement is that the class of fuzzy control algorithms we have considered may be used as nonlinear compensators for instance in relatively simple but common situations where the plant may be described by $\dot{x} = Ax + g(x)u$ and where u can be designed to compensate for the nonlinearity in the input channel stated in terms of g. The resulting system would be considered *feedback linearized*[8].

A simple instance of this situation is in process control where the plant may be described by

$$\ddot{y} + a_1(y,\dot{y})\dot{y} + a_2(y,\dot{y})y = u. \quad (31)$$

Then u defined as

$$u = u_c + (\hat{a}_1(y,\dot{y})\dot{y} + \hat{a}_2(y,\dot{y})y), \quad (32)$$

where $\hat{a}_1$ and $\hat{a}$ are estimates of a_1 and a_2 respectively, renders the following system

$$\ddot{y} = u_c, \quad (33)$$

provided $\|a_1 - \hat{a}_1\|$ and $\|a_2 - \hat{a}_2\|$ are negligible. Now given the earlier argument that a fuzzy control algorithm of the type we have considered can *emulate* any continuous function of its arguments over a finite interval, it is easy to see that u given by (32) can indeed be implemented in terms of some fuzzy control algorithm. In the same vein there are numerous situations where fuzzy control can be and is used to compensate for some dominant nonlinearity in the plant. There are limits though as can be expected. For instance, if only the plant output and its derivative are directly accessible in the implementation of the fuzzy control algorithm, then correspondingly limited set of functions can be emulated and as such limited class of nonlinearities in the plant may be compensated for. This is not a lim-

---

1. It is not difficult to see that as the characteristic functions approach their limiting triangular form, F_j and F'_l approach zero and in fact become identically zero in the limit.

itation of fuzzy control per se but only of the manner of specification of the rules and, at least in principle, can be dealt with using the framework described above.

5 CONCLUSION

This article presents a framework for analyzing the functional behavior of a common class of fuzzy control algorithms. In particular, it shows that in principle fuzzy control may be used to compensate for nonlinear plant behavior and as such can be viewed as a feedback linearization strategy, albeit using rules rather than differential-algebraic relationships. The consequence of this statement is that fuzzy control does indeed accomplish more than its counterparts such as PID in that it not only compensates for dynamic deficiencies in the plant as PID does, but it also linearizes some nonlinear plants and as such provides uniform dynamic behavior, which PID for instance can not.

REFERENCE

[1]   Langari G., and Tomizuka, M., Fuzzy linguistic control of arc welding, in *Sensors and Controls for Manufacturing*, ASME, 1988.

[2]   Tang, K. L. and Mulholland, R. J., Comparing fuzzy logic with classical controllers, *IEEE Transactions on Systems, Man, and Cybernetics*, 17(6), November/December 1987.

[3]   Kickert, W. J. M. and Mamdani, E. H., Analysis of a fuzzy logic controller, *Fuzzy Sets and Systems*, 1(1), 1978.

[4]   Buckley, J. J. and Ying. H., Fuzzy controller theory: Limit theorems for linear fuzzy control rules, *Automatica*, 25(3), 1989.

[5]   Langari, R., A nonlinear formulation of a class of fuzzy linguistic control systems, in *Proceedings of the 1992 American Control Conference, Chicago, Illinois*, June 1992.

[6]   Dubois, D. and Prade, H., *Fuzzy Sets and Systems*, Academic Press, 1980.

[7]   Royden, H. L., *Real Analysis*, MacMillan, 1988.Isidori, A., *Nonlinear Control Systems*, Springer Verlag, 1989.

[8]   Isidori, A., *Nonlinear Control Systems*, Springer Verlag, 1989.

Chapter 13
Fuzzy Controls under Product-Sum-Gravity Methods and New Fuzzy Control Methods

Introduction, *276*
Min-Max-Gravity Method, *276*
Point at Issue of Min-Max-Gravity Method, *278*
Product-Sum-Gravity Method, *280*
Comparison of Min-Max-Gravity Method and Product-Sum-Gravity Method, *282*
Realization of PID Controllers by Product-Sum-Gravity Method, *285*
Fuzzy Control Results, *288*
New Fuzzy Reasoning Methods and Control Results, *291*
Conclusion, *292*
References, *294*

This chapter shows that Mamdani's *min-max-gravity* method, which is widely used as a fuzzy control method, does not necessarily fit our intuition and, therefore, is not the best fuzzy control method. Several new fuzzy control methods are proposed, such as the *product-sum-gravity* method, *(min/product)-max-gravity* method, and the *bounded product-bounded sum-gravity method*. It is shown that the *product-sum-gravity* method is intuitive and simple in nature and that this method can be used to construct a PID controller, in contrast with the *min-max-gravity* method. Indeed, it is found from computer simulations that the proposed new fuzzy control methods can get better control results than the *min-max-gravity* method.

FUZZY CONTROLS UNDER PRODUCT-SUM-GRAVITY METHODS AND NEW FUZZY CONTROL METHODS

M. Mizumoto
Division of Information and Computer Sciences
Osaka Electro-Communication University
Neyagawa, Osaka 572, Japan

1. INTRODUCTION

Most of the existing fuzzy logic controllers are based on the fuzzy reasoning method called "min-max-gravity method" by Mamdani [1].

This chapter shows that the min-max-gravity method does not necessarily fit our intuition and is not the best fuzzy reasoning method for fuzzy control. This chapter also proposes several new fuzzy control methods named "product-sum-gravity method" [2], "(min/product)-max-gravity method", "bounded product-bounded sum-gravity method", and so on for fuzzy control. It is shown that the product-sum-gravity method is intuitive and simple in nature and this method can construct a PID controller, though the min-max-gravity method can not realize it. Moreover, it is found from computer simulations that these new fuzzy control methods can get better control results than the min-max-gravity method.

2. MIN-MAX-GRAVITY METHOD

We shall consider the following multiple fuzzy reasoning form:

$$\begin{array}{ll} \text{Rule 1:} & A_1 \text{ and } B_1 \Rightarrow C_1 \\ \text{Rule 2:} & A_2 \text{ and } B_2 \Rightarrow C_2 \\ \multicolumn{2}{c}{\cdots\cdots\cdots\cdots\cdots\cdots} \\ \text{Rule n:} & A_n \text{ and } B_n \Rightarrow C_n \\ \underline{\text{Fact:}} & \underline{x_o \text{ and } y_o} \\ \text{Cons:} & C' \end{array} \qquad (1)$$

where A_i is a fuzzy set in a set X; B_i in Y; and C_i in Z and $x_o \in X$, $y_o \in Y$. Each fuzzy rule 「A_i and $B_i \Rightarrow C_i$」, $i=1,\cdots,n$, is defined as

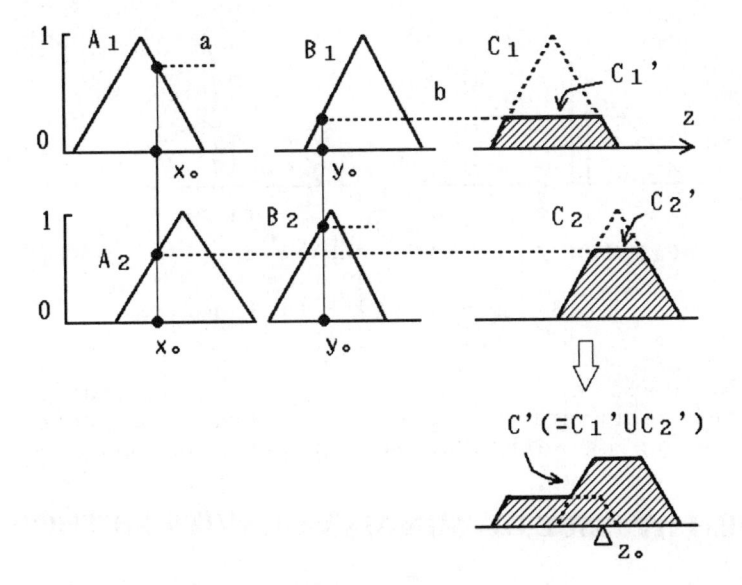

Fig.1 Min-Max-Gravity Method Using (3),(5) and (6)

$$\mu_{Ai \text{ and } Bi => Ci}(x,y,z) = \mu_{Ai}(x) \wedge \mu_{Bi}(y) \wedge \mu_{Ci}(z) \qquad (2)$$

where ∧ stands for min.
 The inference result Ci' which is inferred from the fact ⌈x₀ and y₀⌋ and the fuzzy rule ⌈Ai and Bi => Ci⌋ is given as

$$\mu_{Ci'}(z) = \mu_{Ai}(x_o) \wedge \mu_{Bi}(y_o) \wedge \mu_{Ci}(z) \qquad (3)$$

The final consequence C' of (1) is aggregated by taking the union (U) of C1', C2', ⋯, Cn' obtained in (3). Namely,

$$C' = C1' \cup C2' \cup \cdots \cup Cn' \qquad (4)$$

that is,

$$\mu_{C'}(z) = \mu_{C1'}(z) \vee \cdots \vee \mu_{Cn'}(z) \qquad (5)$$

where ∨ stands for max.
 The representative point z_o for the resulting fuzzy set C' is obtained as the center of gravity of C':

$$z_o = \frac{\int z \cdot \mu_{C'}(z) dz}{\int \mu_{C'}(z) dz} \qquad (6)$$

This fuzzy reasoning method is known as Mamdani's method [1] and called the *"min-max-gravity method"* (see Fig. 1).

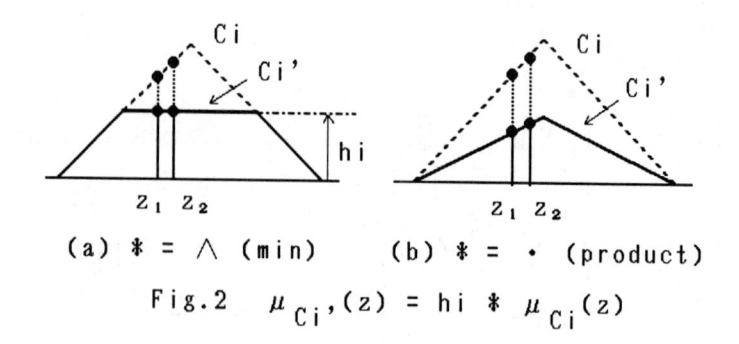

(a) $* = \wedge$ (min)   (b) $* = \cdot$ (product)

Fig. 2 $\mu_{C_i'}(z) = h_i * \mu_{C_i}(z)$

We shall next show that the min-max-gravity method does not necessarily fit our intuition and propose a new fuzzy reasoning method called "product-sum-gravity method".

3. POINT AT ISSUE OF MIN-MAX-GRAVITY METHOD

At first, we shall consider the way of obtaining C_i' in (3). In general, C_i' is obtained as follows by using $\wedge$ (=min).

$$\mu_{C_i'}(z) = h_i \wedge \mu_{C_i}(z) \tag{7}$$

where $h_i = \mu_{A_i}(x_o) \wedge \mu_{B_i}(y_o)$ in (3).

In Fig. 2(a), the degree of contribution of z_1, which corresponds to the grade of membership of z_1, becomes the same as that of z_2 in the resulting fuzzy set C_i' because of using "$\wedge$" (=min), though z_2 had higher degree of contribution than z_1 in the original C_i. It seems to be quite natural to expect that the order of degrees of contribution of z_1 and z_2 is preserved also in C_i'. In order to realize the order preserving for the contributions of z_1 and z_2, the operation of algebraic product ($\cdot$) is recommended to use in (7) instead of $\wedge$ (=min). Namely,

$$\mu_{C_i'}(z) = h_i \cdot \mu_{C_i}(z) \tag{8}$$

As is shown in Fig. 2(b), the order of degrees of contribution of z_1 and z_2 is preserved under the algebraic product, though these degrees become lower.

We shall next consider the case of obtaining

$$h_i = \mu_{A_i}(x_o) \wedge \mu_{B_i}(y_o) \tag{9}$$

in (3). As a simple example, let $\mu_{A_i}(x_o) = 0.8$ and $\mu_{B_i}(y_o) = 0.3$, then we have $h_i = 0.3$ by taking the min ($\wedge$) of 0.8 and 0.3. In this case, $h_i = 0.3$ even if $\mu_{A_i}(x_o) = 0.6$ and, in general, as long as $\mu_{A_i}(x_o) \geq 0.3$, which does not fit our intuition because h_i should change as $\mu_{A_i}(x_o)$

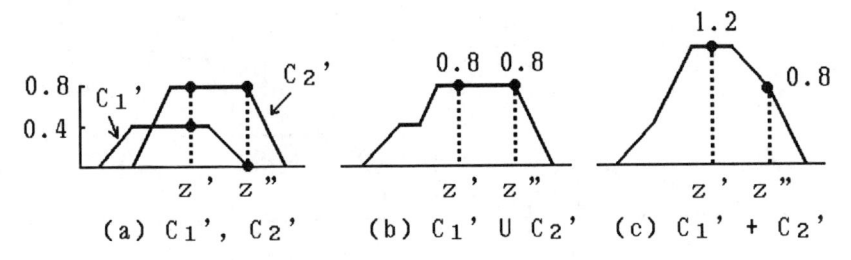

Fig.3 Aggregation of Fuzzy Sets C_1' and C_2'

and $\mu_{Bi}(y_o)$ change. If we use algebraic product $(\cdot)$ instead of $\wedge$ (min) in (9), hi changes as $\mu_{Ai}(x_o)$ and $\mu_{Bi}(y_o)$ change. Thus, we can define hi as

$$hi = \mu_{Ai}(x_o) \cdot \mu_{Bi}(y_o) \qquad (10)$$

In the sequel, it is found that the min operator ($\wedge$) is not necessarily a suitable one in obtaining the inference results Ci' in (3) and that the algebraic product seems to be more suitable than the min.

We shall next discuss the method of obtaining the consequence C' in (5) which is aggregated from $C1', \cdots, Cn'$.

As a simple example, consider the two inference results $C1'$ and $C2'$ in Fig.3(a). The union $C1' \cup C2'$ of $C1'$ and $C2'$ is given as follows by using v (=max) (see Fig.3(b)).

$$C' = C1' \cup C2' \qquad (11)$$

$$\mu_{C'}(z) = \mu_{C1'}(z) \vee \mu_{C2'}(z) \qquad (12)$$

We shall consider two points z' and z''. In the aggregated fuzzy set $C1' \cup C2'$, the same degrees of contribution, 0.8, are obtained at the points z' and z'' as in Fig.3(b). However, z' had originally 0.4 degree of contribution in $C1'$ and 0.8 degree of contribution in $C2'$ as in Fig.3(a). On the other hand, z'' had 0 and 0.8 degrees of contribution in $C1'$ and $C2'$, respectively. Namely, z' has 0.4 and 0.8 as degrees of contribution and z'' has 0 and 0.8. Therefore, z' should have higher degree of contribution than z'' after the aggregation. But the same degrees of contribution are obtained as in Fig.3(b) because of using v (max).

To reflect the difference of degrees of contribution of the two points z' and z'', it is suggested to use the "sum" (+, plus) operator in the aggregation of $C1'$ and $C2'$ instead of max, namely:

$$C' = C1' + C2' \qquad (13)$$

$$\mu_{C'}(z) = \mu_{C1'}(z) + \mu_{C2'}(z) \qquad (14)$$

(a) Many inference results on the right

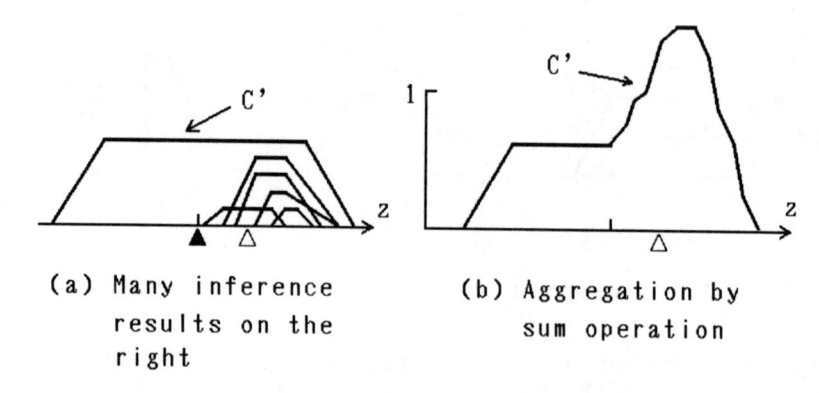

(b) Aggregation by sum operation

Fig.4 Aggregation of Inference Results

For example, as in Fig.3(c), z' has 1.2 degree of contribution by summing up 0.4 and 0.8, whereas z" has 0.8 (= 0 + 0.8). Hence, z' can have higher degree of contribution than z".

We shall inquire further into the aggregation problem by using another example. Fig.4(a) shows that a number of inference results exist on the right side. When the max (v) is used in the aggregation as in (5) and (12), the final consequence C' from these inference results coincides with the outside inference result. Thus the representative point z_o of C' will be at its center (▲). However, it would be quite natural to consider that the representative point would exist on the right (△) near the center (▲) because of the existence of many inference results on the right.

To avoid the ignorance of a number of inference results on the right side, the "sum" (plus) operator is suggested to be used in the aggregation of (5) instead of max. Using (14), we can obtain a final consequence C' by summing up these inference results upwards and the representative point z_o of C' will be slightly to the right (△) of the center, which seems to be quite reasonable.

It should be noted that, as shown in the examples of Fig.3(c) and Fig.4(b), the resulting fuzzy set C' sometimes exceeds 1. To avoid such a fact, we can use arithmetic mean [3] instead of using the sum operation. However, the same center of gravity can be obtained even if we use either the sum or arithmetic mean in the aggregation.

4. PRODUCT-SUM-GRAVITY METHOD

In the sequel, we propose a new method of fuzzy reasoning called "*product-sum-gravity method*" [2] which can be obtained by replacing min by algebraic product, and max by sum in the min-max-gravity method in (3) and (5). Namely, the product-sum-gravity method is defined as follows (see Fig.5):

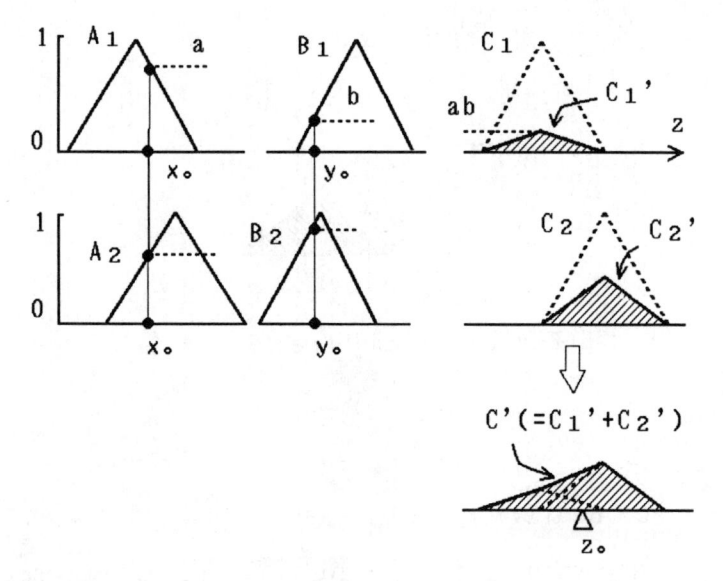

Fig.5 Product-Sum-Gravity Method Using
(15),(17) and (6)

The consequence of C' of (1) by the product-sum-gravity method is obtained as follows:

At first, the inference result Ci' from the fact 「x_0 and y_0」 and the fuzzy rule 「Ai and Bi => Ci」 in (1) is given as

$$\mu_{Ci'}(z) = \mu_{Ai}(x_0) \cdot \mu_{Bi}(y_0) \cdot \mu_{Ci}(z) \tag{15}$$

The consequence C' of (1) is aggregated from C1', C2', ⋯, Cn' by using sum (+):

$$C' = C1' + C2' + \cdots + Cn' \tag{16}$$

Namely,

$$\mu_{C'}(z) = \mu_{C1'}(z) + \cdots + \mu_{Cn'}(z) \tag{17}$$

The representative point z_0 of the C' is obtained by using the center of gravity method (6). However, the center of gravity z_0 of C' in the case of product-sum-gravity method is discussed in the following:

Let z_i be the center of gravity of the inference result Ci' of (15) and Si be the area of Ci' (see Fig. 6), then we have:

$$zi = \frac{\int z \cdot \mu_{Ci'}(z) dz}{\int \mu_{Ci'}(z) dz} = \frac{\int z \cdot \mu_{Ci'}(z) dz}{Si} \tag{18}$$

which leads to

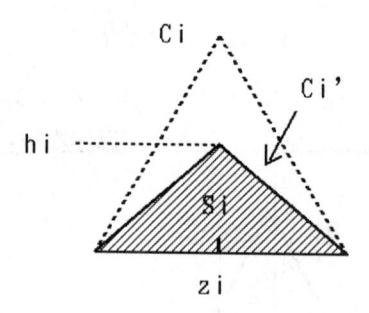

Fig.6 Center zi and Area Si of C_i'

$$\int z \cdot \mu_{C_i'}(z) dz = S_i \, z_i \qquad (19)$$

Therefore, the center of gravity z_o of the final consequence C' of (17) is given as

$$z_o = \frac{\int z \cdot \mu_{C'}(z) dz}{\int \mu_{C'}(z) dz} = \frac{\int z \cdot [\mu_{C_1'}(z) + \cdots + \mu_{C_n'}(z)] dz}{\int [\mu_{C_1'}(z) + \cdots + \mu_{C_n'}(z)] dz}$$

$$= \frac{\int z \cdot \mu_{C_1'}(z) dz + \cdots + \int z \cdot \mu_{C_n'}(z) dz}{\int \mu_{C_1'}(z) dz + \cdots + \int \mu_{C_n'}(z) dz}$$

$$= \frac{S_1 z_1 + S_2 z_2 + \cdots + S_n z_n}{S_1 + S_2 + \cdots + S_n} \qquad (20)$$

Therefore, the center of gravity z_o is given as the weighted average of z_i's (center of gravity of C_i') with weights S_i's. (area of C_i'). This method corresponds to a defuzzification method called *"Area Methods"* in [3].

5. COMPARISON OF MIN-MAX-GRAVITY METHOD AND PRODUCT-SUM-GRAVITY METHOD

We shall compare the two fuzzy reasoning methods of min-max-gravity method and product-sum-gravity method by giving the following example of fuzzy reasoning:

Example: Let z_1, z_2, z_3, and z_4 be the heights at the lattice points (x_1, y_1), (x_1, y_2), (x_2, y_1), (x_2, y_2), respectively, with $x_1 < x_2$ and $y_1 < y_2$, and consider the following fuzzy reasoning:

$$\begin{array}{llllll}
\text{Rule 1:} & \underline{x_1} & \text{and} & \underline{y_1} & \Rightarrow & \underline{z_1} \\
\text{Rule 2:} & \underline{x_1} & \text{and} & \underline{y_2} & \Rightarrow & \underline{z_2} \\
\text{Rule 3:} & \underline{x_2} & \text{and} & \underline{y_1} & \Rightarrow & \underline{z_3} \\
\text{Rule 4:} & \underline{x_2} & \text{and} & \underline{y_2} & \Rightarrow & \underline{z_4} \\
\text{Fact:} & x_o & \text{and} & y_o & & \\
\hline
\text{Cons:} & & & & & z_o
\end{array} \qquad (21)$$

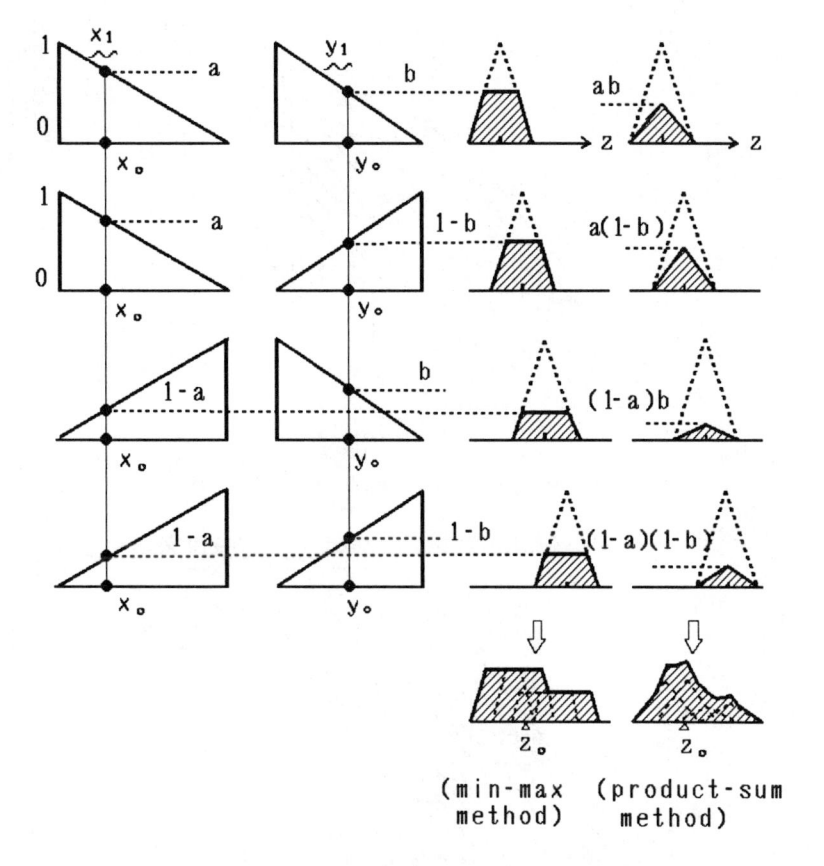

Fig.7 Inference Results z_o of (21) Under
Min-Max-Gravity Method and
Product-Sum-Gravity Method

where $\underset{\sim}{x1}$ is a fuzzy set representing "about x1". Fuzzy sets $\underset{\sim}{x1}$ and $\underset{\sim}{x2}$ are of the triangular type and intersect at the height 0.5 each other (see Fig.7). The same holds for the fuzzy sets $\underset{\sim}{y1}$ and $\underset{\sim}{y2}$. Fuzzy sets $\underset{\sim}{z1}$, $\underset{\sim}{z2}$, $\underset{\sim}{z3}$ and $\underset{\sim}{z4}$ are also of the same width, but the width is not restricted.

We can obtain the height z_o at the point (x_o, y_o) by using the min-max-gravity method and product-sum-gravity method as in Fig.7. Fig. 8 indicates the inference result z_o at x_o and y_o in the case that the points z1, z2, z3 and z4 are assumed to be on the same plane.

The min-max-gravity method does not obtain a plane as the inference result (Fig.8(a)), and the curved surface changes slightly as the width of the fuzzy sets $\underset{\sim}{z1}$, $\underset{\sim}{z2}$, $\underset{\sim}{z3}$, and $\underset{\sim}{z4}$ changes [4].

In contrast, the product-sum-gravity method can obtain a plane in the case that the points z1, z2, z3 and z4 are on the same plane. In general, for any heights z1, z2, z3 and z4, the product-sum-gravity method obtains the following inference result:

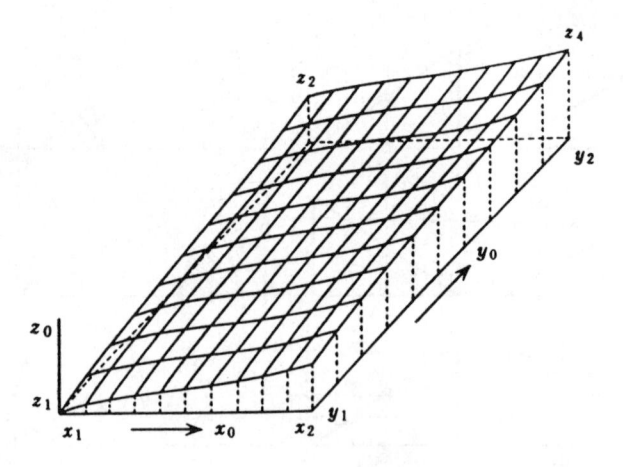

(a) Case of min-max-gravity method

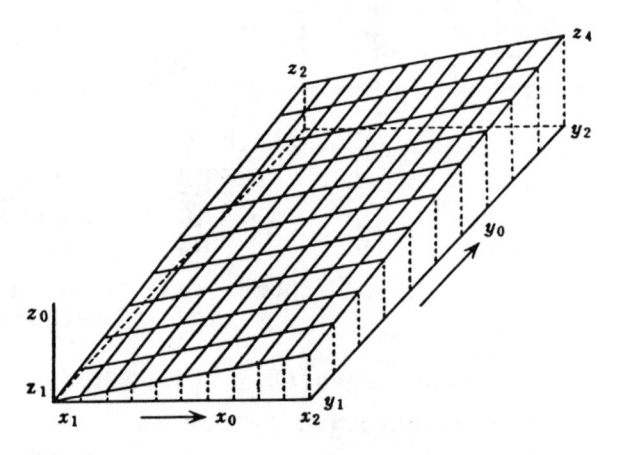

(b) Case of product-sum-gravity method

Fig. 8 Inference Results $z_∘$ from $x_∘$ and $y_∘$ in (21)

Let S be the areas of the fuzzy sets z1, z2, z3 and z4, then the area Si of each inference result zi' inferred from each rule is given as (see Fig. 6).

$$S1 = abS, \quad S2 = a(1-b)S$$
$$S3 = (1-a)bS, \quad S4 = (1-a)(1-b)S \quad (22)$$

where

$$a = \mu_{x1}(x_∘) = (x2-x_∘)/(x2-x1)$$
$$b = \mu_{y1}(y_∘) = (y2-y_∘)/(y2-y1) \quad (23)$$

Therefore, the center of gravity $z_∘$ is given by using (20) as follows:

$$z_o = \frac{S1z1 + S2z2 + S3z3 + S4z4}{S1 + S2 + S3 + S4}$$

$$= \frac{abSz1 + a(1-b)Sz2 + (1-a)bSz3 + (1-a)(1-b)Sz4}{abS + a(1-b)S + (1-a)bS + (1-a)(1-b)S}$$

$$= \frac{abz1 + a(1-b)z2 + (1-a)bz3 + (1-a)(1-b)z4}{ab + a(1-b) + (1-a)b + (1-a)(1-b)}$$

$$= abz1 + a(1-b)z2 + (1-a)bz3 + (1-a)(1-b)z4 \quad (24)$$

It is noted that the denominator of the above equation is equal to 1.

It follows that the product-sum-gravity method is intuitive in nature and obtains a more simple inference result than the min-max-gravity method. The inference result by the min-max-gravity method is not simple and is represented by a fractional expression whose denominator and numerator are of 2nd order form of a and b [4].

6. REALIZATION OF PID CONTROLLERS BY PRODUCT-SUM-GRAVITY METHOD

It is found in Fig. 8(b) that the product-sum-gravity method can generate a plane. This fact suggests the construction of a PID controller by the product-sum-gravity method.

As is well-known, a control action u of *PID controller* is a linear combination of the error e, the change in error Δe, and the integral of the error ∫ edt, namely,

$$u = \alpha e + \beta \Delta e + \gamma \int e dt \quad (25)$$

where α is the proportional coefficient, β is the derivative coefficient, and γ is the integral coefficient for e.

For simplicity, we shall first consider the case of *PD controller*:

$$u = \alpha e + \beta \Delta e \quad (26)$$

Let e1 and e2 be the minimal and maximal values of the possible error e, and let Δe1 and Δe2 be the minimal and maximal values of Δe, that is

$$e1 \leq e \leq e2 ; \quad \Delta e1 \leq \Delta e \leq \Delta e2 \quad (27)$$

Then we define fuzzy sets e1 and e2 for the error e as in Fig. 9(a). Similarly, fuzzy sets Δe1 and Δe2 for the change in error Δe are given in Fig. 9(b).

Then, we can construct fuzzy control rules for the PD controller by the product-sum-gravity method as follows:

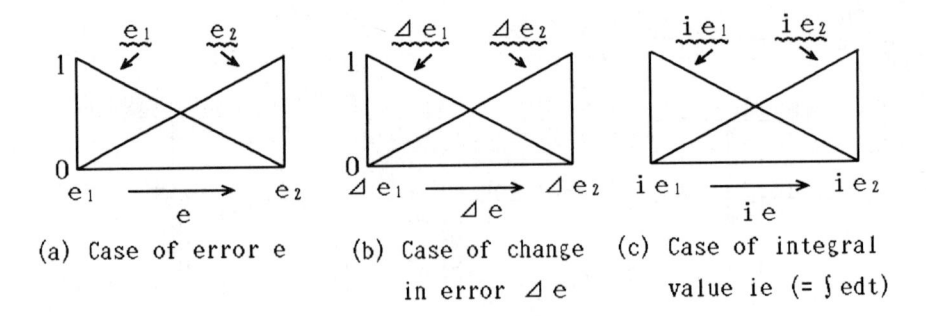

Fig. 9 Fuzzy Sets of Error e, Change in Error Δe and Integral Value ∫ edt

$$\begin{array}{ll}
\text{Rule 1:} & \underline{e1} \text{ and } \underline{\Delta e1} \Rightarrow \underline{u1} \\
\text{Rule 2:} & \underline{e1} \text{ and } \underline{\Delta e2} \Rightarrow \underline{u2} \\
\text{Rule 3:} & \underline{e2} \text{ and } \underline{\Delta e1} \Rightarrow \underline{u3} \\
\text{Rule 4:} & \underline{e2} \text{ and } \underline{\Delta e2} \Rightarrow \underline{u4} \\
\text{Fact:} & e \text{ and } \Delta e \\
\hline
\text{Cons:} & u
\end{array} \qquad (28)$$

where u1, u2, u3 and u4 are real numbers such that

$$\begin{aligned}
u1 &= \alpha\ e1 + \beta\ \Delta e1 \\
u2 &= \alpha\ e1 + \beta\ \Delta e2 \\
u3 &= \alpha\ e2 + \beta\ \Delta e1 \\
u4 &= \alpha\ e2 + \beta\ \Delta e2
\end{aligned} \qquad (29)$$

and $\underline{u1}$ (about u1), $\underline{u2}$, $\underline{u3}$ and $\underline{u4}$ are fuzzy sets of triangular type of the same width.

Then we can obtain the inference result u of (28) from (24) in the following form:

$$\begin{aligned}
u &= \frac{abu1 + a(1-b)u2 + (1-a)bu3 + (1-a)(1-b)u4}{ab + a(1-b) + (1-a)b + (1-a)(1-b)} \\
&= abu1 + a(1-b)u2 + (1-a)bu3 + (1-a)(1-b)u4 \qquad (30)
\end{aligned}$$

where a and b are given as

$$a = \mu_{e1}(e) = \frac{e2-e}{e2-e1} \qquad (31)$$

$$b = \mu_{\Delta e1}(\Delta e) = \frac{\Delta e2 - \Delta e}{\Delta e2 - \Delta e1} \qquad (32)$$

The u1, u2, u3 and u4 of (29) are the values of (26) at the lattice points (e1, Δe1), (e1, Δe2), (e2, Δe1) and (e2, Δe2), and are on the same plane.

Therefore, the inference result (i.e., control action) u can be derived as follows by inserting (29), (31) and (32) into (30):

$$u = \qquad (30)$$
$$= \alpha\ e + \beta\ \Delta e \qquad (33)$$

which indicates the construction of PD controller (26) by means of the product-sum-gravity method.

Note that the min-max-gravity method can not realize such PD controller since it does not generate a plane as shown in Fig. 8(a).

We shall next consider the case of the *PID controller* of (25). Let ie be the integral value $\int e dt$ of an error e, and its minimum and maximal values be ie1 and ie2, respectively. Namely,

$$ie1 \leq ie \leq ie2 \qquad (34)$$

Fuzzy sets $\underline{ie1}$ and $\underline{ie2}$ are given in Fig. 9(c).

Then the fuzzy control rules for the PID controller are given as:

$$
\begin{array}{llll}
\text{Rule 1:} & \underline{e1} \text{ and } \underline{\Delta e1} \text{ and } \underline{ie1} & \Rightarrow & \underline{u1} \\
\text{Rule 2:} & \underline{e1} \text{ and } \underline{\Delta e1} \text{ and } \underline{ie2} & \Rightarrow & \underline{u2} \\
\text{Rule 3:} & \underline{e1} \text{ and } \underline{\Delta e2} \text{ and } \underline{ie1} & \Rightarrow & \underline{u3} \\
\text{Rule 4:} & \underline{e1} \text{ and } \underline{\Delta e2} \text{ and } \underline{ie2} & \Rightarrow & \underline{u4} \\
\text{Rule 5:} & \underline{e2} \text{ and } \underline{\Delta e1} \text{ and } \underline{ie1} & \Rightarrow & \underline{u5} \\
\text{Rule 6:} & \underline{e2} \text{ and } \underline{\Delta e1} \text{ and } \underline{ie2} & \Rightarrow & \underline{u6} \\
\text{Rule 7:} & \underline{e2} \text{ and } \underline{\Delta e2} \text{ and } \underline{ie1} & \Rightarrow & \underline{u7} \\
\text{Rule 8:} & \underline{e2} \text{ and } \underline{\Delta e2} \text{ and } \underline{ie2} & \Rightarrow & \underline{u8} \\
\text{Fact:} & e \text{ and } \Delta e \text{ and } ie & & \\
\hline
\text{Cons:} & & & u
\end{array} \qquad (35)
$$

where u1, u2, ..., u8 are real numbers such that

$$
\begin{aligned}
u1 &= \alpha\ e1 + \beta\ \Delta e1 + \gamma\ ie1 \\
u2 &= \alpha\ e1 + \beta\ \Delta e1 + \gamma\ ie2 \\
u3 &= \alpha\ e1 + \beta\ \Delta e2 + \gamma\ ie1 \\
u4 &= \alpha\ e1 + \beta\ \Delta e2 + \gamma\ ie2 \\
u5 &= \alpha\ e2 + \beta\ \Delta e1 + \gamma\ ie1 \\
u6 &= \alpha\ e2 + \beta\ \Delta e1 + \gamma\ ie2 \\
u7 &= \alpha\ e2 + \beta\ \Delta e2 + \gamma\ ie1 \\
u8 &= \alpha\ e2 + \beta\ \Delta e2 + \gamma\ ie2
\end{aligned} \qquad (36)
$$

which are the values of (25) at the points $(e1, \Delta e1, ie1), \cdots, (e2, \Delta e2, ie2)$. Thus, the control action u for e, Δe and ie is given as

$$u = \frac{abc\,u1 + ab(1-c)u2 + a(1-b)c\,u3 + a(1-b)(1-c)u4 + (1-a)bc\,u5 + (1-a)b(1-c)u6 + (1-a)(1-b)c\,u7 + (1-a)(1-b)(1-c)u8}{abc + ab(1-c) + a(1-b)c + a(1-b)(1-c) + (1-a)bc + (1-a)b(1-c) + (1-a)(1-b)c + (1-a)(1-b)(1-c)}$$

$$\begin{aligned}
&= abcu1 + ab(1-c)u2 + a(1-b)cu3 + a(1-b)(1-c)u4 \\
&\quad + (1-a)bcu5 + (1-a)b(1-c)u6 + (1-a)(1-b)cu7 \\
&\quad + (1-a)(1-b)(1-c)u8
\end{aligned}$$

$$= \alpha\, e + \beta\, \Delta e + \gamma \int edt \tag{37}$$

where

$$c = \mu_{ie1}(ie) = \frac{ie2-ie}{ie2-ie1} \tag{38}$$

Therefore, it is shown that the PID controller can be constructed by the product-sum-gravity method.

7. FUZZY CONTROL RESULTS

In this section we shall compare the control results under the min-max-gravity method and product-sum-gravity method. It is shown in the computer simulations that the product-sum-gravity method can get better control results than the min-max-gravity method.

Now consider a plant model $G(s) = e^{-2s}/(1+20s)$ with first order delay. Fuzzy control rules for the plant model are shown in Table 1 [5] and interpretted as

$$\begin{aligned}
&R1: \quad e \text{ is NB and } \Delta e \text{ is ZO} \Rightarrow \Delta u \text{ is PB} \\
&R2: \quad e \text{ is NM and } \Delta e \text{ is ZO} \Rightarrow \Delta u \text{ is PM} \\
&\quad \cdots \cdots \cdots \cdots \cdots \cdots \\
&R13: \quad e \text{ is ZO and } \Delta e \text{ is PB} \Rightarrow \Delta u \text{ is NB}
\end{aligned} \tag{39}$$

where e is the error, Δe is the change in error and Δu is the change in action. NB, NM, $\cdots$, PB are fuzzy sets as shown in Fig.10. When $e = e_0$ and $\Delta e = \Delta e_0$ are given to a fuzzy controller as premises of (39), each control rule Ri ($i=1,\cdots,13$) infers a fuzzy set Ci' for Δu by using

Table 1 Fuzzy Control Rules e, $\Delta e \Rightarrow \Delta u$

	Δu	Δe						
		NB	NM	PS	ZO	PS	PM	PB
e	NB				PB			
	NM				PM			
	NS				PS			
	ZO	PB	PM	PS	ZO	NS	NM	NB
	PS				NS			
	PM				NM			
	PB				NB			

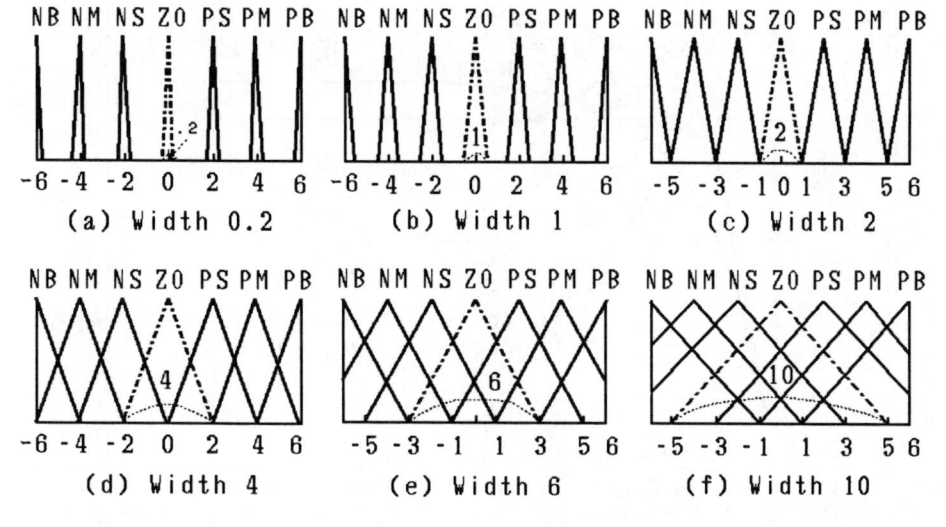

Fig.10 Fuzzy Sets in Case of Changing the Width W

$$\mu_{C_i'}(\Delta u) = \mu_{A_i}(e_o) \wedge \mu_{B_i}(\Delta e_o) \wedge \mu_{C_i}(\Delta u) \quad (40)$$
[min-max-gravity method]

$$\mu_{C_i'}(\Delta u) = \mu_{A_i}(e_o) \cdot \mu_{B_i}(\Delta e_o) \cdot \mu_{C_i}(\Delta u) \quad (41)$$
[product-sum-gravity method]

where Ai, Bi and Ci are fuzzy sets NB, NM, ⋯, PB in Fig.10 and i=1,⋯,13.

The resulting fuzzy set C' is aggregated from C1', ⋯, C13' by using (5) or (17). The actual change in action $\Delta u = \Delta u_o$ to the plant is obtained as the center of gravity of the fuzzy set C'.

Fig.11 shows control results when using the min-max-gravity method, where W stands for the width of fuzzy sets NB, NM, ⋯, PB in Fig. 10. When the width of the fuzzy sets is small (W = 0.2, 1), that is, the fuzzy sets are completely separated, the control results are bad and do not converge on the set point 40. On the other hand, when the width is large (W = 8, 10), the control results are satisfactory but their overshoots become rather big. When W = 4, 6, we get good control results. Namely, good control results are obtained when the fuzzy sets are not isolated but do not overlap too much.

Fig.12 indicates the control results using the product-sum-gravity method. Similar control results are observed as in Fig. 11, but their overshoots are smaller than those by the min-max-gravity method in Fig. 11.

Fig.13 compares control results by the min-max-gravity method and product-sum-gravity method in the case that the width W of the fuzzy sets is 8 (see Fig.10).

It is found from the computer simulations that the product-sum-gravity method gives better control results than the min-max-gravity method.

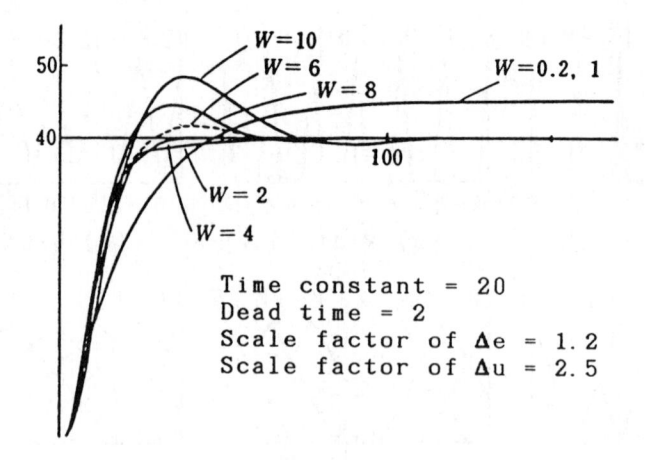

Fig. 11 Control Results by Min-Max-Gravity Method

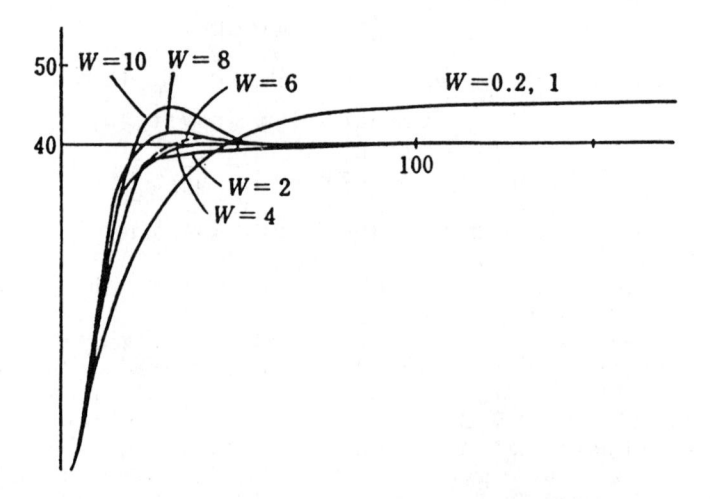

Fig. 12 Control Results by Product-Sum-Gravity Method

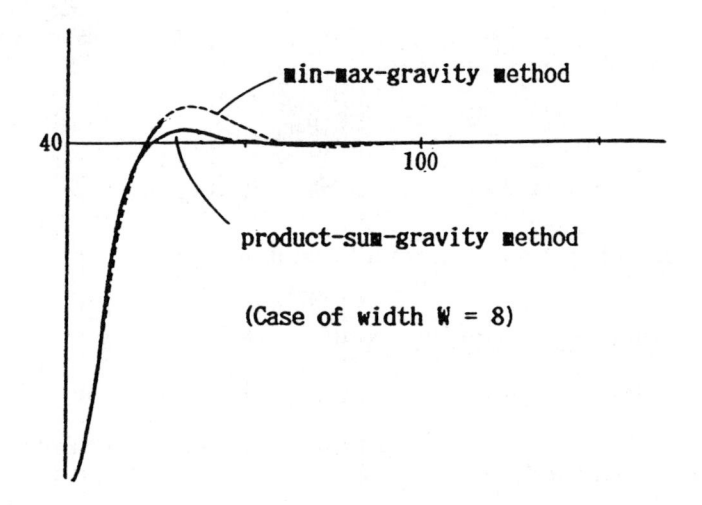

```
                 ■in-■ax-gravity ■ethod
       40
                        100

             product-sum-gravity ■ethod

             (Case of width W = 8)
```

Fig. 13 Comparison of Control Results

8. NEW FUZZY REASONING METHODS AND CONTROL RESULTS

As was seen in the definition of the product-sum-gravity method, the method can be obtained by replacing min with algebraic product and max with sum in the min-max-gravity method. This fact indicates the possibility of proposing new fuzzy reasoning methods by using appropriate operations in place of min and max.

For example, we can propose new fuzzy reasoning method of *min-sum-gravity method* by using (3) and (17).

Moreover, the *(min/product)-max-gravity method* is obtained by the following: The inference result Ci' is given by using ∧ (=min) and then · (=product) in (3). Namely, we have:

$$\mu_{Ci'}(z) = [\mu_{Ai}(x_\circ) \wedge \mu_{Bi}(y_\circ)] \cdot \mu_{Ci}(z) \qquad (42)$$

In the min-max-gravity method, the operations of min and max are dual in nature, but product and sum used in the product-sum-gravity method are not dual. Therefore, it is possible to propose the *product-algebraic sum-gravity method* in the following:

$$\mu_{Ci'}(z) = \mu_{Ai}(x_\circ) \cdot \mu_{Bi}(y_\circ) \cdot \mu_{Ci}(z) \qquad (43)$$

$$\mu_{C'}(z) = \mu_{C1'}(z) \dotplus \cdots \dotplus \mu_{Cn'}(z) \qquad (44)$$

where algebraic sum $\dotplus$ is defined as

$$\text{Algebraic Sum:} \quad x \dotplus y = x + y - xy \qquad (45)$$

We can also propose a *bounded product-bounded sum-gravity method* as follows:

$$\mu_{Ci'}(z) = \mu_{Ai}(x_\circ) \ominus \mu_{Bi}(y_\circ) \ominus \mu_{Ci}(z) \tag{46}$$

$$\mu_{C'}(z) = \mu_{C1'}(z) \oplus \cdots \oplus \mu_{Cn'}(z) \tag{47}$$

where bounded product $\ominus$ and bounded sum $\oplus$ are dual and are given as

$$\underline{\text{Bounded Product}}: \quad x \ominus y = 0 \vee (x + y - 1) \tag{48}$$

$$\underline{\text{Bounded Sum}}: \quad x \oplus y = 1 \wedge (x + y) \tag{49}$$

Furthermore, we can define a *drastic product-drastic sum-gravity method*, where

$$\underline{\text{Drastic Product}}: \quad x \triangle y = \begin{cases} x & \cdots \; y = 1 \\ y & \cdots \; x = 1 \\ 0 & \cdots \; x, y < 1 \end{cases} \tag{50}$$

$$\underline{\text{Drastic Sum}}: \quad x \triangledown y = \begin{cases} x & \cdots \; y = 0 \\ y & \cdots \; x = 0 \\ 1 & \cdots \; x, y > 0 \end{cases} \tag{51}$$

In the same way, we can propose a number of other fuzzy reasoning methods for fuzzy control, say, *bounded product-algebraic sum-gravity method, min-bounded sum-gravity method, (product/min)-algebraic sum-gravity method, (min/drastic product)-sum-gravity method, (product /drastic product)-sum-gravity method* and so on by combining several operations of t-norms and s-norms [6].

Fig. 14 compares control results by min-max-gravity method, product-sum-gravity method, min-sum-gravity method and (min/product)-max-gravity method. It is found from their computer simulations that product-sum-gravity method gives the best control results compared with the other methods and that the min-sum-gravity method and (min/product)-max-gravity method are found to be better than the min-max-gravity method.

Finally, Fig. 15 shows control results by min-max-gravity method, product-algebraic sum-gravity method and bounded product-bounded sum-gravity method in which the operations used are dual to each other. The min-max-gravity method gives also a bad control result.

9. CONCLUSION

The operations of min and max used in the min-max-gravity method have very strong nonlinearity and thus the reasoning method does not seem to fit our intuition. However, min and max are easy to implement in hardware and thus the min-max-gravity method will be used as the useful

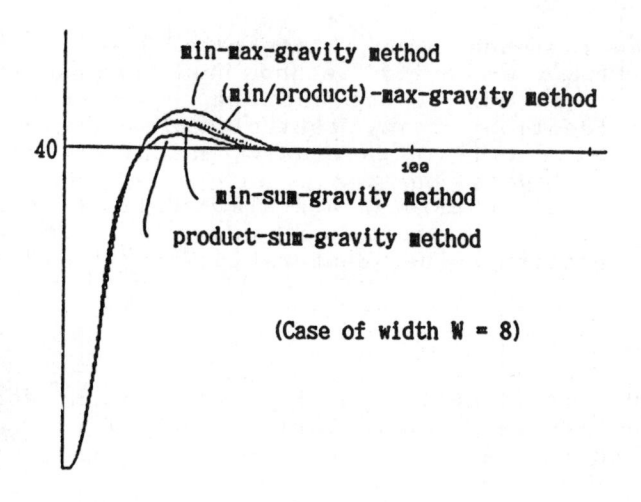

Fig. 14 Control Results by New Fuzzy Reasoning Methods

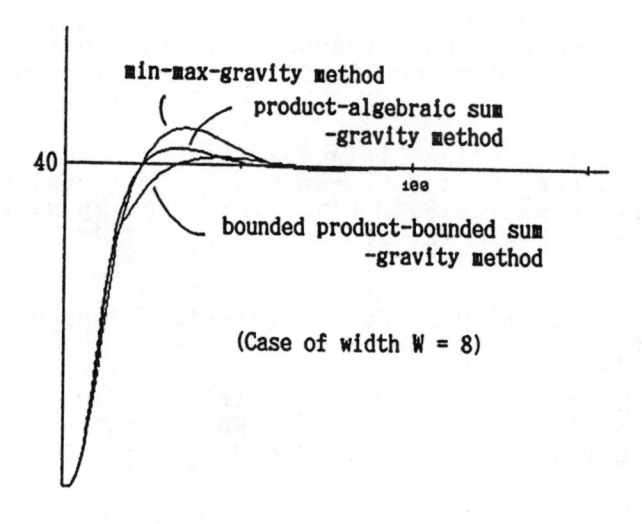

Fig. 15 Cases of Fuzzy Reasoning Methods with Dual Operations

method for the implementation in hardware.

The product-sum-gravity method has some advantages because of the use of product and sum. For example, in this method, identical fuzzy control rules can be used twice or more at a time in the fuzzy reasoning process so that emphatic effects on fuzzy inference results are realized. Suppressive effects are also realized by employing fuzzy control rules whose consequence part is characterized by a negative valued membership function [7].

REFERENCES

1. Mamdani, E. H., "Applications of Fuzzy Algorithms for Control of a Simple Dynamic Plant", *Proc. of IEEE*, 121, 1585-1588, 1974.

2. Mizumoto, M., "Fuzzy Controls by Product-Sum-Gravity Method", *Advancement of Fuzzy Theory and Systems in China and Japan* (Liu and Mizumoto Eds.) (Proc. of Sino-Japan Joint Meeting on Fuzzy Sets and Systems, Oct. 15-18, 1990, Beijing, China), pp. c1.1-c1.4, International Academic Publisher, 1990.

3. Mizumoto, M., "Improvement methods of fuzzy controls", *Proc. of 3rd IFSA Congress* (Seattle, Aug. 6-11, 1989), pp. 60-62, 1989.

4. Mizumoto, M. & Yamada, Y., "Interpolation by Fuzzy Reasoning Methods (Part 2) -Case that conditional part consists of two variables", *Proc. of 8th Fuzzy System Symposium* (Hiroshima, May 26-28, 1992), 217-220, 1992 (in Japanese).

5. Yamazaki, T. & Sugeno, M., "Fuzzy controls", *Systems and Controls*, 28, 442-446, 1984 (in Japanese).

6. Mizumoto, M., "Pictorial Representations of Fuzzy Connectives, Part I: Cases of t-norms, t-conorms and averaging operators", *Fuzzy Sets and Systems*, 31, 217-242, 1989.

7. Mizumoto, M., "Fuzzy Controls with Fuzzy Rules of Emphatic and Suppressive Types", *Proc. of 1992 Int. Fuzzy Systems and Intelligent Control Conference*, 134-140, March 15-18, 1992.

Chapter 14
Fuzzy Modeling for Adaptive Process Control

Introduction, *296*
Modeling Methodology, *297*
Preliminaries, *298*
Modeling Strategy, *299*
Hyperellipsoidal Clustering, *301*
Consequence Modeling, *306*
Premise Modeling, *307*
Fuzzy Dynamic Model, *309*
Predictive Control, *310*
Conclusion, *313*
References, *313*

This chapter introduces some techniques to build auto-regressive exogenous models and to carry out the simulation of predictive control interactively and convincingly. To build a fuzzy model, an interactive approach is emphasized in which knowledge or intuition can play an important role. The main proposal in this chapter is a clustering technique, called the *hyperellipsoidal clustering method*, which assists modelers in finding fuzzy subsets suitable for building a fuzzy model. Other problems in fuzzy modeling are also considered such as the effect of data standardization, the selection of conditional and explanatory variables, the shape of a membership function and its tuning problem, the manner of evaluating weights of rules, and the simulation technique for verifying a fuzzy model. Also proposed in this chapter is the design of predictive control using a fuzzy auto-regressive exogenous model in which the future weights of the rules are given by the reference trajectory.

FUZZY MODELING FOR ADAPTIVE PROCESS CONTROL

Yoshiteru Nakamori
Department of Applied Mathematics, Konan University
8-9-1 Okamoto, Higashinada-ku, Kobe 658, Japan

1 INTRODUCTION

The auto-regressive model based on the input-output data is usually used to design a process control because the complexity of phenomena makes it difficult to build a theoretical model. However, due to the high non-linearity of the process, the order of a model often becomes very high so that past effects are taken into account, even if that is physically unrealistic. One way to cope with such a difficulty is to develop a number of submodels which are simple, understandable, and responsible for respective sub-domains. The idea of multi-model approach[1] is not new, but the idea of fuzzy modeling[2] using the concept of the fuzzy sets theory[3] offers a new technique to build multi-models of a process based on the input-output data.

The fuzzy model, or the fuzzy implication inference model[2] consists of a number of rules: in each rule the premise is described by a fuzzy proposition and the consequence is given by a linear model, or an auto-regressive exogenous model in case of the process system. The output from a fuzzy model is given by the weighted sum of outputs from individual consequences where the weights are calculated by the membership grades of inputs to respective premises. In this chapter some new techniques will be proposed to build fuzzy auto-regressive exogenous models for process systems[4][5].

In order to determine a sequence of controls based on an auto-regressive model, the predictive control method[6] received much attention recently. Its main feature is to obtain a control sequence which gives an output sequence close to a reference trajectory, by compensating the inaccuracy of the model based on the measurement of the process. Another proposal in this chapter is to design such a predictive control using a fuzzy auto-regressive exogenous model, in which the future weights of the rules are given by the reference trajectory[4][5].

The developed software[7] offers the facilities to set reference trajectories, the length of control and coincidence horizons, and the number of impulse steps of a fuzzy impulse response model which is transformed from the fuzzy auto-regressive model. The software allows to modify the model in such a way that one can adjust the parameters in the premises of the fuzzy model, or return to the modeling phase to re-build new models with different structures. By such an interaction between modeling and simulation based on the actual process data, one can finally find a better model and a better predictive control for a process system.

2 MODELING METHODOLOGY

We sometimes call measurement data "samples" because an imaginary population and random or systematic errors in measurement flash across our mind. In an assumption that allows unlimited nonlinearity or orders of a model, we can build a model close to the data that is nothing more than just samples from an imaginary population. The determination coefficient adjusted for the degrees of freedom[8] or Akaike's information criterion[9] puts a penalty on decrease in the degrees of freedom. The group method of data handling[10] copes with this problem by introducing an unbiased criterion.

But many of these works hardly satisfy modelers who are facing complex and nonlinear systems. We have to recognize that modeling is art[11] and it is important to understand that the diffusion of this thought has promoted the fuzzy sets theory[3]. Apart from pattern recognition, etc., system modeling is generally an act to understand things directly rather than by computer. And we can understand at most a linear combination like a fuzzy model. Modeling is an act that clears up our vague knowledge, not an act that tries to express vagueness imprecisely.

The fuzzy model proposed by Takagi and Sugeno[2] is a nonlinear model consisting of a number of rule-based linear models and membership functions that will determine the degrees of confidence of the rules. Fuzzy modeling has some interdependent subproblems such as fuzzy partition of a data space, identification of membership functions and linear models. To obtain a satisfactory model from among an unlimited number of combinations, we should decide two things in advance: which sphere we will examine and how we obtain a satisfactory result.

It is generally a good incentive for us to determine a criterion and a searching algorithm because it is logically satisfying to accept a solution found by them. This is also a big temptation in building a fuzzy model. The algorithm developed by Sugeno and Kang[12][13] follows a thoughtful procedure, but it is theoretically impossible to come across an ideal model by a normative approach only. Unless we use intuition to find a way to the goal, it is difficult to obtain a convincing model that can be used in an actual situation. The model proposed by Takagi and Sugeno should be called a fuzzy model in this sense; otherwise it is not different from a neural network model from the point of view of nonlinear modeling.

The fuzzy sets theory does not merely provide interpolation techniques to analyze nonlinear systems. It is something to join logic and intuition together. A modeling algorithm should have a strategy for increasing the chances of finding a better model through the modeler's judgment. In this respect, we recommend an interactive approach with computer assistance[14], where the theme is how to analyze data in order to summarize it to a certain level at which we can understand the nature of the data. This is the philosophical background of this study.

The main technical proposal in this chapter is a clustering algorithm that will search fuzzy subsets based on our desire about their shapes, where heuristics is very important when creating a balance between the continuity and the linearity of data distribution within clusters[15]. After developing a number of fuzzy subsets, we identify linear substructures of the system under study using the interactive modeling support system[14]. Other technical proposals are related to the integration of rules: selection of conditional variables, identification of membership functions, and evaluation of a fuzzy model[15].

3 PRELIMINARIES

The fuzzy model proposed by Takagi and Sugeno[2] is a nonlinear model which consists of a number of rules such as

$$\text{Rule } R^k : \begin{cases} \text{if} \quad x_p \text{ is } A_p^k, \text{ and } x_q \text{ is } A_q^k, \text{ and } \cdots, \\ \text{then} \quad y^k = c_0^k + \sum_{i=1}^{m} c_i^k \cdot x_i. \end{cases} \quad (1)$$

Here, variables x_p, x_q, $\cdots$ in the conditional sentence are called premise variables, and variables x_i ($i = 1, 2, \cdots$) in the linear equation of the concluding part are called consequence variables. A_p^k, A_q^k, $\cdots$ are fuzzy subsets with membership functions $A_p^k(x_p)$, $A_q^k(x_q)$, $\cdots$ that have some tuning parameters called premise parameters. The coefficients c_i^k ($i = 0, 1, 2, \cdots$) of the linear equation are called consequence parameters.

A prediction of the output y by the fuzzy model is given by

$$y_* = \frac{\sum_k w^k \cdot y_*^k}{\sum_k w^k}, \qquad w^k = \prod_p A_p^k(x_{p*}), \quad (2)$$

where x_{p*} denotes an input value, and y_*^k the output value from the rule R^k. The weight w^k of rule R^k is given by the product of membership grades corresponding to input values of all premise variables. Thus, a predicted value is the weighted average of outputs from all rules. This implicitly means that at least one rule should be activated.

Why do we need such a model? Because it is hard enough to express complicated phenomena, but using complicated expressions well is still harder. We often divide an object into some understandable small parts that are usually linear substructures and manage the total nonlinearity by joining them together. In fact, we can recognize that the original form of the fuzzy model is the piecewise linear model in which the whole set is divided into a number of crisp subsets. In the field of control theory, the multi-model control has been examined for a long time[1][16].

How do we build a fuzzy model? An answer given by Sugeno and Kang[12][13] is an iterative algorithm that takes account both of the following problems at the same time:

- Selection of consequence variables and identification of regression coefficients.
- Selection of premise variables and identification of membership functions.

Letting

$$z_i^k = \frac{w^k \cdot x_i}{\sum_k w^k}, \quad i = 0, 1, 2, \cdots, m; \quad k = 1, 2, \cdots, \quad (3)$$

where $x_0 = 1$, we have

$$y = \sum_k (c_0^k z_0^k + c_1^k z_1^k + \cdots + c_m^k z_m^k). \qquad (4)$$

The new variable z_i^k depends on weights of rules that are unknown at this moment. But if they are fixed the above equation can be identified by the least square method; actually it is a weighted least square method. Therefore, a nonlinear optimization algorithm can be applied by fixing the premise and consequence parts by turns.

If we attach importance to the fact that a set of numerical data is a part (probably biased) of a population, we have to use the logic of reasoning and data processing based on common sense to understand an object intuitively as well as analytically. In this chapter we take an interactive approach instead of a normative one, dividing the modeling process into some tractable parts instead of formulating the problem synthetically. A technical contribution in this chapter is a clustering method to find linear substructures of a system. In this respect, it is important to mention two related works and difficulties in applying them to our problem.

A method to obtain a number of linear regression models from heterogeneous data is developed in Jajuga[17]. But there is no idea of prediction, that is, even if an input value is given, the output cannot be predicted. This cannot be resolved by introducing membership functions because the membership values of the output are necessary to predict itself. This is a difficult problem in the case of using a clustering method to build a fuzzy model. We take great pains with this point in this chapter.

When we think about the clustering to discover linear substructures of data distribution, we have to refer to the work of Bezdek and others. They extend Fuzzy c-Means (FCM) method[18] and consider deeply Fuzzy c-Lines (FCL) method[19] and a generalized Fuzzy c-Varieties (FCV) method[20]. Bezdek and others develop a clustering method from necessary and sufficient conditions of this optimization problem. Since a linear variety is extended to infinity, there is a possibility that some cluster contains two widely-separated groups of data units. To get over this problem, they consider a trade-off with c-Means method[20].

There are the following difficulties when applying their method to our problem:

- The membership grades of points without data in the space are not fixed.

- Values of the criterion depend on the number and density of data units.

The above-mentioned difficulties are the same in a sense. That is, for us, the given data units are not all objects for analysis. Assuming an imaginary data distribution, we want to interpolate (or extrapolate, if possible) data units which do not actually exist. We have to find a new idea.

4 MODELING STRATEGY

We have devised the following strategy for developing a fuzzy model with the intention of joining logic and intuition together[15].

- Stage 1: fuzzy partition of the data space. We will find a number of fuzzy subsets by a clustering method that is the main technical contribution of this work. By this clustering, which is in fact a covering of all data units, we would like to find linear substructures of the system under study. Thoughtful consideration is needed here in the selection of variables to be used for clustering.

- Stage 2: identification of linear substructures. Using divided data units, we will identify linear models: selection of explanatory variables and determination of their coefficients. But, it sometimes happens that we cannot build a good linear model for some fuzzy subspace because of a strong nonlinearity or a crowd of data units in a small region. For such a subspace we will build a possibility model expressed by a membership function. Note that the latter case should be avoided in building an auto-regressive exogenous model.

- Stage 3: integration of rules. Introducing an evaluating criterion of the model (an external criterion of clustering), we will carry out the premise modeling: selection of premise variables and identification of their membership functions. The procedure to select premise variables is another technical contribution of this work. Actually, we prepare several sets of linear models at Stage 2 and examine them after premise modeling to determine the final model.

This strategy is not necessarily the best way to build a fuzzy model from measurement data alone. We emphasize an interactive approach here again to utilize the knowledge of persons who understand the problem very well. By the way, when we consider partition of the data space containing an objective variable, we face difficulties in the following cases:

- If the objective variable is a multi-valued function of some explanatory variables such as in [17], we must find variables which can distinguish the rules of the model.

- If a time-series model is being built, it is often desirable to divide the data set by variables that are not explanatory ones, such as differentiated or integrated variables.

The set of premise variables is a subset of the set of consequence variables in usual fuzzy modeling. To extend the applied sphere, we introduce the following definitions:

(a) Objective variables: output variables or future state variables of the system. If the number of objective variables is more than one, we will build a fuzzy model for each objective variable; of course it is possible to process them simultaneously by introducing an integrated criterion[21].

(b) Explanatory variables: input variables or past state variables of the system. The explanatory variable and consequence variable have the same meaning.

(c) Conditional variables: quantitative variables which can contribute to calculating confidence of rules. The set of conditional variables and the set of explanatory variables usually have an intersection. Conditional variables and premise variables with a wide meaning are the same.

When the linear variety clustering is used for fuzzy modeling, we sometimes encounter a case in which we cannot separate the data space including an objective variable effectively, as mentioned above. In other words, there is a danger of getting quite different outputs from the several rules with similar premises. One way of treating such a situation is to introduce the following definition:

(d) Meta-conditional variables: qualitative variables that can control meta-rules. To find meta-conditional variables is important in an actual application, but is not an object of consideration here.

Since our purpose is to discover linear varieties from the data, we have to be careful to carry out a fuzzy partition which does not lead to similar premises. This is the greatest reason that a definition called conditional variables is introduced instead of premise variables. We want to try to use variables which are not included in the consequence as conditional variables. We carry out clustering by choosing variables that can be important conditional or explanatory ones. We add the following definition:

(e) Clustering variables: the union of an objective variable and a subset of explanatory and conditional variables chosen carefully, which determine the axes of a clustering space. Selection of clustering variables requires the intuition of people who understand the problem under study.

5 HYPERELLIPSOIDAL CLUSTERING

Let us introduce a criterion of the clustering (an inner criterion) which suits the above-mentioned purpose, and propose a clustering algorithm based on it, and then explain the heuristics needed in carrying out the clustering successfully.

5.1 Criterion

Suppose that a set of clustering variables $\{x_1, x_2, \cdots, x_m\}$ that includes an objective variable is prepared, and that the data vectors are given as follows:

$$\alpha_i = \begin{bmatrix} x_{i1} \\ x_{i2} \\ \vdots \\ x_{in} \end{bmatrix}, \quad i = 1, 2, \cdots, m; \quad m \ll n. \tag{5}$$

Assume that the data of each variable is transformed as follows:

$$\sum_{j=1}^{n} x_{ij} = 0, \quad \| \alpha_i \|_{R^n}^2 = \sum_{j=1}^{n} x_{ij}^2 = n \cdot s_i^2, \quad i = 1, 2, \cdots, m. \tag{6}$$

Here, s_i (> 0) are design parameters reflecting the modeler's will. That is, setting a larger value on s_i corresponds to attaching the more importance to x_i. In this sense, we call s_i the degree of importance. Denote the j-th data unit by β_j:

$$\beta_j = (x_{1j}, x_{2j}, \cdots, x_{mj}), \qquad j = 1, 2, \cdots, n \tag{7}$$

which is identified with a point in the m-dimensional Euclidean space R^m.

Introducing a criterion (shown later), we divide the set of data units { β_1, β_2, $\cdots$, β_n } into p subsets (clusters) C^1, C^2, $\cdots$, C^p. The number of clusters p will be determined by a stopping rule which will be introduced later, but actually it should be judged by thinking of the structure of a fuzzy model.

Let S^k be the variance-covariance matrix obtained by all data units of C^k, and $\lambda_1^k, \lambda_2^k, \cdots, \lambda_m^k$ the eigenvalues of S^k such that

$$\lambda_1^k \geq \lambda_2^k \geq \cdots \geq \lambda_m^k \quad (\geq 0). \tag{8}$$

Denote by $|C^k|$ the number of data units in C^k. If $|C^k| \leq m$, there exists an l such that $\lambda_l^k = 0$. Even when $|C^k| > m$ it is possible that some eigenvalues are 0 or close to 0. We replace such eigenvalues with ε, a positive small number. The regularizing parameter ε is quite important in creating a balance between the continuity and the linearity of data distribution. In the criterion which will be introduced later, taking a larger value of ε corresponds to setting more importance on continuity than linearity. Therefore, we call ε the degree of continuity.

Denote the eigenvector corresponding to λ_l^k by e_l^k ($\| e_l^k \|_{R^m} = 1$). In case some eigenvalues are 0, we determine the corresponding eigenvectors arbitrarily but orthogonally to others. Let E^k be a hyperellipsoid with the center being the center of gravity of data units in C^k and the axes being made up of $e_1^k, e_2^k, \cdots, e_m^k$:

$$E^k: \quad \frac{(z_1^k)^2}{\lambda_1^k} + \frac{(z_2^k)^2}{\lambda_2^k} + \cdots + \frac{(z_m^k)^2}{\lambda_m^k} = d^k. \tag{9}$$

Here, d^k is the smallest positive number so that E^k contains all data units of C^k inside or on the boundary. If we decide to ignore extraordinary values of the minority, we can design hyperellipsoids containing some percentage of data units.

Thus, we get p hyperellipsoids $E^1, E^2, \cdots, E^p$ that cover n data units $\beta_1, \beta_2, \cdots, \beta_n$. The volume of hyperellipsoid E^k is given by

$$V^k = \frac{\pi^{\frac{m}{2}} \cdot (d^k)^{\frac{m}{2}}}{\Gamma(\frac{m+3}{2})} \cdot \prod_{l=1}^{m} (\lambda_l^k)^{\frac{1}{2}} \tag{10}$$

where $\Gamma(\cdot)$ denotes the Gamma function. We define the criterion for clustering as the sum of volumes of hyperellipsoids, neglecting constants:

$$SVH(p) = \sum_{k=1}^{p} (d^k)^{\frac{m}{2}} \cdot \prod_{l=1}^{m} (\lambda_l^k)^{\frac{1}{2}}. \tag{11}$$

We will find the data partition $C^1, C^2, \cdots, C^p$ which minimizes the SVH criterion.

The reasons why this SVH criterion suits our purpose are summarized as follows:

- The theoretical minimum of the SVH criterion is 0 because $\lambda_l^k \geq 0$ ($k = 1, 2, \cdots, p$; $l = 1, 2, \cdots, m$). In this case the dimension of each hyperellipsoid is less than m; this implies that it is possible to find a linear relation between $x_1, x_2, \cdots, x_m$. The SVH criterion can also find a group of data units that gather together in a small region, which is one of the objectives of our clustering.

- There is resistance to a biased data distribution because the SVH criterion does not directly depend on the number of data. But, the bias of using hyperellipsoids with centers that are the centers of gravity of data units is impossible to avoid.

- A hyperellipsoid covers all data units in each cluster, not data units within the limits of the standard deviations in the directions of principal components. Because of this, for extraordinary values of the minority, a big increase in the SVH criterion is caused. That is, we have a function for detecting extraordinary data.

- The whole data units are divided into clusters C^1, C^2, $\cdots$, C^p, but a covering of data units is made by hyperellipsoids E^1, E^2, $\cdots$, E^p. This hints toward the repeated use of some data units when the rules of a fuzzy model are assembled. That is, using data units contained in E^k, we can build the rule R^k.

5.2 Algorithm

The problem here is how to attain the minimum of the SVH criterion. The best way is, of course, to study all combinations. But we must develop an algorithm when the number of data units does not permit it. Our algorithm has the following tactics:

Tactics 1: initial clusters. By using the Ward method, we form a set of initial clusters of the hyperellipsoidal clustering. The reasons for using the Ward method are the following:

- If the number of data units is great or if a large number of data units are close to each other, the group of initial clusters should be built by gathering data units within a limit.

- In each stage of clustering, we have to compute eigenvalues and eigenvectors of a number of variance-covariance matrices. If the number of initial clusters is smaller, it is better for reduction of calculation time.

- To avoid having a cluster consisting of two data units which are located far away from each other, it is better to put close data units into a cluster in advance. This helps with an easy setting of the parameter ε because one of the eigenvalues is definitely 0 in case of two data units.

We introduce two stopping parameters of the Ward method. That is, we set the number of initial clusters less than or equal to n_w, or the lowest limit of distances between the centers of gravity of clusters d_w.

Tactics 2: hierarchical procedure. Starting with the initial clusters, at each step we lump two clusters that give the minimum increase in the SVH criterion. Here, we consider lumping of two clusters with a distance between the centers of gravity of less than a parameter δ_u, which is a stopping parameter of the hyperellipsoidal clustering and an important design parameter like the regularizing parameter ε when creating a balance between the continuity and the linearity of data distribution within clusters. Another stopping parameter v_h restricts the rate of increase in the SVH criterion. A rapid increase of the SVH criterion means that we are failing to obtain a good linear model using a subset of the data.

Tactics 3: non-hierarchical procedure. At each step of the hierarchical procedure, we introduce a non-hierarchical one that changes some data units between clusters so as to reduce the SVH value. If we move two or more data units at the same time, we may get a better result. But we only repeat movement of one data unit to avoid combinatorial explosion. A parameter δ_m is introduced here to limit the sphere of searching in replacing data units. We should set δ_m larger than δ_u to allow replacement of data units at the final step. To set δ_u and δ_m large implies attaching importance to linearity. In this sense they are called the degrees of linearity.

5.3 Design Parameters

In the above algorithm a number of design parameters are introduced and they have great influences on the result. We summarize the heuristics necessary in setting those parameters in the following:

(a) The degree of importance s_i $(i = 1, 2, \cdots, m)$: standard deviations. Normalization of variances is usually recommended in the clustering. But, it is desirable that the modeler assigns weights of variables by using judgment and understanding about the problem. For the purpose of building fuzzy models, the objective variable should have a larger weight than explanatory variables because we want to divide the data set so as to find the trend of the objective variable. Moreover, by giving different weights to explanatory variables depending on the purpose of model use, we can distinguish variables which we need in the model from those that are less important.

(b) The stopping parameters n_w, d_w in the Ward method. When there is a large number of data units or many of them gather together in a small region, the initial clusters of the hyperellipsoidal clustering are prepared by the Ward method. In this case, we set the number of initial clusters less than or equal to n_w, or the lowest limit of distances between the centers of gravity of clusters d_w. Since the distance of the centers of gravity of two clusters depends on the values of variances s_i^2 $(i = 1, 2, \cdots, m)$, determination of d_w requires trial and error. If the density of the data distribution is biased, at the final stage of clustering by the Ward method, we can replace all data units within each cluster with a representative point.

(c) The degree of continuity ε: the regularizing parameter. This parameter is introduced in order to consider full dimensional hyperellipsoids at every step of the clustering. That is, this parameter is needed when a cluster consists of less than $m+1$ data units, or when an eigenvalue of the variance-covariance matrix is (nearly) equal to 0. To set ε small implies laying stress on linearity in the clustering. On the other hand, to make its value large corresponds to putting importance on continuity of data distribution. We have to find suitable ε to avoid the case that separated data units belong to the same cluster at early stages of the clustering.

(d) The degrees of linearity δ_u, δ_m: regulating parameters in lumping clusters and replacing data units. The parameter δ_u is a stopping parameter of the hyperellipsoidal clustering, and an important design parameter like the regularizing parameter ε when creating a balance between the continuity and the linearity of data distribution within clusters. The parameter δ_m limits the sphere of search in replacing data units. We should set δ_m larger than δ_u to allow replacement of data units at the final step. To set δ_u and δ_m large implies attaching importance to linearity.

(e) The stopping parameter v_h in the hyperellipsoidal method. We can let a hyperellipsoid contain some percentage of data units within a cluster. But here we let it contain all data units. Because of this, the clustering process has the function of detecting extraordinary values. That is, the data which is a great distance from the center of the distribution is left behind in the clustering, by limiting the rate of increase in the SVH criterion under v_h. But we need care to handle v_h because it may happen that we just miss data units reaching the extraordinary level. Not only the problem of extraordinary data, the rapid increase in the SVH criterion implies that we are failing to obtain a good linear model using a subset of the data.

(f) In terms of technical problems, we sometimes have to give small fluctuations to data units in the following cases: When we do not use the Ward method, we should let the identical data units move slightly by giving random fluctuations in order to avoid irregular conditions when lumping clusters. When there are plural data units on or near the boundary of a hyperellipsoid, they obstruct the replacement of data units even though movement of data units is desirable.

5.4 Membership Functions

We divide the data space by the proposed clustering technique and build a rule-based model. At this time, if we think a great deal about the fact that conditional variables are correlated to each other, we should identify multi-dimensional membership functions with hyperellipsoidal contour-lines. However, we adopt the idea to construct one-dimensional membership functions for individual variables for the following reasons:

- The clustering is done in the variable space including the objective variable, but when running the model it is necessary to use membership functions of conditional variables only.

- It is theoretically possible to identify a multi-dimensional membership function related to a set of conditional variables, but this is a little difficult because of the existence of a large number of tuning parameters.

- It is easy to understand the model if the premises are expressed through language, and it is further understandable if expressions are made by use of a small number of important conditional variables.

According to this plan, we will identify membership functions $A_i^k(x_i)$ for all prepared variables (not only the clustering variables) corresponding to the hyperellipsoid E^k. Put

$$X_i^k = \{x_{il} \mid \beta_l \in E^k\}, \quad k = 1, 2, \cdots, p \tag{12}$$

and denote the first, second (median) and third quartiles of X_i^k by q_{i1}^k, q_{i2}^k and q_{i3}^k, respectively. If two of them are equal, give one of them a small fluctuation to keep the restriction that $q_{i1}^k < q_{i2}^k < q_{i3}^k$. Let us define a membership function of x_i related to the fuzzy subset A_i^k by

$$A_i^k(x_i) = \begin{cases} inf\{exp\{-\frac{(x_i-q_{i2}^k)^2}{2(t_{i1}^k)^2(q_{i1}^k-q_{i2}^k)^2}\}, \chi_i(x_i)\}, & x_i \leq q_{i2}^k \\ inf\{exp\{-\frac{(x_i-q_{i2}^k)^2}{2(t_{i2}^k)^2(q_{i3}^k-q_{i2}^k)^2}\}, \chi_i(x_i)\}, & x_i \geq q_{i2}^k \end{cases} \tag{13}$$

where $\chi_i(x_i)$ is the characteristic function corresponding to the domain of x_i, and t_{i1}^k, t_{i2}^k (> 0) are tuning parameters with the unit default[21]. The function $A_i^k(x_i)$ is an asymmetrical curve with two inflection points that are internally or externally dividing points between the median and the first (resp. the third) quartile in the ratio $t_{i1}^k : 1 - t_{i1}^k$ (resp. $t_{i2}^k : 1 - t_{i2}^k$).

There are several reasons to use such a membership function. It is not easily influenced by extraordinary data units because quartiles are robust statistics. It is always positive for every value of x_i, therefore, theoretically accepts any input. But it is important to declare the effective sphere of the model by introducing domains of individual variables.

6 CONSEQUENCE MODELING

6.1 Linear Substructures

When we build a linear model with each subset of the data, we must examine how many components of a hyperellipsoid we should use. If we have a conclusion that we should use q-components, we will try building linear models about all combinations of $q - 1$ explanatory variables and compare the average of square errors. If there exists a conditional variable which is not thought of as an explanatory one, we can reduce the number of explanatory variables used.

About the decision concerning the numbers of components, even statistics cannot provide a clear answer, but we consider the following procedures[22]:

- In the cluster C^k, we choose q components such that the rate of cumulative contribution $\sum_{l=1}^{q} \lambda_l^k / \sum_{i=1}^{m} (S^k)_{ii}$ is greater than some level, for example, 80%. This idea is a response to the need to explain most of the fluctuations by selected components. We think this guide is appropriate to our problem.

- We can choose components that have a rate of contribution more than the average in the cluster C^k, that is, $\lambda_l^k \geq \frac{1}{m}\sum_{i=1}^{m}(S^k)_{ii}$. In this procedure each selected component must have more information than one variable.

- On the assumption that variables $x_1, x_2, \cdots, x_m$ follow a multi-dimensional normal distribution in the cluster C^k, we carry out a hypothesis test $H_0 : \lambda_{q+1}^k = \cdots = \lambda_m^k$, and if it is accepted, we will not adopt components after q-th. This hypothesis means that if we take q components, the rest of $m - q$ components are equivalent in all directions and we cannot choose some of them.

In an actual application, having in mind these standards together with the knowledge on interdependencies between variables, we will identify the consequences of a fuzzy model. We can also think of variables not used in the clustering as explanatory variables, if necessary. Here, we would like to give important notes on the consequence modeling.

In linear modeling, we use not only data units contained in each cluster but also data units contained in each hyperellipsoid. In this way, by allowing repeated data use, we can expect improvement in model precision. By repeated data use, we can add the effect of the weighted least square method to the usual least square one.

A fuzzy model consisting of a number of rule-based linear models can be verified only after premise modeling. Therefore, we prepare a number of possible linear models for individual subsets of the data at this stage. Often, we have to come back to this stage after premise modeling.

6.2 Modeling Support

Development of linear models is based on the use of observed values of the variables together with statistical techniques for the evaluation of parameters of the model. But often experts developing the model may have additional knowledge with regard to interdependencies between variables. This knowledge interpreted as a mental model of the system considered can be used to make the model a more adequate representation of the given process or system.

On the other hand, this knowledge is often of a subjective nature and cannot be fully formalized. Therefore, the use of this knowledge in developing a desired model necessitates the direct participation of experts in the modeling process itself. This in turn necessitates the development of computer-aided modeling systems that facilitate such participation. We have an interactive modeling support system[14] to carry out linear modeling with the assistance of experts.

Including the above mentioned facility, we have developed a fuzzy modeling support system in which the Ward method can be used interactively to obtain fuzzy subspaces and the simulation can be carried out using the developed fuzzy model. This time we have developed new facilities to help the hyperellipsoidal clustering and the premise modeling that will be explained in the next section. With this support system based on logic, we can utilize the intuition of experts effectively.

7 PREMISE MODELING

7.1 Conditional Variables

It is not necessary to use all variables used in the clustering as premise variables. This is the same idea as that of selecting explanatory variables. We can also use variables not used in the clustering as premise variables. Here we calculate the degrees of separation using defined membership functions and build premises by a set of variables with a large degree of separation. That is, we construct premises by fixing the consequences. It is a complex optimization problem of selecting a combination of variables and tuning parameters t_{i1}^k, t_{i2}^k.

As mentioned in Section 5.4, we construct one-dimensional membership functions for individual variables instead of a multi-dimensional one. This is an important approach here. Another proposal is that we construct all the premises of a fuzzy model using the same combination of conditional variables. Let I be the set of variables that can be conditional. We define the degree of separation $S(I_c)$ of a subset $I_c \subset I$ as follows:

First calculate the following quantities:

$$w^k(I_\alpha) = \sum_{\substack{l \\ x_{il} \in X_i^k}} \prod_{\substack{i \\ x_i \in I_c}} A_i^k(x_{il}), \quad k = 1, 2, \cdots, p \tag{14}$$

$$\bar{w}^k(I_\alpha) = \sum_{h \neq k} \sum_{\substack{l \\ x_{il} \in X_i^k}} \prod_{\substack{i \\ x_i \in I_c}} A_i^h(x_{il}), \quad k = 1, 2, \cdots, p \tag{15}$$

where X_i^k is a subset of data defined in Section 5.4, and p denotes the number of rules. Let us define the degree of separation by

$$S(I_c) = \min_k \left\{ \frac{w^k(I_c)}{\bar{w}^k(I_c)} \; ; \; k = 1, 2, \cdots, p \right\}. \tag{16}$$

For possible subsets I_c's with relatively large degrees of separation, we try to optimize parameters t_{i1}^k, t_{i2}^k by the complex method[23] with the criterion of the sum of square errors, and then adopt one I_c. In order to find a better subset I_c, we can use a forward selection method though there is no assurance of obtaining the best one.

7.2 Model Evaluation

If we determine a criterion, it is easy to evaluate the model. But especially when we think there is an extrapolation problem, we need a tool to check quality of the model by intuition. We usually plot predicted and measured outputs together along with the time axis and check the behavior of the model. But this is not enough to ascertain the degree of confidence of the rules and of the model outputs.

Input variables for the fuzzy model are explanatory and conditional variables. To check the model, we fix the values of some of them and free the rest. Since input variables are correlated to each other, it is natural that the admissible input ranges of free variables will change. Let I_f be the set of variables whose values are fixed and $\bar{I}_f$ the set of free variables. For $x_j \in \bar{I}_f$, define an admissible input distribution (a possibility distribution)[21] by

$$B_j(x_j) = \frac{\sum_{k=1}^{p} u^k \cdot A_j^k(x_j)}{\sum_{k=1}^{p} u^k}, \qquad u^k = \prod_{x_i \in I_f} A_i^k(x_{i*}) \tag{17}$$

where x_{i*} is a fixed input value of $x_i \in I_f$. In order to cope with the case where $I_f = \phi$, and to define $B_i(x_i)$ for $x_i \in I_f$, let us define the default as $u^k = 1$ $(k = 1, 2, \cdots, p)$. For the output variable y, we can define a possibility distribution $B(y)$ similarly.

The simulation is carried out by the Monte-Carlo method. We give n sets of inputs to the fuzzy model successively. Here, for $x_i \in I_f$, n inputs are equal and for $x_j \in \bar{I}_f$, n inputs are random numbers from the distribution:

$$N[B_j(x_j)] := \frac{B_j(x_j)}{\|B_j(x_j)\|} \tag{18}$$

where $\|\cdot\|$ indicates the integral operation on the domain.

The prediction of y by the l-th set of inputs is given by

$$y_l = \frac{\sum_{k=1}^{p} w^k \cdot y_l^k}{\sum_{k=1}^{p} w^k}, \qquad w^k = \prod_{x_i \in (I_f \cup \bar{I}_f) \cap I_c} A_i^k(x_{il}). \tag{19}$$

Define the degree of confidence of y_l by

$$C(y_l) = \frac{\prod_{x_i \in I_f \cup \bar{I}_f} N[B_i(x_{il})]}{\max_l \left\{ \prod_{x_i \in I_f \cup \bar{I}_f} N[B_i(x_{il})] \right\}}. \tag{20}$$

We give a large degree of confidence to an estimate y_l if it is obtained by the combination of inputs that is highly possible to occur based on the observation of the past data by which the model is constructed. We can observe quality of the model by plotting $(y_l, C(y_l))$ $(l = 1, 2, \cdots, n)$ and comparing with $B(y)$ on the screen of a computer.

8 FUZZY DYNAMIC MODEL

Using the above-mentioned techniques, we can build a fuzzy ARX (Auto-Regressive eXogenous) model which consists of several rules such as

$$\text{Rule } R^k : \begin{cases} \text{if } y_\alpha(t - l_\alpha) \text{ is } A_\alpha^k, \text{ and } y_\beta(t - l_\beta) \text{ is } A_\beta^k, \text{ and } \cdots, \\ \text{then } Y_t^k = \sum_{l=1}^{T} C_l^k \cdot Y_{t-l} + \sum_{l=1}^{T} D_l^k \cdot U_{t-l} + V^k \end{cases} \tag{21}$$

where $A_\alpha^k, A_\beta^k, \cdots$ are fuzzy subsets with respective membership functions, C_l^k, D_l^k coefficient matrices, and

$$Y_t = (y_1(t), y_2(t) \cdots, y_O(t))^\mathsf{T} \tag{22}$$

$$U_t = (u_1(t), u_2(t) \cdots, u_I(t))^\mathsf{T} \tag{23}$$

$$V^k = (v_1^k, v_2^k \cdots, v_O^k)^\mathsf{T} \tag{24}$$

are output variables, input variables, and constants, respectively.
The estimate of Y_t can be obtained by the following formula:

$$Y_t^* = \frac{\sum_{k=1}^{p} w_t^k \cdot Y_t^k}{\sum_{k=1}^{p} w_t^k}, \qquad w_t^k = \prod_\alpha A_\alpha^k(y_\alpha^*(t - l_\alpha)), \tag{25}$$

where p is the number of rules, and Y_t^k the output from the rule R^k. The weight w_t^k is given by the product of membership grades of all conditional variables.

9 PREDICTIVE CONTROL

The multi-model approach[1] is an idea to use a number of process models that are effective around the respective operating points and to unify the respective optimal controls by reasonable weights. The difficulties of this approach are how to unify the outputs from individual submodels to predict future outputs of the process, and how to lead the optimal controls in the universe of discourse.

An attempt to design controllers using fuzzy models was done in Sugeno and Kang[12], where the linear control theory is used to design optimal regulators based on the respective rule-based models, and then the optimal control in the universe of discourse is obtained by taking account of the degrees of confidence of rules. However, the weighted sum of respective optimal regulators is not necessarily optimal in the universe of discourse.

We will present an algorithm of designing controllers based on the idea of model predictive control. First we transform the ARX model to the fuzzy impulse response model:

$$\text{Rule} \quad R^k: \qquad Y_t^k = \sum_{l=1}^{q} H_l^k \cdot U_{t-l} + \sum_{l=1}^{q} G_l^k \cdot V^k. \tag{26}$$

The number of impulse steps q is determined by observing impulse response curves. H_l^k is an impulse response matrix, where $H_l^k = 0$ for $l > q$. Putting

$$H_l^t = \frac{\sum_{k=1}^{p} w_t^k \cdot H_l^k}{\sum_{k=1}^{p} w_t^k}, \qquad G_l^t = \frac{\sum_{k=1}^{p} w_t^k \cdot G_l^k \cdot V^k}{\sum_{k=1}^{p} w_t^k}, \tag{27}$$

we can write the output of fuzzy dynamic model as

$$Y_t^m = \sum_{l=1}^{q} H_l^t \cdot U_{t-l} + \sum_{l=1}^{q} G_l^t. \tag{28}$$

We assume that the series of inputs $U_{t-1}, U_{t-2}, \cdots$ were added to the process, and the series of inputs $U_t, U_{t+1}, \cdots, U_{t+M-1}$ will be added from the current time t, and assume that

$$U_{t+M+l} = U_{t+M-1} \quad for \quad l \geq 0. \tag{29}$$

We can further rewrite the output equation as follows:

$$Y_M = Y_{M_0} + H_F U_F + H_0 U_0 + G_0, \tag{30}$$

where

$$Y_M = [(Y_{t+L}^m)^\mathsf{T}, (Y_{t+L+1}^m)^\mathsf{T}, \cdots, (Y_{t+L+P-1}^m)^\mathsf{T}]^\mathsf{T} \tag{31}$$

$$Y_{M_0} = [(Y_t^m)^\mathsf{T}, (Y_t^m)^\mathsf{T}, \cdots, (Y_t^m)^\mathsf{T}]^\mathsf{T} \tag{32}$$

$$U_F = [(U_t)^\mathsf{T}, (U_{t+1})^\mathsf{T}, \cdots, (U_{t+M-1})^\mathsf{T}]^\mathsf{T} \tag{33}$$

$$U_0 = [(U_{t-1})^\mathsf{T}, (U_{t-2})^\mathsf{T}, \cdots, (U_{t-q})^\mathsf{T}]^\mathsf{T} \tag{34}$$

$$H_F = \begin{bmatrix} H_L^{t+L} & \cdots & H_{L-M+2}^{t+L} & \sum_{l=1}^{L-M+1} H_l^{t+L} \\ H_{L+1}^{t+L+1} & \cdots & H_{L-M+3}^{t+L+1} & \sum_{l=1}^{L-M+2} H_l^{t+L+1} \\ \vdots & & \vdots & \vdots \\ H_{L+P-1}^{t+L+P-1} & \cdots & H_{L-M+P+1}^{t+L+P-1} & \sum_{l=1}^{L-M+P} H_l^{t+L+P-1} \end{bmatrix} \tag{35}$$

$$H_0 = \begin{bmatrix} H_{L+1}^{t+L} - H_1^t & \cdots & H_{L+q}^{t+L} - H_q^t \\ H_{L+2}^{t+L+1} - H_1^t & \cdots & H_{L+q+1}^{t+L+1} - H_q^t \\ \vdots & & \vdots \\ H_{L+P}^{t+L+P-1} - H_1^t & \cdots & H_{L+q+P-1}^{t+L+P-1} - H_q^t \end{bmatrix} \tag{36}$$

$$G_0 = \begin{bmatrix} \sum_{l=1}^{q} G_l^{t+L} - \sum_{l=1}^{q} G_l^t \\ \sum_{l=1}^{q} G_l^{t+L+1} - \sum_{l=1}^{q} G_l^t \\ \vdots \\ \sum_{l=1}^{q} G_l^{t+L+P-1} - \sum_{l=1}^{q} G_l^t \end{bmatrix}. \tag{37}$$

Predicted outputs may differ from the measured data because of identification errors or the existence of disturbances. The following equation is provided to reduce the difference:

$$Y_P = Y_M + \{Y - Y_{M_0}\}, \tag{38}$$

where Y_P is the vector of predicted outputs, Y the vector of measured data:

$$Y_P = [(Y_{t+L}^p)^T, (Y_{t+L+1}^p)^T, \cdots, (Y_{t+L+P-1}^p)^T]^T, \tag{39}$$

$$Y = [(Y_t)^T, (Y_t)^T, \cdots, (Y_t)^T]^T. \tag{40}$$

Let us introduce a reference trajectory Y_R:

$$Y_R = [(Y_{t+L}^r)^T, (Y_{t+L+1}^r)^T, \cdots, (Y_{t+L+P-1}^r)^T]^T \tag{41}$$

which is given by

$$Y_{t+l}^r = \alpha^{(l-L+1)} Y_t + (1 - \alpha^{(l-L+1)}) Y^*, \tag{42}$$

where Y^* is the vector of targets, and α a tuning parameter.

It is desirable to move the output along a smooth trajectory to the target than to move it drastically. When a first order delay curve is chosen as a reference trajectory, the time constant can be used as a parameter to adjust the response speed.

The control action at the current time should be determined in order to minimize the difference between the predicted and reference trajectories during some future time period. Therefore an optimal control can be obtained by minimizing a cost function, for instance,

$$J = \| Y_P - Y_R \|^2 + \beta \| U_F \|^2. \tag{43}$$

This function depends on U_F nonlinearly through membership functions, that is, the future values of confidence of rules are required in evaluating H_F, H_0 and G_0; this fact causes the problem to be very difficult. But, if we use membership grades obtained by the reference trajectory, the solution to the problem is given by

$$U_F = (H_F^T H_F + \beta I)^{-1} H_F^T (Y_R - Y - H_0 U_0 - G_0). \tag{44}$$

Though U_F includes M time steps of control actions, just one-step control will be put into the process.

The measured outputs at the next time step $t+1$ may be different from the predicted values at the current time t. Then we repeat the same steps by resetting $t+1$ as the current time.

In this chapter we formulated the model predictive control using the impulse response model, but we can formulate it using the parametric models, that is, the state space model or the auto-regressive exogenous model[24].

10 CONCLUSION

It is very difficult to develop effective models for a complex process based on input-output data. Because of this, the problems are how to design controllers using poor models as well as how to build process models. This chapter proposed the algorithms to develop fuzzy dynamic process models and to design model predictive controllers using the developed models. We have carried out an application study on a stable incineration problem of a rotary-kiln process treating excess sludge from a municipal wastewater treatment plant, in which almost operations have been left to the skilled operators.

The introduced techniques with an interactive computer software[7] realize the interaction between modeling and simulation of the process control. With this software, the fuzzy auto-regressive exogenous models are built interactively and convincingly, and the simulation of predictive control is carried out in such a way that the user can determine the design parameters and the reference trajectories freely, and if necessary, the user can modify the parameters in the model. It is also possible to return back to the modeling phase in order to examine other structures of the model to find a better predictive control based on the actual process data.

REFERENCES

[1] Binder, Z. et. al, "About a multimodel control methodology, algorithm, multi-processors, implementation and application." In *Proc. of the 8th IFAC World Congress*. Kyoto, 1991, pp. 981-986.

[2] Takagi, T. and Sugeno, M., "Fuzzy identification of systems and its applications to modeling and control." *IEEE Trans. on Syst. Man and Cybern.*, Vol. 15, No. 1, pp. 116-132, 1985.

[3] Zadeh, L. A., "Fuzzy sets." *Information and Control*, Vol. 8, pp. 338-353, 1965.

[4] Nakamori, Y., Suzuki, K. and Yamanaka T., "Model predictive control using fuzzy dynamic models." In *Proc. of IFSA'91 Brussels*, Vol.Engineering. Brussels, July 7-12, 1991, pp. 135-138.

[5] Suzuki, K. and Nakamori, Y., "Model predictive control based on fuzzy dynamic models." In *Proc. of 4th Int. Sympo. on Process Systems Engineering*, Vol. 2. Waterloo, August 5-9, 1991, pp. 18.1-15 .

[6] Oshima, M., "Model predictive control." In *Text of Process Control Techniques 1989*. Tokyo, The Society of Chemical Engineers, 1989.

[7] Suzuki, K., Nakamori, Y. and Yamanaka, T., "An interactive support system for fuzzy modeling and fuzzy model predictive control system design." In *Proc. of Int. Fuzzy Engineering Sympo*. Yokohama, November 13-15, 1991, pp. 1102-1103.

[8] Theil, H., *Economic Forecasts and Policy*, 2nd ed. Amsterdam, North-Holland, 1961.

[9] Akaike, H., "A new look at the statistical model identification." *IEEE Trans. on Automatic Control*, Vol. AC-19, No. 6, pp. 716-723, 1974.

[10] Ivakhnenko, A. G., "The group method of data handling, a rival of the method of stochastic approximation." *Soviet Automatic Control*, Vol. 13, No. 3, pp. 43-55, 1968.

[11] Majone. G., "The craft of applied systems analysis." In Tomlinson, R. and Kiss, I. (eds): *Rethinking the Process of Operational Research and Systems Analysis*. Oxford, Pergamon Press, 1984, pp. 143-157.

[12] Sugeno, M. and Kang, G. T., "Fuzzy modelling and control of multilayer incinerator." *Fuzzy Sets and Systems*, Vol. 18, pp. 329-346, 1986.

[13] Sugeno, M. and Kang, G. T., "Structure identification of fuzzy model." *Fuzzy Sets and Systems*. Vol. 28, pp. 15-33, 1988.

[14] Nakamori, Y., "Development and application of an interactive modeling support system." *Automatica*, Vol. 25, No. 2, pp. 185-206, 1989.

[15] Nakamori, Y. and Ryoke, M., "Exploratory data analysis for fuzzy modeling and decision support." In *Proc. of 10th Int. Conf. on Multiple Criteria Decision Making*, Vol. 2. Taipei, July 19-24, 1992, pp. 281-290.

[16] Lainiotis, P. G. et al., "Optimal adaptive control: a nonlinear separation theorem." *Int. J. Control*, Vol. 15, No. 5, pp. 877-888, 1972.

[17] Jajuga, K., "Linear fuzzy regression." *Fuzzy Sets and Systems*, Vol. 20, pp. 343-353, 1986.

[18] Dunn, J., "A fuzzy relative of the ISODATA process and its use in detecting compact well-separated clusters." *J. Cybernetics*, Vol. 3, pp. 32-57, 1974.

[19] Bezdek, J. C. et al., "Detection and characterization of cluster substructure I. linear structure: fuzzy c-lines." *SIAM J. Appl. Math.*, Vol. 40, No. 2, pp. 339-357, 1981.

[20] Bezdek, J. C. et al., "Detection and characterization of cluster substructure II. fuzzy c-varieties and convex combinations thereof." *SIAM J. Appl. Math.*, Vol. 40, No. 2, pp. 358-372, 1981.

[21] Kainuma, M., Nakamori, Y. and Morita, T., "Integrated decision support system for environmental planning." *IEEE Trans. on Syst. Man and Cybern.*, Vol. SMC-20, No. 4, pp. 777-790, 1990.

[22] Tanaka, Y. and Wakimoto, K., *Methods of Multivariate Statistical Analysis*, Gendai-Sugakusya, (in Japanese), 1983.

[23] Box, M. J., Davies, D. and Swann, W. H., *Non-Linear Optimization Techniques*, Obiver & Boyd, 1969, pp. 52-54.

[24] Nishitani, H. et al., "Model description for model predictive control." In *Proc. of 29th SICE Annual Conference*. Tokyo, July 24-26, 1990, pp. 737-740.

Chapter 15
Fuzzy Controller with Matrix Representation

Introduction, *316*
Fuzzy Control Statements, *317*
Basic Concept of Matrix Representation, *318*
Reduction of Matrix Representation, *322*
Reduction by Simple Self-Tuning, *322*
Self-Tuning by Modified Simplex Method, *324*
Neural Networks for Fuzzy Controller, *328*
Conclusion, *334*
References, *335*

This chapter proposes a methodology for converting fuzzy control statements into a matrix representation which provides fast computation of the fuzzy controller outputs. Various methods for improving the value of the matrix representation are proposed. These include some self-tuning methods as well as using neural networks. To illustrate the matrix representation method, it is combined with a neural network and applied to the problem of traffic signal control. It is shown that the over all delays of vehicles are reduced substantially with this method.

FUZZY CONTROLLER WITH MATRIX REPRESENTATION

M. Nakatsuyama   J. H. Yan   H. Kaminaga
Department of Electronic Engineering, Yamagata University
Yonezawa 992 Japan

1. INTRODUCTION

Many researchers have been researching the architecture of fuzzy controllers and a large number of manufacturers have been producing many kinds of the fuzzy controllers. Most controllers have a large number of parallel processors and are faster than even a supercomputer in computing the fuzzy inference rules. Many types of the computer boards of the fuzzy controller have been made for personal computers and used for home electric appliances, but they are not so fast. Yamakawa proposed a controller in which the values of the membership functions must be fixed before computation and controls successfully the inverted pendulum [1]. At first Zadeh proposed a fuzzy algorithm and showed an example of fuzzy automatic control of an automobile which moves as if a human controls it [2]. Pappis proposed the fuzzy control of traffic signals by using fuzzy control statements and got a better result than by other methods [3]. It is the basic and important idea that Fuzzy control statements are easily converted into a matrix.

In general, fuzzy control is supposed to be nondeterministic. A fuzzy control program consists of many fuzzy control statements. The final decision is made by choosing the maximum values of these statements using a simple calculation or the gravity method [4]. Though a fuzzy program is based on fuzzy sets or ambiguous concepts, fuzzy control programs must provide only one determined value when some fixed inputs are applied to the fuzzy control system. If multiple results are selected, then exact and stable control can not be expected. Then the fuzzy control must become deterministic. Nakatsuyama showed that the fuzzy program can be converted into matrices that we call the matrix representation [5]. One of the advantages of the matrix representation is faster computation than in the fuzzy control statements. The disadvantage is that it takes more memory, but nowadays the price of memory is very cheap so this is not a serious obstacle. Since the value of the matrix is not adequate initially, fine adjustment of the value is required in the next stage.

We use simple self-tuning to get a better value for the matrix and derive successfully the reduction of the matrix [6]. We trained a neural network to get the adequate value of the matrix representation. Then, combining the matrix representation, the neural network, and simple heuristic technique we get the good result for controlling the traffics [7]. We also get a good result by using the modified simplex method for computing the adequate value of the matrix [8].

In this chapter, we outline methods for improving the values of the matrix. The simple self-tuning is very easy but not so effective to get the better value. The simplex method seems to be the best since it gets the smallest over all delays among these methods. However, we believe that neural networks will get the best result because a

2. FUZZY CONTROL STATEMENTS

Fuzzy control statements are the basic concept in fuzzy control and describe clearly the ambiguous nature of fuzzy theory by using the linguistic representation. The linguistic representation is the strongest tool for illustrating fuzzy theory. The fuzzy control statements consist of the assembly of statements such as "if .. then .. else .. ". We show an example of a fuzzy program which controls the traffic signals as follows:

if   T = medium and A = mt(medium) and Q = lt(small)
                                          then E = medium
else if T = long   and A = mt(many)   and Q = lt(medium)
                                          then E = long

The terms mt. and lt denote "more than" and "less than" respectively. The symbols T, A, Q and E are the fuzzy variables of time, arrived vehicles, queue of vehicles and the signal duration, respectively. In fact, the real calculation is executed as follows:

$$\mu = \max(\mu_R(t,a,q,e),\ \mu_S(t,a,q,e)...) \tag{1}$$

The symbol μ denotes the membership function. If $\mu_R = \mu_S = ...$, we are able to choose any number as the most suitable value as follows:

$$\text{The set of optimum values} = \{\ (t_1,a_1,q_1,e_1),\ (t_2,a_2,q_2,e_2),.....(t_n,a_n,q_n,e_n)\ \} \tag{2}$$

So the fuzzy control is nondeterministic from this point of view. However, it is necessary to select only one value to guarantee stable and precise control, and we do that by using the gravity method. Then, the fuzzy control becomes deterministic and we get the following equation.

$$FP:\ f(t,a,q,e)\ \rightarrow\ d \tag{3}$$

If the value of the term e is constant, Eq.(3) can be written as follows:

$$FP:\ f(t,a,q)\ \rightarrow\ d \tag{4}$$

The terms t, a, and q are not an integer, but are always normalized. Let M be a matrix, then this equation will become

$$dm = M(k_1,k_2,k_3,...k_n). \tag{5}$$

The term k_i is an integer. The fuzzy control statements are represented as a matrix which we call the matrix representation.

3. BASIC CONCEPT OF MATRIX REPRESENTATION

It is rather difficult to calculate the computing time of the fuzzy control statement or the matrix representation. We use the traffic control problem for comparing the fuzzy control statement with the matrix representation.

In fuzzy control programs for traffic control, we use 5×5 statements. Each statement contains 3 comparison operations to calculate the minimum value. For obtaining maximum value of statements, $5 \times 5 - 1$ comparisons are necessary. So there must be $3 \times 5 \times 5 + 5 \times 5 - 1 = 99$ comparisons. Each fuzzy set has a table to get its membership function. One example of these tables is shown in Table 1.

Table 1 Fuzzy Sets on Extension

Fuzzy sets	Extension time (sec)									
	1	2	3	4	5	6	7	8	9	10
short	1	0.8	0.1	0	0	0	0	0	0	0
medium	0	0	0.1	0.8	1	0.8	0.1	0	0	0
long	0	0	0	0	0.1	0.8	1	0.8	0.1	0

In each statement, it is necessary to get a value from the table before comparison. Then each statement needs 4 adding to calculate the address of a necessary data in the table. The total number of additions is $5 \times 5 \times 4$. The execution time of addition or subtraction or comparison may be τ and the execution time of multiply is probably 10 τ in most of the latest computers. Therefore, fuzzy control programs needs 199 τ in computing time.

To obtain a value from M(t,a,q), it is necessary to calculate the address of e in the computer memory. Let N be the size of the matrix. The address r is determined by the equation such as

$$r = o + NN \times (t - 1) + N \times (a - 1) + q - 1 \qquad (6)$$

where $NN = N \times N$ and o denotes the origin of the matrix. It needs 2 multiply and 6 addition or subtraction operations to get the address. So the matrix representation needs only 25 τ. Therefore, the matrix representation is calculated much faster than the fuzzy control program.

The fuzzy program needs about $N \times N$ memory units for storing the table which is used for calculating "more" or "medium" or "small". On the other hand, the matrix representation generally needs about even $N \times N \times N$ memory units. Then the memory for the matrix representation is larger by N times than the fuzzy program. By considering the latest price of the memory, it is not a serious obstacle to use the large memory. Indeed, we estimate that the size of the memory is only 1600 bytes for the matrix representation of traffic control. However, once we can degenerate the matrix as we will describe afterwards, the memory size of the matrix representation is comparable with the fuzzy program. It is clear that the fuzzy program which controls 16 traffic junctions requires long computing time.

We assume that there are 8 one-way roads and 16 traffic intersections as shown in Fig. 1. A fuzzy program which controls 16 traffic intersections is time-consuming. We apply the matrix representation to control them. The matrix is rather sparse, but contains N × N × N elements. In this case, N is 20. We show a part of the matrix in Fig. 2. If a = 13, q = 2 and t = 5, then we get the value e = 4 seconds. The term a = 13 means that the number of arrived vehicles v is between 420/hour and 640/hour. The term q=2 means that the queue of vehicles r is 1. The term t=5 means that the green time of the traffic signal g is between 6 and 8 seconds. The term e means the extension of the time. These terms are illustrated in Table 2. We need 245 seconds to calculate the simulation of 16 traffic signals for 3000 seconds by using a personal computer NEC PC-9801V.

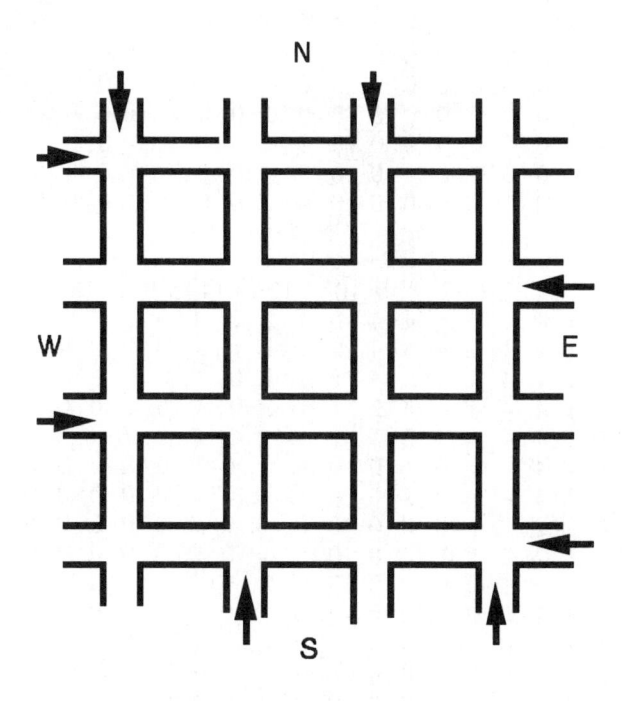

Fig. 1. Road Map

```
t = 0  1  2  3  4  5  6  7  8  9 10 11 12 13 14 15 16 17 18 19   q
    4  4  4  4  4  4  4  4  2  2  1  0  0  0  0  0  0  0  0  0   0
    4  4  4  4  4  4  4  2  1  1  0  0  0  0  0  0  0  0  0  0   1
    4  4  4  4  4  4  2  1  1  0  0  0  0  0  0  0  0  0  0  0   2
    4  4  4  4  4  2  2  1  0  0  0  0  0  0  0  0  0  0  0  0   3
    4  4  4  4  2  2  1  0  0  0  0  0  0  0  0  0  0  0  0  0   4
    4  4  4  2  2  2  1  0  0  0  0  0  0  0  0  0  0  0  0  0   5
    4  4  4  2  2  2  0  0  0  0  0  0  0  0  0  0  0  0  0  0   6
    4  4  4  2  2  1  0  0  0  0  0  0  0  0  0  0  0  0  0  0   7
    4  4  2  2  1  0  0  0  0  0  0  0  0  0  0  0  0  0  0  0   8
    4  2  2  1  0  0  0  0  0  0  0  0  0  0  0  0  0  0  0  0   9
    2  2  1  0  0  0  0  0  0  0  0  0  0  0  0  0  0  0  0  0  10
    2  2  1  0  0  0  0  0  0  0  0  0  0  0  0  0  0  0  0  0  11
    2  1  0  0  0  0  0  0  0  0  0  0  0  0  0  0  0  0  0  0  12
    1  1  0  0  0  0  0  0  0  0  0  0  0  0  0  0  0  0  0  0  13
    1  1  0  0  0  0  0  0  0  0  0  0  0  0  0  0  0  0  0  0  14
    0  0  0  0  0  0  0  0  0  0  0  0  0  0  0  0  0  0  0  0  15
    0  0  0  0  0  0  0  0  0  0  0  0  0  0  0  0  0  0  0  0  16
    0  0  0  0  0  0  0  0  0  0  0  0  0  0  0  0  0  0  0  0  17
    0  0  0  0  0  0  0  0  0  0  0  0  0  0  0  0  0  0  0  0  18
    0  0  0  0  0  0  0  0  0  0  0  0  0  0  0  0  0  0  0  0  19
                                    (a)
```

```
t = 0  1  2  3  4  5  6  7  8  9 10 11 12 13 14 15 16 17 18 19   q
    4  4  4  4  4  4  4  4  4  4  2  2  1  0  0  0  0  0  0  0   0
    4  4  4  4  4  4  4  4  4  2  2  1  0  0  0  0  0  0  0  0   1
    4  4  4  4  4  4  4  4  4  2  1  1  0  0  0  0  0  0  0  0   2
    4  4  4  4  4  4  4  4  2  1  1  0  0  0  0  0  0  0  0  0   3
    4  4  4  4  4  4  4  2  1  1  0  0  0  0  0  0  0  0  0  0   4
    4  4  4  4  4  4  2  1  0  0  0  0  0  0  0  0  0  0  0  0   5
    4  4  4  4  4  2  2  1  0  0  0  0  0  0  0  0  0  0  0  0   6
    4  4  4  2  2  2  1  0  0  0  0  0  0  0  0  0  0  0  0  0   7
    4  4  4  2  2  2  1  0  0  0  0  0  0  0  0  0  0  0  0  0   8
    4  4  4  2  2  1  0  0  0  0  0  0  0  0  0  0  0  0  0  0   9
    4  4  2  2  1  0  0  0  0  0  0  0  0  0  0  0  0  0  0  0  10
    4  2  2  1  0  0  0  0  0  0  0  0  0  0  0  0  0  0  0  0  11
    2  2  1  0  0  0  0  0  0  0  0  0  0  0  0  0  0  0  0  0  12
    2  2  1  0  0  0  0  0  0  0  0  0  0  0  0  0  0  0  0  0  13
    2  1  0  0  0  0  0  0  0  0  0  0  0  0  0  0  0  0  0  0  14
    1  1  0  0  0  0  0  0  0  0  0  0  0  0  0  0  0  0  0  0  15
    1  1  0  0  0  0  0  0  0  0  0  0  0  0  0  0  0  0  0  0  16
    0  0  0  0  0  0  0  0  0  0  0  0  0  0  0  0  0  0  0  0  17
    0  0  0  0  0  0  0  0  0  0  0  0  0  0  0  0  0  0  0  0  18
    0  0  0  0  0  0  0  0  0  0  0  0  0  0  0  0  0  0  0  0  19
                                    (b)
```

Fig. 2. The Matrix of (a) a = 13 and (b) a = 14

Table 2 Normalized Data

a, t, q	vehicles / hour	time	queue
0	$5 > v \geq 0$	$g = 0$	$r = 0$
1	$10 > v \geq 5$	$g = 1$	$r = 1$
2	$14 > v \geq 10$	$g = 2$	$r = 1$
3	$20 > v \geq 14$	$g = 3$	$r = 2$
4	$30 > v \geq 20$	$6 > g \geq 4$	$r = 3$
5	$40 > v \geq 30$	$8 > g \geq 6$	$r = 4$
6	$60 > v \geq 40$	$12 > g \geq 8$	$r = 5$
7	$80 > v \geq 60$	$16 > g \geq 12$	$r = 6$
8	$120 > v \geq 80$	$24 > g \geq 16$	$r = 7$
9	$160 > v \geq 120$	$32 > g \geq 24$	$11 > g \geq 8$
10	$220 > v \geq 160$	$42 > g \geq 32$	$16 > g \geq 11$
11	$320 > v \geq 220$	$55 > g \geq 42$	$20 > g \geq 16$
12	$420 > v \geq 320$	$67 > g \geq 55$	$26 > g \geq 20$
13	$640 > v \geq 420$	$80 > g \geq 67$	$30 > g \geq 26$
14	$940 > v \geq 640$	$90 > g \geq 80$	$36 > g \geq 30$
15	$1280 > v \geq 940$	$100 > g \geq 90$	$43 > g \geq 36$
16	$1880 > v \geq 1280$	$110 > g \geq 100$	$51 > g \geq 43$
17	$2560 > v \geq 1880$	$120 > g \geq 110$	$60 > g \geq 51$
18	$3600 > v \geq 1560$	$130 > g \geq 120$	$70 > g \geq 60$
19	$v \geq 3600$	$g \geq 130$	$11 > g \geq 70$

We propose the matrix representation instead of the fuzzy control statements. The matrix representation requires large memory. Nowadays the price of the memory is not expensive, so we may use a large memory. We propose the fuzzy controller with the matrix representation shown in Fig. 3 which is fast in computation.

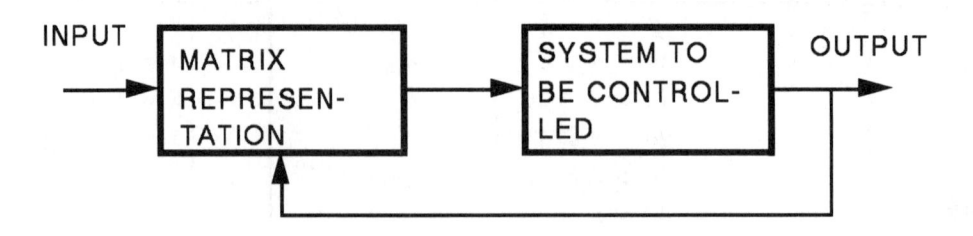

Fig. 3. Fuzzy Controller with Matrix Representation

4. REDUCTION OF MATRIX REPRESENTATION

Though the matrix is considerably sparse, it is rather difficult to determine all the precise value of elements of the matrix. If the dimension of the matrix can be reduced, then the process of determining the element values may become easier. In Fig. 2, we determined that the maximum value of the extension time is 4 seconds. If the value of the extension time is a constant except 0, the size of the matrix becomes only $N \times N$. The matrix shown in Fig.2 becomes the one shown in Fig. 4, by using the constant value 1 for e. The terms tm and tr depict the duration of the traffic signals and the traffics, respectively. In Fig. 4, the element depicts the reference value of q. If q is less than the value shown in Fig. 4, then the green light will be extended 1 second. The execution time may be bigger in the $N \times N$ matrix than the $N \times N \times N$ matrix. It requires 255 seconds to simulate the traffic signal control, but the increase is only about 4.0 % in the PC-9801V.

5. REDUCTION BY SIMPLE SELF-TUNING

A number of researches about self-tuning have been made [9],[10],[11]. We propose a simple self-tuning of the values of the matrix elements. At every 1000 seconds, we calculate the mean delay time of the vehicles both at each traffic junction and at each traffic direction. At first, a matrix check(tr,tm) is prepared and initialized. If the newly calculated delay time is greater than the mean delay time calculated in the last period, the value of the check(tr,tm) is decreased by 1; if the delay time is smaller, the value is increased by 1. Finally, if the value of check(tr,tm) is negative, the value of the matrix deg(tr,tm) is added by 1; if positive, the value is decreased by 1. Then we get the modified matrix mdeg(tr,tm) as shown in Fig. 5.

The matrix can be reduced by two rules. The first reduction rule is simple: Adjacent rows where elements are all the same are unified into one row. The second reduction rule is to eliminate the rows which do not affect the control. In Fig. 5,

tm =	0	1	2	3	4	5	6	7	8	9	10	11	12	13	14	15	16	17	18	19	tr
	1	1	1	1	1	1	1	1	1	1	1	1	1	1	1	1	1	1	1	1	0
	1	1	1	1	1	1	1	1	1	1	1	1	1	1	1	1	1	1	1	1	1
	2	1	1	1	1	1	1	1	1	1	1	1	1	1	1	1	1	1	1	1	2
	2	1	1	1	1	1	1	1	1	1	1	1	1	1	1	1	1	1	1	1	3
	2	1	1	1	1	1	1	1	1	1	1	1	1	1	1	1	1	1	1	1	4
	3	1	1	1	1	1	1	1	1	1	1	1	1	1	1	1	1	1	1	1	5
	3	1	1	1	1	1	1	1	1	1	1	1	1	1	1	1	1	1	1	1	6
	4	2	1	1	1	1	1	1	1	1	1	1	1	1	1	1	1	1	1	1	7
	6	3	2	1	1	1	1	1	1	1	1	1	1	1	1	1	1	1	1	1	8
	8	4	2	1	1	1	1	1	1	1	1	1	1	1	1	1	1	1	1	1	9
	10	5	3	2	1	1	1	1	1	1	1	1	1	1	1	1	1	1	1	1	10
	12	6	4	2	1	1	1	1	1	1	1	1	1	1	1	1	1	1	1	1	11
	14	7	5	3	2	1	1	1	1	1	1	1	1	1	1	1	1	1	1	1	12
	16	11	9	7	6	5	4	3	2	1	1	1	1	1	1	1	1	1	1	1	13
	17	15	13	11	10	9	8	7	6	5	4	3	2	1	1	1	1	1	1	1	14
	18	16	14	14	12	12	11	10	9	5	7	6	5	4	3	2	1	1	1	1	15
	19	18	17	16	15	14	13	12	11	10	9	8	7	6	5	4	3	2	1	1	16
	19	19	19	19	18	18	17	17	16	16	14	14	12	12	11	10	8	6	4	3	17
	19	19	19	19	19	19	19	19	18	18	16	16	14	14	13	12	10	10	8	6	18
	19	19	19	19	19	19	19	19	19	19	19	19	19	18	17	16	15	14	12		19

Fig. 4. First Reduced Matrix deg(tr,tm)

tm =	0	1	2	3	4	5	6	7	8	9	10	11	12	13	14	15	16	17	18	19	tr
	1	1	1	1	1	1	1	1	1	1	1	1	1	1	1	1	1	1	1	1	0-1
	2	1	1	1	1	1	1	1	1	1	1	1	1	1	1	1	1	1	1	1	2-4
	3	1	1	1	1	1	1	1	1	1	1	1	1	1	1	1	1	1	1	1	5-6
	4	2	1	1	1	1	1	1	1	1	1	1	1	1	1	1	1	1	1	1	7
	6	3	2	1	1	1	1	1	1	1	1	1	1	1	1	1	1	1	1	1	8
	8	4	2	1	1	1	1	1	1	1	1	1	1	1	1	1	1	1	1	1	9
	10	5	3	2	1	1	1	1	1	1	1	1	1	1	1	1	1	1	1	1	10
	12	6	4	2	1	1	1	1	1	1	1	1	1	1	1	1	1	1	1	1	11
	14	7	5	3	2	1	1	1	1	1	1	1	1	1	1	1	1	1	1	1	12
	16	11	9	7	5	4	3	2	1	1	1	1	1	1	1	1	1	1	1	1	13
	17	15	13	10	8	7	6	4	4	4	3	2	1	1	1	1	1	1	1	1	14
	18	16	15	14	12	12	8	7	6	6	6	6	5	4	3	2	1	1	1	1	15
	19	18	17	16	16	16	14	9	8	8	8	8	7	6	5	4	3	2	1	1	16
	19	19	19	19	18	18	17	17	16	16	14	14	12	12	11	10	8	6	4	3	17
	19	19	19	19	19	19	19	19	18	18	16	16	14	14	13	12	10	10	8	6	18
	19	19	19	19	19	19	19	19	19	19	19	19	19	18	17	16	15	14	12		19

Fig. 5 The Second Modified Matrix mdeg(tr,tm)

tm =	0	1	2	3	4	5	6	7	8	9	10	11	12	13	14	15	16-19	tr
	3	1	1	1	1	1	1	1	1	1	1	1	1	1	1	1	1	0-6
	4	2	1	1	1	1	1	1	1	1	1	1	1	1	1	1	1	7
	6	3	2	1	1	1	1	1	1	1	1	1	1	1	1	1	1	8
	8	4	2	1	1	1	1	1	1	1	1	1	1	1	1	1	1	9
	10	5	3	2	1	1	1	1	1	1	1	1	1	1	1	1	1	10
	12	6	4	2	1	1	1	1	1	1	1	1	1	1	1	1	1	11
	14	7	5	3	2	1	1	1	1	1	1	1	1	1	1	1	1	12
	16	11	9	7	5	4	3	2	1	1	1	1	1	1	1	1	1	13
	17	15	13	10	8	7	6	4	4	4	3	2	1	1	1	1	1	14
	18	16	15	14	12	12	8	7	6	6	6	6	5	4	3	2	1	15
	19	18	17	16	16	16	14	9	8	8	8	8	7	6	5	4	3	16-19

Fig. 6. Final Reduced Matrix

rows 0-6 are unified by the first rule and columns 16-19 and rows 18-19 are unified by the second rule. Then we get the final modified matrix shown in Fig. 6.

We need 230 seconds to calculate the simulation of 16 traffic signals at 1440 vehicles/hour for 3600 seconds by using the final reduced matrix, while we need 233 seconds to calculate the same simulation by using the modified matrix with PC-9801V. The memory size is improved considerably.

6. SELF-TUNING BY MODIFIED SIMPLEX METHOD

In general, the simplex method is effective in determining the maximum value or the minimum value in linear programming [12]. The fuzzy rules are sometimes adjusted by using the simplex method.

The vertex X_k is defined as

$$X_k = (x_1^k, x_2^k, \ldots, x_n^k), \quad k = 1, 2, \ldots m.$$

The evaluation function P_k of the vertex X_k is represented as $P_k = P(X_k)$. If $P(X_w)$ is the largest (or smallest) value among $P_1, P_2, \ldots P_n$, we define the new vertex X'_w by the following equation.

$$X'_w = X_g + a(X_g - \beta X_w) \tag{7}$$

where,

$$X_g = \frac{1}{m-1} \sum_{k=1, k \neq w}^{m} X_k.$$

The vertex X'_w is expected to have a better value. We propose to add the term β to the simplex method. This term takes a role in reducing the computing time. If $n = 2$, $m = 3$ and $\beta = 1$, we show the relation between Xi in Fig. 7.

We repeat to calculate the new X'_w and to adjust the value of the weight until we get the smallest value of the evaluation function. Let the value ε be an appropriate small number. We stop the calculation when:

$$\text{DIF} = \left[\frac{1}{m} \sum_{k=1}^{m} (P_k - \bar{P})^2\right]^{\frac{1}{2}} < \varepsilon,$$

where the term $\bar{P}$ is the mean value of P_K.

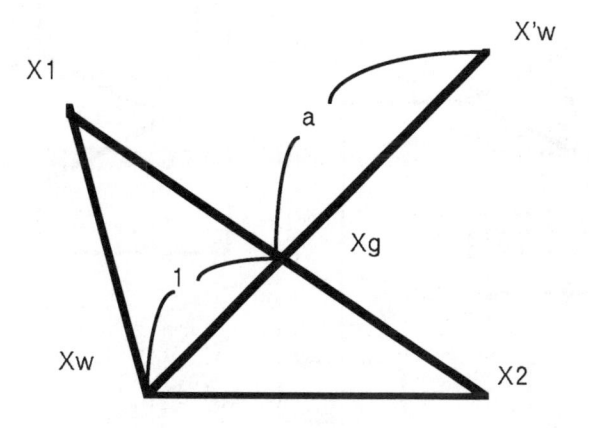

Fig. 7. The Modified Simplex Method

We apply the simplex method to the control of the traffic signals. There are 8 one-way roads and therefore there are 16 traffic junctions as shown in Fig. 1. In this case, the evaluation function P_k is determined as the overall delay time of the vehicles. The 16 traffic junctions can be controlled by the fuzzy controller with matrix representation which is operated by only one personal computer. The original matrix representation is also shown in Fig. 4. The figure of the matrix corresponds to the value of the queue of the vehicles. The values of the original matrix representation are not the appropriate ones, so we must adjust these values for efficient control. We evaluate the fuzzy control system by using the overall delays of the vehicles. We illustrate the flow chart of the self-tuning algorithm in Fig. 8. We get the new values of a part of the matrix representation from the input signal, the output signal and the evaluation function. It is cumbersome to adjust all the values of the matrix representation at once, so we calculate only a part of the matrix representation at each time. We continue this process until we get the optimum values of the matrix.

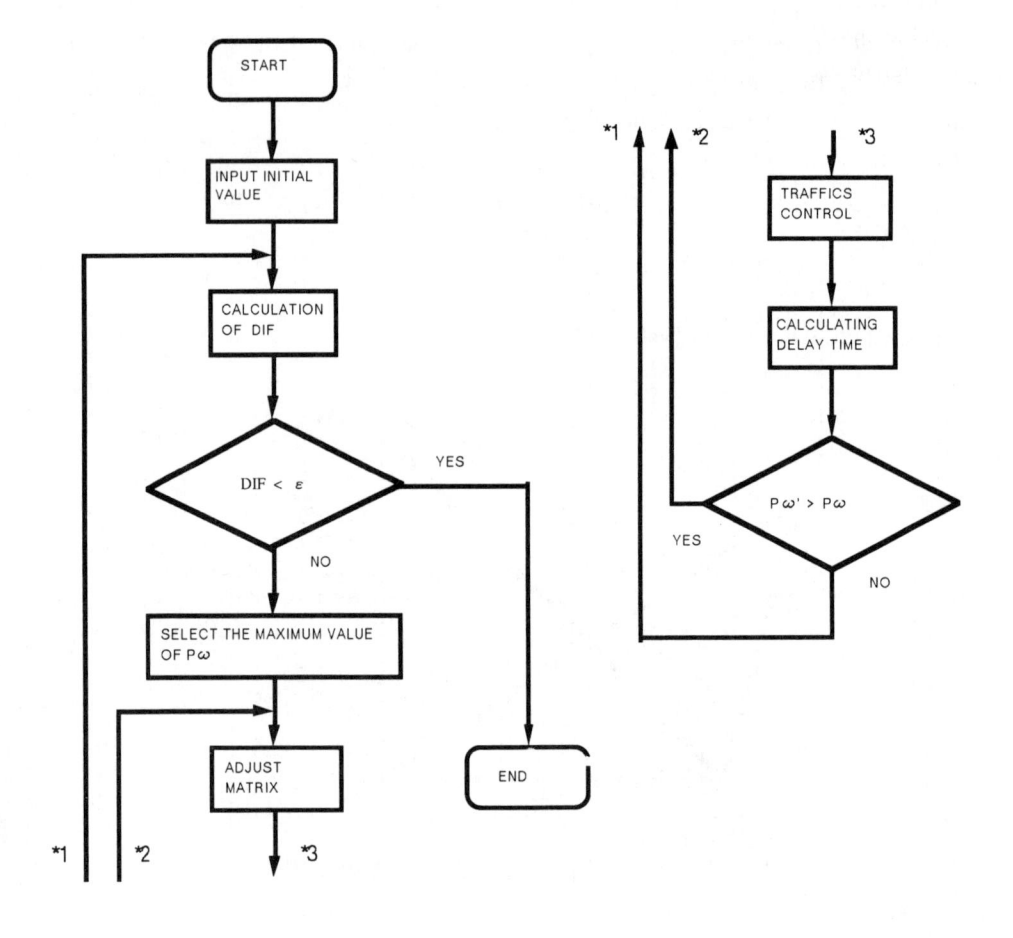

Fig. 8. The Self-tuning Algorithm

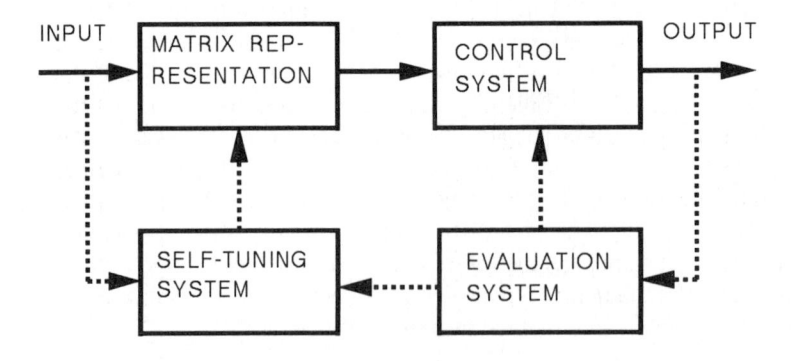

Fig. 9. Fuzzy Control System with Modified Simplex Method

We show the fuzzy control system with the modified simplex method in Fig. 9. The input signals are the traffics and the output signal is the flow of vehicles at each traffic junction. In the evaluation system, we calculate the overall delays of the vehicles, the queue of the vehicles and the duration of the traffic signals. We improve substantially the value of the overall delay by using the self-tuning method. The improved result is shown in Table 3. The improved value relies too much on the value ε.

Table 3 Overall Delays

| Traffics(vehicles/hour) | | ε | Overall Delays | | |
NS	EW		origin	simplex	improved (%)
720	720	0.2	10.26	8.44	17.8
720	720	0.04	10.26	8.31	19.1
940	940	0.5	18.81	12.45	33.8
940	940	0.05	18.81	12.43	33.9
1280	1280	0.5	58.47	16.41	71.9
1280	1280	0.05	58.47	14.66	74.9

7. NEURAL NETWORKS FOR FUZZY CONTROLLER

The original matrix representation obtained from the fuzzy program is also shown in Fig. 1, but we use the matrix shown in Fig. 4 as the matrix representation. The terms tr and tm are the traffics and the time duration of traffic signals respectively. At first, the value of the elements of the matrix representation are supposed not to be optimal, so the fine adjustment of the values should be done before using the matrix for control. The adjustment will be done by the neural network.

7.1 Neural Network

We proposed the method for matrix modification which modifies the values of the matrix by repetition of the control and by using the simple self-tuning. We got good results, but found that the correction is not sufficient. To get a more precise adjustment, we adopt the neural network as shown in Fig. 10.

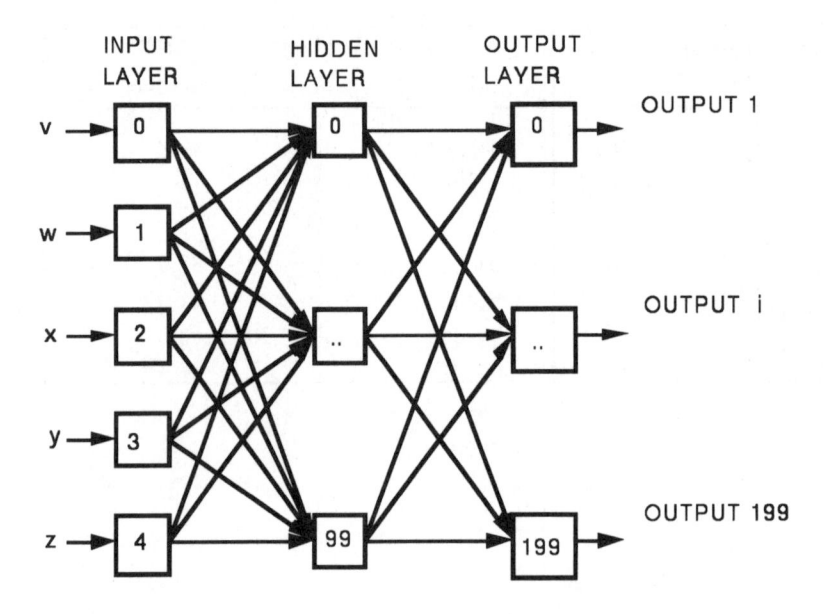

Fig. 10. Neural Network

After several hundred training iterations, we get adequate values for the weights. The output of the output layer is almost equivalent to the value of the matrix representation. The input terms v, w, x, y and z are the overall delay of the present and the past, the mean traffic, the mean value of the signal duration, and the mean value of the queue, respectively. An input data set just corresponds one element of the matrix, but we determine the values of 5×5 or 3×3 elements at each input data set. More than one element can be calculated approximately, because the value of the matrix keeps the continuity. To get more precise values of 5 × 5 or 3 × 3 elements, we use the mask matrix shown in Fig. 11.

$$\begin{matrix} 0.6 & 0.7 & 0.8 & 0.7 & 0.6 \\ 0.7 & 0.8 & 0.9 & 0.8 & 0.7 \\ 0.8 & 0.9 & 1.0 & 0.9 & 0.8 \\ 0.7 & 0.8 & 0.9 & 0.8 & 0.7 \\ 0.6 & 0.7 & 0.8 & 0.7 & 0.6 \end{matrix}$$

Fig. 11. The Mask Matrix

If a data set corresponds only to one neuron, it does not need the mask matrix. To calculate the weight, we use the following function:

$$\text{out} = 1 / (1 + \exp (- p)), \tag{8}$$

which is called a simple sigmoidal activation function [13]. The term p is the value of the output layer. After about 1000 training iterations, we get almost the same value as the matrix representation by using 49 data sets which include the traffics {32, 64, 128, 256, 512, 1024, 1280} × {32, 64, 128, 256, 512, 1024, 1280}. The value of the matrix represents the queue q. If the mean values of tr, tm and q are 15.18, 13.15 and 4.38, then we get the trained data shown in Fig. 12a. Fig. 12b is the original data. The calculated time delay of the vehicles is 56.08. Both the trained data and the original data are normalized by 20.

$$\begin{matrix} 0.43 & 0.35 & 0.24 \\ 0.67 & 0.66 & 0.65 \\ 0.90 & 0.85 & 0.80 \end{matrix}$$
(a)

$$\begin{matrix} 0.50 & 0.40 & 0.35 \\ 0.70 & 0.60 & 0.60 \\ 0.80 & 0.80 & 0.80 \end{matrix}$$
(b)

Fig. 12. (a) The Trained Data (b) The Original Data

7.2 Heuristic Algorithm

The trained data simulate only the original matrix. Therefore it does not improve the performance of the control by itself. Since the neural network itself has not the ability of reasoning, it is necessary to get this ability by using the training algorithm. Generally speaking, it is supposed that there is no explicit goal in fuzzy control. It is required only to get a better result than before.

We adopted the so-called heuristic algorithm to get the adequate value of the neural network. The outline of the heuristic algorithm is as follows.

STEP 1
Determine the matrix representation and temporary goal. swt = 0. cv = 0.

STEP 2
Simulation or experiment. If swt = i, then the result i is obtained.

STEP 3
If cv is 0, then swt = 1 and go to STEP 5.

STEP 4
If swt is 3 and the result i is better than result j, the temporary goal is result i and the weight i of the neural network is selected and swt becomes 1 and cv = cv +1.

STEP 5
Calculate the output of the neural network and adjust the weight by using the temporary goal and result i.

STEP 6
Stop to repeat when the difference between the new over-all delay and the last over-all delay is less than ϵ. The term ϵ is any small number.

STEP 7
If swt is 1, the weights 1 of the neural network are to be subtracted by the weight adjustment. If swt is 2, the weights 2 of the neural network are to be added by the weight adjustment.

STEP 8
Determine the matrix representation by using the modified neural network. swt = swt + 1. Go to STEP 2.

This algorithm is easily executed on a workstation.

```
INPUT  →  [MATRIX REPRESENTATION]  →  [SYSTEM TO BE CONTROLLED]  →  OUTPUT
                    ↑                              │
            [NEURAL NETWORKS] ←───────────────────┘
                    ↑
            [HEURISTIC ALGORITHM]
```

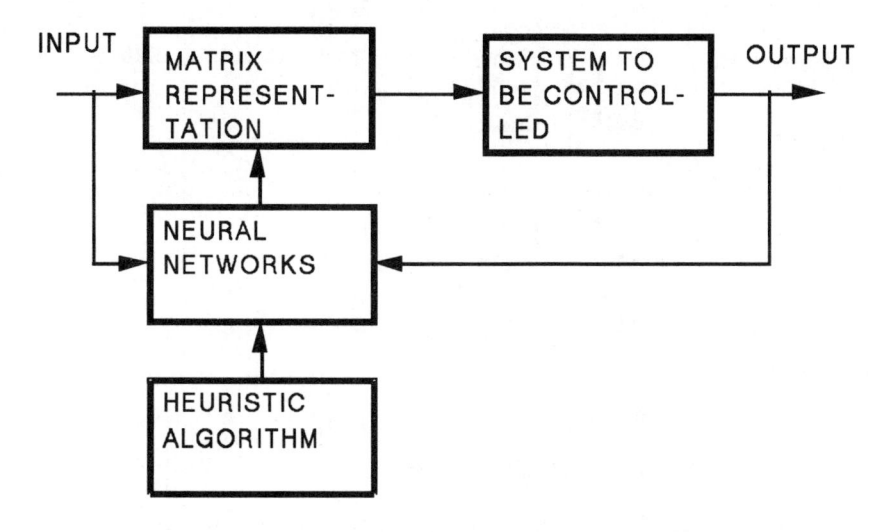

Fig. 13. Fuzzy Controller with Neural Network

Our new proposition is based on the matrix representation modified by the neural network which is adjusted by simulation or experimental data. The decision process is executed by the heuristic algorithm. Its architecture is shown in Fig. 13. The values of the matrix representation will be modified by the repetition of the experiment or simulation.

7.3 Application to Traffic Control

We adopt the fuzzy controller with matrix representation modified by neural network to the traffic control. We suppose there are 8 one-way roads and there are 16 traffic junctions shown in Fig. 1. Each traffic junction may be controlled by a simple fuzzy controller which provides a good result. In fact, these 16 traffic signals are controlled by a single 20×20 matrix which is shown in Fig. 4.

If some one-way roads are supposed to be the artery road, the phase control is very effective. We illustrate one example of an artery road in Fig. 14. The time chart of the traffic signal for phase control is also shown in Fig. 15. The term τ is the signal duration. The terms L and L' are the quantities for controlling the phase. We showed that a personal computer can control 16 traffic junctions easily. We use the data sets shown in Table 4 for training the neural network. It is not very effective to improve the time delay if the traffics are less than the 1280 vehicles / hour, since the values have already been precisely adjusted.

Traffic junction i

Traffic junction j

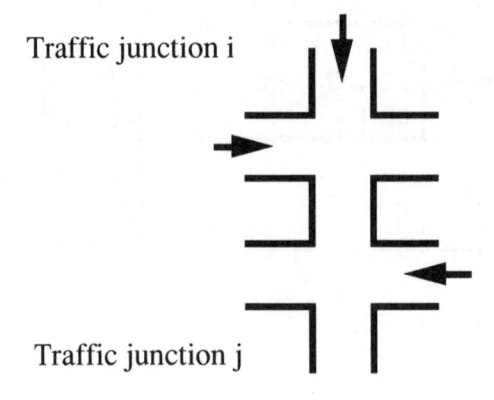

Fig. 14. Traffic Junctions

| GREEN | AMBER | RED |

$\leftarrow \tau-8 \rightarrow \leftarrow L \rightarrow$ $\leftarrow \tau-8 \rightarrow \leftarrow L' \rightarrow$ (a)

| RED | GREEN | AMBER |

(b)

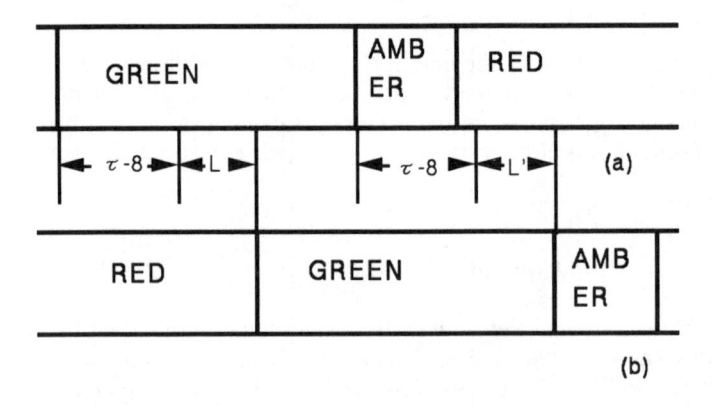

Fig. 15. (a) Traffic Junction i and (b) Traffic Junction j

So we try to improve the delay of the traffics 1280 vehicles / hour. At first, we calculate the overall delays of the vehicles shown in Table 4. All the traffic junctions are controlled by a simple fuzzy controller with the matrix representation.

Table 4 Overall Delays

tr1	tr2	delay	tr1	tr2	delay	tr1	tr2	delay	tr1	tr2	delay
32	32	3.14	32	64	3.06	32	128	2.73	32	256	1.67
32	512	1.19	32	1024	1.6	32	1280	2.21			
64	32	3.23	64	64	3.64	64	128	3.62	64	256	2.61
64	1280	5.27									
128	32	2.44	128	64	3.26	128	128	4.15	128	256	4.46
128	512	4.45	128	1024	6.32	128	1280	8.32			
256	32	1.68	256	64	3.01	256	128	4.37	256	256	5.25
256	512	5.90	256	1024	8.36	256	1280	10.51			
512	32	4.01	512	256	5.90	512	512	6.96	512	1024	9.49
512	1280	12.44									
1024	32	1.87	1024	64	4.32	1024	128	6.22	1024	256	8.41
1024	512	9.45	1024	1024	11.48	1024	1280	27.22			
1280	32	2.21	1280	64	5.21	1280	128	8.50	1280	256	10.78
1280	512	12.58	1280	1024	40.33	1280	1280	56.08			

Table 5 Error

tr1	tr2	error
1280	32	0.00000644183
1280	64	0.00000143224
1280	128	0.00000130032
1280	256	0.00000029831
1280	512	0.00000067673
1280	1024	0.00000218982
1280	1280	0.00000441928

The terms tr1 and tr2 are the traffic in the directions of east-west and north-south, respectively. It requires about 500 training iterations to get the adequate weights of the neural network. The output p of the neural network is calculated by Eq. 9:

$$p(i,j) = \sum_k wk(i,j,k) \times (\sec(k,0) \times v + \sec(k,1) \times w + \sec(k,2) \times x + \sec(k,3) \times y + \sec(k,4) \times z) \tag{9}$$

The terms wk and sec are both the weights of the neural network. The terms v, w, x, y, and z are the present delay time, the past delay time, the traffics, the queue, and the signal duration, respectively. The weight adjustment del and the error are calculated by the following equations:

$$del(i,j) = \sum_k h \times (temp - mean) \times (\sec(k,0) \times v + \sec(k,1) \times x + \sec(k,3) \times y + \sec(k,4) \times z) \tag{10}$$

$$error = \sum_{i=cr-1}^{cr+1} \sum_{j=cm-1}^{cm+1} del(i,j) \times del(i,j) \tag{11}$$

The terms cr and cm are the mean traffics and the mean signal duration respectively. The term h is 0.005. The term temp is the temporal goal of the fuzzy control. The term mean is the present delay time. We show a part of the error of the training data in Table 5. By using the matrix representation modified by the neural network, we get the delay time 43.63 sec / vehicle at the traffics tr1 = tr2 = 1280, while the original delay time is 56.08. The delay time is improved by about 22%.

8. CONCLUSION

The matrix representation is very convenient and fast in computation. The values of the matrix, however, are not optimum at first. Hitherto we determine the values of the matrix by the simple self-tuning, the simplex method and the neural network. The performance of the fuzzy controller is substantially improved by using the simplex method. So the simplex method seems to be superior. However, we guess the neural network may be superior to the other methods, because the simple heuristic algorithm is very flexible. After improving the heuristic algorithm, the neural network may have a better result than other methods. We are trying to use another neural network instead of the simple heuristic algorithm. We expect that this neural network will work better than the heuristic algorithm. We applied this method to the control of traffic signals and got good results. This method may be easily applied to any other control system. We gratefully acknowledge the help of Mr. Mizunuma and Ms. Sugimoto.

REFERENCES

[1] Yamakawa T., "Stabilization of an inverted pendulum by a high-speed fuzzy logic controller hardware system", *Fuzzy Sets and Systems*, 32, 2, pp.161-180., 1989

[2] Zadeh L.A., "Fuzzy algorithm", *Information and Control*, 12,pp.91-102, 1968

[3] Pappis C.P. and E.H. Mamdani, "A fuzzy logic controller for a traffic junction", *IEEE Trans. Syst. Man Cybern.*, SMC-7,10,pp.701- 717, 1977

[4] Mizumoto M., "Min-max-gravity method versus product-sum-gravity method for fuzzy controls", *IFSA'91 Brussels*, (Engineering) pp.127-130, 1991

[5] Nakatsuyama M., H. Nagahashi, N. Nishizuka and K. Watanabe, "Matrix representation for fuzzy program and its application to traffic control", *11th IFAC WORLD Congress in Tallinn*, 7, pp.83-88, 1990

[6] Nakatsuyama M., H. Nagahashi and K. Watanabe, "Matrix representation for fuzzy programming and reconstitution of fuzzy program from modified matrix", *Proc. of Sino-Japan Joint Meeting on Fuzzy Sets and Systems*, Beijing, 1990

[7] Nakatsuyama M. and H. Kaminaga, "Fuzzy controller modified by neural networks", *SICICA'92*, Malaga, Spain, 1992

[8] Yan J. H., M. Nakatsuyama and H. Kaminaga, "Fuzzy controller with matrix representation optimized by using modified simplex method", *Korea-Japan Joint Conference for Fuzzy Systems and Engineering* (to appear), 1992

[9] Baakini N., Automatic learning control using fuzzy logic, Ph.D Thesis, London Univ., 1976

[10] Yamazaki T., M. Sugeno, "Self-organizing fuzzy control", *Trans. of SICE*, 20, 8, pp. 720-726, 1984

[11] Yamamoto T., S. Omatsu, H. Ishihara, "A construction of self-tuning PID control system", *Trans. of SICE*, 25, 10, pp.1069-1075, 1989

[12] Gass, S.I., *Linear programming*, McGraw Hill Kogakusha, 1975

[13] Wasserman P.D., *Neural Computing Theory and Practice*, Van Nostrand Reinold, New York, 1989

Chapter 16
A Self Tuning Method of Fuzzy Reasoning by Genetic Algorithm

Introduction, *338*
A Conventional Self-Tuning Method, *339*
Optimization of the Inference Rules by Genetic Algorithm, *341*
Numerical Examples, *346*
Conclusion, *353*
References, *353*

This chapter proposes a new self-tuning method for fuzzy reasoning based on the application of a genetic algorithm. Simplified fuzzy inference rules whose consequent parts are expressed in terms of real numbers are employed in this method. Using the genetic algorithm, the number of inference rules and the shapes of the membership functions in the antecedent parts are determined so as to optimize an information criterion. The consequent parts of the inference rules are optimized by the descent method from input-output data. Using the proposed method, the inference rules are determined so as to optimize the learning and generalization capabilities of fuzzy reasoning. Numerical examples are provided to illustrate the effectiveness of the proposed method.

A SELF-TUNING METHOD OF FUZZY REASONING BY GENETIC ALGORITHM

H. Nomura, I. Hayashi and N. Wakami
Central Research Laboratories,
Matsushita Electric Industrial Co., Ltd. Osaka JAPAN

1. INTRODUCTION

In order to provide fuzzy reasoning with learning function, various learning methods have been proposed. These methods include the self-tuning fuzzy controller employing a descent method[1-3] and the neural network driven fuzzy reasoning[4], which can optimize the shape of membership functions in inference rules from input-output data. However, these methods have suffered from the inherent prerequisite problems, such as an advanced setting of the number of inference rules, which have to be derived by trial and error.

The result of the learning depends on the number of inference rules. When the number of inference rules is small, the inference rules cannot express the input-output relation of given data well. On the contrary, when the number is large, the generalization capability of the inference rules is sacrificed because of the overfitting. Therefore, the number of inference rules has to be determined from a standpoint of overall learning capability and generalization capability. The work to determine the number of inference rules requires to designer a large number of experiments by trial and error.

In order to solve such problem, a new learning method of fuzzy reasoning by means of a genetic algorithm is proposed here. The genetic algorithm[5, 6] is an optimization method developed from the theory of biological evolution.

In this method, a simplified fuzzy reasoning in which the consequent parts are expressed in real numbers is employed. The number of inference rules and the shapes of membership functions in the antecedent part are determined by applying the genetic algorithm, and the real numbers of the consequent parts are derived by using the descent method.

In this chapter, the conventional learning method of fuzzy reasoning employing the descent method, which constitutes the foundation of the proposed method, is described first. Then, an algorithm to determine the structure of the antecedent parts by using the genetic algorithm is explained. Finally, in order to demonstrate the effectiveness of the present method, some numerical examples are reported.

2. A CONVENTIONAL SELF-TUNING METHOD

2.1. Learning Algorithm Using Descent Method

Simplified fuzzy reasoning in which the consequent parts are expressed by real numbers is employed in this method. Expressing input variables by $x_j, (j = 1, \ldots, m)$ and an output variable by y, the inference rules of the simplified fuzzy reasoning can be expressed by the following:

$$\text{Rule } i: \quad \text{If} \quad x_1 \text{ is } A_{i1} \text{ and}, \ldots, \text{ and } x_m \text{ is } A_{im} \quad \text{then} \quad y \text{ is } w_i \tag{1}$$

wherein $i, (i = 1, \ldots, n)$ is the number of the inference rules, $A_{i1}, \ldots, A_{im}$ are the membership functions in the antecedent part, and w_i is a real number in the consequent part. The output of the simplified fuzzy reasoning, y, can be derived by using the following equations:

$$\mu_i = \prod_{j=1}^{m} A_{ij}(x_j) \tag{2}$$

$$y = \frac{\sum_{i=1}^{n} \mu_i \cdot w_i}{\sum_{i=1}^{n} \mu_i} \tag{3}$$

where μ_i is a membership value of i-th inference rule.

By using a descent method, the real numbers w_i of the consequent parts are optimized from the input-output data[1, 2]. The descent method can alter the tuning parameters so as to minimize an objective function H, which is expressed by the following equation in this case:

$$H = \frac{1}{2}(y^{rp} - y^p)^2 \tag{4}$$

where y^{rp} is a desirable output data for the p-th input data $(x_1^p, \ldots, x_m^p)$, and y^p is an output of the fuzzy reasoning corresponding to the same p-th input data $(x_1^p, \ldots, x_m^p)$. The objective function H means the squared inference error.

Using the descent method, the learning rule of the real numbers in the consequent parts can be expressed by the following:

$$\begin{aligned} w_i(t'+1) &= w_i(t') - K \cdot \frac{\partial H}{\partial w_i} \\ &= w_i(t') - K \cdot \frac{\mu_i^p}{\sum_{i=1}^{n} \mu_i^p}(y^p - y^{rp}) \end{aligned} \tag{5}$$

where t' is the number of iteration of learning, μ_i^p is a membership value of the i-th inference rule corresponding to the p-th input-output data, and K is a constant.

By applying this learning rule to the input-output data repeatedly, the consequent parts are updated so as to minimize the objective function. In this case, the result of learning doesn't converge into a local optimum, but into a global optimum, because $\partial^2 H/\partial w_i^2 \geq 0$ is obtained for all i [7].

2.2. Problems of Conventional Self-Tuning Method

The input-output data for learning (Training Data: TRD), and the input-output data for evaluation (Checking Data: CHD), are expressed by the following:

$$\begin{array}{lll} \text{TRD} & : & (x_1^p, \ldots, x_m^p, y^{rp}), \quad p = 1, \ldots, P \\ \text{CHD} & : & (x_1'^q, \ldots, x_m'^q, y'^{rq}), \quad q = 1, \ldots, Q. \end{array}$$

The inference errors for these two types of input-output data are specified respectively by the following equations:

$$E_{TRD} = \frac{1}{P} \sum_{p=1}^{P} (y^p - y^{rp})^2 \tag{6}$$

$$E_{CHD} = \frac{1}{Q} \sum_{q=1}^{Q} (y'^q - y'^{rq})^2. \tag{7}$$

Figure 1 shows generalized relations between the number of inference rules and the inference errors E_{TRD}, E_{CHD} derived by the optimized inference rules using the descent method. In this method, the larger the number of inference rules, the smaller the inference error E_{TRD} obtained. However, the inference error E_{CHD} becomes larger for a larger number of inference rules after it exceeded a certain threshold value as shown in Figure 1. This phenomenon is caused by an excessive learning applied to the TRD.

The generalization capability of the inference rules can be expressed by the inference error E_{CHD}. Therefore, it can be said that the generalization capability of the inference rules would be lower if an excessive number of inference rules were applied for the learning.

In this conventional self-tuning method, the designer has to search the optimal number of inference rules by trial and error. The work to search the optimal number of inference rules requires the designer to perform a large number of experiments.

The descent method can optimize not only the real numbers in the consequent parts, but also the membership functions in the antecedent parts[1]. For the learning of only the consequent part, the result of learning converges into an global optimum in general. However, the result of learning the antecedent part can converge into a local optimum.

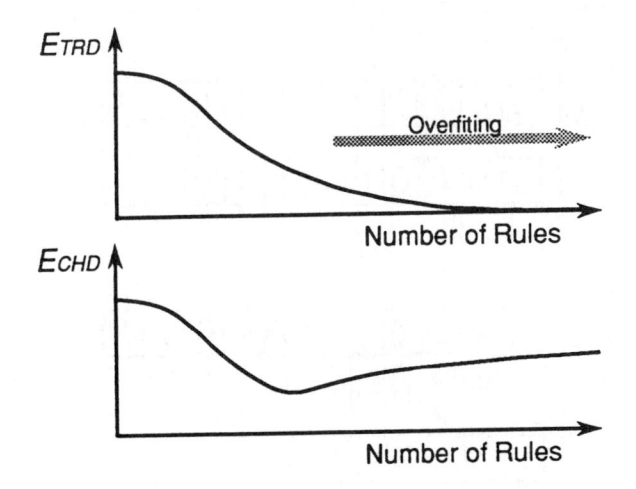

Figure 1: Transition of Inference Errors

3. OPTIMIZATION OF THE INFERENCE RULES BY GENETIC ALGORITHM

The proposed method is to optimize the number of inference rules and the shapes of the membership functions in the antecedent parts by a genetic algorithm.

3.1. Genetic Algorithm

A genetic algorithm (abbreviated GA) [5, 6] is a method to obtain an optimal solution by applying a theory of biological evolution. The most advantageous feature of the GA is a possibility of escaping from local optimum because of probabilistic operations such as *crossover* and *mutation*. In the GA, a solution candidate s_r which maximizes an objective function $E(s_r)$ called *fitness*, is searched. The solution candidate is expressed by the string, called *individual*, which is expressed by the following:

$$s_r = L_{r1}L_{r2}\ldots L_{rG} \tag{8}$$

where $L_{rg}, (g = 1, \ldots, G)$ is a variable taking a value of either "1" or "0". For instance, an example of the individual s_r with $G = 13$ is expressed by the following string:

$$s_r = 1001000110011. \tag{9}$$

A set of individuals, S, called *population*, is expressed as follows:

$$S = \{s_1, s_2, \ldots, s_R\}. \tag{10}$$

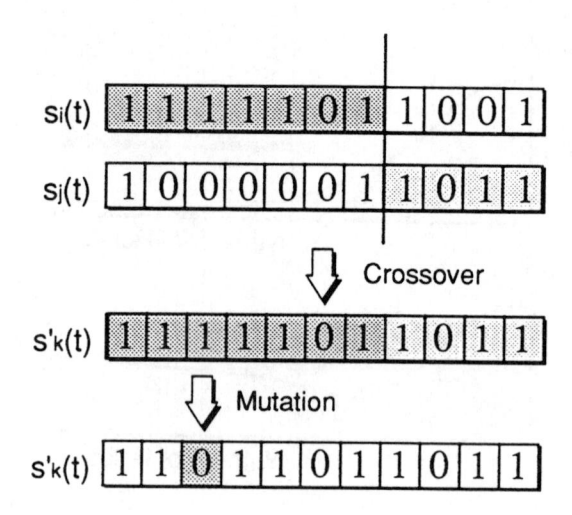

Figure 2: Operators in Genetic Algorithm

3.2. A Learning Procedure of GA

An optimal solution is searched by applying the following procedure:

1) The individuals $s_1, s_2, \ldots, s_R$ which constitute a population $S(t)$ of the 0-th generation $(t = 0)$ are determined by uniform random numbers.

2) The fitness $E(s_r)$ for each individual s_r is derived to determine a selection probability $P_{sr}(t)$, which is expressed by the following:

$$P_{sr}(t) = \frac{E(s_r(t))}{\sum_{r=1}^{R} E(s_r(t))} \qquad (11)$$

where $r = 1, \ldots, R$.

And, then, the number of the subsequently produced individual, k, is initialized at 1.

3) Two individuals $s_i(t)$ and $s_j(t)$ are selected out from the population $S(t)$ in accordance with the selection probabilities $P_{si}(t)$ and $P_{sj}(t)$.

4) An operation called crossover, shown in Figure 2, is applied to the individuals $s_i(t)$ and $s_j(t)$. The crossover operation selects a boundary in the strings with probability of $1/(G-1)$, and exchanges the blocks of strings about the boundary. One of the two individuals produced by this operation is selected at random, and is nominated as the new individual $s'_k(t)$.

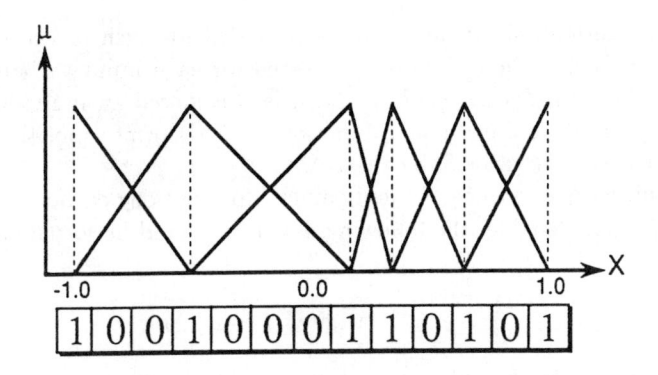

Figure 3: Membership Functions

5) An operation called mutation is applied to $s'_k(t)$. By this, each element of the individual $s'_k(t)$ is reversed according to a mutation probability P_m. Taking the example showing Figure 2, the element positioned third from the left end is reversed by the mutation.

6) The number of newly produced individuals, k, is compared with the total number of individuals, R, and if $k < R$, k is incremented by one, and steps (3) to (6) are repeated. Otherwise, the algorithm proceeds to the next step.

7) The new population, $S'(t) = \{s'_1(t), \ldots, s'_R(t)\}$, produced in steps (3) to (6), substitute into the population on the next generation $S(t+1)$.

8) The generation t is incremented by one, and the steps (2) to (8) are repeated until the terminating conditions are satisfied.

3.3. Algorithm to Optimize Inference Rules

The optimization of the number of inference rules and the shapes of membership functions in the antecedent part by means of the GA is described here.

Figure 3 shows the membership functions in the antecedent part employed in the simplified fuzzy reasoning. The membership function takes triangular shape, and the width of each membership function is defined to be the length between the centers of neighboring membership functions.

The number and shapes of membership functions can be expressed in terms of strings consisting of "0" and "1" as shown in Figure 3, wherein the center position of each membership function is expressed by "1". The string can be expressed as follows:

$$L_{j1}L_{j2}\ldots L_{jG} \quad \text{where} \quad j = 1, \ldots, m. \tag{12}$$

$L_{jr}, (r = 1, \ldots, G)$ is a variable taking a value of either "0" or "1".

In this method, since the string is provided for each of the input variables x_j, a string combining the strings L_{jr} provided for each input variable x_j, which is expressed by $L_{11}...L_{1G}L_{21}...L_{2G}...L_{m1}...L_{mG}$, is considered an individual in the GA. The optimal number of membership functions and the center positions of these are searched for each input variable by the GA.

Assuming the existence of the membership functions on both sides of the domain of each input variable, the following equations could be formulated.

$$\begin{cases} L_{j1} = 1 \\ L_{jG} = 1 \end{cases} \text{where} \quad j = 1, ..., m. \qquad (13)$$

As the fitness in the GA, an information criterion[8] shown below is employed here:

$$C = N \cdot \log(E_{TRD}) + 2(\text{The number of parameters}) \qquad (14)$$

The information criterion C shows the overall capability for learning and generalization of fuzzy reasoning. A smaller value of the information criterion is considered to mean better inference rules. Expressing the total number of membership functions allocated to the input variable x_j by N_j, the number of parameters shown in Eq. (14) can be expressed by the following:

$$\text{The number of parameters} = \sum_{j=1}^{m}(N_j - 2) + \prod_{j=1}^{m} N_j. \qquad (15)$$

The fitness $E(s_r)$ in the GA is defined by the formula:

$$E(s_r) = \max_r C(s_r) - C(s_r). \qquad (16)$$

In this method, the number of membership functions and the center positions of the membership functions maximizing the fitness $E(s_r)$ are derived by using the GA.

Although the information criterion expressed by Eq. (14) is employed as a fitness in this case, the use of some other objective function such as the unbiasedness criterion could be better depending on the actual problem.

3.4. Self-Tuning Procedure

The procedures to obtain the optimal inference rules using the GA is shown below:

[Step 1] All of the individuals $s_r(t)$ where $r = 1, ..., R$ on the 0-th generation ($t = 0$) are determined by uniform random numbers. In a concrete form, because of the conditions given by Eq. (13), $L_{j2}, ..., L_{jG-1}$ values of individuals $s_r(t)$ are set at either "0" or "1" by the uniform random numbers.

[Step 2] The learning by the descent method is applied according to [Step 2-1] to [Step 2-5] in order to determine the real numbers of the consequent parts. These processes have to be applied to all individuals $s_r(t)$, where $r = 1, ..., R$.

To begin with, the number of individual, r, is initialized at 1.

[Step 2-1] The number and the shapes of membership functions in the antecedent parts are determined according to the string of the individual $s_r(t)$. Then, the number of the input-output data, p, is initialized at 1, and the number of iterations of learning by descent method, t', is initialized at 1.

[Step 2-2] The fuzzy reasoning is applied to the p-th input data $(x_1^p, ..., x_m^p)$ by using Eqs. (2) and (3), in order to determine the membership value u_i^p of each inference rule and to obtain the output of the fuzzy reasoning y^p.

[Step 2-3] Based on the output of the fuzzy reasoning y^p, the membership value u_i^p and the output data y^{rp}, the real numbers in the consequent part, w_i, are updated by using Eq. (5).

[Step 2-4] Comparing the number of input-output data, p, with the total number of input-output data, P. If $p < P$, the algorithm is returned to [Step 2-2] after adding 1 to p, otherwise, the algorithm proceeds to next step.

[Step 2-5] In this step, a change of the inference error, $E_{TRD}(t') - E_{TRD}(t'-1)$, is derived. When the change satisfies the following formula, learning by the descent method is terminated:

$$|E_{TRD}(t') - E_{TRD}(t'-1)| < \delta \qquad (17)$$

where δ is a threshold value to judge the convergence of the inference error E_{TRD}, which has to be set in advance.

If Eq. (17) is not satisfied, after adding 1 to t' and initializing p to 1, the process is returned to [Step 2-2].

If Eq. (17) is satisfied, the fitness $E(s_r)$ is derived from the values of converged inference errors E_{TRD} by applying Eqs. (14) to (16), and the selection probability $P_{sr}(t)$ are derived by Eq. (11). Then, if $r < R$, the algorithm is returned to [Step 2-1] after adding 1 to r, otherwise, the number of the individual produced next, k, is initialized at 1, and the algorithm proceeds to next step.

[Step 3] Two individuals $s_i(t)$ and $s_j(t)$ are selected from $S(t)$ according to the selection probabilities $P_{si}(t)$ and $P_{sj}(t)$.

[Step 4] The crossover is applied to these two selected individuals to derive a new individual $s'_k(t)$. In this case, since the fuzzy reasoning has m input variables, m-points crossover[5] is applied.

[Step 5] The mutation is applied to each of the elements of individual $s'_k(t)$ according to the mutation probability P_m.

[Step 6] The series of processes, [Step 3] to [Step 5], is repeated until the number of new individual, k, becomes R.

[Step 7] A new population $S'(t) = \{s'_1(t), s'_2(t), \ldots, s'_R(t)\}$ produced by the processes from [Step 3] to [Step 6] is defined as $S(t+1)$.

[Step 8] The number of the generation, t, is incremented by one, and the processes from [Step 2] to [Step 8] are repeated until a convergence of the population S is obtained.

An individual having the highest fitness in the converged population is defined as the final solution.

4. NUMERICAL EXAMPLES

In order to demonstrate the usefulness of the GA method, two simple identification problems of nonlinear systems are discussed here.

The two nonlinear systems having single input and single output are shown below:

$$\text{SYSTEM 1} : y = \begin{cases} -x & x \leq 0 \\ x^2 & x > 0 \end{cases} \quad (18)$$

$$\text{SYSTEM 2} : y = 1 - x - x^2 + x^3 + R_g \quad (19)$$

where R_g is a function that generates the random numbers with gaussian distribution having an average value of 0.0 and a standard deviation of 1.0.

By changing the input x in Eqs. (18) and (19) in the range [-1, 1] according to the uniform random numbers, input-output data are generated respectively. The produced input-output data are divided into the TRD and the CHD in each system.

By using the GA method, each of the input-output relations given by Eqs. (18) and (19) is identified in terms of fuzzy inference rules from the produced TRD.

The initial conditions for this GA method are shown in the following:

The total number of individuals	:	$R = 20$,
The length of individual	:	$G = 13$,
The mutation probability	:	$P_m = 0.01$,
The number of generations	:	$t = 50$,
The threshold value	:	$\delta = 1.0 \times 10^{-5}$.

Figure 4 shows the input-output relation of SYSTEM 1. In this example, 40 TRD and 40 CHD are used. The square dots in Figure 4 show the TRD.

Figure 5 shows the membership functions obtained by applying the GA to the TRD of SYSTEM 1. The number of inference rules is determined at 7 by the GA. In the region ($x > 0$) having non-linear input-output relation, more inference rules are assigned than in the region ($x \leq 0$) having linear input-output relation. This result indicates that this self-tuning method can arrange the membership functions in accordance with the nonlinearity of the given TRD.

Figure 6 shows the output of fuzzy reasoning using the optimized membership function of Figure 5. In Figure 6, the output of fuzzy reasoning coincides with the true output of Eq. (18) approximately. Therefore, it can be said that this self-tuning method acquires the input-output relation of given TRD well. Using the inference rules obtained by the GA, the E_{TRD} and E_{CHD} became 2.1×10^{-6} and 5.7×10^{-6}, respectively.

In order to show the higher learning capability of this method, a comparison with a conventional self-tuning method was conducted. As the conventional method, the self-tuning controller by the descent method[1] only was employed, and the inference rules were constructed from the same TRD.

In Figure 7, relations of the conventional method between the number of inference rules and the inference errors E_{TRD}, E_{CHD} are shown. The numbers of inference rules in the conventional method were varied from 2 to 27 manually. Figure 7 proves that a lower inference error E_{TRD} is obtained for a higher number of inference rules, while a higher inference error E_{CHD} is obtained for a higher number of inference rules exceeding a certain threshold. As shown in Figure 7, the number of inference rules which gives a minimum E_{CHD} is 7. The values of E_{TRD} and E_{CHD} at 7 inference rules are 3.0×10^{-5} and 6.3×10^{-5}, respectively, both of which are larger than those obtained by the GA. This shows that the inference rules obtained by the GA have a higher learning capability than those obtained by the conventional self-tuning method.

In order to demonstrate that the solution obtained by the GA doesn't converge into a local optimum but to a global optimum, a result of searching in total space is shown next. In this example, the total number of solution candidates is $2^{11} (= 2048)$. The learning of the consequent parts by the descent method is applied to all of the 2048 solution candidates in order to derive the value of the information criterion C. The results are shown in Table 1, where they are sorted in terms of the value of the information criterion. Table 1 shows that the optimal solution obtained by the total search is 1000001111101. This solution coincides with the one obtained by the GA. This example shows that the optimal inference rules minimizing the value of the information criterion is indeed obtained by the GA.

The result of learning applied to SYSTEM 2 is shown next. It is very often that noise component is included in observed input-output data. This example simulates the case where input-output data include noise component. Through this example, the generalization capability of the GA method is shown.

Figure 8 shows the input-output relation of SYSTEM 2, the thick line shows

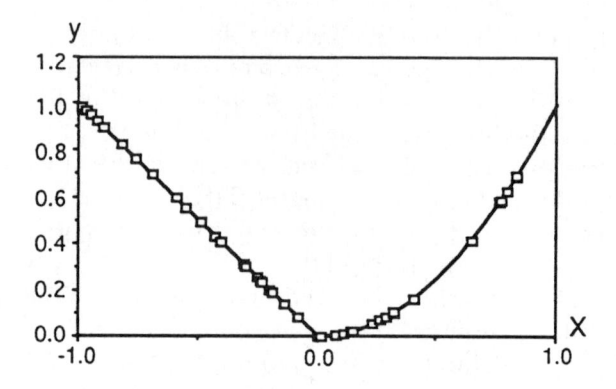

Figure 4: Input-output Relation (SYSTEM 1)

| 1 | 0 | 0 | 0 | 0 | 0 | 1 | 1 | 1 | 1 | 1 | 0 | 1 |

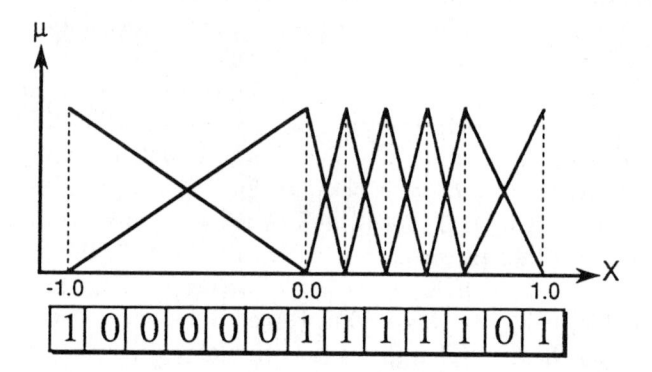

Figure 5: Optimized Membership Functions (SYSTEM 1)

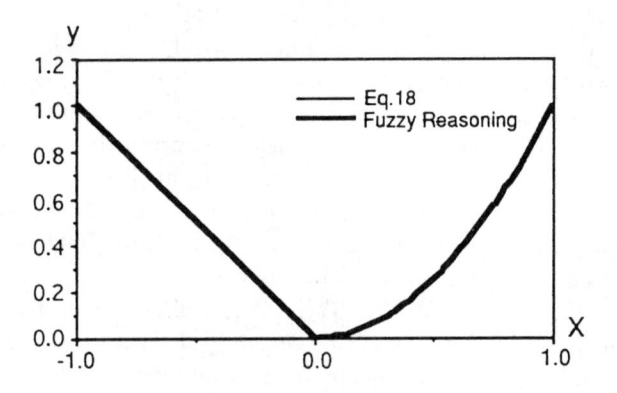

Figure 6: Output of Fuzzy Reasoning (SYSTEM 1)

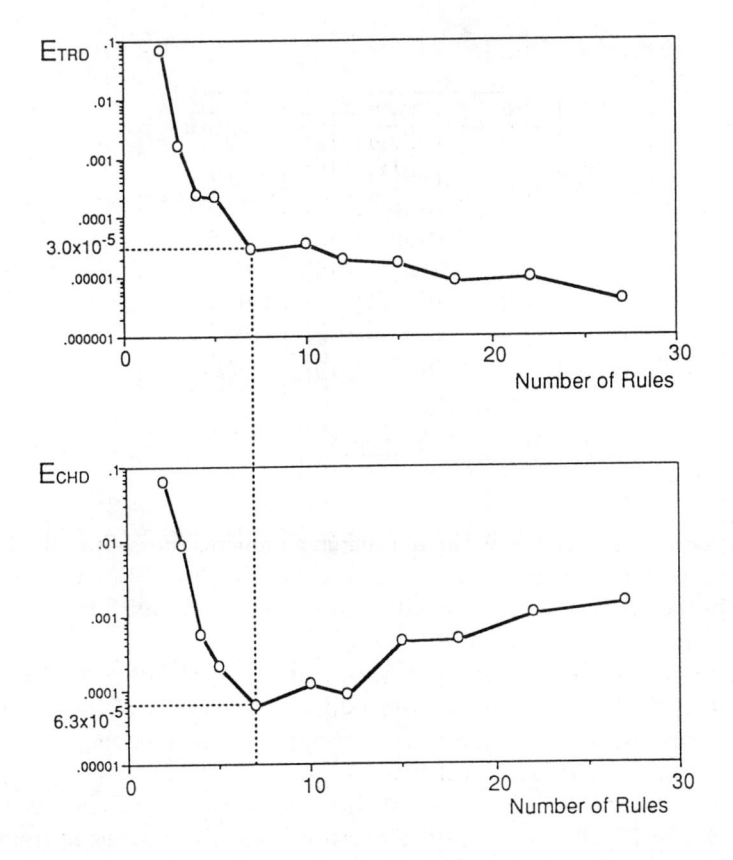

Figure 7: Transition of Inference Errors (SYSTEM 1)

Table 1: Result of Searching in Total Space (System 1)

No.	Strings	C
1	1000001111101	-496.9
2	1000011111101	-494.7
3	1000101111101	-494.3
4	1001001111101	-493.8
5	1010001111101	-493.5
6	1100001111101	-493.1
7	1000001111111	-490.7
8	1001011111101	-490.7
	......	
2048	1000000000011	-94.0

the true input-output relation without random numbers, the square dots show the TRD. Since a term having random numbers is included in Eq. (19), a noise component is included in the produced TRD and CHD. In this example, 400 TRD and 400 CHD are used.

If the conventional self-tuning method were applyed for such TRD by giving a large number of inference rules, the generalization capability of the inference rules could be particularly low. Therefore, the searching of an optimal number of inference rules become more important in such case.

Figure 9 shows the membership functions obtained by applying the GA to the TRD of SYSTEM 2 . The number of inference rules is determined at small number 4. This result shows that the number of inference rules is determined so as to avoid overfitting by the GA.

Figure 10 shows the output of the fuzzy reasoning optimized by the GA, the thick line means the output of the optimized fuzzy reasoning, the thin line shows the true output without random numbers. Since Figure 10 shows that the output of the fuzzy reasoning is not constrained by the noise of the TRD, and coincides with the true output approximately, it can be said that the GA method has high generalization capability.

The results of a total search similar to the one carried out for SYSTEM 1, are shown in Table 2. The optimum solution obtained by the total search coincides with the solution 1000010001001 obtained by the GA. This example shows also that optimum inference rules can be obtained by the GA.

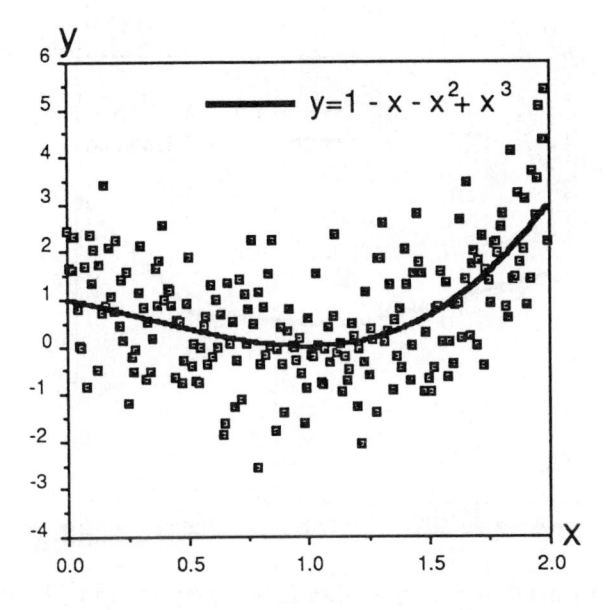

Figure 8: Input-output Relation (SYSTEM 2)

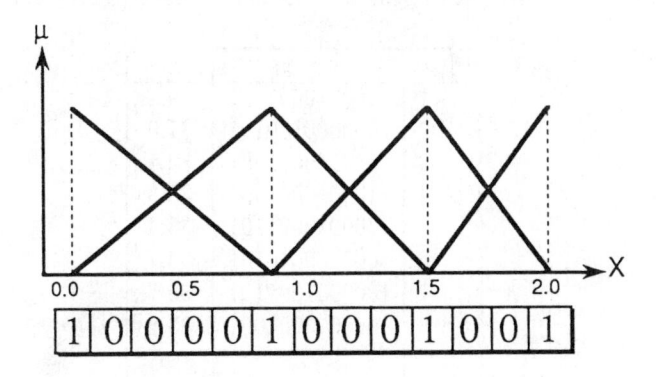

Figure 9: Optimized Membership functions (SYSTEM 2)

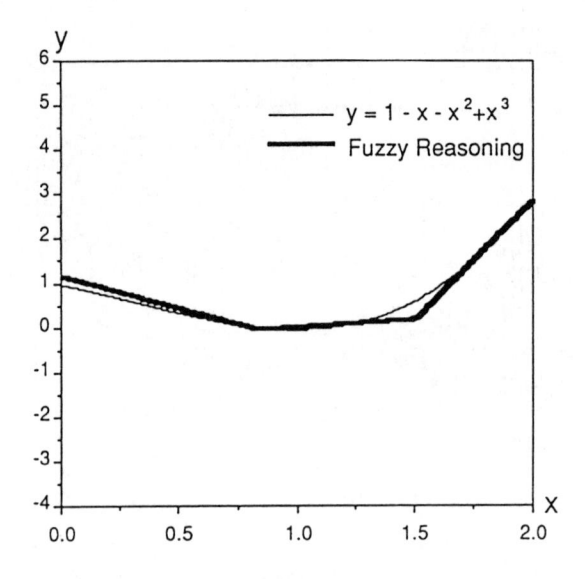

Figure 10: Output of optimized fuzzy reasoning (SYSTEM 2)

Table 2: Result of Searching in Total Space (SYSTEM 2)

No.	Strings	C
1	1000010001001	29.2
2	1000000100101	31.0
3	1000100001001	32.4
4	1100010001001	33.3
5	1000010001101	33.4
6	1000100010101	33.7
7	1000001001101	33.7
8	1010010001001	33.8
......		
2048	1000000000001	391.4

5. CONCLUSION

A new self-tuning method of fuzzy reasoning by the genetic algorithm is proposed here. With this method, the number of inference rules and the shapes of membership functions in the antecedent part are determined by the genetic algorithm so as to optimize an information criterion expressing the quality of the inference rules.

Using two numerical examples, we have shown that this GA method has the higher learning capability than the conventional self-tuning method using the descent method.

As a future problem, the effectiveness of the proposed method, when it is applied to more complicated systems having multi-inputs, should be established.

REFERENCES

[1] Nomura, H., Hayashi, I., Wakami, N., "A self-tuning method of fuzzy control by descent method". *Proceedings of 4th IFSA Congress*, Engineering, pp. 155-158, 1991.

[2] Nomura, H., Hayashi, I., Wakami, N., "A learning method of fuzzy inference rules by descent method". *Proceedings of IEEE International Conference on Fuzzy Systems*, pp. 203-210, 1992.

[3] Nomura, H., Hayashi, I., Wakami, N., "A self-tuning method of fuzzy reasoning by delta rule and its application to a moving obstacle avoidance". *Journal of Japan Society for Fuzzy Theory and Systems*, Vol. 4, pp. 379-388, 1992. (in Japanese)

[4] Hayashi, I., Nomura, H., Yamasaki, H., Wakami, N., "Construction of fuzzy inference rules by NDF and NDFL". *International Journal of Approximate Reasoning*, Vol.6, pp. 241-266, 1992.

[5] Goldberg, D. E., *Genetic Algorithms in Search, Optimization and Machine Learning*, Addison Wesley Publishing Company, 1989.

[6] Holland, J. H., *Adaptation in Natural and Artificial Systems* , The University Michigan Press, 1975.

[7] Ichihashi, H., Watanabe, T., "Learning control by fuzzy models using a simplified fuzzy reasoning". *Journal of Japan Society for Fuzzy Theory and Systems*, Vol. 2, No. 3, pp. 429-437, 1990. (in Japanese)

[8] Akaike, H., "A new look at the statistical model identification". *IEEE Transactions on Automatic Control*, AC-19, No. 6, 1974.

[9] Nomura, H., Hayashi, I., Wakami, N., "A self-tuning method of fuzzy reasoning by genetic algorithm". *Proceedings of the 1992 International Fuzzy Systems and Intelligent Control Conference*, pp. 236-245, 1992.

[10] Nomura, H., Hayashi, I., Wakami, N., "A learning method fusing fuzzy reasoning and genetic algorithm". *Proceedings of the IMACS/SICE International Symposium on Robotics, Mechatronics and Manufacturing Systems*, pp. 155-160, 1992.

Chapter 17
Hybrid Neural-Fuzzy Reasoning Model with Application to Fuzzy Control

Introduction, *356*
Fuzzy Models, *357*
Hybrid Neural-Fuzzy Reasoning Model, *365*
Conclusions, *370*
References, *372*

The hybrid neural-fuzzy reasoning model (HNFRM) proposed in this chapter combines the computational paradigms of neural networks and the new fuzzy reasoning model (NFRM). The HNFRM is applied to fuzzy control using a direct-current series motor, representing an example of a simple controlled process, that serves as a vehicle for the evaluation of the performance of the various controllers discussed in this chapter. It is shown that the hybrid system performs much better than the NFRM component standing alone. Furthermore, the HNFRM reveals two important features. First, the neural network component of the HNFRM is single-layered. Second, the interconnection weights in the neural network are interpreted as fuzzy relation degrees between the input and output linguistic variables, so that the inference process of the neural network is not unintelligible.

HYBRID NEURAL-FUZZY REASONING MODEL WITH APPLICATION TO FUZZY CONTROL

Daihee Park
Department of Computer Science
Florida State University
Tallahassee, FL 32306

Abraham Kandel
Department of Computer Science and Engineering
University of South Florida
Tampa, FL 33620

Gideon Langholz
Department of Electrical Engineering
Florida State University *and* Tel-Aviv University
Tallahassee, FL 32306

1. INTRODUCTION

Since the introduction of the basic methods of fuzzy reasoning by Zadeh [1,2], and the success of their first application to fuzzy control [3], the Fuzzy Reasoning Method (FRM) has been widely studied and used successfully in a number of control problems [4]. Nevertheless, certain challenging open problems are associated with the FRM, including (1) the question of the completeness of the fuzzy rule base; (2) the subjective definitions of the fuzzy subsets; and (3) the choice of fuzzy implication operators.

Recently, Cao *et al.* [5] have proposed a New Fuzzy Reasoning Method (NFRM) which turned out to be superior to Zadeh's FRM. Kandel *et al.* [6] have applied the NFRM to the cart-pole system and compared its performance with the FRM's, whereas Yu *et al.* [7] have applied the NFRM successfully to the control of an activated sludge plant.

However, the NFRM still retains some of the problems associated with the FRM. In particular, the fuzzy relation matrix in the NFRM, determined by the human operator according to his experience, plays an important role, but may be difficult to extract optimally from the operator particularly as the system increases in complexity. Therefore, it is important to establish a mechanism for finding the optimal fuzzy relation matrix autonomously. One way in which this may be accomplished is by incorporating event-driven neural network learning mechanisms into the NFRM.

In this chapter, we propose a Hybrid Neural-Fuzzy Reasoning Model (HNFRM) that combines the computational paradigms of neural networks and the NFRM. (Note that we consider here *only* the NFRM because of its advantages over the FRM.) We apply the HNFRM to fuzzy control using a DC (direct current) series motor as an example of a simple controlled process that can serve as a vehicle for the evaluation of the performance of the various controllers discussed in this chapter.

The particular DC series motor that we use is the one investigated by Kiszka *et al.* [8,9]. Figure 1 reproduces the steady-state characteristic of the motor, obtained by Kiszka *et al.* from actual measurements. In a DC series motor, the field winding is connected in series with the armature. Since the magnetic flux in the motor is proportional to the field current, which varies, the relationship between the current and the speed at which the motor's shaft rotates is nonlinear in general.

We show, by studying the application of the HNFRM to fuzzy control, that the hybrid system performs much better than the NFRM component standing alone.

The chapter is organized as follows: In Section 2, we discuss the FRM and NFRM controllers. In Section 3, we propose the HNFRM controller and then apply it to the DC series motor. Finally, Section 4 provides a comparative summary of the results obtained.

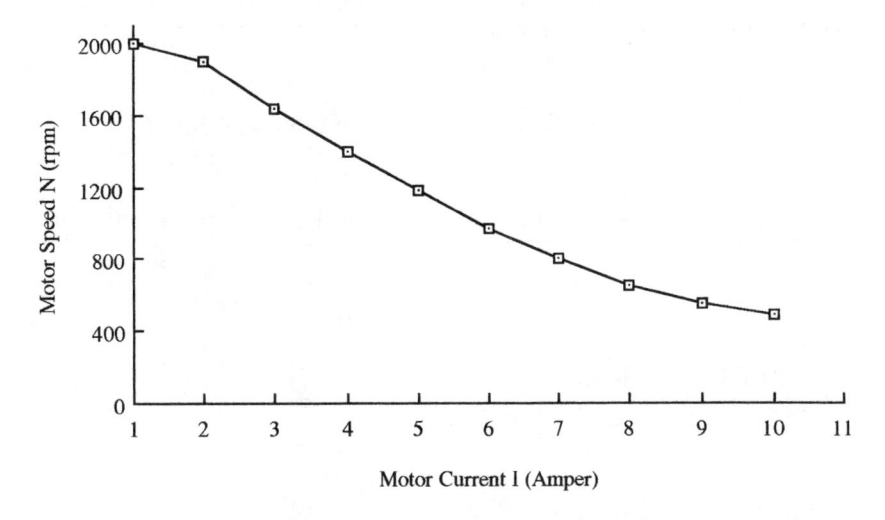

Figure 1. Actual Characteristic of the DC Series Motor (*Source:* [8], Figure 2.2. Reproduced by permission of *Elsevier Science Publishers*)

2. FUZZY MODELS

2.1. Fuzzy Reasoning Method

A block diagram depicting the application of the Fuzzy Reasoning Method (FRM) [2] to process control is shown in Figure 2. The fuzzy control system comprises four principal components: Fuzzification interface; Knowledge base; Decision-making logic; and Defuzzification interface [10].

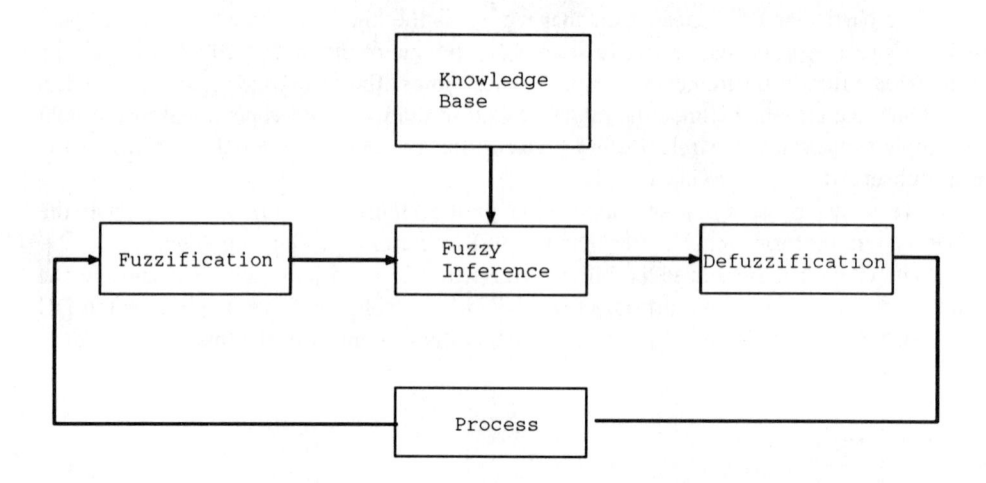

Figure 2. Structure of a Fuzzy Rule-based Control System

Throughout this chapter, we adopt the following fuzzy conditional statements to describe a certain situation:

IF X is A(1)   THEN   Y is B(1)
   ALSO
IF X is A(2) THEN   Y is B(2)   (2.1)
   ALSO
   .
   .

IF X is A(n)   THEN   Y is B(n)

where X and Y are two variables, and A(1), ..., A(n) and B(1), ..., B(n) are linguistic descriptions of X and Y, respectively, which can be quantified by the fuzzy subsets A^*_i and B^*_i for each i in the ranges of X and Y.

The fuzzy conditional statements (2.1) can be formalized in the form of the fuzzy relation R(X,Y) [2]:

$$R(X,Y) = ALSO\ (R_1, R_2, ..., R_i, ..., R_n)$$

where ALSO represents a sentence connective which combines the R_i's into the fuzzy relation $R(X,Y)$, and R_i denotes the fuzzy relation between X and Y determined by the *ith* fuzzy conditional statement, namely:

$$R_i = \int \mu_{A*i}(x) \otimes \mu_{B*i}(y) / (x,y), \qquad i = 1, 2, ..., n$$

where $\otimes$ is a fuzzy implication operator.

Once we establish a fuzzy relation $R(X,Y)$ between two variables X and Y, we can then apply the compositional rule of inference to infer the fuzzy subset B for Y, given the fuzzy subset A for X:

$$B = A \circ R(X,Y)$$

where o is a fuzzy compositional operator.

Thus far, the output of the inference process is a fuzzy set. However, since a deterministic (nonfuzzy) control action is required in many practical applications, a defuzzification strategy is employed to produce a nonfuzzy control action that best represents the possibility distribution of the inferred fuzzy control action. We use the following defuzzification method [8,9]:

$$y = \sum_{k=1}^{m} y_k / m$$

where y is a particular value of the variable Y, y_k is the support value in which the membership function $\mu_{B*k}(y)$ reaches its maximum grade of membership, and m is the number of such support elements.

In Table 1, we list several fuzzy implication operators which will be used in the sequel. Note that the sentence connective ALSO in the linguistic description is interpreted as intersection (also denoted as ^ or the "minimum").

Consider now the DC series motor introduced in Section 1. Observing the relationship between the armature current I and the speed N of the motor's shaft (Figure 1), the process operator may formulate the following linguistic descriptions to specify the static characteristic of the motor (*Source:* [8], Equation (2.2). Reproduced by permission of *Elsevier Science Publishers*):

$$\begin{aligned}
&\text{IF } I = \text{null} \quad \text{THEN } N = \text{very large} \\
&\quad \text{ALSO} \\
&\text{IF } I = \text{zero} \quad \text{THEN } N = \text{large} \\
&\quad \text{ALSO} \\
&\text{IF } I = \text{small} \quad \text{THEN } N = \text{medium} \qquad (2.2)\\
&\quad \text{ALSO} \\
&\text{IF } I = \text{medium} \quad \text{THEN } N = \text{small}
\end{aligned}$$

ALSO
IF I = large   THEN N = zero
ALSO
IF I = very large THEN N = zero

where I and N are the linguistic variables corresponding to the current and speed, respectively. Notice that the first linguistic statement in (2.2) is not apparent in Figure 1. It reflects the fact that, in a DC series motor, if the load is disconnected from the motor shaft, high speeds would result because of the small armature current that flows [11].

Table 1. Fuzzy Implication Operators

$$\otimes_2 = \int_{U \times V} [\mu_A(a) \rightarrow_2 \mu_B(b)] / (a,b)$$
$$\text{where } \mu_A(a) \rightarrow_2 \mu_B(b) = 1 \text{ if } \mu_A(a) \leq \mu_B(b)$$
$$= 0 \text{ otherwise}$$

$$\otimes_4 = \int_{U \times V} [1 \wedge (\mu_B(b) / \mu_A(a))] / (a,b)$$

$$\otimes_5 = \int_{U \times V} [1 \wedge (1 - \mu_A(a) + \mu_B(b))] / (a,b)$$

$$\otimes_{27} = \int_{U \times V} [\mu_A(a) \rightarrow_{27} \mu_B(b)] / (a,b)$$
$$\text{where } \mu_A(a) \rightarrow_{27} \mu_B(b) = 1 \quad \text{if } \mu_A(a) \leq \mu_B(b)$$
$$= \mu_B(b) \text{ otherwise}$$

$$\otimes_{28} = \int_{U \times V} [1 \wedge [\mu_B(b)*(1-\mu_A(a)) / (\mu_A(a)*(1-\mu_B(b)))]] / (a,b)$$

Notes: (1) a and b are particular values of the variables A and B, respectively.
(2) $\mu_A(a)$ and $\mu_B(b)$ are the grades of membership of a and b, respectively.
(3) x denotes the Cartesian product, and * denotes the arithmetic product.
(4) U and V are the universes of discourse of A and B, respectively.

From the verbal descriptions (2.2), we see that the variable I has six linguistic values: *Null, Zero, Small, Medium, Large,* and *Very Large,* whereas the variable N has five: *Zero, Small, Medium, Large,* and *Very Large.* The membership functions of the fuzzy subsets for the motor current I and speed N are shown in Figures 3 and 4, respectively.

Kiszka *et al.* [8,9] applied the FRM to the DC series motor using the implication operator $\otimes_2$, $\otimes_4$, $\otimes_5$, $\otimes_{27}$, and $\otimes_{28}$ (see Table 1), and obtained the results shown in Figure 5. Each of these operators resulted in the same, smallest mean square error e^2, that estimates the discrepancy between the motor's actual (measured) characteristic (Figure 1) and that produced through the application of the FRM:

$$e^2 = \sum_{i=1}^{v} (n_{ri} - n_{mi})^2 / \sum_{i=1}^{v} n_{ri}^2 \qquad (2.3)$$

Here, n_{ri} is the actual speed value for some current value i, n_{mi} is the corresponding FRM-produced value, and v is the number of discretization intervals of N.

While the FRM offers model-free estimation of the control system characteristics, and may prove more robust and easier to modify than a mathematical model of the process, the results obtained depend on a number of factors, such as: (1) the question of completeness of the fuzzy rule base; (2) the subjective definitions of the fuzzy subsets; (3) the choice of fuzzy implication operators; and (4) the defuzzification procedure that calculates the deterministic value of a fuzzy set.

2.2. New Fuzzy Reasoning Method

The New Fuzzy Reasoning Method (NFRM) [5] is illustrated schematically in Figure 6. Let $\mathbf{x} = [x_1, x_2, ..., x_n]$ designate a fuzzy vector consisting of the membership degrees of the real-valued input x; let $\mathbf{y} = [y_1, y_2, ..., y_m]$ be a fuzzy vector consisting of the membership degrees of the output variable y; and let $\mathbf{R}$ be an $n \times m$ fuzzy relation matrix whose elements w_{ij} (i=1,2,...,n; j=1,2,...,m) indicate the relation degree between the *ith* linguistic description for X and the *jth* linguistic description for Y.

The y_i's are determined by:

$$y_i = \Sigma_j x_j * w_{ji}$$

whereas the real number y in the universe of Y is evaluated using the moment method [5]:

$$y = \Sigma_j (f_j * y_j) / \Sigma_j y_j \qquad (2.4)$$

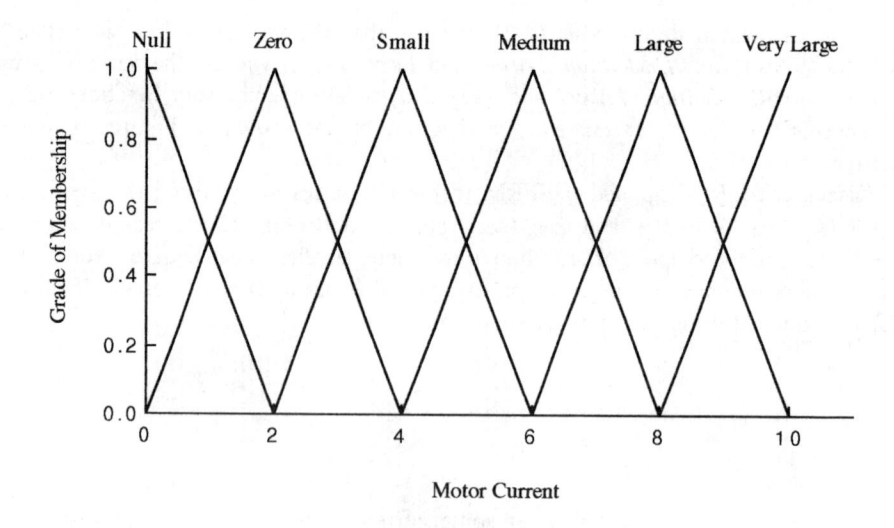

Figure 3. Membership Functions of the Fuzzy Subsets for Motor Current I

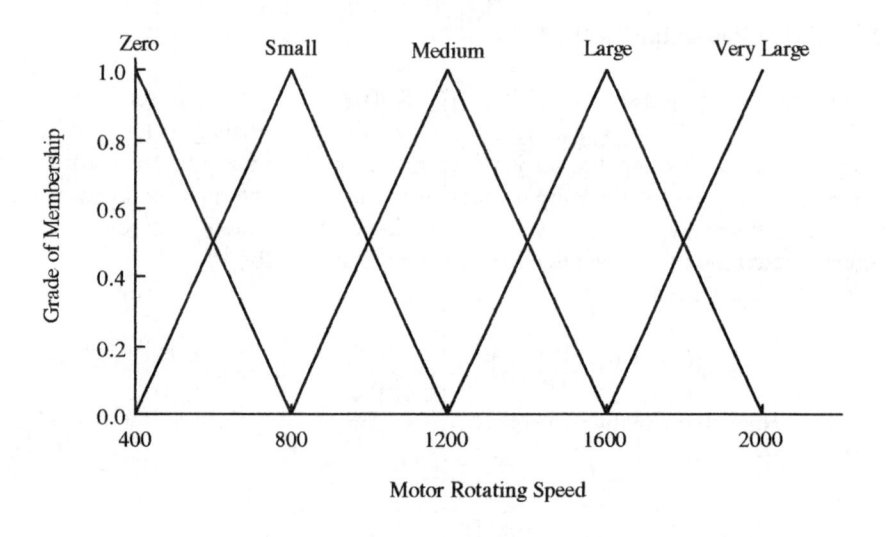

Figure 4. Membership Functions of the Fuzzy Subsets for Motor Speed N

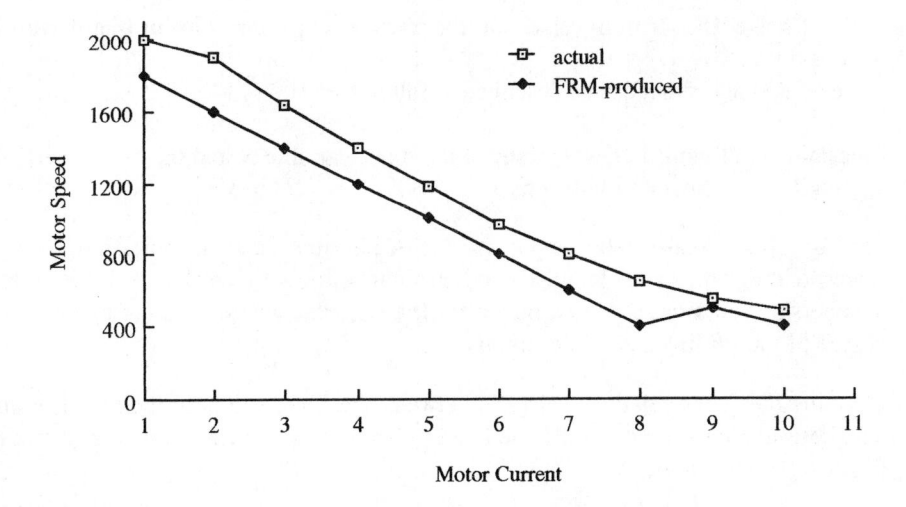

Figure 5. Actual and FRM-produced Motor Characteristic ($e^2 = 2.504 \ast 10^{-2}$)

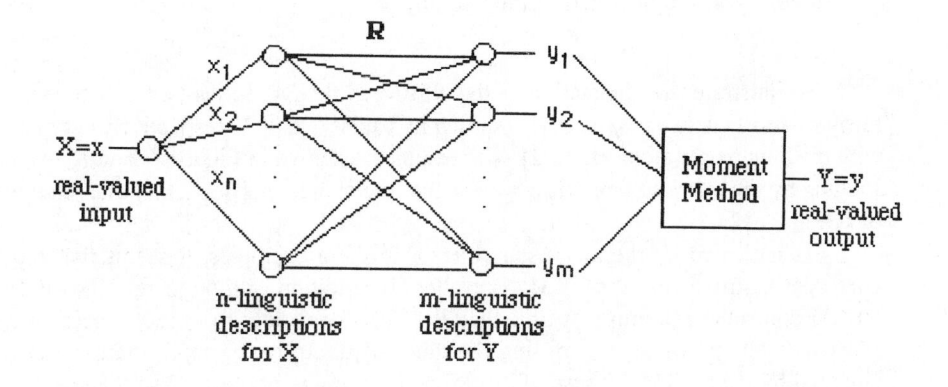

Figure 6. NFRM Schema

where the f_j's are the central values of the membership functions of the linguistic descriptions for Y.

The NFRM algorithm can be described as follows:

1: Determine the linguistic descriptions of the input variable X and the output variable Y, and their membership functions.

2: For a given real-valued input $X=x$, determine the membership degree corresponding to each linguistic description of X by means of the fuzzy membership functions. That is, obtain $x=[x_1, ..., x_n]$ where x_i is the membership degree of the ith linguistic description of X.

3: Determine the fuzzy relation matrix **R** between the linguistic descriptions of X and Y. These relations are based on the expert's knowledge which is given in terms of linguistic descriptions.

4: Calculate the vector $y=[y_1, y_2, ..., y_m]$ in terms of the vector x and the fuzzy relation matrix **R**:

$$y = x \circ R$$

Note that, in the sequel, we use the sum-of-product operator for the fuzzy compositional operator in the NFRM.

5: Using the moment method (2.4), transform the vector y to the corresponding real number y in the universe of the variable Y.

To illustrate the application of the NFRM to the DC series motor, we consider two fuzzy relation matrices R_1 and R_2 (shown in Tables 2 and 3, respectively) commensurate with the linguistic statements (2.2). The results are shown in Figures 7 and 8, respectively. Judging by the mean square error, we see that the choice of R_1 is a bad one relative to that of R_2.

As indicated by the mean square error in these examples, it seems that the NFRM controller outperforms the FRM controller. In addition, although not shown here, the NFRM controller requires less time than the FRM controller to execute the control tasks. Moreover, the problem of choosing the most adequate fuzzy implication operator is not inherited by the NFRM because of its different reasoning process. Nevertheless, the results obtained through the application of the NFRM still depend on the same factors as in the FRM case, *except* for the choice of fuzzy implication operators, but in addition, they also depend on the choice of the fuzzy relation matrix **R**.

Table 2. Fuzzy Relation Matrix **R₁**

I \ N	Zero	Small	Medium	Large	Very large
Null	0.0	0.0	0.0	0.0	1.0
Zero	0.0	0.0	0.0	1.0	0.0
Small	0.0	0.0	1.0	0.0	0.0
Medium	0.0	1.0	0.0	0.0	0.0
Large	1.0	0.0	0.0	0.0	0.0
Very large	1.0	0.0	0.0	0.0	0.0

Table 3. Fuzzy Relation Matrix **R₂**

I \ N	Zero	Small	Medium	Large	Very large
Null	0.0	0.0	0.0	0.0	1.0
Zero	0.0	0.0	0.0	0.5	0.9
Small	0.0	0.0	0.5	0.9	0.3
Medium	0.5	1.0	0.2	0.0	0.0
Large	1.0	0.5	0.0	0.0	0.0
Very large	1.0	0.2	0.0	0.0	0.0

3. HYBRID NEURAL-FUZZY REASONING MODEL

Some of the problems associated with NFRM-based controllers can be alleviated by incorporating neural network learning mechanisms into the fuzzy controller. A system of this type, referred to as Hybrid Neural-Fuzzy Reasoning Model (HNFRM), is proposed in this section.

In the NFRM, the fuzzy relation matrix plays an important role. However, the process of selecting an adequate fuzzy relation matrix is subjective and most time consuming, generally completed by trial-and-error. Furthermore, it is often impractical to obtain the fuzzy relation matrix from the process operator, particularly if the system is complex and/or if the fuzzy conditional statements have more than two variables. Since the process operator usually has only a general idea of the fuzzy relation matrix in a given region, the process of making that general idea precise is the most difficult task in the design of a finely-tuned fuzzy controller.

It is therefore important to establish a mechanism for adjusting the fuzzy relation matrix autonomously in order to make the controller perform robustly. To this end we take advantage of the learning capability of the neural network in the context of the NFRM.

In the following sections, we propose the HNFRM system and then apply it to the DC series motor.

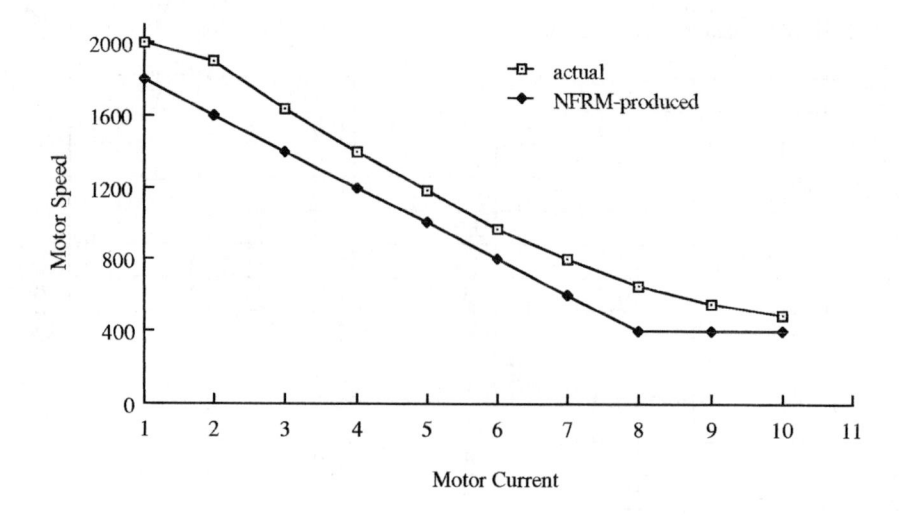

Figure 7. Actual and NFRM-produced Motor Characteristic Corresponding to R_1 of Table 2   ($e^2 = 2.633*10^{-2}$)

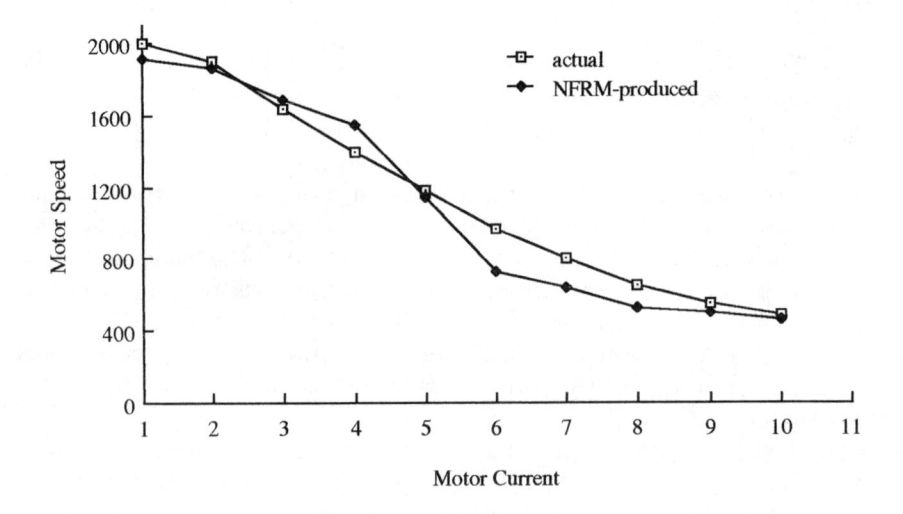

Figure 8. Actual and NFRM-produced Motor Characteristic Corresponding to R_2 of Table 3   ($e^2 = 8.556*10^{-3}$)

3.1. HNFRM Design

The transfer of knowledge from the NFRM to the neural network component of the HNFRM initially breaks the rule base clauses into two classes: antecedents clauses requiring input from the user, and consequent clauses which yield the final output decisions. Each fuzzy degree element in the fuzzy relation matrix is identified with interconnection weights in the neural network component, and antecedents and consequents are associated with input and output nodes, respectively. These classes give rise to a mapping of the fuzzy inference component of the NFRM onto a single-layer linear feedforward neural network which constitutes the neural network component of the HNFRM.

The mapping is carried out as follows:

1. Fuzzy rules are mapped into interconnections in the neural network.
2. Linguistic descriptions for the inputs are mapped into neurons in the input layer.
3. Linguistic descriptions for the outputs are mapped into neurons in the output layer.
4. The elements of the fuzzy relation matrix are mapped into corresponding interconnection weights in the neural network.
5. The fuzzy compositional operator used is mapped into the sum-of-products operator.

The correspondence between the neural network component of the HNFRM and the fuzzy inference component of the NFRM is summarized in Table 4.

In Figure 9, the HNFRM control system is represented in a block diagram form. The inputs to the network, as well as the elements of the fuzzy relation matrix, are fuzzy degrees, each ranging between 0 and 1. The outputs of the neural network component are also fuzzy degrees and, therefore, the activation functions in this case are the identity functions.

The HNFRM consists of two phases: a *learning* phase and an *operation* phase. After carrying out the mapping, the HNFRM adjusts the fuzzy relation matrix during the learning phase by training the neural network using the Widrow-Hoff algorithm [13] (see below). The new fuzzy relation matrix thus obtained is then transferred back to the NFRM.

Consequently, the HNFRM is based on the premise that the transfer of knowledge between the fuzzy control system and the neural network is *bidirectional*. Initially, the fuzzy control system is invoked with a known base of knowledge. The neural network then takes the knowledge from the fuzzy control system and modifies it through learning. Since the fuzzy control system supplies the meta-knowledge to begin the learning process, learning in the neural network is implemented more efficiently. Finally, the fuzzy control system makes use of knowledge acquired from the neural network to process its task.

Table 4. Neural Network-NFRM Correspondence

NEURAL NETWORK	NFRM
Interconnections	Fuzzy rules
Neurons in the input and output layer	Linguistic descriptions
Weights	Fuzzy relation degrees
Sum-of-products	Fuzzy compositional operator

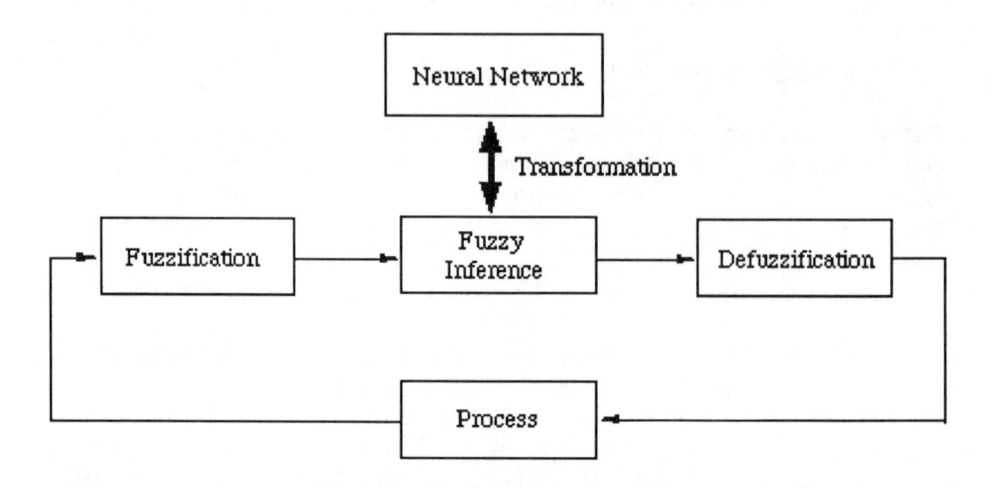

Figure 9. Structure of The HNFRM Control System

The Widrow-Hoff algorithm, used by the neural network during the *learning phase*, is a variant of the gradient descent method. The algorithm proceeds as follows:

1: Present the real-valued input $X = x$ to the HNFRM. By means of the membership functions of each of the linguistic descriptions for X, obtain the input pattern $x=[x_1,x_2,...,x_n]$ corresponding to a given real value of X, where each x_i is a certain membership degree of x in the ith linguistic description for X.

2: Compute the HNFRM output pattern $y=[y_1,y_2,...,y_m]$ using $y_j=\Sigma_i x_i * w_{ij}$. The pattern y is a fuzzy set of the set of all linguistic descriptions of the object Y, and w_{ij} is the fuzzy relation degree between ith linguistic description for X and jth linguistic description for Y.

3: Compute the training error.

3.1: Compute a real output value by means of the moment method [5]:

$$y = \Sigma_j (f_j * y_j) / \Sigma_j y_j$$

where the f_j's are the central values of membership functions of the linguistic descriptions for Y.

3.2: Given a desired scalar output value **d**, find a desired output vector $\mathbf{y^d} = [y^d_1, y^d_2, ..., y^d_m]$ by means of the fuzzy membership functions for Y.

3.3: Compare $\mathbf{y^d}$ and $\mathbf{y}$:

$$\varepsilon_j = y_j^d - y_j = y_j^d - \Sigma_i x_i * w_{ij}$$

3.4: Compute the total learning squared error:

$$LE = 0.5 * \Sigma_j \varepsilon_j^2 \qquad (3.1)$$

4: If LE is sufficiently small, go to Step **7**.

5: Apply the Widrow-Hoff algorithm to modify the weights:

$$w_{ij}(t+1) = w_{ij}(t) + \Delta w_{ij}(t+1) \qquad (3.2)$$

where,

$$\Delta w_{ij}(t+1) = \beta * \varepsilon_j * x_i / \Sigma_k (x_k)^2 + \alpha * \Delta w_{ij}(t)$$

The coefficients ß and α are the learning coefficient and the momentum coefficient, respectively.

6: Return to Step **1**.

7: Stop.

3.2. HNFRM Control of DC Motor

The neural network component of the HNFRM DC motor controller consists of six neurons in the input layer and five neurons in the output layer, corresponding to the six linguistic descriptions of the input variables and five linguistic descriptions of the output variables in accordance with the linguistic statements (2.2).

We employed $\mathbf{R_2}$ of Table 3 as the initial fuzzy relation matrix, and assigned the values 0.1 and 0.01 to the learning coefficient ß and the momentum coefficient α, used in (3.2) for weight adjustments, respectively.

Table 5 depicts the HNFRM-generated fuzzy relation matrix after 100 learning iterations. With this matrix transferred back into the NFRM, Figure 10 shows the results obtained when the HNFRM was used to control the DC series motor after 100 learning iterations.

To see how well the HNFRM controller generalizes to solve arbitrary problem instances, we presented to it 46 test data, which included the 10 data used for training. As can be seen in Figure 11, it generates a very smooth curve.

Note that, although the training algorithm for the NFRM minimizes the total learning squared error LE [Eqn. (3.1)], we converted LE into e^2 [Eqn. (2.1)] in Figures 10 and 11 so that the results would be consistent and comparable throughout this chapter.

Since the fuzzy relation matrix has now been adjusted autonomously to minimize the error, the errors produced by the HNFRM controller are much smaller than those obtained by the NFRM controller (compare Figure 10 with Figure 8).

Table 5. HNFRM-generated Fuzzy Relation Matrix for the DC Motor

I \ N	Zero	Small	Medium	Large	Very large
Null	0.000	0.000	0.000	0.000	1.000
Zero	0.000	0.000	0.000	0.454	0.660
Small	0.000	0.000	0.622	0.569	0.000
Medium	0.000	0.631	0.515	0.000	0.000
Large	0.292	0.807	0.000	0.000	0.000
Very large	0.817	0.151	0.000	0.000	0.000

4. CONCLUSIONS

We studied several fuzzy controllers and analyzed their performance using a DC motor as the controlled process. We have shown that the performance of fuzzy rule-based controllers can be improved considerably if we combine the two complementary knowledge representation techniques, neural networks and fuzzy rule bases, in a hybrid system.

Tables 6 and 7 summarize our results. Table 6 provides model descriptions, whereas Table 7 compares the results obtained for each controller. In these tables, NFRM-1 and NFRM-2 refer to the NFRM with R_1 (Table 2) and R_2 (Table 3), respectively.

The HNFRM system reveals several important features which merit further investigation. Among them, we have: (1) The neural network component of the HNFRM is single-layered; and (2) the interconnection weights in the neural network are interpreted as fuzzy relation degrees between the input and output linguistic variables, so that the inference process of the neural network is not unintelligible.

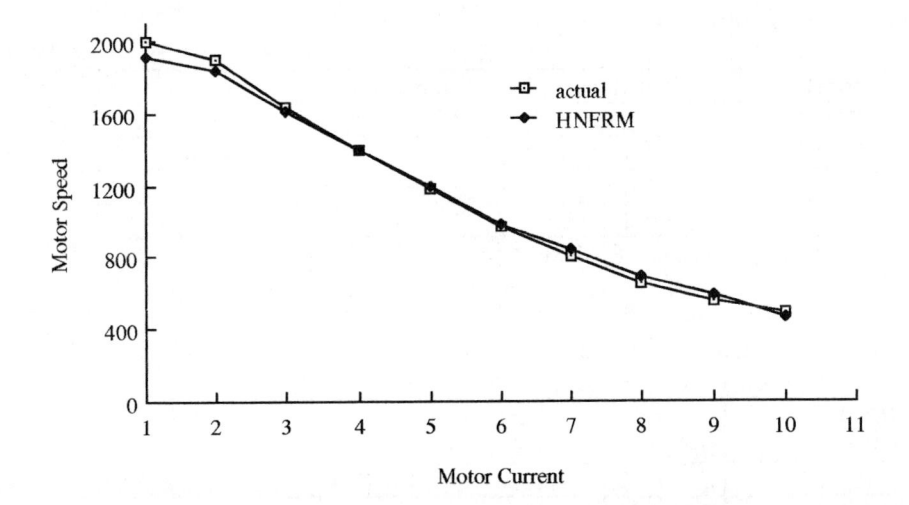

Figure 10. Learning Phase: Actual and HNFRM-produced Motor Characteristic After 100 Iterations ($e^2 = 1.098*10^{-3}$)

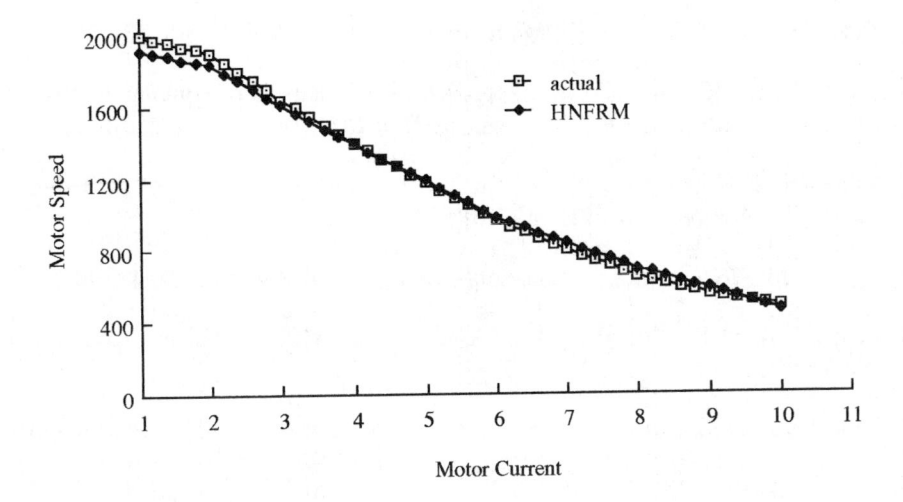

Figure 11. Generalization Phase: Actual and HNFRM-produced Motor Characteristic with 46 Test Data ($e^2 = 2.670*10^{-4}$)

Table 6. Model Descriptions

Model	Learning	Operator	AF
FRM	N/A	$\otimes_2, \otimes_4, \otimes_5$, etc	N/A
NFRM-1	N/A	SOP	Identity
NFRM-2	N/A	SOP	Identity
HNFRM	Widrow-Hoff	SOP	Identity

(N/A = not applicable; AF = activation function; SOP = sum-of-products)

Table 7. Comparison of Results

Model	e^2	Learning Iterations	LR	MR
FRM	$2.504*10^{-2}$	N/A	N/A	N/A
NFRM-1	$2.633*10^{-2}$	N/A	N/A	N/A
NFRM-2	$8.556*10^{-3}$	N/A	N/A	N/A
HNFRM	$1.098*10^{-3}$	100	0.1	0.01

(LR = learning rate ß; MR = momentum rate α; N/A = not applicable)

REFERENCES

[1] Zadeh, L.A., "Fuzzy sets." *Information and Control*, Vol. 8, pp. 338-353, 1965.

[2] Zadeh, L.A., "Outline of a new approach to the analysis of complex systems and decision processes." *IEEE Trans. Systems Man Cybernet*, Vol. 3, pp.28-44, 1973.

[3] Mamdani, E.H., "Application of fuzzy algorithms for the control of a dynamic plant." *IEEE Proc.*, Vol. 121, pp. 1585-1588, 1974.

[4] Sugeno, M., *Industrial Applications of Fuzzy Control*. North-Holland, 1985.

[5] Cao, Z., Kandel, A., and Li, L., "A new model of fuzzy reasoning." *Fuzzy Sets and Systems*, Vol. 36, pp. 311-325, 1990.

[6] Kandel, A., Li, L., and Cao, Z., "Fuzzy Inference and its Applicability to Control Systems." To appear in *Fuzzy Sets and Systems*, 1992.

[7] Yu, C., Cao, Z., and Kandel, A., "Application of fuzzy reasoning to the control of an activated sludge plant." *Fuzzy Sets and Systems*, Vol. 38, pp.1-14, 1990.

[8] Kiszka, J.B., Kochanska, M.E., and Sliwinska, D.S., "The influence of some fuzzy implication operators on the accuracy of a fuzzy model - Part I." *Fuzzy Sets and Systems*, Vol. 15, pp. 111-128, 1985.

[9] Kiszka, J.B., Kochanska, M.E., and Sliwinska, D.S., "The influence of some fuzzy implication operators on the accuracy of a fuzzy model - Part II." *Fuzzy Sets and Systems*, Vol. 15, pp. 223-240, 1985.

[10] Efstathiou, J., *Expert Systems in Process Control*. The Institute of Measurement and Control, United Kingdom, 1989.

[11] Del Toro, V., *Electric Machines and Power Systems*. Prentice Hall, 1985.

[12] Wasserman, P.D., *Neural Computing: Theory and Practice*. Van Nostrand Reinhold, 1989.

Chapter 18
Learning Fuzzy Control Rules from Examples

Introduction, *376*
The SC-Net Approach, *377*
Learning Fuzzy Motor Control, *381*
Summary, *390*
References, *394*

A fuzzy controller is normally generated by developing an appropriate set of rules in conjunction with some expert in the area of the control problem. The rules and, equally important, the fuzzy membership functions may need to be revised several times during the process. The process is one of trial-and-error mixed with insight and expertise. This chapter concentrates on learning fuzzy control rules from examples. A hybrid connectionist, symbolic learning system which uses a mixture of instance based and inductive learning is described and used to learn fuzzy control rules and fuzzy membership functions used in the rules. An example of the control of a DC motor is given and compared to other fuzzy controllers for the same problem to show that good performance is possible through learning.

LEARNING FUZZY CONTROL RULES FROM EXAMPLES

Steve G. Romaniuk and Lawrence O. Hall
Department of Computer Science and Engineering
University of South Florida
Tampa, FL 33620

1 INTRODUCTION

Fuzzy controllers are normally built with the use of fuzzy rules [9, 16]. These fuzzy rules are obtained either from domain experts or by observing the people who are currently doing the control. The membership functions for the fuzzy sets will be derived from the information available from the domain experts and/or observed control actions. The building of such rules and membership functions requires tuning. That is, performance of the controller must be measured and the membership functions and rules adjusted based upon the performance. This process will naturally be somewhat ad hoc and time intensive. This has spawned efforts to start with an initial set of fuzzy rules and use reinforcement learning to refine the rules and/or the fuzzy set used by the rules over time [1, 2, 3]. Some impressive results have been achieved.

This chapter describes the use of a hybrid connectionist, symbolic machine learning system, SC-net [4, 14], to generate the fuzzy rules and membership functions for control. A training set of examples is given to the system in the learning phase. After learning, rules can be extracted for use in fuzzy control. Membership functions for the defined fuzzy sets are also learned during the training process with the dynamic plateau modification feature of SC-net [13]. The system is given no initial set of fuzzy control rules. A different approach to learning strictly from data is given in [10]. They determine the rule antecedent membership functions by learning the center value and width of triangular shaped functions with the use of backpropagation. A real-valued consequent is learned similarly. In [5], rules are extracted from a neural expert system which uses modified backpropagation training. Other work along these lines has been reported in [6], but is not done in the control domain.

The rest of this chapter will consist of a description of the relevant features of the SC-net learning system, a description of an example control domain (a DC motor), and results of learning to control the motor. An analysis of the results and the possibilities for success in other domains will also be presented.

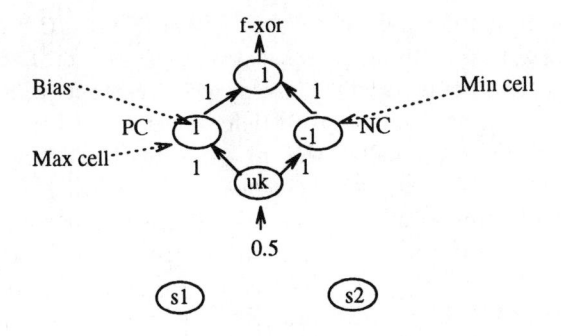

Figure 1: SC-net Structure

2 THE SC-NET APPROACH

SC-net configures its connectionist architecture based upon the training examples presented to it. The learning algorithm responsible for the creation of the network topology is the Recruitment of Cells algorithm (RCA) [4, 13]. RCA is an incremental, instance-based algorithm that requires only a single pass through the training set. Every training instance is presented to the network for a single feedforward pass. After the pass has been completed, the actual and the expected activation for every output are compared. Three possible conditions may result from this comparison.

- The example was correctly identified (error is below some epsilon). No modifications are made to the network.
- The example is similar to at least one previously seen and stored instance (error within 5 epsilon). For those output cells that have an activation within 5 epsilon of the expected output, a bias is adjusted to incorporate the new instance.
- The example could not be identified by the network. This results in the recruitment of a new cell (referred to as an information collector cell, ICC). Network inputs which are positive (> 0.5) are directly connected to the IC cell with a weight of $\frac{1}{output\ value}$, where **output value** is the expected value for the output to which the IC cell is connected. If the input value is < 0.5, it is sent through a negation cell converting it to the 0.5-1 range for the IC cell, which is a min cell. The connection to the negation cell has a weight of 1 and from the negation cell to ICC has a link weight of $\frac{1}{1-output\ value}$. The ICC cell itself is connected to either the positive (PC) or negative collector (NC) cell. The PC is used to collect positive evidence, whereas the NC accumulates negative evidence. The initial empty network structure for a two input (one output) fuzzy exclusive-or is presented in Figure 1. Note that the uk cell always takes an activation of 0.5.

To improve on the generalization capabilities of the RCA generated SC-net network two forms of post training generalization can be used. The generalization is most important in crisp domains which have large percentages of nominal (symbolic as opposed to numeric) attributes. We briefly cover them here for completeness. One generalization method is called the min-drop feature. Whenever a test pattern is presented to the system, which cannot be identified by any of the output cells, the min-drop feature is applied. Given a pattern that cannot be recognized by the network, all output cells will be in an inactive state (an unknown response of 0.5 is returned). In this case the min-drop feature is applied to find the nearest corresponding output for the current pattern. It is based on the fact that new patterns are stored in the network through recruitment of IC cells (and possibly some negation cells). These IC cells are essentially min-cells, which return the minimum of the product formed from the incoming activation, the bias and the weight on the corresponding connection. The min-drop feature works by dropping (ignoring) the next piece of evidence which is below some threshold. The process of ignoring evidence corresponds to dropping connections to IC cells in the network. The min-drop process is repeated until one or more output cells enter an active state (fire). The final number of connections dropped indicates the degree of generalization required to match the newly presented pattern. In a second mode, a bound may be placed on the min-drop value, preventing an unwarranted over-generalization. RCA and post training generalization in the form of the min-drop feature provide good generalization. However, several problems can be associated with the RCA learning phase.

- Network growth can be linear in the number of training examples.
- As a direct consequence of the first problem storage and time (to perform a single feedforward pass) requirements may increase beyond the networks physical limitations.
- Generalization on yet unseen patterns is limited, and requires use of min-drop feature.

To address the above problems a network pruning algorithm was developed. The GAC (Global attribute Covering) algorithm's [13] main purpose is to determine a minimal set of cells and links, which is equivalent to the network generated by RCA. That is, all previously learned information should be retained in the pruned network. In the particular example studied in this chapter, the GAC algorithm and min-drop prove to be unnecessary. The fuzzy sets of the patterns in the conditions of the rule provide the flexibility for smooth control. Since, GAC has no effect in this domain it will not be further discussed.

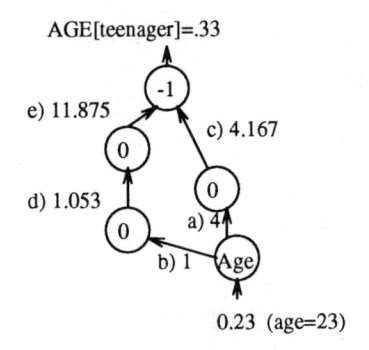

Figure 2: The fuzzy Variable Teenager.

2.1 Dynamic Plateau Modification of Fuzzy Membership Functions

All fuzzy membership functions in SC-net are represented as trapezoidal fuzzy sets [13, 15]. They are represented in the network by a group of cells as shown in Figure 2 for the fuzzy variable teenager. Teenager takes membership values of 1 in [13..19], of course. In this implementation the membership function goes linearly to 0 at the ages of 5 and 25. In the network ages are translated into [0,1] from the [0,100] year range. So the age of 22 is translated to 0.22. Figure 3 shows the actual graph of the membership function for the fuzzy teenager variable.

The dynamic plateau modification function (DPM) is designed to bring in or extend the arms of the fuzzy membership function. In general, we allow the range of the membership function for unknown functions to initially be the range of the fuzzy variable. The range in which the function obtains a value of 1 is at least one point (all fuzzy sets in SC-net are normal in the sense that they contain at least one full member) and usually much smaller than the function range. Hence, for the teenage example with a 100 year range the right arm of the trapezoidal membership function would initially go to 0 at age 100, if we had no information on constructing the membership function other than where it is crisp (attains a membership value of 1). We always assume that the crisp (normal) portion of the membership function is known. The DPM function allows us to arbitrarily set the arms too wide and then adjust them during the learning process. Clearly, in our example it is impractical for someone 99 or 100 years old to have membership in the fuzzy set teenager.

A high-level description of the DPM method is as follows. When it is determined that the fuzzy membership value has caused an incorrect output, the maximal membership that will not cause an error is determined. This value for the set element given and the nearest element at which the membership function takes a value of 1 are used to specify the linear arm of the function. This provides a new upper or lower plateau value (point at which the function goes to 0) for the fuzzy membership function which is used to update the weights labeled **a** thru **e** in Figure 2 [15].

Fuzzy Membership function for teenager.

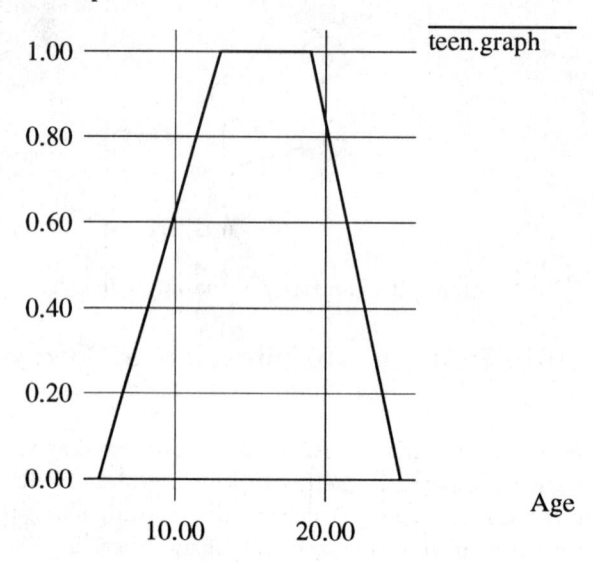

Figure 3: Graph of Membership Function for Fuzzy Variable Teenager.

Table 1: Sample data points used for DC motor control

Current	1.0	2.0	3.0	8.0	10.0
Speed	2000.0	1900	1650	700.0	500.0

3 LEARNING FUZZY MOTOR CONTROL

3.1 Description of Problem

The task to be studied is the control of a DC series motor [7, 11, 12] that has the following ratings:

- Power Rating: $P_N = 0.7kW$

- Voltage Rating: $U_N = 110V$

- Rated Current: $I_N = 8.84A$

- Rated Speed: $n_N = 1500 r.p.m$

The actual measurements of the current I versus the rotating speed value n during steady states were obtained from [11, 12]. Performance accuracy is measured in terms of root-mean-square (RMS) error of the motors real static characteristic and the predicted fuzzy characteristic generated by the trained SC-net network. The RMS error is calculated by

$$\Delta^2 e = \frac{\sum_{l=1}^{v}(n_{rl} - n_{ml})^2}{\sum_{l=1}^{v} n_{rl}^2} \tag{1}$$

where n_{rl} is the actual value for rpm at point l, n_{ml} is the answer from the fuzzy rules at point l, and v is the number of discrete points at which measurements are made. In this case v=100 as measurements are made at every 0.1 amp from 0 to 10 amps.

The intent of the learned control rules is to provide a smooth change in rotating speed as the current changes.

Table 1 contains a representative subset of sample points used for training. Note, that the current ranges between 1 and 10 amperes.

3.2 Description of Experiments

During training only samples at each amp between 1 and 10 inclusive were used, yielding a total of 10 training patterns. Increments of 0.1 amps across the [1, 10] amps interval were used for testing. Unlike other studies conducted using this data

Table 2: Initial definition of membership functions

Name	Lower-Bound	Upper-Bound	Lower-Plateau	Upper-Plateau
Zero	0.0	0.0	-20.0	30.0
One	1.0	1.0	-20.0	30.0
Two	2.0	2.0	-20.0	30.0
Three	3.0	3.0	-20.0	30.0
Four	4.0	4.0	-20.0	30.0
Five	5.0	5.0	-20.0	30.0
Six	6.0	6.0	-20.0	30.0
Seven	7.0	7.0	-20.0	30.0
Eight	8.0	8.0	-20.0	30.0
Nine	9.0	9.0	-20.0	30.0
Ten	10.0	10.0	-20.0	30.0

[7, 8, 11, 12], the selection of the 0.1 amps increment was required in order to determine SC-net's generalization ability. In this domain, SC-net will store all 10 examples exactly for some error bounds. It is important to point out that no a priori knowledge in the form of rules was used during the training process. The initial fuzzy membership partitions for the input current were chosen at every full amp. The initial fuzzy membership definitions are displayed in Table 2. The reason they are chosen with such wide bounds has to do with the meaning of the output in SC-net. A value of 0.5 on an output is used to indicate **unknown**, a value of 0 indicates the output is off and a value of 1 indicates the output is on. As the output value increases from 0.5 to 1 there is increasing belief in it and likewise there is increasing belief in its negation as the output goes from 0.5 to 0.

In this problem, the output is speed which ranges from 0 to 2000 r.p.m. This presented a data encoding problem since SC-net has been mostly used as a classification learning system. However, the problem is solved by mapping [0, 2000] to [0.5, 1]. The necessary precision will still exist, as floating point numbers are handled using exponentials. Hence, both intervals will have the same amount of precision, albeit in different forms.

We now turn attention to the actual experimental results. The epsilon value chosen for use during learning will play an important role. If an example is within 5ϵ of the expected output, only a bias value is changed during the cell recruitment process. Otherwise, the example is stored via a new set of cells in the network.

The other important effect of epsilon is on the dynamic plateau modification process. Our fuzzy membership functions appear clearly **over general** at the start. After an example is correctly learned some cell(s) will have been added to the network and connected to the positive collector cell shown in Figure 1. The positive collector is a maximum cell. It is possible that for a given example the output value will be too large and since we have only positive evidence the only method to reduce it is to bring in the "offending" arm of the appropriate fuzzy membership function or functions if more than one contribute. This can be directly traced back from the positive collector cell, since it is the cell with maximum input whose value will propagate further.

The DPM function is only applied **after** all the training examples have been seen. At this time assignments to inputs, taken from the train set, necessary to activate paths to outputs in the network are determined by tracing the network backwards from the output. These input assignments are applied to the network. In the case that the fuzzy membership functions cause paths to be active which were not intended to be active the appropriate arm of the membership function is adjusted as discussed earlier.

The fuzzy rules that are learned are shown in Figure 4. In this domain, with 10 training examples, 10 rules are always learned. The definitions of the fuzzy membership functions will vary with the choice of epsilon and this is discussed and shown in the proceeding. Essentially, lower ϵ values lead to membership functions which allow for better generalizability. A smaller ϵ causes more tuning of the fuzzy membership functions, because it is used in deciding the tolerance within which an output value is acceptable. In Figure 5, we show the actual learned neural network. Each box indicates a set of cells with appropriate link weights to represent the corresponding fuzzy variable. All the cells on the layer labeled ICC for information collector cells are minimum cells.

In Figure 6, SC-net's response curve versus the expected response curve for the speed of the DC motor as the current changes is displayed. The first curve was obtained using a value of $\epsilon = 0.01$, whereas the second is based on a value of $\epsilon = 0.008$. For these initial values of ϵ SC-net's performance is poor in that it learned a simple linear approximation for the expected response curve.

For the values of $\epsilon = 0.006$ and $\epsilon = 0.005$ in Figure 7 the initial linear approximation attempt is replaced by two linear functions, which give a good first fit of the training data and improve the predicted curve with respect to the expected response curve.

This behavior continues throughout Figures 8 thru 10. As ϵ reaches a value of 0.001 and smaller the predicted curve is very similar to the expected response curve. This result clearly indicates that SC-net possesses the necessary smoothness during generalization that is essential for modeling control problems such as a DC motor.

In Table 3 the corresponding RMS ($\Delta^2 e$) error is shown for the various values of the learning parameter ϵ. In correspondence to the overall increase of generalization on yet unseen data points, the $\Delta^2 e$ error also declines (general tendency) as ϵ is decreased. Finally, as $\epsilon = 0.00001$ the actual and expected motor responses are identical, causing the error to be zero.

```
Rule 1  if ( fuzzy(Current[Ten])   = 1.000 ) then Speed ( 0.625 ).

Rule 2  if ( fuzzy(Current[Nine])  = 1.000 ) then Speed ( 0.650 ).

Rule 3  if ( fuzzy(Current[Eight]) = 1.000 ) then Speed ( 0.675 ).

Rule 4  if ( fuzzy(Current[Seven]) = 1.000 ) then Speed ( 0.700 ).

Rule 5  if ( fuzzy(Current[Six])   = 1.000 ) then Speed ( 0.750 ).

Rule 6  if ( fuzzy(Current[Five])  = 1.000 ) then Speed ( 0.800 ).

Rule 7  if ( fuzzy(Current[Four])  = 1.000 ) then Speed ( 0.850 ).

Rule 8  if ( fuzzy(Current[Three]) = 1.000 ) then Speed ( 0.913 ).

Rule 9  if ( fuzzy(Current[Two])   = 1.000 ) then Speed ( 0.975 ).

Rule 10 if ( fuzzy(Current[One])   = 1.000 ) then Speed ( 1.000 ).
```

Figure 4: Fuzzy rules

Table 3: Learning parameter ϵ versus $\Delta^2 e$ error.

ϵ	$\Delta^2 e$
0.01	$2.68 * 10^{-2}$
0.008	$2.57 * 10^{-2}$
0.006	$2.60 * 10^{-3}$
0.005	$2.42 * 10^{-3}$
0.004	$1.20 * 10^{-3}$
0.003	$6.90 * 10^{-4}$
0.002	$5.90 * 10^{-4}$
0.001	$1.50 * 10^{-5}$
0.0001	$1.30 * 10^{-7}$
0.00001	0.00

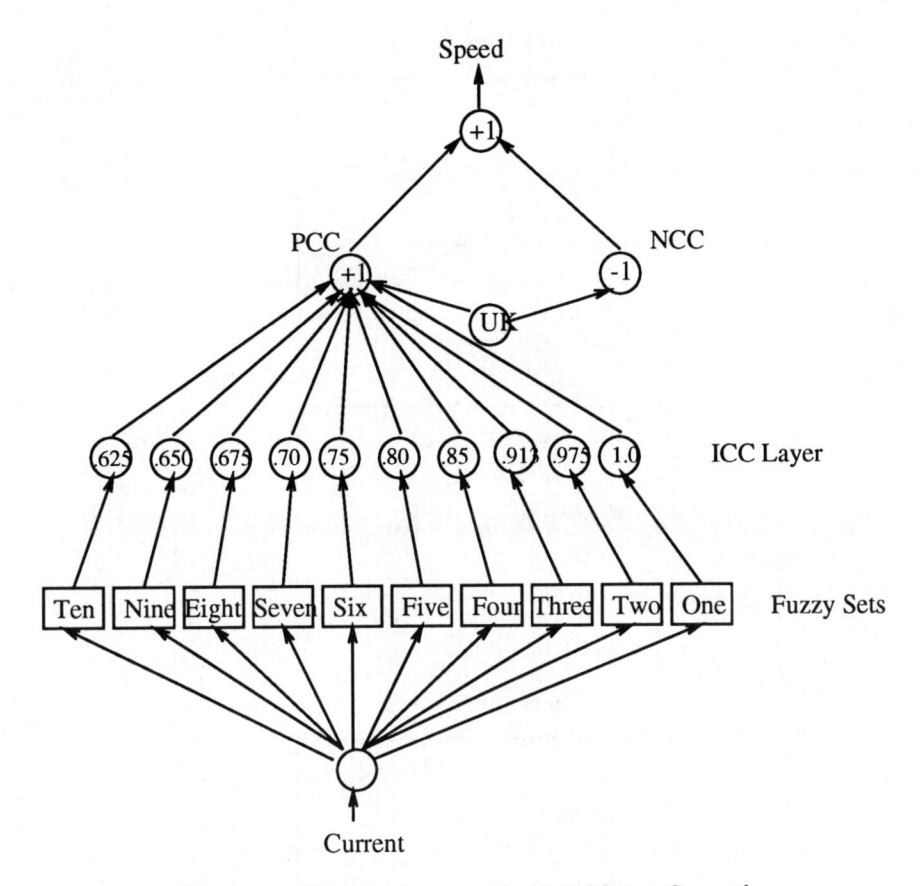

Figure 5: SC-net Structure for DC Motor Control

Current vs. Speed

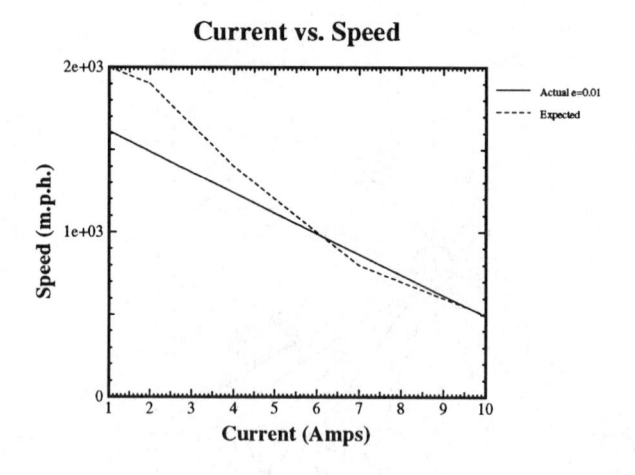

a) $\epsilon = 0.01$

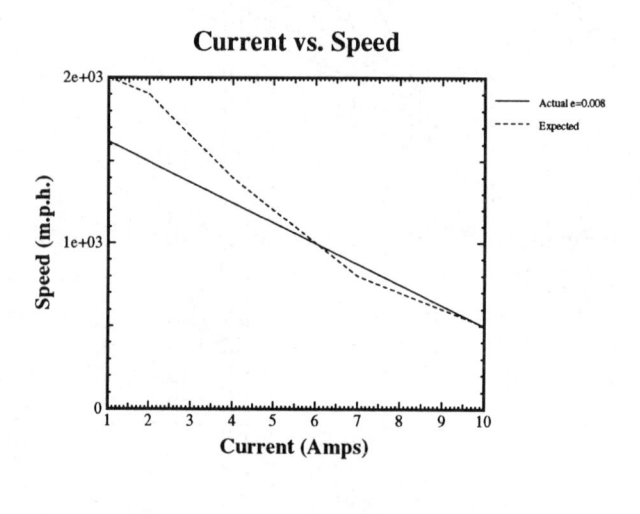

Current vs. Speed

b) $\epsilon = 0.008$

Figure 6: Actual vs. Expected Motor Response

Current vs. Speed

(a) $\epsilon = 0.006$

Current vs. Speed

(b) $\epsilon = 0.005$

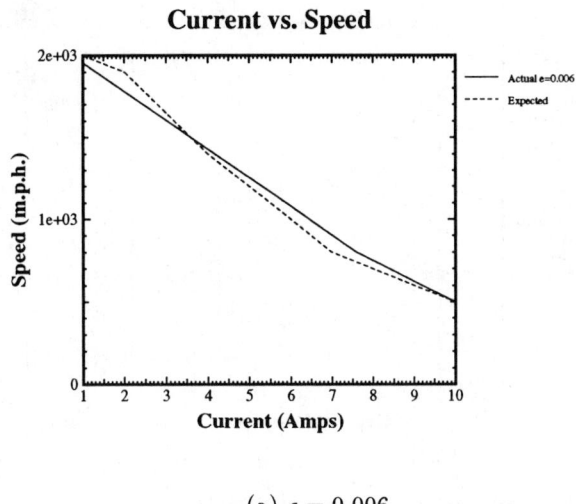

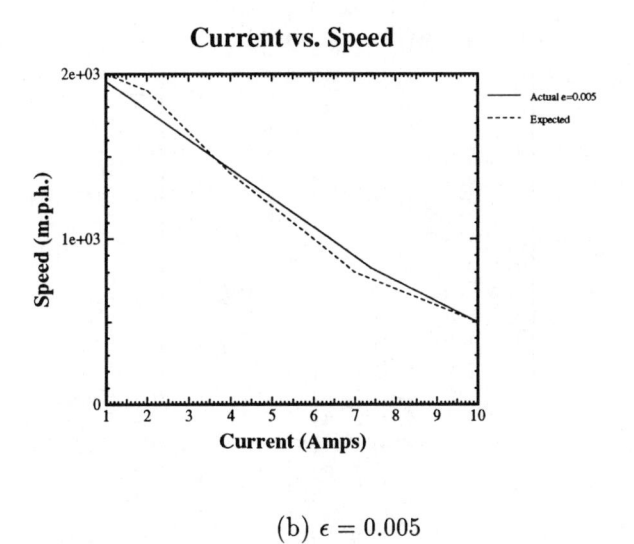

Figure 7: Actual vs. Expected Motor Response

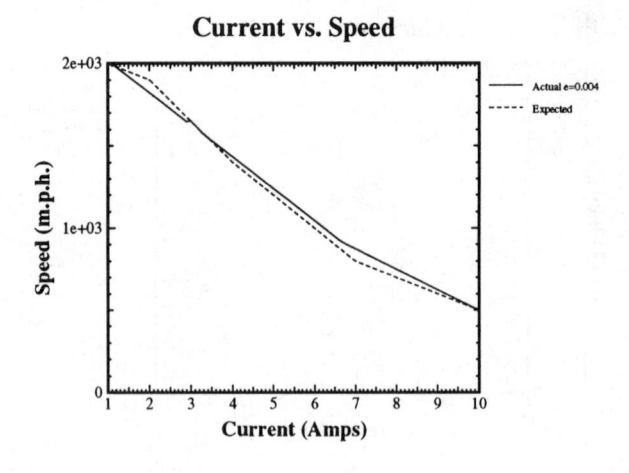

Current vs. Speed

(a) $\epsilon = 0.004$

Current vs. Speed

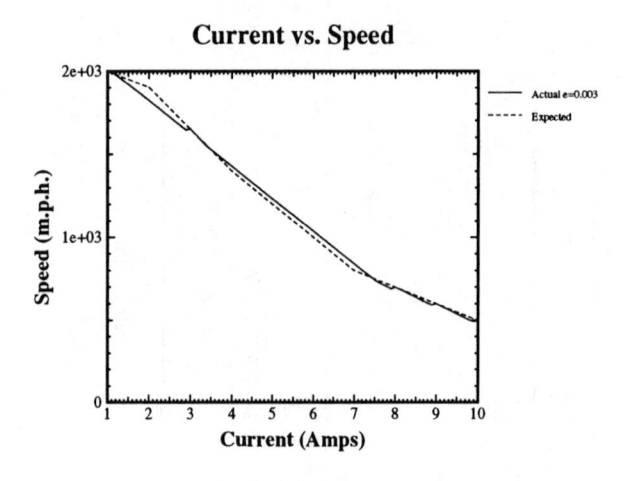

(b) $\epsilon = 0.003$

Figure 8: Actual vs. Expected Motor Response

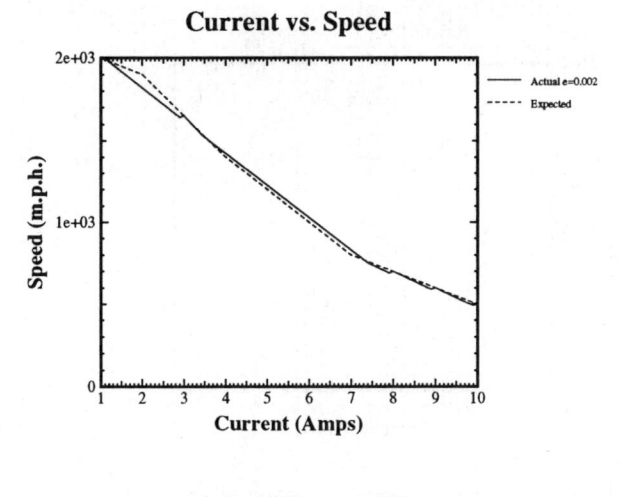

Current vs. Speed

(a) $\epsilon = 0.002$

Current vs. Speed

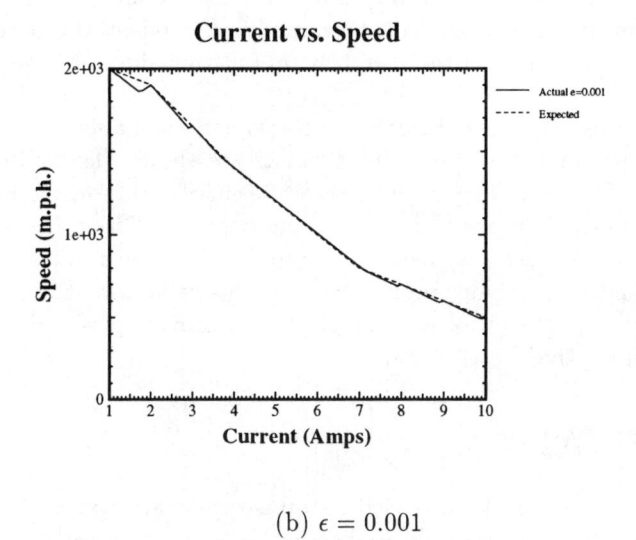

(b) $\epsilon = 0.001$

Figure 9: Actual vs. Expected Motor Response

Current vs. Speed

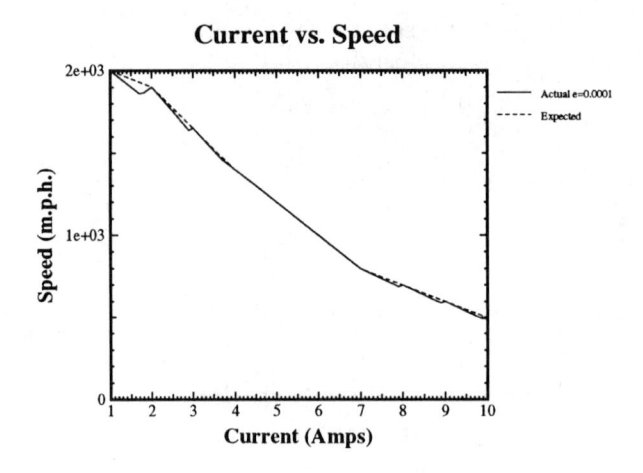

Figure 10: Actual vs. Expected Motor Response

Figure 11 shows the initial scaled membership functions for the fuzzy labels Zero through Ten. In Figure 12, the DPM adapted membership functions are presented with a mapping on the ampere scale of $-20 \rightarrow 0$ and $30 \rightarrow 1$. As expected several of the membership arms have been modified to reflect the actual train data. Essentially, the right arms of the membership functions for 1-6 have been brought in.

To further test the generalization capability of SC-net a second experiment was conducted. This time only five training points are chosen (first all odd amps are trained on then all even). The resulting response curves are shown in Figure 13. The response is quite good even with half the training points. The only major difference is that the learned control is poor at 1 amp for the even train set. This is not unreasonable since the only example given near this value was at 2 amps. The good results indicate that the fuzzy sets and DPM can provide powerful generalizations in SC-net with relatively little data.

4 SUMMARY

We have shown that an instance-based neural network can learn effective fuzzy control rules. The particular example studied is the smooth control of a direct current (DC) motor. With just 10 examples the system learns to provide smooth control over a current range of 1-10 amps. In this case the most important feature of the

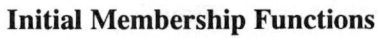

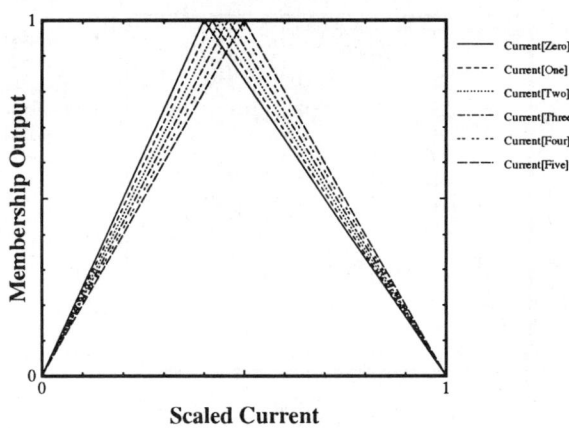

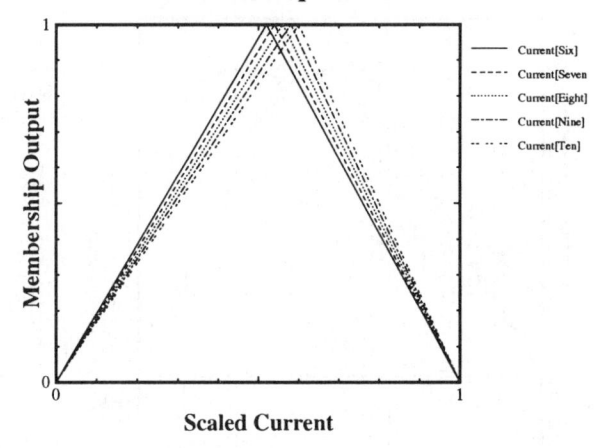

Figure 11: Initial membership functions (before DPM)

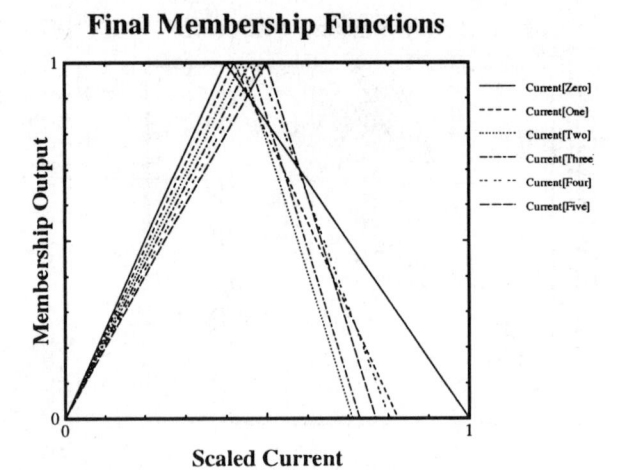

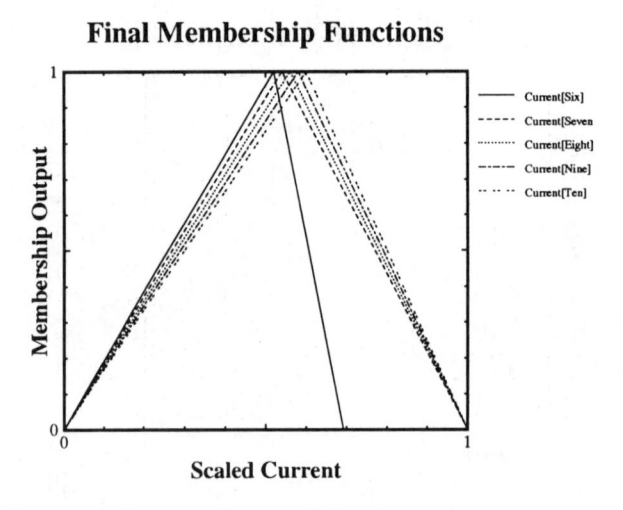

Figure 12: Final membership functions (after DPM)

Current vs. Speed

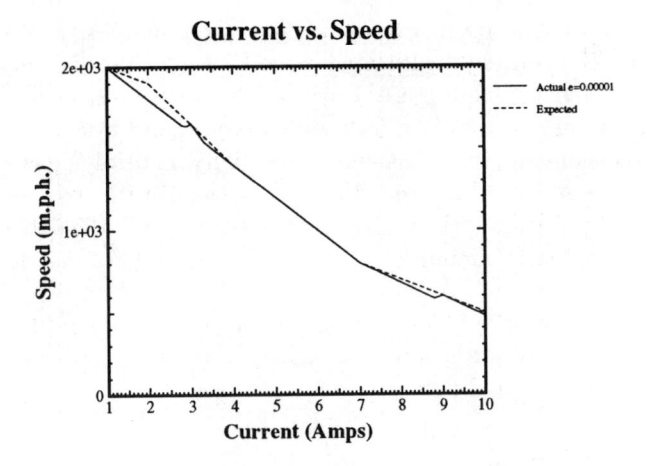

(a) Odd amps training set

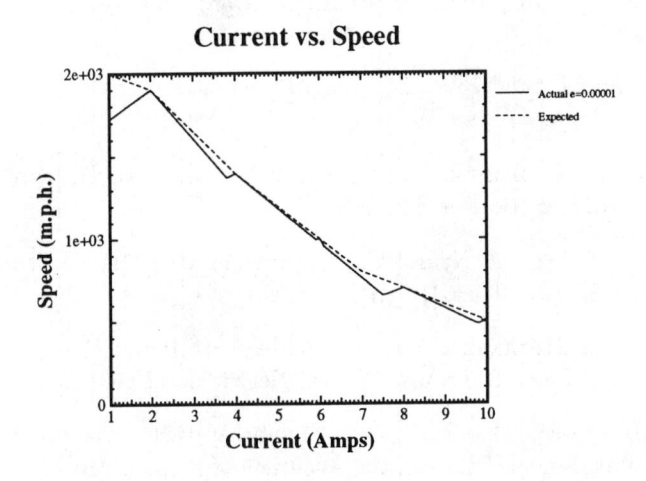

Current vs. Speed

(b) Even amps training set

Figure 13: Response Curves from limited training sets

system is its ability to incorporate fuzzy membership functions. These membership functions are modified during the learning process by dynamic plateau modification. They provide the generalizability of the system.

The rules generated could be used in a fuzzy expert system that calculated its actions in the same fashion as SC-net. Essentially, multiplying the membership function for the premise by the encoded speed associated with the rule to determine the actual speed by decoding the multiplied result back into the 0-2000 rpm range. Only the rule with the maximum membership in the premise match is allowed to fire.

The results are promising for learning the fuzzy sets used in fuzzy control and the fuzzy rules. Future work is needed on allowing multiple fuzzy rules to fire and learning in more complicated domains.

REFERENCES

[1] Berenji, H.R. "Strategy Learning in Fuzzy Logic Control", *NAFIPS '91*, pp. 301-306, Columbia, Missouri, 1991.

[2] Burkhardt, D.G. and Bonissone, P.P. "Automated Fuzzy Knowledge Base Generation and Tuning", *IEEE International Conference on Fuzzy Systems*, pp. 179-188, 1992.

[3] Chen, Y., Lin, K, and Hsu, S. "A Self-Learning Fuzzy Controller", *IEEE International Conference on Fuzzy Systems*, pp. 189-196, 1992.

[4] Hall, L.O. and Romaniuk, S.G., " A hybrid connectionist, symbolic learning system", *AAAI-90*, Boston, Ma. 1990.

[5] Hayashi, Y. (1992), A Neural Expert System Using Fuzzy Teaching Input, FUZZ-IEEE'92, pp. 485-491, San Diego, CA.

[6] Keller, J.M. and Tahani, H. (1991), Implementation of Conjunctive and Disjunctive Fuzzy Logic Rules with Neural Networks, IJAR.

[7] Kiska, J.B., Kochanska, M.E., and Sliwinska, D.S. "The Influence of some Fuzzy Implication operators on the Accuracy of a Fuzzy Model - Part I", *Fuzzy Sets and Systems*, V. 15, N. 2, pp. 111-128, 1985.

[8] Kiska, J.B., Kochanska, M.E., and Sliwinska, D.S. "The Influence of some Fuzzy Implication operators on the Accuracy of a Fuzzy Model - Part I", *Fuzzy Sets and Systems*, V. 15, N. 2, pp. 223-240, 1985.

[9] Lee, C.C. "Fuzzy Logic in Control Systems: Fuzzy Logic Controller – Part I and Part II", *IEEE Transactions on Systems, Man and Cybernetics*, V. 20, No. 2, pp. 404-435, 1990.

[10] Nomura, H., Hayashi, I., and Wakami, N., "A Learning Method of Fuzzy Inference Rules by Descent Method", *IEEE International Conference on Fuzzy Systems*, pp. 203-210, 1992.

[11] Park D. "Intelligent Fuzzy Reasoning Models with Application to Fuzzy Control", Ph.D. Dissertation, Department of Computer Science, Florida State University, Tallahassee, Fl. 1992.

[12] Park, D., Kandel, A., and Langholz, G. "Hybrid Neural-Fuzzy Reasoning Model with Application to Fuzzy Control", In *Fuzzy Control Systems* (A. Kandel and Langholz, G. Eds.), CRC Press

[13] Romaniuk, S.G. " Extracting Knowledge from a Hybrid Symbolic, Connectionist Network", Ph.D. Dissertation, Department of Computer Science and Engineering, University of South Florida, 1991.

[14] Romaniuk, S.G. and Hall, L.O. "SC-net: A Hybrid Connectionist, Symbolic Network", *Information Sciences*, To Appear, 1993.

[15] Romaniuk, S.G. and Hall, L.O. "Learning Fuzzy Information in a Hybrid Connectionist, Symbolic Model", *IEEE International Conference on Fuzzy Systems 1992*, pp. 309-312., San Diego, Ca. 1992.

[16] Sugeno, M., Industrial Applications of Fuzzy Control, Amsterdam, North-Holland, 1985.

Chapter 19
A Computational Approach to Fuzzy Logic Controller Design and Analysis Using Cell State Space Methods

Introduction, *398*
Control Problem, *400*
Cell State Space Optimal Control, *402*
Fuzzy Logic Controller Tuning, *408*
Angular Position Control of an Inverted Pendulum, *411*
Extensions to Three and Four Variables, *418*
Summary and Conclusions, *425*
References, *425*

This chapter presents a systematic approach to fuzzy logic controller design based on computational methods. This approach provides algorithms for the automated design, calibration, and analysis of fuzzy logic control systems. Although fuzzy logic controllers (FLC) have shown promise in controlling many complex, nonlinear, or ill-defined systems, FLCs can be difficult to formulate, analyze, and fine-tune. In many cases, a mathematical model of the plant is available but it is strongly nonlinear, making it difficult to analytically derive a closed form solution for the optimal control. Using *cell state space* based algorithms, a discrete approximation to the desired optimal control can be generated. The process state space is quantized into discrete units called *cells*. The continuous dynamical behavior of the system is approximated with discrete-valued functions called *cell-to-cell mappings*. Given a controller performance function, a numerical optimal control algorithm is used to find the desired control action for each cell in the system's state space. The set of control actions constitutes the control table, and using it as a set of input-output data, a gradient descent algorithm is used to fine-tune the output of the FLC.

A COMPUTATIONAL APPROACH TO FUZZY LOGIC CONTROLLER DESIGN AND ANALYSIS USING CELL STATE SPACE METHODS

Samuel M. Smith
Florida Atlantic University
Boca Raton FL 33431

Brent R. Nokleby
Brigham Young University
Provo UT 84602

David J. Comer
Brigham Young University
Provo UT 84602

1. INTRODUCTION

In the process of designing controllers for industrial processes or guided vehicles, nonlinear functions are often encountered. For example, it is not unusual for the equations that govern some controlled process or plant to be nonlinear [1]. Even with a plant that can be considered linear, if the response of the system is to be optimal in some way, practical limitations on the actuator may lead to a required drive signal that is highly nonlinear. Intentional nonlinearity may then be introduced to achieve the desired performance [1].

Nonlinear control design is generally difficult to apply to practical systems. Over the years this fact has led to the application of linear methods to problems best suited to nonlinear design. The resulting response may be far less desirable than the optimal response, but the design methods used are far less complex than nonlinear approaches. Because of the complexities of nonlinear analysis, procedures such as describing function methods are not amenable to systems of third or higher order.

Two popular methods used in the design of nonlinear plant controllers have been piecewise linearization of characteristics or the application of *p-i-d* (proportional-integral-derivative) techniques if the system is not highly nonlinear. If the plant cannot be accurately linearized or if certain aspects of performance are critical, nonlinear controllers must be used. The *Bang-Bang* controller was introduced over thirty years ago for time-optimal performance. In this controller, the actuator drives the plant with maximum signal to accelerate, then drives with minimum signal to decelerate while approaching the target position. The actuator drive signal may become a linear combination of functions only as the target position is neared. This type of controller may be difficult to design and may result in deficiencies that detract from overall performance.

Fuzzy logic has successfully been applied to the design of many complex control problems [2,3]. As this field develops, more theoretical methods are being used to replace trial and error approaches to the design procedure. While one appealing feature of fuzzy logic systems is the control of unknown plants, another area of potential application is that of complex, nonlinear plant control. This chapter reports an improved method of designing fuzzy logic set-point controllers [4,5,6], such as that shown in Fig. 1, for nonlin-

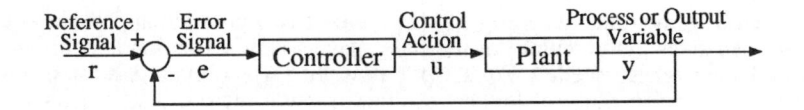

```
   Reference     Error                Control        Process or Output
   Signal   +   Signal                Action         Variable
     r      ─○──── e ──→ [Controller] ──── u ──→ [Plant] ──── y ──→
             - │                                                   │
               └───────────────────────────────────────────────────┘
```

Fig. 1: Block Diagram of Typical Set-point Control System.

ear, but known plant models. These models are typically expressed in terms of a set of nonlinear equations or state equations.

Hsu [7,8,9] introduced the concept of a *cell state space* that can be used in the optimal design of nonlinear controllers. This approach discretizes the state space of the system. The result of applying Hsu's method is an *optimal control table* (OCT) that contains the most appropriate actuator output or plant input signal for each discrete cell in the state space. If this drive signal is plotted in the state space, it generates the discrete control surface for optimal control.

The use of state space or phase space in the description and design of nonlinear control systems has been used since the 1960s [10]. While these early studies contributed to the understanding of nonlinear systems, few practical design methods resulted, partly because of the limited computing power then available. In 1980, Hsu [7] proposed a method of discretizing the state space into cells which could be used to map the response of the system for various input values. He further developed a scheme for selecting the sequence of drive signals of a system that minimizes a cost function as the system moves to a desired set point [9]. The result of Hsu's method is an OCT that contains an input drive value for each cell in the state space. As the system trajectory moves from one cell to another, the appropriate drive signal that contributes to the minimization of the overall cost is obtained from the OCT and applied to the system. This scheme is referred to as direct digital control.

A problem addressed by this chapter is that of extending Hsu's method to a system with three or four state variables. With such a system, depending on cell size, it is easy to exceed the memory storage capability of the computer. For direct digital control of a system with four state variables, the number of bytes required for storage of the OCT can be astronomically high. The computation time can also become unacceptably large for these systems. As an example, the generation of an OCT to result in minimal settling time was considered for a two-, three-, and four-dimensional system. The OCT was generated with the use of an RS6000 workstation for the two-dimensional system with estimates for the higher-dimensional systems based on this computer (see Table 1).

If large cell sizes are used, insufficient resolution results, which leads to increased cumulative error problems when applying the methods of [4,5,6]. The use of small cell sizes increases the accuracy but can lead to excessive memory requirements.

Several methods have been employed to solve this problem with varying degrees of

Table 1: Memory and Time Requirements.

No. of State Variables	No. of Cells Per Variable	Total No. of Cells	Memory Required (kbytes)	Approximate CPU Time (hours)
2	101	10,201	5,379	0.25
3	101	1,030,301	543,323	25.25*
4	101	104,060,401	54,875,602	2550.25*

*predicted value

success [11]. The methods generally fall into one of two possible approaches. The first uses larger cell sizes in the regions of the state space where resolution is unimportant and smaller cell sizes where needed. For set-point systems, the region near the origin of the state space (zero error) requires higher resolution, while regions distant from the origin may not. In a system designed to minimize response time, the regions far from the zero error point often dictate a maximum positive or minimum negative drive signal. Only as the error nears zero does the drive signal vary sharply with movement in the state space.

A second method applies larger cell sizes, but considers error magnitudes in moving from cell to cell. The OCT is found from a combination of cost minimization and error minimization. The procedures reported in this chapter extend Hsu's method and allow its application to systems with as many as four state variables.

The process then matches a fuzzy logic control surface, produced by a *fuzzy logic controller* (FLC), to the discrete control surface of the OCT. The advantages of the FLC are several. A major advantage is the minimization of storage required by the controller. The FLC reduces the required run-time storage by several orders of magnitude. Also, the chattering that is characteristic of a discrete drive signal is eliminated by the FLC. The transition from a drive signal specified by one cell to that of an adjacent cell can be very abrupt in the direct digital controller. The FLC smoothes these transitions and rids the system of chattering. This can be an important consideration in some applications.

2. CONTROL PROBLEM

The problem to be addressed is that of achieving the optimal set-point control of a process driven by a nonlinear controller. The plant is modeled by the nonlinear autonomous system

$$\dot{\mathbf{x}} = \mathbf{h}(\mathbf{x}, u), \tag{1}$$

where $\mathbf{x}$ is the state vector and u is the control signal applied to the plant. An autonomous system requires that the control signal u be solely a function of the process state, that is $u = g(\mathbf{x})$.

The computational methods used for both the design and implementation of the controller require a computer. Thus the system must eventually be expressed in a discrete time format. Assuming a sufficiently high sample rate with a sample period of T, a discrete time approximation of the plant is given by the difference equation

$$\mathbf{x}(k+1) = \mathbf{f}(\mathbf{x}(k), u(k)), \tag{2}$$

where k is the number of the time sample.

2.1 Set Point Control

In a simple set-point control system (see Fig. 1), the process variable x is measured and compared to a reference signal r, thereby creating an error signal e. Current and previous values of e are used by the controller, in accordance with its control policy, to develop the control signal, u, which is applied to the plant. More generally, the set-point control system may simultaneously control several process variables, with respect to a vector set point, $\mathbf{r}$, thereby generating a vector error signal $\mathbf{e}$. It may be possible to express the system dynamics in terms of the translated state vector, $\mathbf{e} = \mathbf{r} - \mathbf{x}$. Thus,

$$\dot{\mathbf{e}} = \frac{d}{dt}(\mathbf{r} - \dot{\mathbf{x}}) = -\mathbf{h}(\mathbf{r} - \mathbf{e}, u), \tag{3}$$

where the set point, **r**, becomes the origin of the translated error system. Driving the process variable, *x*, to rest at the set point, *r*, is equivalent to driving the state of the translated system to the origin. For autonomous systems, the output of the controller, *u*, is some general nonlinear function, *g*, of the vector error signal given by

$$u(k) = g(\mathbf{e}(k)). \tag{4}$$

The controller output, *u*, when plotted over the controller input space, **e**, generates what is called the *control surface*. The problem facing the designer is that of finding the control surface that optimizes some given performance index and then selecting a control strategy that will best implement the desired control surface.

2.2 Optimal Control

The optimal control problem involves choosing, from a set of allowable control actions, those control signals that drive the system from a given initial state to the desired set point while minimizing the given cost function or performance index, **L**.

Let $\mathbf{e}(k)$ be the state vector of the error system at any time, *kT*, with initial state, $\mathbf{e}(0)$, and the desired final state, $\mathbf{e}(\cdot) = \mathbf{0}$. The next state of the system after one sample period is given by

$$\mathbf{e}(k+1) = f(\mathbf{e}(k), u(k)). \tag{5}$$

The incremental cost function at time, *k*, is

$$\mathbf{L}_k = Q(\mathbf{e}(k), u(k), \delta(k)), \tag{6}$$

where $\mathbf{e}(k)$ is the state of the system, $u(k)$ is the applied control signal, and $\delta(k)$ is the time over which the control is applied. If it is assumed that the optimal control of the system requires *m* time steps to drive the system from the initial to the final state, then the total cost, **L**, is the sum of the costs at each step; that is,

$$\mathbf{L} = \sum_{k=0}^{m} \mathbf{L}_k. \tag{7}$$

The sequence of control actions that minimize the total cost, **L**, for a given initial state, is the optimal vector for that initial state:

$$\mathbf{u}_{opt} = [u_0, u_1, \ldots, u_m], \tag{8}$$

where u_k is the optimal control signal at time, *k*.

In order to generate the global optimal control of the system, an optimal control signal must be found for every allowable initial condition. For an autonomous system, every state along a trajectory can be considered a new initial condition; thus, the optimal control at any time is a function of only the current system state. The global optimal control therefore, consists of a single control action for every allowable system state.

Finding an analytical expression for the optimal control, with respect to the cost function, is generally impossible for most nonlinear systems and can be problematic for many linear ones. In many of these cases, however, numerical data processing methods may be able to directly generate a discrete approximation to the optimal control surface [9,12,13]. The next section presents one such method.

3. CELL STATE SPACE OPTIMAL CONTROL

Hsu [9] proposed an efficient cell state space based algorithm tailored for set-point control systems. In Hsu's method, the process state space is quantized into discrete units called cells, which form the *cell state space* of the system. The continuous dynamic behavior of the system is approximated with discrete valued functions termed *cell-to-cell mappings* [7,8,14]. Propagation of state trajectories is approximated with the propagation of cell trajectories. The center point of each cell is used as the point from which the next cell transition is computed, based on the underlying point-to-point system.

3.1 Cell-to-Cell Mapping

Consider the system governed by

$$\mathbf{x}(t_{k+1}) = \mathbf{h}(\mathbf{x}(t_k), u(t_k)) \text{ and } u(t_k) = g(\mathbf{x}(t_k)), \tag{9}$$

where $\mathbf{h}(\cdot)$ is a point mapping defined on $\mathbf{X} \times \mathbf{U} \to \mathbf{X}$, and $\mathbf{X}$ is the n-dimensional state space of the process with state vector $\mathbf{x} = [x_1, \cdots, x_n]^T$, $x_i \in X_i$, $i = 1, \ldots, n$, and $\mathbf{U}$ is the one-dimensional region of interest for the fuzzy controller output, u. The separation between two time indices, $\Delta t_k = t_k - t_{k-1}$, is the time interval over which a mapping of the system state occurs. The values of Δt_k are not necessarily uniform in duration but are integer multiples of the sample period T.

To obtain the cell mapping from the point mapping, the system's state space must first be partitioned into cells. In a practical system, only a finite region of the state space is of interest. Cells are formed by dividing the region of interest of each state space variable, x_i, into equal intervals of size, s_i. Each of these intervals is denoted by the integer valued index, z_i and contains all the points, x_i, satisfying

$$(z_i - 1/2)s_i \le x_i \le (z_i + 1/2)s_i. \tag{10}$$

A cell is the n-tuple of intervals, $\mathbf{z} = [z_1, \ldots, z_n]^T$, $\mathbf{z} \in \mathbf{Z}$. The remainder of state space extending to infinity is lumped together into a special cell called the *sink cell*. The evolution of the system in cell space is described by the *cell-to-cell mapping*, $\mathbf{C}(\mathbf{z}) : \mathbf{Z} \to \mathbf{Z}$, such that $\mathbf{z}(k+1) = \mathbf{C}(\mathbf{z}(k))$. The cell mapping is derived from the underlying point mapping by propagating the corresponding center point, $\mathbf{x}^c(t_k)$, of each cell, $\mathbf{z}(k)$, until a cell transition occurs; that is,

$$\text{IF } h(\mathbf{x}^c(t_k)) \in \mathbf{z}', \text{ THEN } \mathbf{z}(k+1) = \mathbf{C}(\mathbf{z}(k)) = \mathbf{z}'. \tag{11}$$

A cell-to-cell trajectory is generated by recursive application of the cell mapping, $\mathbf{C}$; thus, cells along a trajectory are given by

$$\mathbf{z}(0) \to \mathbf{C}(\mathbf{z}(0)) \to \mathbf{C}^2(\mathbf{z}(0)) \to \cdots, \tag{12}$$

where $\mathbf{z}(r) = \mathbf{C}^r(\mathbf{z}(0))$ denotes the rth cell along a trajectory.

An *equilibrium cell*, z^*, of the cell mapping, **C**, satisfies $z^* = C(z^*)$, and is analogous to a fixed point in a point mapping. A *periodic motion* of period K (P–K motion) of a mapping **C** is a set of K distinct cells, $z^*(0),\ldots,z(K-1)$, that satisfy

$$z^*(0) = C^K(z^*(0)) \text{ and } z^*(m) = C^m(z^*(0)), \; m = 1,\ldots,K-1. \tag{13}$$

A periodic motion is analogous to a limit cycle in a point mapping. An equilibrium cell is a P–1 motion. The r-step *domain of attraction* of a periodic motion is the set of all cells that map in r steps, or less, to one of the periodic cells in that periodic motion.

3.2 The Unraveling Algorithm

Hsu and Guttalu developed an efficient algorithm for finding all the periodic motions and their respective domains of attraction called the *unraveling algorithm* [8]. A major result of Hsu's extensive work is that cell mappings preserve the qualitative nature of the underlying point mapping systems to the resolution allowed by the cell sizes. Indeed, in the limit, as cell size shrinks, the cell mapping and point mapping converge [14]. Thus, the number and type of fixed points and limit cycles in a point mapping, for even strongly nonlinear systems, is reflected by the number and type of p-k motions in a cell map. In a set point control system, for example, a cell-to-cell map of the closed loop system directly indicates the global absolute stability of the system about the set point.

Hsu's original algorithm used a fixed time interval for cell transitions which could result in the creation of spurious equilibrium cells. The following summarizes the development, by the authors, of a modified unraveling algorithm that solved this problem with variable time intervals [4,5].

If N_i is the number of intervals along each axis, x_i, of the cell state space, then the total number of regular cells is given by

$$N_c = \prod_{i=1}^{n} N_i. \tag{14}$$

It is assumed that trajectories entering the sink cell cannot leave. Because the sink cell maps to itself, it is designated as an equilibrium cell. The total number of cells, including the sink cell, is $N_c + 1$. It is always possible to devise a scheme for sequentially ordering all the cells in **Z**. For the sake of computational efficiency, instead of using a vector format, a scalar format is possible where each cell **z** is denoted by its corresponding position in the sequential ordering. The cells are labeled with an integer valued index, c, from 0 to N_c, with the sink cell numbered N_c. This generates the scalar cell mapping, C, where

$$\begin{cases} c(k+1) = C(c(k)), & \text{for } c(k) < N_c \text{ and } c(k+1) \leq N_c \\ C(N_c) = N_c. \end{cases} \tag{15}$$

In the original unraveling algorithm, the time interval over which a cell mapping occurs is fixed, that is, $\Delta t_k = T$. This however requires that the system sample rate and cell size be matched closely, otherwise the system state may not move far enough during time period, T, to leave the cell even though it would have left eventually. This leads to the creation of spurious equilibrium cells.

Variable Time Interval The authors' solution is to use a variable time interval for cell transitions that is dynamically determined. The point-to-point trajectory, starting at the center point of a cell, is recursively propagated (each stage of the point-to-point trajectory is evaluated over the time interval, T) until one of the following is true:

a) The point-to-point trajectory enters another cell. In this case

$$\Delta t_k = mT, \qquad (16)$$

where m is the number of stages needed to enter another cell.

b) Δt_k becomes very large without entering another cell; that is,

$$\Delta t_k > MT, \quad M \gg 1, \qquad (17)$$

where M is a positive integer. This indicates the presence of a fixed point or small limit cycle within the cell. In this case, the originating cell is deemed to be an equilibrium cell and the time period for a cell transition is forced to $\Delta t_k = T$.

c) The distance traveled in each dimension is too small to exit the cell before Δt_k would become large; that is,

$$\left| x_i(t_{k-1} + T) - x_i(t_{k-1}) \right| < \frac{1}{2M} s_i, \quad i = i, \ldots, n. \qquad (18)$$

In this case the originating cell is also deemed an equilibrium cell and $\Delta t_k = T$. This saves computational effort.

The global properties of each cell, c, relative to the cell mapping, C, are characterized by the following five numbers: the image cell, $C(c)$, the group number, $G(c)$, the periodicity number, $P(c)$, the step number, $S(c)$, and the step increment number, $I(c)$. Each periodic motion is sequentially assigned a group number as each periodic motion is discovered. The periodicity number is the period of a given period motion. All the cells in the r-step domain of attraction of a periodic motion are assigned the same group and periodicity numbers. The step number indicates the total number of times steps, each of duration, T, it takes to map that cell to a periodic cell. The step increment number, m, indicates the number of time steps it takes to complete a cell transition.

Four conclusions result from this cell mapping setup. First, by definition, the sink cell is a P-1 motion. Second, among the regular cells, the number of periodic cells cannot exceed N_c. Third, the length of the longest period cannot exceed MN_c. And last, the evolution of the system starting at any regular cell, c, can lead to only three possible outcomes: (a) cell, c, is itself a periodic cell of a periodic motion; (b) cell, c, is mapped into the sink cell in r steps; or (c) cell, c, is mapped into a periodic cell of a periodic motion in r steps.

Assignment of Properties The conclusions above lead directly to the unraveling algorithm. During processing, each cell will be in one of three stages: *not yet processed*, *in processing*, and *completely processed*. A cell is completely processed if all its properties have been assigned. First, the image cell and step increment number for each cell are computed and assigned. Then the sink cell's properties are assigned to be

$$C(N_c) = N_c \quad G(N_c) = 0 \quad P(N_c) = 1 \quad S(N_c) = 0 \quad I(N_c) = 0. \tag{19}$$

The sink cell is then labeled as *completely processed,* while the rest of the regular cells are assigned the stage of *not yet processed.* Next, the regular cells are processed one at a time in the order, c = 0,1,...,N_C - 1. A *processing sequence* for a given cell, c, is the sequence of cells generated by recursive application of the cell mapping starting at c. At each step of a processing sequence, $C^j(c)$, there are three possibilities:

(a) $C^j(c)$ is a new unprocessed cell. In this case the stage of $C^j(c)$ is set to *in processing* and the processing sequence is continued to the next cell.

(b) $C^j(c)$ is a *completely processed* cell. In this case the processing sequence is terminated. The group and periodicity numbers of all the cells in the processing sequence are assigned to be those of $C^j(c)$. The step number of each cell is computed as

$$S(C^l(c)) = \sum_{i=l}^{j-1} I(C^i(c)) + S(C^j(c)), \quad l = 0,\ldots,j-1. \tag{20}$$

(c) $C^j(c)$ is an *in processing* cell and has already been encountered in this processing sequence. This means a new periodic motion has been discovered. The processing sequence is terminated. A new group number is generated and assigned to each cell in this sequence. If the first time that cell $C^j(c)$ appeared in this processing sequence was at step k, $k<j$, then the periodicity number is given by

$$P(\cdot) = \sum_{i=k}^{j-1} I(C^i(c)), \tag{21}$$

and the step number by

$$S(C^l(c)) = \begin{cases} \sum_{i=l}^{k-1} I(C^i(c)), & l = 0,\ldots,k-1 \\ 0, & l = k, k+1,\ldots,j-1. \end{cases} \tag{22}$$

Once a processing sequence has been terminated, a new sequence is begun with the next unprocessed cell.

3.3 Nonuniform Quantization

Given a fixed total number of cells, both the accuracy of the cell mapping is increased and the amount of computation is reduced by placing smaller cells where the system dynamics are slower and larger cells where the system dynamics are faster. For example, in a set-point control system, precise measurement of convergence to the set point is important. This research has developed a nonuniform quantization scheme that concentrates small cells near the set point and larger cells farther away [6]. This is done through a nonlinear one-to-one mapping of each axis of the state space onto the respective axis of a *compactified* space. A uniform quantization of the compactified space results in a nonuniform quantization of the original state space. Suppose **x** and **w** represent respectively points in the original and compactified state spaces. The transformations between the ith elements is given by

$$x_i = q_i^{-1}(w_i) = \begin{cases} \alpha_i\left(\lambda_i w_i^2 + (1-\lambda_i)|w_i|\right)\text{sign}(w_i) + \delta_i, & 0 < \lambda_i \le 1 \\ \alpha_i w_i + \delta_i, & \lambda_i = 0, \end{cases} \qquad (23)$$

$$w_i = q_i(x_i) = \begin{cases} \left\{\dfrac{\lambda_i - 1}{2\lambda_i} + \left(\dfrac{(1-\lambda_i)^2}{\lambda_i^2} + \dfrac{4|x_i - \delta_i|}{\lambda_i \alpha_i}\right)^{1/2}\right\}\dfrac{\text{sign}(x_i - \delta_i)}{2}, & 0 < \lambda_i \le 1 \\ \dfrac{x_i - \delta_i}{\alpha_i}, & \lambda_i = 0, \end{cases} \qquad (24)$$

where δ_i is the offset between the origins of the two systems, λ_i is a parameter determining the degree of nonlinearity in the transformation, (if $\lambda_i = 1$ the transformation is quadratic and if $\lambda_i = 0$, linear) and $\alpha_i = \max(|a_i - \delta_i|, |\delta_i - b_i|)$.

3.4 Optimal Control Algorithm

Based on the cell state space concept, Hsu proposed an optimal set-point control algorithm [9] that computes a discrete approximation to the minimizing control for a given cost function. Hsu's algorithm is a simplified form of dynamic programming that exploits both the spatially and temporally discrete nature of a cell space expression of a set-point control system. This research has enhanced Hsu's original algorithm. The modified algorithm is summarized below [5,6].

Consider the *cell mapping pair*, (See Fig. 2(a))

$$\mathbf{z}(k+1) = \mathbf{C}(\mathbf{z}(k), u(k), \Delta(k)), \qquad (25)$$

where $\mathbf{z}(k+1)$ is the *image cell* that $\mathbf{C}$ generates over the time interval $\Delta(k)$, given *do-*

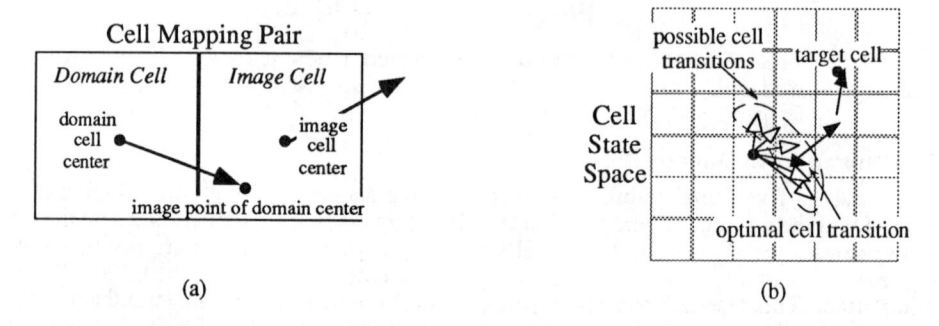

Fig. 2: Cell State Space System. (a) Cell Mapping Pair Illustrating the Center-Point Mapping Method. (b) Small Region of Cell State Space About the Set point (Target Cell). The Possible Cell Transition Images Are the Result of Different Control Actions from the Center Point of the Indicated Cell. The Optimal Transition Maps to the Trajectory with Minimum Total Cost.

main cell, $\mathbf{z}(k)$, and intervalwize constant control action, $u(k)$. The allowable control actions are limited to the finite set

$$u(k) \in U = \{u_1, \ldots, u_{N_U}\}. \tag{26}$$

The goal of the optimal control action is to find a sequence of controls that will drive the system state, starting at any cell, to a target cell(s), while minimizing some cost function. In an autonomous system, any given cell transition is not a function of when in a trajectory a cell is encountered; therefore, each cell along a trajectory can be considered a new initial condition. Thus, the solution for the optimal control consists of finding, for each cell, the one control action that maps that cell to the best possible image cell.

The cost increment accrued by any cell transition is given by

$$L_k = Q(\mathbf{z}(k), u(k), \Delta(k)). \tag{27}$$

The total cost to drive any cell to the target cell is the sum of the cost increments along the cell trajectory. Consider a cell trajectory that starts at cell $\mathbf{z}(0)$ and ends N_T cell transitions later at the target cell. The total cost of that trajectory is given by

$$L = \sum_{k=0}^{N_T} L_k = \sum_{k=0}^{N_T} Q(\mathbf{z}(k), u(k), \Delta(k)). \tag{28}$$

A list associating each cell and its optimal control action (i.e. minimizes L) is called the optimal control table (OCT). The procedure for constructing the OCT follows.

Procedure At any stage in the algorithm, each cell belongs to one of three sets. These are the set of *controllable cells*, the set of *uncontrollable cells*, and the set of *pending cells*. A cell is uncontrollable if no sequence of allowable control actions can map that cell to the target cell. Initially, the sink cell is added to the set of uncontrollable cells, the target cell which contains the set point is placed in the set of controllable cells, and the rest of the cells are place in the set of pending cells. Due to the discrete number of cells, control actions, and time intervals, L is discrete. The minimum change in L is called the basic cost unit. Associated with each iteration of the algorithm is a current cost level which is an integer multiple of the basic cost unit. The current cost level is initialized to 0. The steps to the algorithm are as follows [5,6,9]:

(a) For each cell, determine the image cell for each of the allowable control actions. The image cells are computed by recursively applying the control action, $u(k)$, until one of the conditions in Eqs. 16, 17, or 18 is met. For each image cell, compute the cost increment incurred by the cell transition. By dynamically determining the time interval for a cell transition, the number of images that must be computed is greatly reduced over Hsu's original algorithm [5,6,9].

(b) For each pending cell, determine the stage of its image cells. If all possible image cells are uncontrollable, then add that pending cell to the list of uncontrollable cells. Otherwise, for each controllable image cell, compute the total cost for the associated control sequence by adding the cost increment for the transition to that image cell to the image cell's total cost . If the total cost is less than or equal to the current cost level, then the associated control action is part of an optimal control sequence. Assign the control action with the lowest total cost as the optimal control for that pending cell and add the cell to the list of controllable cells. When more that one control action results in the same total cost, a set of discriminating functions are used to select the optimal control. A graphical illustration of this step is shown in Fig. 2(b).

(c) Increment the current cost level.

(d) Repeat steps (b) and (c) until the set of pending cells is either empty or the current cost level becomes large with no reduction in the set of pending cells.

Because it considers every possible cell-to-cell transition at every cost level, the optimal control algorithm finds the globally minimizing set of control actions. The set of control actions thus found constitutes the OCT mentioned previously. The OCT provides a spatially discrete approximation to the optimal control surface.

Cost Functions for Minimum-Time Control In minimum-time control, the goal is to drive the system to the set point as quickly as possible. The cost function increment is the time interval over which a cell transition occurs. The total cost is

$$L_k = f(\mathbf{z}(k), u(k), \Delta(k)) = \Delta(k) = m_k T \quad \text{and} \quad L = \sum_{k=0}^{N_T} \Delta(k). \tag{29}$$

4. FUZZY LOGIC CONTROLLER TUNING

The optimal control table (OCT) could be used directly at run time as a controller. However, the outputs provided by the OCT are not continuous, and these quantized outputs can introduce oscillations, chattering, and steady state errors. The OCT may also contain inaccuracies stemming from the discrete nature of its derivation. Finally, a very large number of cells may be needed to accurately approximate the desired control policy, resulting in impractical on-line memory requirements for the controller. Instead of using the OCT directly, this work implements the desired control policy with a fuzzy logic controller (FLC), where the FLC is tuned to model the control policy embodied by the OCT. Because the FLC can express its output as a smooth, nonlinear function of its inputs, the FLC is more robust to variations in system parameters [15,16] and avoids the problems caused by the quantized output of the OCT [17]. The FLC smoothes out isolated errors in the OCT as well. In comparison to the OCT, the memory needed for the FLC is negligible. Moreover, the FLC can be further modified to reflect additional information the controller designer may have about controller behavior, and is also amenable to run time scaling and adaptation.

4.1 Fuzzy Logic Controller Format

The FLC used in this work is based on a type of controller proposed by Sugeno and his colleagues [18,19]. The essential difference between Sugeno's controller and the more conventional FLC of Mamdani [20] lies in the way the outputs to the control rules are formed. In the conventional FLC, the compositional rule of inference [21] is used to generate fuzzy sets as the rule consequents. Some form of weighting is used to combine rule outputs and produce a defuzzified drive signal.

In contrast, the rule consequents of Sugeno's controller are expressed as functions of the input variables and rule combination is done by forming a weighted convex sum of the output function values. To achieve a given level of performance, this type of FLC requires significantly fewer rules than does the conventional FLC [19].

The functions used for the rule consequents are parameterized functions of the controller inputs. This choice makes the controller amenable to modification or optimization with numerical parameter identification methods [4]. The format for a single control rule is

$$R_j \equiv \text{IF } x_1 \text{ is } A_{j1} \text{ and } x_2 \text{ is } A_{j2} \text{ and} \ldots \text{and } x_n \text{ is } A_{jn}, \text{ THEN } y_j = g_j(x_1,\ldots,x_n,\mathbf{c}_j), \quad (30)$$

where x_i, $i = 1,\ldots,n$, are the n inputs to the controller, A_{ji} is a fuzzy set with membership function, $\mu_{A_{ji}}(x_i)$, that restricts the ith input to the jth rule, and y_j is the output of the jth rule given by the parameterized function $g_j(\cdot)$ with output-parameter vector, $\mathbf{c}_j$.

Assuming that the inputs $(x_1, x_2, \ldots, x_n)$ to the FLC have the crisp (non-fuzzy) values given by $(x'_1, x'_2, \ldots, x'_n)$, the truth value, β_j, of each rule is given by

$$\beta_j = \min\left(\mu_{A_{j1}}(x'_1), \mu_{A_{j2}}(x'_2), \ldots, \mu_{A_{jn}}(x'_n)\right). \quad (31)$$

The min operator in Eq. 31 could be replaced by the product of another suitable t-norm. The rules are combined to form the FLC output, y*, by taking the weighted average of the rule outputs. This is given by

$$y^* = \frac{\sum_{j=1}^{m} \beta_j y_j}{\sum_{j=1}^{m} \beta_j}, \quad (32)$$

where m is the number of rules and β_j is the truth value of the jth rule.

4.2 Parameter Identification

The Widrow-Hoff *least-mean-square* (LMS) algorithm [17,22] is a gradient following technique used to minimize mean-squared error through recursive parameter updates. This research uses a novel derivation [6,17] of the LMS algorithm to calculate the FLC-rule output-parameter vectors of Eq. 30 that minimize the average-squared error between the output surface generated by the FLC and the output *surface* of the OCT.

The average-squared error, to be minimized, is defined as

$$\xi = \frac{1}{N}\sum_{k=1}^{N} \varepsilon^2(k), \quad (33)$$

where $\varepsilon(k) = d(k) - y^*(k)$. The values $d(k)$ and $y^*(k)$ are the outputs of the OCT and FLC, respectively, for a given input, $\mathbf{x}(k)$, and k is an index into the N-point input-output data set.

In this research, the output of the jth rule is

$$y_j = g_j(\mathbf{x}, \mathbf{c}_j) = c_{j0} + c_{j1}x_1 + \cdots + c_{jn}x_n. \quad (34)$$

Defining $\mathbf{v} = [v_0 \quad v_1 \quad \cdots \quad v_n]^T = [1 \quad x_1 \quad \cdots \quad x_n]^T$, Eq. 34 becomes

$$y_j = c_{j0}v_0 + c_{j1}v_1 + \cdots + c_{jn}v_n = \sum_{l=0}^{n} c_{jl}v_l, \qquad (35)$$

which allows the output of the FLC to be expressed in terms of the output parameters:

$$y^* = \frac{\sum_{j=1}^{m}\beta_j y_j}{\sum_{j=1}^{m}\beta_j} = \frac{\sum_{j=1}^{m}\beta_j \sum_{l=0}^{n} c_{jl}v_l}{\sum_{j=1}^{m}\beta_j}. \qquad (36)$$

The error gradient-descent algorithm recursively updates the values of the parameters, c_{jl}, to move them in the direction of the greatest decrease in ξ. This direction is given by the negative gradient of ξ with respect to the parameters, c_{jl}. The Widrow-Hoff LMS algorithm approximates the gradient of ξ with the gradient of the squared error at a single point. The jlth element of the estimated gradient is given by

$$\hat{\nabla}_{jl}\xi \equiv \frac{\partial \varepsilon^2(k)}{\partial c_{jl}} = -2\varepsilon(k)\frac{\partial(y^*(k))}{\partial c_{jl}} = -2\varepsilon(k)\left(\frac{\beta_j(k)v_l(k)}{\sum_{i=1}^{m}\beta_i(k)}\right). \qquad (37)$$

The Widrow-Hoff single-parameter update equation can now be written as

$$c_{jl}(k+1) = c_{jl}(k) - \alpha\frac{\partial \varepsilon^2(k)}{\partial c_{jl}} = c_{jl}(k) + \frac{2\alpha\varepsilon(k)\beta_j(k)v_l(k)}{\sum_{i=1}^{m}\beta_i(k)}. \qquad (38)$$

where α is some positive constant.

As mentioned previously, k is an index to the set of input-output points, and not the time step. In order to avoid adapting to localized regions only, the points are randomized before the algorithm is applied. The total average-squared error between the OCT and the FLC surface is calculated after each pass through *all* the data points. When the change in error, from the previous pass, is below a certain threshold, the process is terminated.

In order for a set-point control system to be absolutely stable, the set point must be a stable equilibrium point. This requires that the control at the set point cause the system to remain at the set point. Therefore, after the parameters have been updated to minimize the average-squared error, one more update, called the zero-error update [5], is performed using the scale factor value of

$$\alpha = \frac{\left(\sum_{j=1}^{m}\beta_j\right)^2}{2\sum_{j=1}^{m}\beta_j^2 \sum_{l=0}^{n}v_l^2}, \qquad (39)$$

which guarantees that the parameter vector results in zero error, between the OCT and FLC surface, at the set point.

4.3. Stability and Performance Analysis of the Closed-Loop FLC System

After the FLC has been calibrated, a cell-to-cell map of the entire closed-loop system is generated to cross check overall system stability and performance. From the cell-to-cell map, a global indication is obtained of the absolute stability of the system about the set point. An indication of relative stability of the system to parameter variations can be obtained from a succession of cell maps. The cell mapping algorithm provides a computational method of guaranteeing the stability of the closed-loop FLC system. The OCT serves as a valuable guide to the controller designer as to the nature of the optimal control surface should the designer wish to make further modifications to the FLC.

One of the major criticisms of conventional FLC design is that there is no way to analytically guarantee stability. Common practice has been to use testing and simulation to empirically demonstrate stability, but usually with some trepidation. Cell mapping, however, provides a formal computational tool for demonstrating system stability and is founded in the extensive work of Hsu and his colleagues [7,8,9,14]. Indeed, cell mapping can be thought of as an efficient means of performing a rigorously exhaustive simulation of system behavior. The more widespread use of formal computational methods, a la cell mapping, should assuage some of the fears industry has expressed about fuzzy control. The major restriction to the use of cell mapping techniques is that a computer simulation model of the plant must be available, although the model may be nonlinear, fuzzy, or an artificial neural net.

5. ANGULAR-POSITION CONTROL OF AN INVERTED PENDULUM

Using the techniques and algorithms of the previous sections, the time-optimal angle-position control of the inverted pendulum system will now be derived.

5.1 Design Procedure

The controller design procedure can be summarized as follows:

(a) Build the simulation model of the plant

(b) Define the cost function

(c) Generate the optimal control table

(d) Formulate the initial fuzzy rule base

(e) Identify the FLC output function parameters

(f) Evaluate the FLC's stability and performance

(g) Perform additional scaling and adaptation

Inverted Pendulum System The inverted pendulum system is shown in Fig. 3. The goal is to balance the pole vertically with horizontal forces on the cart. The cart travels along a frictionless track. The pole is connected to the cart by a frictionless hinge that rotates in the vertical plane aligned with the track. The angle dynamics of the inverted pendulum are given in Eq. 40 [23], where θ is the angular position of the pole with respect to the vertical, F is the driving force on the cart, l is the half-length of the pole, m_p is the mass of the pole, m_c is the mass of the cart, and g is the gravitational acceleration constant.

$$\ddot{\theta}(t) = \frac{g\sin\theta(t) - \dfrac{\cos\theta(t)}{m_p + m_c}\left(m_p l \dot{\theta}^2(t)\sin\theta(t) + F(t)\right)}{\dfrac{4l}{3} - \dfrac{m_p l \cos^2\theta(t)}{m_p + m_c}} \qquad (40)$$

Fig. 3: Inverted Pendulum System (cart and pole).

This example is only concerned with the control of the angular position of the pole and not the position or velocity of the cart. Due to the nonlinear unstable nature of the inverted pendulum system, previous attempts at control were primarily concerned with stabilizing control [23,24,25] and were restricted to small initial angles of the pole. In contrast, the cell space based optimal control algorithm enables us to strive for minimum time control for even large initial angles of the pole.

The angular velocity and position are computed by numerical integration of Eq. 40, where the time increment between successive states is the sample period, T. The parameter values are: $l = 0.5$ m, $m_p = 0.1$ kg, $m_c = 0.9$ kg, and $T = 0.01$ seconds.

A diagram of the fuzzy control system is shown in Fig. 4. To regularize the expression of the fuzzy control rules, each control input is scaled and then amplitude limited. This gives $x_1(k) = K_1 \theta(k)$ and $x_2(k) = K_2 \dot{\theta}(k)T \cong K_2(\theta(k) - \theta(k-1))T$. In addition, the controller output is amplitude limited and scaled; that is, $u(k) = K_u y(k)$. These scaling coefficients and amplitude limits are given in Table 2.

Optimal Control Table Generation The set of allowable control actions consists of $U = \{0, \pm 2, \pm 4.67, \pm 8, \pm 12, \pm 16.67, \pm 22, \pm 28, \pm 34.67, \pm 50\}$. The number of cells is $51 \times 51 = 2601$. The cells are distributed nonuniformly (see Eqs. 23 and 24) with a nonuniform ratio of 3/2 for both axes. Using the cost function of Eq. 29, the cell state space optimal control algorithm generates the optimal control table (OCT) shown in Fig. 5. Note that the optimal control surface of Fig. 5 is highly nonlinear and looks like a smoothed version of a bang-bang controller.

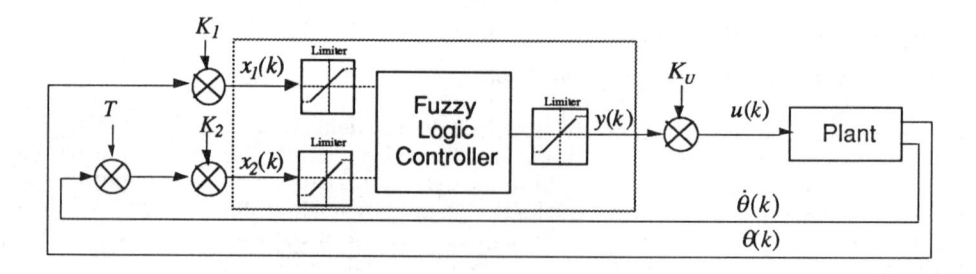

Fig. 4: Fuzzy Logic Controller Block Diagram

Table 2: Regions of Interest for Inverted Pendulum FLC.

Description	State Variable	Scaling Coefficient	FLC Variable
Error Input	$\theta \in [-1.5, 1.5] \, rad$	$K_1 = 1$	$x_1 \in [-1.5, 1.5]$
Error Change Input	$\dot{\theta}T \in [-0.08, 0.08] \, rad$	$K_2 = 1$	$x_2 \in [-1.5, 1.5]$
Force Output	$u \in [-50, 50] \, newtons$	$K_u = 1$	$y \in [-50, 50]$

Fuzzy Logic Controller An initial nine rule fuzzy logic controller (FLC) is formulated such that the rule antecedents cover the complete input space with a high degree of overlap. The membership functions for the rule antecedents are shown in Fig. 6, where the horizontal axis is scaled to the respective regions of interest for each input. Using the LMS algorithm described in section 4.2, and the input-output data provided by the OCT (see Fig. 5), the output function coefficients are then identified. The calibrated FLC rules are given in Table 3. The average-squared error between the FLC and the OCT is 23.53. The total number of controllable cells is 2315 out of 2605.

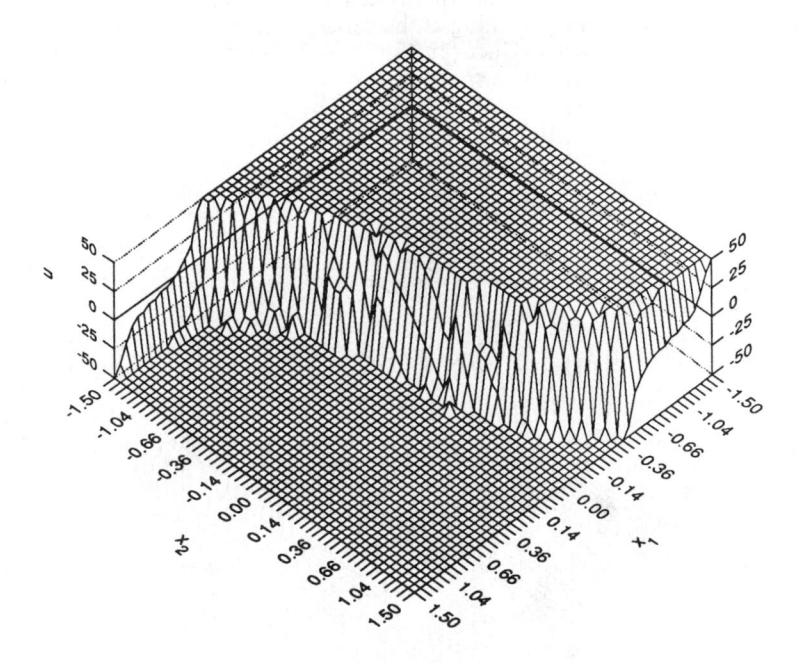

Fig. 5: Inverted Pendulum Minimum-Time Control Table.

Table 3: Rules for Inverted Pendulum FLC.

Antecedents		Output Function Parameters		
Neg	Neg	78.77	-13.88	-23.11
Zero	Neg	106.20	-145.0	-91.83
Pos	Neg	-232.10	149.10	-75.06
Neg	Zero	410.60	-83.49	18.19
Zero	Zero	0.02	-1119.0	-106.10
Pos	Zero	-403.40	-74.48	23.9
Neg	Pos	230.60	150.80	-72.75
Zero	Pos	-105.90	-145.0	-92.65
Pos	Pos	-70.44	-9.11	-17.24

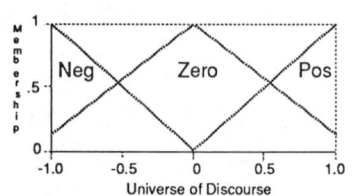

Fig. 6: Fuzzy Set Membership Functions for Fuzzy Control Rules.

Performance and Stability The control surface of the tuned FLC is shown in Fig. 7. This is clearly a smooth approximation to Fig 5. A trajectory map of the response of the system to the FLC is given in Fig. 8, where each arrow represents the end points of the closed-loop system trajectory beginning at the center of each cell. For the sake of clarity, a 25x25 cell grid is used. Also evident in Fig. 8 is the variation in cell sizes due to the nonuniform quantization of the state space.

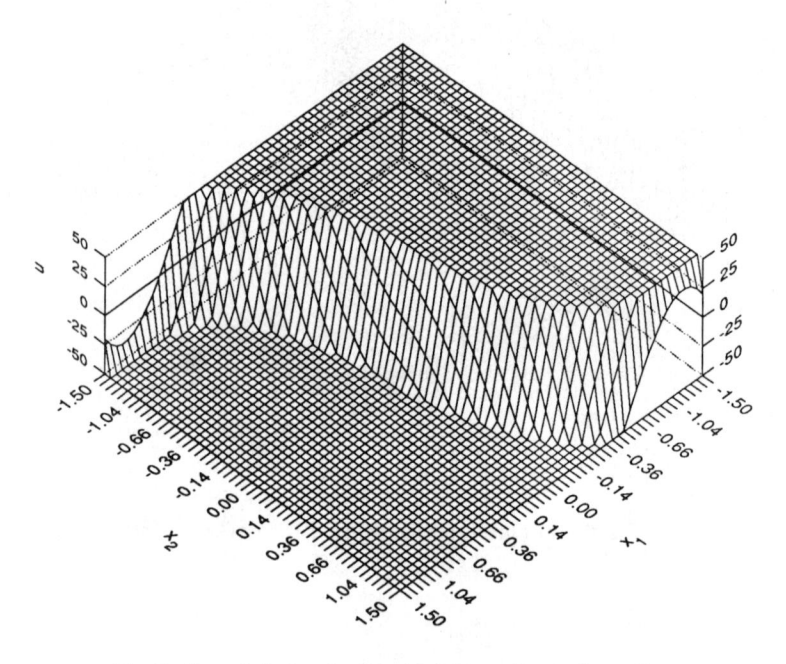

Fig. 7: Control Surface for Nine-Rule Fuzzy Logic Controller.

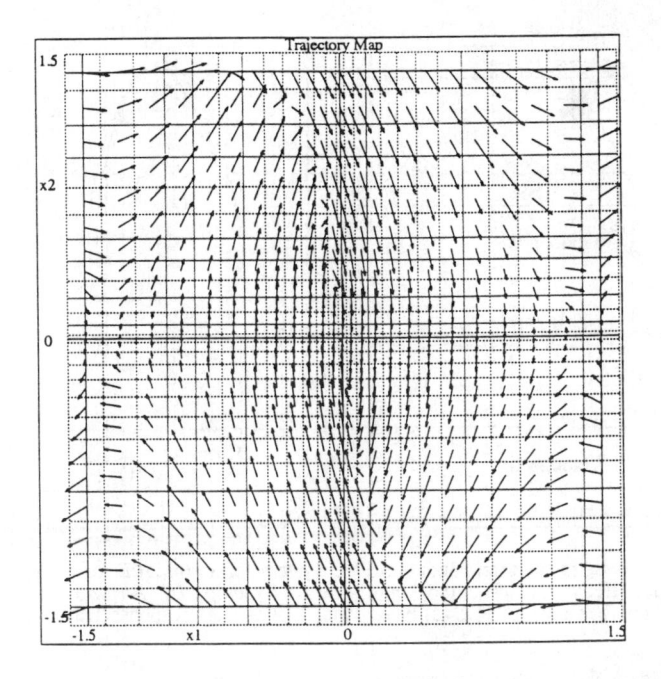

Fig. 8: Trajectory Map of Closed-Loop Control System.

To test the robustness of the FLC to variations in plant parameters, the inverted pendulum's parameters are varied by ±10% (also referred to as *heavy* and *light*). The step-response performance, for an initial error of one radian, is shown in Fig. 9 (*9R* implies the FLC has nine rules). Some time-domain step-response characteristics of the FLC for a one radian initial error are listed in the first three rows of Table 4 (found three pages hence). Note the correspondence between the continuous state trajectories in Fig. 9(c) and the cell based trajectory plot in Fig. 8. Using Fig. 8, it is possible to predict the qualitative behavior of system for any given initial error without having to explicitly generate a different step response for each initial error. Although not shown here, tracing the boundary of the controllable cells indicates the regions of state space that converge to the set point. This is the region of stable operation.

5.2 FLC Scaling and Adaptation

The results at the end of the previous section demonstrate that the FLC is fairly robust to changes in system parameters. Though the performance is degraded, the closed-loop system still converges. It would be desirable to improve the performance without the need to redesign the entire FLC. One approach is to scale the inputs to the FLC [26]. This procedure distorts the FLC's control surface in an attempt to make its shape more similar to a control surface that was tuned for different parameter values.

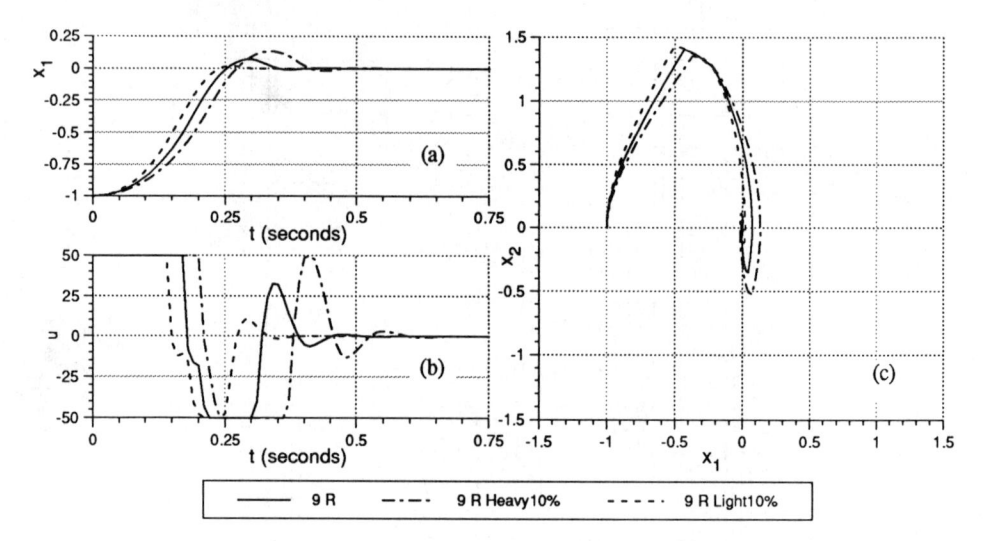

Fig. 9: Step Response of Inverted Pendulum FLC System for an Initial Error of One Radian for Nominal, Heavy (+10%), and Light (-10%) Systems. (a) Angle Error of Inverted Pendulum. (b) Output of FLC (c) Phase Plane Plot of Step-Response Trajectory.

Applying this distorting technique to time-optimal control of *linear* systems is a fairly straightforward task, since a closed-form solution exists. An example would be time-optimal angle-position control of an armature-controlled DC motor. (The techniques of this chapter have also been applied to a DC motor [5,6,17,27].) In this case, it can be seen that scaling the system parameters distorts the switching curve; yet, it retains the same qualitative shape. In fact, the distorted curve is now similar to the optimal switching curve for a different load.

Although a closed-form solution for minimum-time control of the inverted pendulum is not available, the qualitative nature of the inverted pendulum controller surface, as shown in Fig. 7, is similar to that of the DC motor mentioned above. Consequently, compensation for parameter variations in the inverted pendulum might be accomplished in a manner similar to that used for the DC motor. An explicit formula for the DC motor indicates how the optimal switching curve varies with changes in the load, thus determining the scaling that should be used. For the inverted pendulum, an iterative approach is taken to determine the appropriate input scaling.

As shown in Fig. 9, the FLC has some overshoot for the designed parameters and even more for the *heavy* parameters. Possible causes for this overshoot are switching errors introduced by the sampling rate, cell size, and inaccuracies in the FLC approximation. Scaling the inputs to the FLC, so that the transition region is encountered sooner, should reduce this overshoot. This can be done by reducing K_1 and increasing K_2.

For nominal values of the inverted pendulum parameters, K_1 and K_2 are decreased and increased, respectively, by 5%, 10%, and 20% of their original values. The responses are plotted in Fig. 10. The indicated figures show that, as expected, both the overshoot and settling time are reduced, but with a slight increase in rise time. The degree of change is proportional to the scaling.

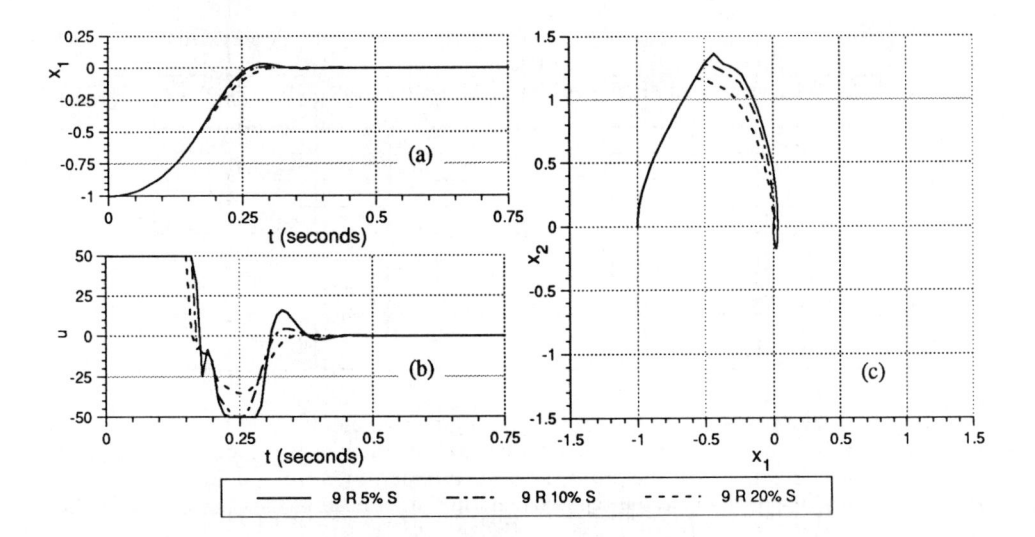

Fig. 10: Step Response of Inverted Pendulum FLC System for an Initial Error of One Radian for Nominal System Parameters. The Scaling Coefficients of the FLC Are Modified by the Percentages Specified in the Legend. (a) Angle Error of Inverted Pendulum. (b) Output of FLC. (c) Phase Plane Plot of Step-Response Trajectory.

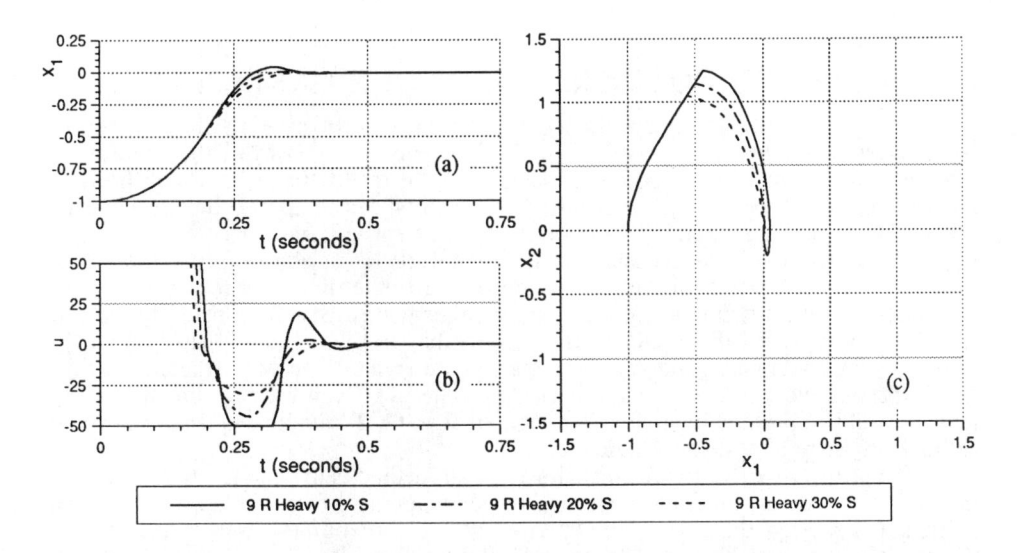

Fig. 11: Step Response of Inverted Pendulum FLC System for an Initial Error of One Radian for Heavy System Parameters. The Scaling Coefficients of the FLC Are Modified by the Percentages Specified in the Legend. (a) Angle Error of Inverted Pendulum. (b) Output of FLC. (c) Phase Plane Plot of Step-Response Trajectory.

Table 4: Inverted Pendulum FLC Time-Domain Step-Response Characteristics for an Initial Error of One Radian.

FLC & Parameters	10-90% Rise Time (sec)	1% Settling Time (sec)	Overshoot %
Nominal	0.14	0.37	6.9
Heavy (+10%)	0.16	0.41	11.5
Light (-10%)	0.13	0.27	1.9
5% S	0.15	0.32	3.2
10% S	0.15	0.26	0.9
20% S	0.17	0.29	0.2
Heavy 10% S	0.16	0.36	4.3
Heavy 20% S	0.17	0.30	0.7
Heavy 30% S	0.19	0.34	0.1

For the heavy values of the inverted pendulum parameters, K_1 and K_2 are decreased and increased, respectively, by 10%, 20%, and 30% of their original values, and the responses plotted in Fig. 11. The input scaling compensates very adequately for the parameter variations of the heavy model. Some time-domain step-response characteristics for the scaled responses are listed in the last six rows of Table 4.

These results strongly suggest that the automated controller design procedure for some systems can be easily augmented by scaling the inputs to compensate for modeling errors or parameter variations. In these cases, the OCT need only be generated once for a nominal set of parameters. A performance-adaptive tuning system that selects, at run time, the input scaling coefficients, could then be used to compensate for run-time parameter variations without modifying the structure of the FLC.

6. EXTENSIONS TO THREE AND FOUR VARIABLES

In previous sections, the accuracy of the optimal control table (OCT) was improved by applying the *adjusted-cost* method [6,17]. However, design of the algorithm is a tedious process and requires a good understanding of the qualitative behavior of the system in question. More specifically, the adjusted-cost method can be applied only if the general shape of the switching curve in two-dimensional state space is known.

The above requirements thus restrict the adjusted-cost algorithm to time-optimal set-point control of two-dimensional systems that are either linear, or are nearly linear in the region of the set point. There are many systems that satisfy these restrictions; yet, it would be more useful if the OCT design technique, and the resulting fuzzy logic controller (FLC), could be extended to systems that are greater than two dimensions, highly nonlinear, or have a different cost criteria (e.g., energy). Even for two dimensional systems it would be preferable to be able to construct an OCT without requiring such an exacting knowledge of system behavior.

The primary motivation for the adjusted-cost method is to decrease the inaccuracies in the time-based cost function caused by the cell mapping errors. However, the mapping errors not only affect the time-optimal performance criteria of the closed-loop system, but can also affect the overall performance and stability.

6.1 Effects of Cell Mapping Errors in Three and Four Dimensions

As the dimension of a system increases, geometry dictates that the potential magnitude of the mapping error within a particular cell also increases. This can cause the accumulated error along any given trajectory to increase greatly, which in turn results in the potential for decreased stability and performance of the closed-loop system. The more nonlinear a system, the more serious the effects.

From the above discussion, it is seen that the adjusted-cost method *cannot* be applied efficiently to higher-dimensional (greater than two) systems, but the remainder of the OCT construction algorithm is applicable. The question now, of course, is to what extent the OCT, and the resulting FLC, are adversely affected by mapping errors.

Three-Dimensional Inverted Pendulum Applying the algorithm, without the adjusted-cost implementation, to a three-dimensional inverted pendulum, illustrates what can occur [11]. The third dimension, or state, is the velocity of the cart; that is, the desired velocity of the cart is zero, but the final location of the cart is not relevant (this would require a fourth dimension). The state equations for the three-dimensional system are shown in Table 5 [28].

The map is 15x15x15 cells with a nonuniform ratio of three, and the controller has 125 rules (five per dimension). The shape of the membership functions for the rule antecedents in this, as well as all following examples, are the same shape as those shown in Fig. 6. The step-response performance of the resulting FLC, shown in Fig. 12, is not satisfactory. The pendulum angle *eventually* converges to zero, but the cart velocity does not, and eventually oscillates. The poor performance results from errors in the OCT. The increase in the system from two to three dimensions introduces large errors that were not present in the corresponding two-dimensional system.

Theoretically, by increasing the density of the cells, satisfactory results can be obtained; although memory limitations and reasonable computation time place limits on the amount resolution can be increased. The maximum amount of memory available for our simulations is approximately 256 megabytes. This allows for increasing the map resolution of the above example from 15 to 75 cells per dimension with a nonuniform ratio of 4/3. Notwithstanding, the performance of the resulting FLC is only mediocre. Fig. 13 shows the step-response performance and the associated drive force. The angle and position eventually converge, but the settling times are only marginal.

Table 5: Three-Dimensional Inverted Pendulum Equations.

State Variables	State Equations
$x_1 = \theta$ $x_2 = \dot{\theta}$ $x_3 = \dot{x}$	$\dot{x}_1 = x_2$ $\dot{x}_2 = \dfrac{g \sin x_1 - \cos x_1 \left[\dfrac{F + m_p l (\sin x_1) x_2^2}{m_c + m_p} \right]}{\dfrac{4}{3} l - \dfrac{m_p l \cos^2 x_1}{m_c + m_p}}$ $\dot{x}_3 = \dfrac{F + m_p l \left[(\sin x_1) x_2^2 - (\cos x_1) \dot{x}_2 \right]}{m_c + m_p}$

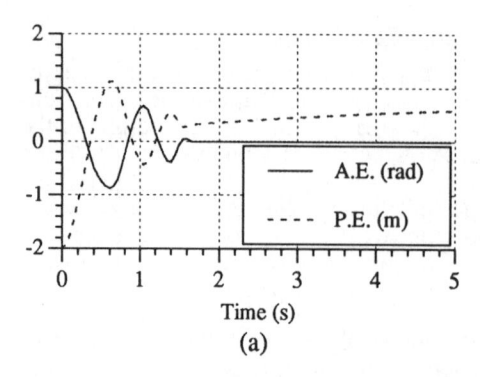

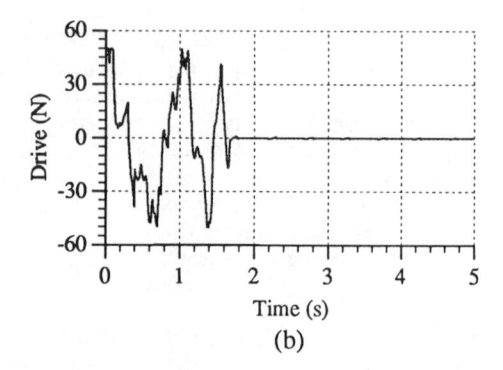

Fig. 12: Step Response Performance of Three-Dimensional Inverted Pendulum. 15 Cells per Dimension, 125 Rules, Time-Optimal Cost Function, and Nonuniform Ratio of 3. (a) Angle and Position Error. (b) Drive Force.

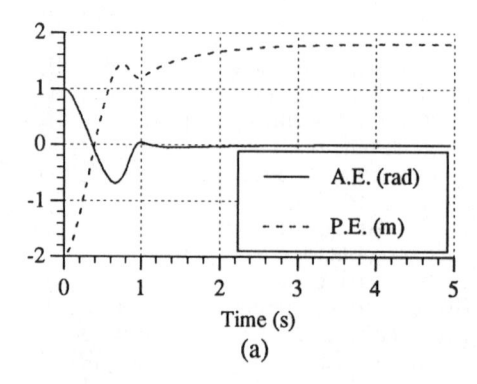

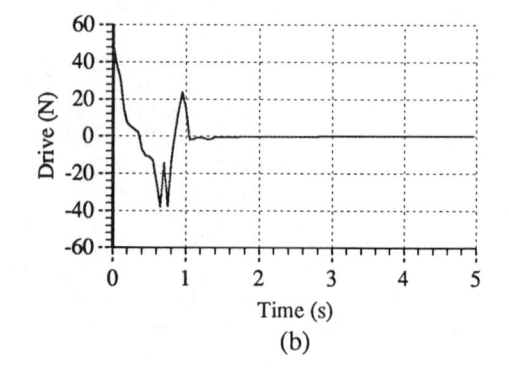

Fig. 13: Step Response Performance of Three-Dimensional Inverted Pendulum. 75 Cells per Dimension, 27 Rules, Time-Optimal Cost Function, and Nonuniform Ratio of 4/3. (a) Angle and Position Error. (b) Drive Force.

Four-Dimensional Inverted Pendulum The *curse of dimensionality* leads to an even greater increase in required memory and computation time when working with four dimensions [11]. Table 6 contains the four-dimensional state equations [28]. Fig. 14 shows the angle and position response for a four-dimensional inverted pendulum with three rules and 25 cells per dimension, and a nonuniform ratio of 4/3. This example requires 205 megabytes of memory; yet, the resolution is so poor that performance is totally unsatisfactory. The FLC cannot adapt properly because of the large mapping errors in the OCT. The trajectories of the resulting closed-loop system diverge, even for very small initial conditions

In summary, the increase in dimensionality not only increases the number of cells, for a given resolution, but also necessitates an increase in map resolution to offset the natural increase in cumulative mapping error. Table 1 further displays the dramatic increase in memory and computation time required as the dimensionality of the system in-

Table 6: Four-Dimensional Inverted Pendulum Equations.

State Variables	State Equations
$x_1 = \theta$ $x_2 = \dot{\theta}$ $x_3 = x$ $x_4 = \dot{x}$	$\dot{x}_1 = x_2$ $\dot{x}_2 = \dfrac{g \sin x_1 - \cos x_1 \left[\dfrac{F + m_p l (\sin x_1) x_2^2}{m_c + m_p} \right]}{\dfrac{4}{3} l - \dfrac{m_p l \cos^2 x_1}{m_c + m_p}}$ $\dot{x}_3 = x_4$ $\dot{x}_4 = \dfrac{F + m_p l \left[(\sin x_1) x_2^2 - (\cos x_1) \dot{x}_2 \right]}{m_c + m_p}$

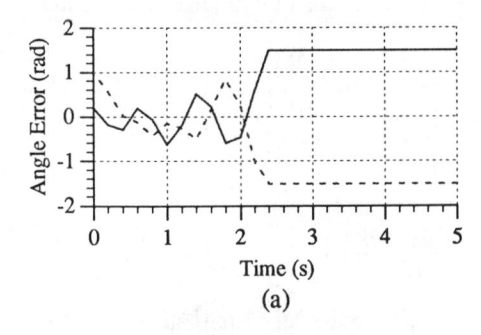

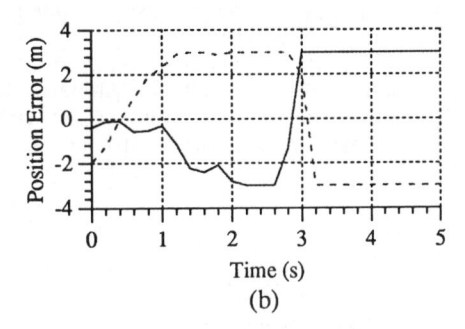

Fig. 14: Step-Response Performance of Four-Dimensional Inverted Pendulum for Two Initial Conditions. 25 Cells per Dimension, 81 Rules, Time-Optimal Cost Function, and Nonuniform Ratio of 4/3. (a) Angle Error. (b) Position Error.

creases. It appears that the conflict between performance, and memory or computation time, cannot be resolved with the present algorithm. The original development of Hsu's algorithm was motivated by the desire to avoid this computational explosion in approaching the resolution of point-to-point mapping [7,8,9].

A nonuniform grid helps reduce the number of needed cells, but as can be seen from the previous examples, it alone is not adequate for higher-dimensional systems.

6.2 The Mapping Error Cost Function

One solution to the problem is to include the mapping error in the cost function [11]. The original algorithm uses mapping error as a discriminating function only; that is, as a tie breaker when discriminating between different trajectories with the same time-based cost. In order to decrease the mapping error adequately, it should be taken into account in the original cost calculation.

For time-optimal control, as used in the original algorithm, the total cost for a trajectory is given by

$$L = \sum_{k=0}^{N_T} m_k T, \qquad (41)$$

where m_k is the number of basic time periods, T, for each cell-to-cell transition, and N_T is the total number of cell-to-cell transitions in the trajectory.

The cell mapping error is defined in each dimension as $E_j = S_j - C_j$, the difference between the exact state, S_j, and the state at the center of the cell, C_j, where j ranges over the dimension of the system. The total cumulative error, E, is then defined for each trajectory as

$$E = \sum_{j=1}^{N_S} \left| \sum_{k=0}^{N_T} E_{jk} \right|, \qquad (42)$$

where j ranges over the number of dimensions of the system and k ranges over the number of cells in the trajectory.

The proposed cost function is based on quadratic regulator theory used in many linear control problems [29]. In standard quadratic regulator design, a practical tradeoff between the rate of convergence of the states to the set point and the energy of the input is sought. The quadratic regulator cost, in the context of discrete optimal control, is given by

$$L = \sqrt{\left(a \sum_{k=0}^{N_T} m_k T\right)^2 + \left(b \sum_{k=0}^{N_T} |U_k|\right)^2}, \qquad (43)$$

where U_k is the actuator output energy, or the energy input to the system, and a and b are constant scale factors. Replacing energy with mapping error, the total cost for the trajectory is given by

$$L = \sqrt{\left(a \sum_{k=0}^{N_T} m_k T\right)^2 + b \left(\sum_{j=1}^{N_S} \left|\sum_{k=0}^{N_T} E_{jk}\right|\right)^2}, \qquad (44)$$

where m_k is the number of sample periods for each cell transition, T is the sample period, E_{jk} is the cell mapping error, N_T is the total number of cell transitions in the trajectory, and a and b are constant scale factors.

The new cost function thus forces a tradeoff between optimality and global stability of the system. From any given initial condition, it is possible that some of the eliminated trajectories may be stable and even more optimal than the chosen trajectory; but, the fact that they have too much accumulated mapping error implies that their actual behavior is not well enough known to assure that the closed loop system will be globally stable. The scale factors, a and b, are presently chosen manually. The following examples use scale values that were determined by trial and error.

Three-Dimensional Inverted Pendulum The OCT for the three-dimensional inverted pendulum system consists of seven cells per dimension, or 343 total cells, with a nonuniform ratio of three. The FLC uses 27 rules (three per dimension), and the error function

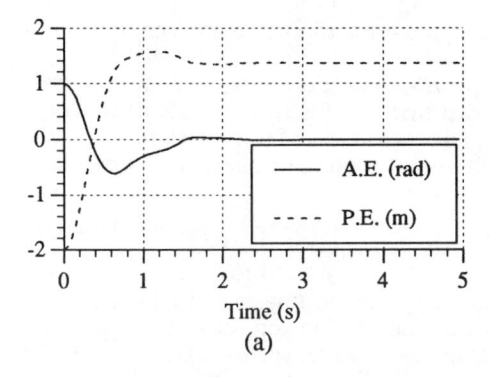

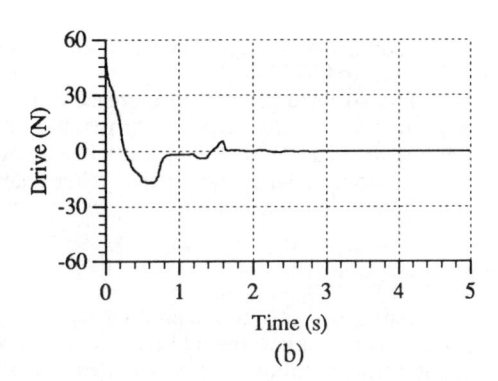

Fig. 15: Step-Response Performance of Three-Dimensional Inverted Pendulum. Seven Cells per Dimension, 27 Rules, Map-Error Cost Function, and Nonuniform Ratio of Three. (a) Angle and Position Error. (b) Drive Force.

scale factors are $a = 1$ and $b = 1000$. The resulting step response and FLC output drive are shown in Fig. 15 [11].

The performance is comparable to that of the 75 cells-per-dimension FLC of Fig. 13; yet, the new method required approximately 1/1000 the memory and computation time. The 1.0 and 0.1 percent settling times of the pendulum angle are improved somewhat with the new algorithm, but the overshoot is increased slightly. However, it should be remembered that time of convergence, and not overshoot, is being optimized.

Four-Dimensional Inverted Pendulum The OCT for the four-dimensional inverted pendulum system consists of 11 cells per dimension, or 14,641 total cells, and a nonuniform ratio of three. The FLC uses 81 rules (three per dimension), and the cost function scale factors are $a = 1$ and $b = 1000$. The step response is shown in Fig. 16 [11].

The performance of the four-dimensional system is excellent, particularly when considering that the map has only 11 cells per dimension. The results seem to be consid-

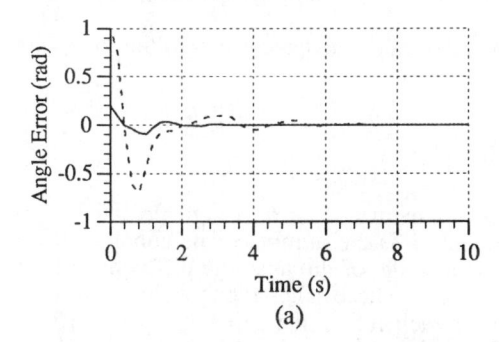

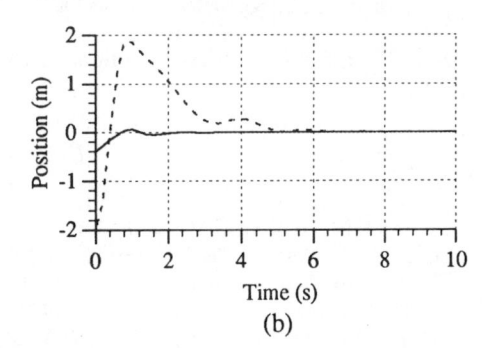

Fig. 16: Step-Response Performance of Four-Dimensional Inverted Pendulum for Two Initial Conditions. 11 Cells per Dimension, 81 Rules, Map-Error Cost Function, and Nonuniform Ratio of Three (a) Angle Error. (b) Position Error

erably better than many other methods used in higher-dimensional nonlinear system controller design [30,31].

If map resolution were increased in the previous two examples, there would obviously be potential for even better results. The fact that the closed-loop systems converge at all, with such low resolution, is a testament to the usefulness of this technique.

It should also be noted that other characteristics could be placed in the cost function, though this research has not yet applied any others.

6.3 Cascading

Another technique, called *cascading*, can be applied to improve steady-state error and speed of convergence [11]. This employs the idea that the control surface generated about the set point of the system is very similar to other smaller scaled regions around the same set point. A scaled version of the controller is applied to a similarly scaled version of the system. Thus, instead of designing several FLCs, each for a smaller region around the set point, one FLC can be designed, and then scaled, to be used consecutively as the system approaches the set point.

With few exceptions, the use of cascaded controllers improves convergence time and steady-state error, though the improvements are sometimes negligible. More research needs to be done in this area to establish guidelines as to when cascading should be applied.

6.4 Map Resolution Enhancement

Another approach to reducing mapping error deals with the grid itself. The previously mentioned nonuniform grid reduces the number of cells needed overall by placing more cells in regions near the center of the state space and less cells in the outer areas. This is based on the assumption that trajectories are moving slower near the center, or set point. In general, though, this is not a very efficient method of allocating the available map cells.

An adaptive method which detects where increased resolution is most effective has been developed [11]. This method increases the resolution in the areas of the state space in which the desired optimal control switches rapidly. These are called *high transition regions*, and can be identified for an OCT, of any dimension, through the following procedure.

First, the amount of transition for each cell is calculated according to the following formula:

$$T_c = \frac{1}{N_k} \sum_{k=1}^{N_k} |C_c - C_k|, \qquad (45)$$

where C_c is the control value of the cell for which the transition is being calculated, C_k is the control value of the kth neighboring cell and N_k is the number of neighboring cells. The amount of transition for a cell is thus the average of the absolute difference of the control value of the cell and all its neighboring cells. These calculations yield a *transition map*, which shows the amount of transition for each cell of the cell state space. A high transition region corresponds to a region in which the transition level of each cell is above a user defined threshold. This threshold is based on the range of possible control actions and the need for increased resolution in certain regions. The transition map is then used to build a new OCT with increased resolution in the high transition regions. The resolution is increased by subdividing the cells in these regions. This process can be applied repeatedly.

This technique appears to be promising but yields inconsistent results. Some satisfactory results can be obtained by adjusting the threshold by trial and error. In some cases the amount of memory required can be decreased by over 100 times while obtaining near comparable performance with the original algorithm [11]. This adaptive cell size method has not yet been applied with the new mapping-error cost function.

Although more work needs to be performed, the performance of the above FLCs demonstrates that the techniques of Section 6 are very useful in constructing three- and four-dimensional OCTs.

7. SUMMARY AND CONCLUSIONS

The reputed high performance and robustness of fuzzy logic controllers (FLC) may in some cases be explained by their smooth and nonlinear nature. The optimal control surface for many systems is nonlinear. Therefore, to achieve high performance levels, a nonlinear controller is needed. In the case of minimum-time control, the optimal controller may require abrupt transitions across nonlinear switching curves. These abrupt transitions can be brittle to parameter variations, whereas the FLC's smoother transitions across switching curves makes FLCs more robust to parameter variations. As demonstrated above, the calibrated FLC for the nonlinear inverted pendulum system is relatively robust to variations in system parameters. Previous attempts at the inverted pendulum problem focused on stabilizing control for large initial angles. The FLC presented here is designed for minimum-time control, even for large initial angles of the pendulum. These results illustrate the viability of this design technique for even strongly nonlinear systems.

The FLC design approach presented in this chapter provides a systematic method for implementing high performance FLCs for a class of two, three, and four variable systems. Although, the cell state space based optimal control algorithm requires a simulation model of the plant to be controlled, the models can be highly nonlinear. Moreover, as demonstrated in this chapter, a controller designed for one set of plant parameters can often be adapted to perform well for a different set of plant parameters by scaling the inputs to the controller, without the need of rewriting the rule base. Thus, the design algorithm is generalizable to a wider range of conditions so long as the qualitative shape of the desired control surface is retained over parameter variations.

Once the output of an FLC is defuzzified, the control is deterministic. In this regard, FLCs can be viewed as smooth variable structure controllers [26]. The performance of the FLC can be verified globally with cell maps over the complete state space. This should lend increasing confidence in the more widespread use of FLCs.

REFERENCES

1. Atherton, D.P., "An overview of nonlinear systems theory." In Billings, S.A., Gray, J.O., and Owens, D.H. (eds): *Nonlinear System Design*. London, UK, Peter Peregrinus Ltd., 1984, pp. 1-29.

2. Self, K., "Designing with fuzzy logic," *IEEE Spectrum*. November 1990, pp. 42-45.

3. Sugeno, M., *Industrial Applications of Fuzzy Control,* M. Sugeno, ed., Elsivier, 1985.

4. Smith, S. M., and Comer, D.J., "Self-Tuning of a Fuzzy Logic Controller Using a Cell State Space Algorithm," *Proceedings of the 1990 IEEE International Conference on Systems, Man, and Cybernetics*, L.A. CA, Nov. 4–7, pp. 445–450, 1990.

5. Smith, S.M. Comer, D.J., "Automated calibration of a fuzzy logic controller using a cell state space algorithm," *IEEE Control Systems Magazine*, August, 1991.

6. Smith S. M., and Comer, D. J., "An algorithm for automated fuzzy logic controller tuning," *IEEE International Conference on Fuzzy Systems*. March 1992, in press.

7. Hsu, C.S., "A theory of cell-to-cell mapping dynamical systems", *Journal of Applied Mechanics*, Vol. 47, December, pp. 931–939, 1980.

8. Hsu, C.S. and Guttalu, R.S., "An unraveling algorithm for global analysis of dynamical systems: an application of cell-to-cell mappings," *Journal of Applied Mechanics*, Vol. 47, December, pp. 940–948, 1980.

9. Hsu, C.S., "A discrete method of optimal control based upon the cell state space concept," *Journal of Optimization Theory and Applications*, Vol. 46. No 4, August 1985.

10. Thaler, G.J. and Brown, R.G., *Analysis and Design of Feedback Control Systems*. New York: McGraw-Hill Book Company, 1960.

11. Cook, B.L., *Fuzzy Logic Controller Design for Higher Dimensional Nonlinear Systems,* Masters Thesis, Brigham Young University 1991.

12. Lapidus, L. and Luus, R., *Optimal Control of Engineering Processes*. Blaisdell, 1967.

13. Bertsekas, D. P., *Dynamic Programming: Deterministic and Stochastic Models*. Prentice-Hall, 1987.

14. Hsu, C.S., "*Cell-to-Cell Mapping*, Springer-Verlag, 1987.

15. Tong, R.M., "A control engineering review of fuzzy systems," *Automatica*, Vol. 13, pp 559–569, 1977.

16. Sugeno, M., "An introductory survey of fuzzy control," *Information Sciences*, Vol. 36, pp. 59–83, 1985.

17. Smith, S.M., *A Computational Approach to Fuzzy Logic Controller Design*, Ph.D. Dissertation, Brigham Young University, 1991.

18. Sugeno, M. and Nishida, M., "Fuzzy control of model car," *Fuzzy Sets and Systems*, Vol. 16 pp. 103–113, 1985.

19. Takagi, T. and Sugeno, M., "Fuzzy identification of systems and its application to modeling and control," *IEEE Transactions on Systems, Man, and Cybernetics*, Vol. SMC-15, No. 1, pp. 116–132, 1985.

20. Mamdani, E. H., "Advances in the linguistic synthesis of fuzzy controllers," *International Journal of Man-Machine Studies*. Vol. 8, 1976, pp. 669-678.

21. Zadeh, L. A., "The concept of a linguistic variable and its application to approximate reasoning, I, II, III," *Information Sciences*. Vol. 8, pp. 199-249, pp. 301-357, Vol. 9, pp. 43-80, 1975.

22. Widrow, B. and Stearns, S.D., *Adaptive Signal Processing*, Prentice-Hall, 1985.

23. Chen Y.Y and Tsao, T.C., "A description of the dynamical behavior of fuzzy systems," *IEEE Transactions on Systems, Man, and Cybernetics*, Vol. SMC-19, No. 4, pp.745–755, 1989.

24. Charles W. Anderson, Learning to Control and Inverted Pendulum Using Neural Networks, IEEE Control Systems Magazine, pp. 31-37, April 1989.

25. Yamakawa, T., "Stabilization of an inverted Pendulum by a High-Speed Fuzzy Logic Controller Hardware System," Fuzzy Sets and Systems, Vol. 32 pp. 161-180, 1989.

26. Braae, M. and Rutherford, D.A., "Selection of parameters for a fuzzy logic controller," *Fuzzy Sets and Systems*, Vol. 2 pp. 185-199, 1979.

27. Smith, S. M., and Comer, D.J., "Automated fuzzy logic controller calibration and performance evaluation," *International Fuzzy Systems and Intelligent Control Conference*. March 1992, in press.

28. Cannon Jr., R.H., *Dynamics of Physical Systems*, McGraw-Hill, 1967.

29. Kailath T., *Linear Systems,* Prentice-Hall, 1980.

30. Anderson, C.W., "Learning to control an inverted pendulum using neural networks," *IEEE Control Systems Magazine,* pp. 31-37, April 1989.

31. Patrikar, A., Provence, J., "A Self-organizing controller for dynamic processes using neural networks," *IEEE International Joint Conference on Neural Networks,* Vol 3, pp. 359-364, 1990.

Chapter 20
An Adaptive Fuzzy Control Model Based on Fuzzy Neural Networks

Introduction, *430*
Building a Fuzzy Control System, *430*
The Implementation of Fuzzy Control, *451*
Controllability and Stability of Fuzzy Control, *452*
The Comparisons between the Fuzzy Control Theory and Modern Control Theory, *454*
Conclusion, *455*
References, *456*

This chapter provides a theoretical foundation for fuzzy control theory based on fuzzy control neural networks. It deals with the problem of deriving a feasible control model when the differential equations characterizing the process cannot be solved easily, or when they are difficult, or even impossible, to get. The chapter discusses the establishment, implementation, and properties (such as controllability, stability, and robustness) of fuzzy control systems in a systematic manner, and compares some methodological aspects of modern control theory with those fuzzy control theory.

AN ADAPTIVE FUZZY CONTROL MODEL BASED ON FUZZY NEURAL NETWORKS

Xinghu ZHANG
Institute of Systems Science
National University of Singapore
Singapore 0511

Peizhuang WANG
Institute of Systems Science
National University of Singapore
Singapore 0511

Zuliang SHEN
Institute of Systems Science
National University of Singapore
Singapore 0511

Xiantu PENG
Aptronix Inc.
2150 North First Street
San Jose, CA 95131

0. INTRODUCTION

To build a control system, the main task is to chose the related factors and proper force forms determined by these factors so that this control system can be controlled under these forces. Because of this, to chose the related factors and the forces becomes the crucial problem to build a control system. In **Modern Control (MC)** theory, the general approach is to build differential equations first, and then to linearize or simulate the differential equations. From the simplified differential equations, we can get the related factors, and chose proper force forms so that the differential equations can be easily solved. But in practice, sometimes it is very difficult, even impossible, to get differential equations for many control systems. Even we can get the differential equations, it is sometimes very difficult to solve them. So it become very important to build a feasible control model without differential equations. **Fuzzy Control (FC)** theory, we will discuss in this paper, is just the control theory to have this feature. Unlike modern control theory, in fuzzy control theory we chose the factors using neural networks, and determine the forces using fuzzy rules. All these processes can be executed in a **Fuzzy Control Neural Network (FCNN)** we will build. In this paper, we also discuss the implementation of FC. Finally we briefly discuss Controllability, Stability, and Robustness of FC, and give the Comparisons between MC and FC.

1. BUILDING A FUZZY CONTROL SYSTEM

Corresponding to the modern control system, a fuzzy control system is also considered in terms of object vector **x**, factor set F, constraint force vector **u**, and target function $\mathbf{T}(t)$. This can be formally stated by the following definition.

Definition: A **Fuzzy Control (FC)** system is a multicomponent model FC(**x**, F, **u**, **T**(t)), which has the following meanings:

(1) **x** is the **object vector**, it represents the objects we want to control;

(2) F is the lattice of all factors [2] which may have effect on **x**;

(3) **u** is the **control force vector**, it represents the control forces constrained to the system;

(4) **T**(t) is the **target vector function**, where t is *time* variable. When **T**(t) is constant, this kind of control is called **fixed point control**; when **T**(t) is not constant, this kind of control is called **fixed trace control**. As a vector function, **T**(t) is usually denoted as a multi-function $\{T_i(t)\}(i=1, 2, \ldots, s)$.

According to the above definition, **x** is a variable in a subspace of the factor space F, and **x** has the same dimension as the vector function **T**(t), and we denote the dimension to be s. Actually **T**(t) is a subset (a point, an area, or a surface, etc.) of the subspace **x** represents.

Example 1: In an inverted pendulum control problem as sketched in Fig.-1, **x** is the angular φ; **u** is the force constrained on the cart; F is the lattice generated by the set of all factors which have effect on the pendulum: the weight of pendulum, the angular φ, the changing rate of φ, the length $2L$ of the stick, the center of the stick gravity, the fiction of the wheels with the ground, the position of the cart C in the y direction, the changing rate of the position in the y direction, etc.; **T**(t) is the constant-value function that **T**(t)≡0.

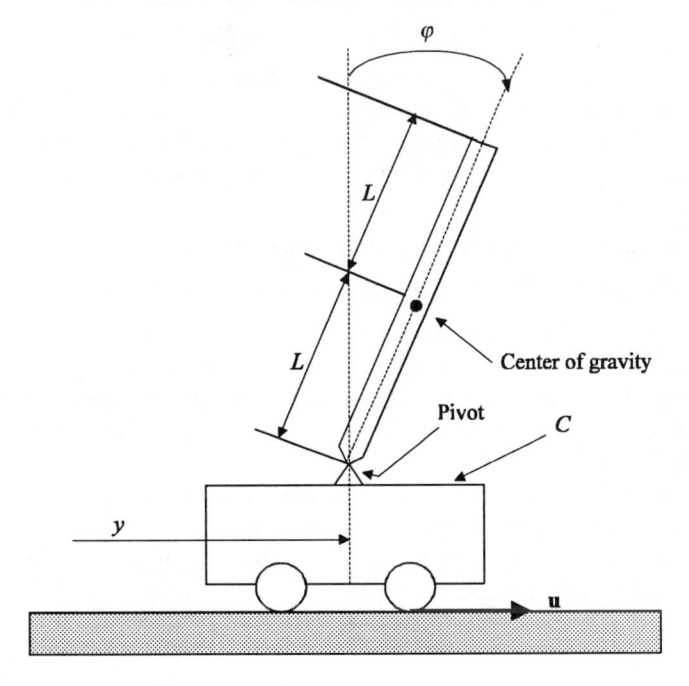

Fig. 1

Example 2: Suppose our control system be a society control system, then our object variable **x** may represent the *living-standard*; the factor lattice F may be generated by the set {*race, religion, living-standard, class, political-system,* ...}; the control force **u** may be chosen as *class-struggle, promoting production, war*, etc.; the target function $T(t)$ may be chosen as *the personal income increases 10% each year*.

Definition: Given a fuzzy control system FC(**x**, F, **u**, $T(t)$), for any $f \in F$, we may define a measure $E(f, \mathbf{x})$: $F \to [0, 1]$, which represented the effect degree of a factor f to **x**. We name this kind of measure as **Effect Measure**, and it satisfies:
(1) Suppose 1 is the biggest element in the lattice F, then we have that $E(1, \mathbf{x})=1$;
(2) For any factor $f \in F$, f^c is the complement element in lattice F, then $E(f^c, \mathbf{x})=1-E(f, \mathbf{x})$;
(3) For any factors $f, g \in F$, if $f \geq g$, then $E(f, \mathbf{x}) \geq E(g, \mathbf{x})$;
(4) For any two independent factors [2] $g, h \in F$,
  if $f=g \wedge h$, then $E(f, \mathbf{x})=E(g, \mathbf{x})E(h, \mathbf{x})$;
  if $f=g \vee h$, then $E(f, \mathbf{x})=E(g, \mathbf{x})+E(h, \mathbf{x})-E(g, \mathbf{x})E(h, \mathbf{x})$.

In practice, according to the experts' experiences and designers' professional knowledge, it is not difficult to chose a proper factor $f=f_1 \vee f_2 \vee \cdots \vee f_n$ such that $E(f, \mathbf{x}) \geq \theta$ for a big enough threshold θ ($0<\theta \leq 1$), where $f_1, f_2, \cdots, f_n$ are independent atomic factors.[2] However, it is difficult to minimize the factor set $\{f_1, f_2, \cdots, f_n\}$, or to say the problem is how we can cancel those unnecessary factors from the factor set $\{f_1, f_2, \cdots, f_n\}$. In order to keep an efficient speed in fuzzy control it is very important to get a minimal factor set. In the following, we will propose a method based on fuzzy neural networks[7]. This fuzzy neural networks is based on neural networks approaches and truth value flow inference[1, 4].

Given a control system FC(**x**, F, **u**, $T(t)$), chose $f=(f_1, f_2, \cdots, f_n)$ such that $f=f_1 \vee f_2 \vee \cdots \vee f_n$ and $E(f, \mathbf{x}) \geq \theta$ for a big enough threshold θ ($0<\theta \leq 1$), where $f_1, f_2, \cdots, f_n$ are independent atomic factors, and suppose $\mathbf{u}=(u_1, u_2, \cdots, u_m)$, $\mathbf{x}=(x_1, x_2, \cdots, x_s)$, then we can build a fuzzy neural networks in terms of f, **u**, and **x** as follows. *See* Fig.-2.

For convenience, we name this neural networks as the **Fuzzy Control Neural Networks (FCNN)** of the system FC(**x**, F, **u**, $T(t)$). We will define this neural networks step by step as following.

In order to write concisely, we give every layer a name. From first layer to last layer, they are named as "**F**actor layer", "**P**redicate layer", "**R**ule layer", "**G**rade layer", "**U**-force layer", "**E**ntombment layer", and "**S**tate layer", and they are denoted by f, p, r, g, u, e, and s respectively as shown in the Fig-2.

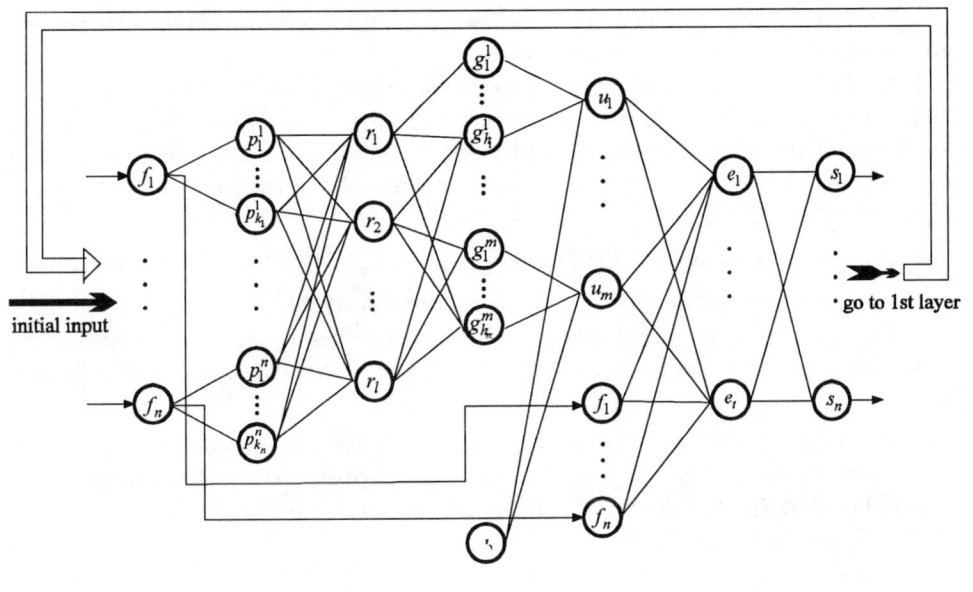

Fig. 2

1.1 The Definitions of Layers

♠ The layer f (first layer) consists of all atomic factors $f_1, f_2, ..., f_n$ that have been chosen.

♠ The layer p consists of all fuzzy predicates that they represent fuzzy quantities in factor spaces. In different factor spaces, we may have different numbers of fuzzy quantities. In general, we suppose that the number of fuzzy quantities in the factor space f_i is k_i, and p^i_j is denoted the jth fuzzy quantities in the factor space f_i, where $i=1, 2, ..., n$, $j=1, 2, ..., k_i$, see Fig-2. For example, suppose the factor space f_1 is the interval [0, 10], then the fuzzy predicates *small*, *middle*, and *large* in it may have the graphic expressions as shown in Fig-3.

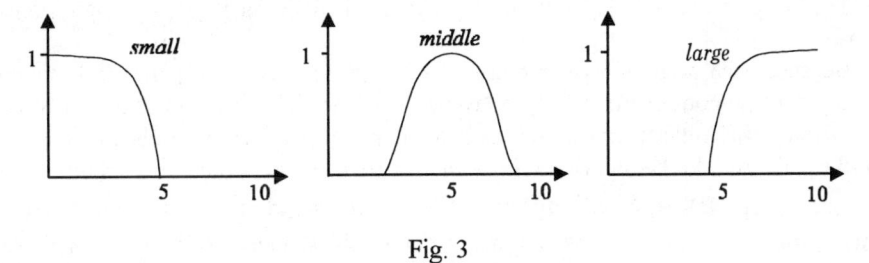

Fig. 3

♠ The layer r consists of all fuzzy rules that represent the relations between factors and control forces.

♠ The layer g consists of all fuzzy grades that represent the fuzzy measures of forces in the force spaces. Similar with layer p, we may suppose that, in different force spaces we may have different numbers of fuzzy grades, and the number of fuzzy grades in the force space u_i is h_i, and g_j^i is denoted the jth fuzzy grades in the force space u_i, where $i=1, 2, \ldots, m, j=1, 2, \ldots, h_i$, see Fig-2.

♠ The layer u consists of all the forces $u_1, u_2, \cdots, u_m$ constrained on the fuzzy control system and all the atomic factors $f_1, f_2, \cdots, f_n$ appeared in layer f.

♠ The layer e consists of a number of entombment nodes $e_1, e_2, \ldots, e_t$, where t is big enough so that we can get a corresponding relation between layer u and layer s through layer e.

♠ The layer s consists of $s_1, s_2, \ldots, s_n$ which represent factors $f_1, f_2, \cdots, f_n$ respectively. However they take the values of the intermediate states of these factors.

♠ In Fig.-2, the node b does not belong to any layer, it is a bias node. The nodes $f_1, f_2, \ldots, f_n$ belong to layer f and layer u in the same time.

1.2 The Definitions of Weights

♥ The weights between layer f and layer p are denoted as $W(f:i, p:j)$ and defined as follows:

Each node (factor) in layer f connects to those nodes (predicates) in layer p which are predicates in its factor space with weights 1, and connects to the other nodes with weights 0. That is

$$W(f:f_i, p:p_k^j) = \begin{cases} 1 & \text{if } i = j, \\ 0 & \text{if } i \neq j. \end{cases}$$

where $i, j=1, 2, \ldots, n; k=1, 2, \ldots, k_j$.

For example, node f_1 only connects to those nodes $p_1^1, \ldots, p_{k_1}^1$ which are predicates in its factor space as shown in Fig.-2.

♥ The weights between layer p and layer r are denoted as $W(p:i, r:j)$ and defined as follows:

Because each node in layer r represents a rule, so each node (rule) in layer r only connects to those nodes (predicates) in layer p which are related with this rule. Moreover, as the inputs of a logic neural networks (*see* section 1.4.1), these predicates connecting with the rule should be indicated in some sequence. For example, in logic formula $(A \wedge \neg B) \vee (\neg A \vee B)$, A, B have different role, so we should indicate them in a sequence. We use numbers $\{1, 2, 3, \ldots\}$ as weights to denote the sequence of inputs. For those not connecting nodes, using 0 as their weights. For example, for rule r_1, we may have the following weights:

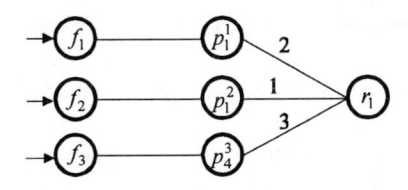

Fig. 4

which means that the number array $(O(p^2{}_1), O(p^1{}_1), O(p^3{}_4))$ is an input pattern of the logic neural network r_1, where $O(p^2{}_1), O(p^1{}_1), O(p^3{}_4)$ are the output values from nodes $p^2{}_1$, $p^1{}_1, p^3{}_4$.

♥ The weights between layer r and layer g are denoted as $W(r:i, g:j)$ and defined as follows:

Because each connection between layer r and layer g is the succeedant of a control rule, the weight attached to it reflects the **credibility, frequency** or **strength** of this rule. In other words, it is the truth value of how much possibility that the rule activates the forces. The initial weights can be got from experts' experiences, as well as by the possibility distribution function, that if the given rule r_1 is "if factor f_1 is $p^1{}_1$ and factor f_3 is $p^3{}_4$ or factor f_2 is $p^2{}_1$, then u_1 is taken $g^1{}_2$", for example, then the r_1 node is the logic formula of disjunction of the proposition "f_2 is $p^2{}_1$" and the conjunction of the proposition "f_1 is $p^1{}_1$" and the proposition "f_3 is $p^3{}_4$", i.e., ("f_2 is $p^2{}_1$" or ("f_1 is $p^1{}_1$" and "f_3 is $p^3{}_4$")), and the possibility distribution function in r_1 may be like this

$$\left\{ \frac{0.8}{g^1_1}, \frac{1}{g^1_2}, \frac{0.7}{g^1_3}, \frac{0.5}{g^1_4}, \frac{0}{g^1_5}, \ldots, \frac{0}{g^1_{h_1}} \right\}$$

This can be shown in the FCNN as following:

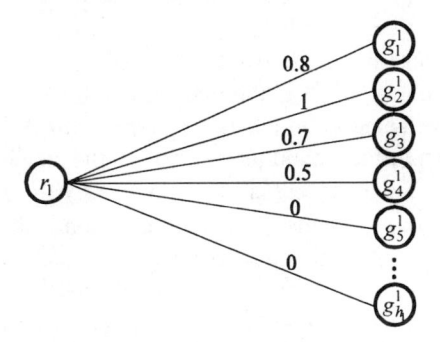

Fig. 5

Continuing the above process, we can build the all initial weights between layer r and layer g.

♥ The weights between layer g and layer u are denoted as $W(g{:}i, u{:}j)$ and defined as follows:

Each node (force) in layer u connects to those nodes (power grade) in layer g which are grades in its force space with weights 1, and connects to the other nodes with weights 0. That is

$$W(g{:}g_k^i, u{:}u_j) = \begin{cases} 1 & \text{if } i = j, \\ 0 & \text{if } i \neq j. \end{cases}$$

where $i, j=1, 2, \ldots, m$; $k=1, 2, \ldots, h_i$.

For example, node u_1 in layer u only connects to those nodes $g_1^1, g_2^1, \ldots, g_{h_1}^1$ in layer g with weights 1, and connects to the other nodes in layer g with weights 0. See Fig-2.

In layer u, the weights between the nodes $f_1, \ldots, f_n$ and the nodes in layer f are denoted as $W(f{:}i, u{:}j)$, and their values are 1 if $i=j$, and their values are 0 if $i \neq j$. For example, f_1 and f_1 have weight 1, f_1 and f_i ($i \neq 1$) have weight 0; f_2 and f_2 have weight 1, and f_2 and $f_i (i \neq 2)$ have weight 0. See Fig-2.

♥ The initial weights between layer u and layer e, and the initial weights between layer e and layer s can be given arbitrarily. Of course, we can assign these initial weights as 0. The weights between layer u and layer e are denoted as $W(u{:}i, e{:}j)$, and the weights between layer e and layer s are denoted as $W(e{:}i, s{:}j)$. Through training the FCNN, we can get some relations between layer u and layer s through the entombment layer e.

1.3 The Definitions of Transfer Functions and Implementation of FCNN

First, we denote the inputs and outputs in layer i as $I(i, j)$ and $O(i, j)$, where $i \in \{f, p, r, g, u, e, s\}$, j goes through the all nodes of layer i.

♦ In layer p, the transfer functions are those corespondent membership functions of fuzzy predicates. For example, if p^1_1 is the fuzzy predicate *very small*, then the transfer function in node p^1_1 is the membership function of *very small*; Similarly, if p^1_2 is the fuzzy predicate *small*, then the transfer function in node p^1_2 is the membership function of *small*. See Fig.-6. In fuzzy control, this process is called the **fuzzifying process**.

♦ In layer r, each node represents the antecedent of a rule, and an antecedent is a combination of propositions through logic operations, therefore each node in layer r is a logic combination of propositions. For example, if a rule is "if A and B are both true, then C takes big", then the node is prepositional combination formula $A \wedge B$; similarly, if a rule is "if A or B is true, and A and B can not be true at the same time, then C takes quite big", then the node is logic formula $(A \vee B) \wedge \neg (A \wedge B)$. About what kind of the transfer functions we should take in layer r and how to implement logic formulae by these transfer functions, we will discuss this problem in section 1.4.1. There we will establish a new kind of function form, logic neural network, to implement all forms of logic formulae. The outputs from layer r are also logic truth values. In this layer, the transfer functions are logic neural networks, and the input values for logic neural networks are obtained from the formula

$$I(r, j) = \text{sgn}(W(p{:}i, r{:}j))O(p, i)$$

where i goes through layer p, j goes through layer r, and sgn(.) is the *symbolic function*.

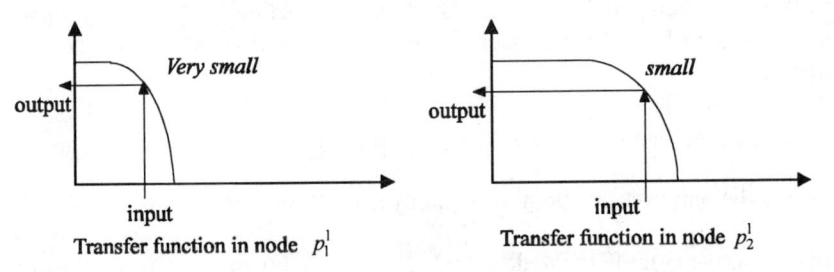

Fig. 6

♦ In layer g, the transfer functions are also those correspondent membership functions of fuzzy grades. For example, if g^1_1 is the fuzzy grade (predicate) *small*, then the transfer function in node g^1_1 is the membership function of *small*; Similarly, if g^1_2 is the fuzzy grade (predicate) *middle*, then the transfer function in node g^1_2 is the membership function of *middle*. See Fig.-7. In fuzzy control, this process is called the **defuzzifying process**.

The input for the transfer function in the node j is $\underset{i}{\oplus}(W(r{:}i, g{:}j) \otimes O(r, i))$, and this is a truth value obtained from the outputs $O(r, i)$ by using the dual operators ($\oplus, \otimes$) on the interval [0, 1]. And the output $O(g, j)$ in node j is the shadow area got from $\underset{i}{\oplus}(W(r{:}i, g{:}j) \otimes O(r, i))$-cut of correspondent transfer function, i.e., membership function, *see* Fig.-7:

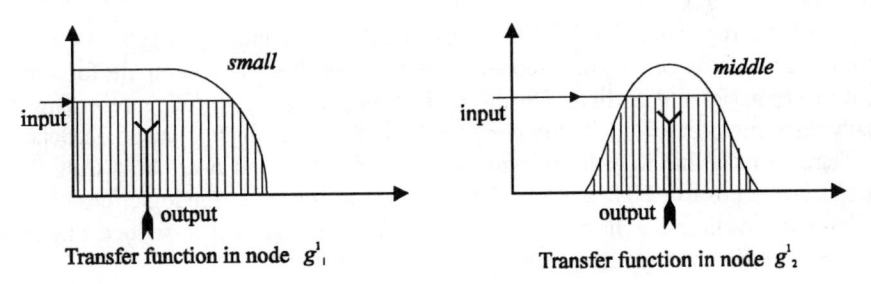

Fig. 7

♦ In layer u, when $j=u_k$ ($k=1, \ldots, m$), the transfer function is the *barycentric function*. The input for the transfer function in the node j is the union of all shadow areas obtained in layer g, i.e., $\bigcup_i (W(g{:}i, u{:}j) * O(g,i))$, where we consider $O(g, i)$ as a fuzzy set and operator $*$ is defined as the multiply of number and fuzzy set. The output in node j is the u_0 measure force if (u_0, y_0) is the barycenter of the set $\bigcup_i (W(g{:}i, u{:}j) * O(g,i))$.

In layer u, when $j=f_k$ ($k=1, \ldots, n$), the transfer function is the *identity function*. Its input is the output of the correspondent node in layer f, and its output has the same value as the input.

♦ In layer e, the input in node e_j ($j=1, \ldots, t$) is $\sum_i W(u{:}i, e{:}j) O(u,i)$ (i goes through layer u), and the output is got from the input by using a Signoid transfer function [11].

♦ In layer s, the input in the node s_j ($j=1, \ldots, n$) is $\sum_i W(e{:}i, s{:}j) O(e,i)$, the output is a value read from sensor. In the computer simulation, we can give estimated output values.

1.4 The Learning Rules and Learning Capacity of the FCNN

The adaptive FCNN is a *time-varying* system. System parameters gradually change as the FCNN processes data. Below we discuss how neural network algorithms can adaptively train FCNN rules from training data. In principle, the training may be performed in three different ways: 1. training the *internal connection strength* among the propositions within a rule and *necessary degrees* of propositions for a rule; 2. training membership functions, i.e., *fuzzy representations* of rules; 3. training the weights, i.e., *inference strengths* of rules. In the following, we will respectively discuss these three problems in the following section 1.4.1, section 1.4.2, and section 1.4.3.

1.4.1 Logic Neural Networks

The logic representation of rules is one of the most important subjects in AI and decision-making science. A good representation can enable us to compute formulae and execute rules easily and feasibly. On the other hand, the problem of representing rule is actually the same problem of representing the logic formula which is the antecedent of the rule. That is to say that in order to represent the rule "if A and B are both true, then the control force is positive big", we actually only need to represent the logic formula $A \wedge B$. Based on the truth value flow inference[1, 4], following we will introduce one type of networks to represent logic formulae.

We first discuss the representation of formulae only with operators $\wedge$ and $\neg$.

If a rule is "if A is true, then the control force is positive big", then the formula is A, and it has the following network representation as shown in Fig.-8 (1).

If a rule is "if A is not true, then the control force is negative big", then the formula is $\neg A$, and it has the representation as shown in Fig.-8 (2).

If a rule is "if A and B are both true, then the control force is positive big", then the formula is $A \wedge B$, and it has the following network representation as shown in Fig.-8 (3).

If a rule is "if A is true and B is not true, then the control force is positive middle", then the formula is $A \wedge \neg B$, and it has the following network representation as shown in Fig.-8 (4).

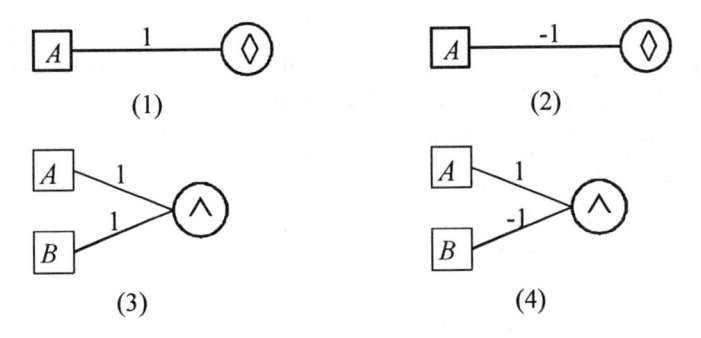

Fig. 8

Further more, if we use "A is true with truth value s" and "B is not true with truth value t" to replace "A is true" and "B is not true" respectively in the above rules, then we can get the following four rules and correspondent network representations.

Corresponding to the first rule, the rule is "if A is true with truth value s, then the control force is positive big", and it has the formula $A[s]$ and the representation as shown in Fig.-9 (1).

Corresponding to the second rule, the rule is "if A is false with truth value s, then the control force is negative big", and it has the formula is $\neg A[s]$ and the representation as shown in Fig.-9 (2).

Corresponding to the third rule, the rule is "if A is true with truth value s and B is true with truth value t, then the control force is positive big", and it has the logic formula $A[s] \wedge B[t]$ and the network representation as shown in Fig.-9 (3).

Corresponding to the fourth rule, the rule is "if A is true with truth value s and B is false with truth value t, then the control force is positive middle", and it has the logic formula $A[s] \wedge \neg B[t]$ and the network representation as shown in Fig.-9 (4).

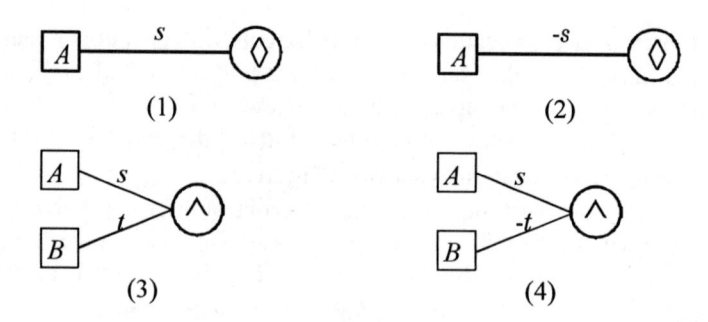

Fig. 9

Let s and t range in the interval $[-1, 1]$, then the eight figures in Fig.-8 and Fig.-9 can be integrated into the Fig.-10 as shown in the following:

$s, t \in [-1, 1]$.

Fig. 10

When $t=0$, just need to use ◊ to replace ∧ in the same place, Fig.-10 becomes Fig.-9 (1), or Fig.-9 (2).

Besides the usual functions such as *symbolic function* sgn, *integer function* int, and *abstract function* abs, we also define the *decimal function* dec and *boundary function* bnd as follows:

Definition: The decimal function dec on the real line $\Re$ is defined as follows:

$$\text{dec}(x) \overset{\Delta}{=} x - \text{int}(x) \qquad (x \in \Re).$$

Definition: The pair boundary function bnd on the real line $\Re$ with α as the lower bound and β as the upper bound is defined as follows:

$$\text{bnd}(x)_\alpha^\beta \overset{\Delta}{=} \begin{cases} \alpha & \text{if } x < \alpha; \\ \beta & \text{if } x > \beta; \\ x & \text{else.} \end{cases}$$

In particular, $\text{bnd}_0^1(.)$ represents the boundary function which has the lower bound 0, and upper bound 1; $\text{bnd}_{-1}^1(.)$ represents the boundary function which has the lower bound -1 and upper bound 1; etc. For simplicity, we use $\text{bnd}(.)$ to denote $\text{bnd}_0^1(.)$.

Using T(A) and T(B) to denote the truth values of A and B, and using in(···) and out(···) to denote the input and output values of each node in a network respectively, then, in reference to Fig.-10, the flowing and transferring of truth values in the network have the following procedures:

In the Fig.-10, the input and output in 'node A' are both defined as T(A), i.e., in(A)=out(A)=T(A); The input and output in 'node B' are both defined as T(B), i.e., in(B)=out(B)=T(B); The inputs in 'node $\wedge$' are T(A) and T(B), i.e., $\text{in}_1(\wedge) = \text{T}(A)$, $\text{in}_2(\wedge) = \text{T}(B)$, and the output value out($\wedge$) in 'node $\wedge$' is defined as

$$\text{out}(\wedge) \stackrel{\Delta}{=} \text{bnd}\left(\frac{\text{dec}(\text{sgn}(s)\text{in}_1(\wedge))}{\text{abs}(s)}\right) \otimes \text{bnd}\left(\frac{\text{dec}(\text{sgn}(t)\text{in}_2(\wedge))}{\text{abs}(t)}\right) \quad \text{(Formula-I)}$$

where $\otimes$ could be any one of the bi-operators that correspond to logic operator $\wedge$. That is, $\otimes \in \{\min, \bullet, \times, \wedge, \text{bounded-product}, \ldots\}$.

When $t=0$, we define $\text{bnd}\left(\dfrac{\text{dec}(\text{sgn}(t)\text{T}(A))}{\text{abs}(t)}\right) = 1$, and use $\Diamond$ to replace $\wedge$, then, in this case, $\text{out}(\Diamond) = \text{bnd}\left(\dfrac{\text{dec}(\text{sgn}(s)\text{T}(A))}{\text{abs}(s)}\right)$, this is the output value in the 'node $\Diamond$' in Fig.-9 (1) or Fig.-9 (2).

In Formula-I, when s and t range in the interval [-1,1], we can get the truth value of any logic formula only containing logic operators $\wedge$ and $\neg$, such as A, $A \wedge \neg B$, $A[s] \wedge B[t]$, etc. For example, taking $s=1$, $t=-1$ and taking $\otimes$ to be <u>min</u>, then

$$\text{bnd}\left(\frac{\text{dec}(\text{sgn}(s)\text{T}(A))}{\text{abs}(s)}\right) \otimes \text{bnd}\left(\frac{\text{dec}(\text{sgn}(t)\text{T}(B))}{\text{abs}(t)}\right)$$

$$= \text{bnd}(\text{dec}(\text{T}(A))) \otimes \text{bnd}(\text{dec}(-\text{T}(B)))$$

$$= \text{T}(A) \otimes (1-\text{T}(B)) = \min\{\text{T}(A), (1-\text{T}(B))\}$$

This is exactly the truth value of logic formula $A \wedge \neg B$ when $\wedge$ takes <u>min</u> and $\neg$ means that T($\neg B$)=1-T(B) for any proposition B.

∀ For those logic formulae only with operators $\vee$ and $\neg$, we have a similar result. That is, for logic formulae $A[s] \vee B[t]$, $A[s] \vee \neg B[t]$, etc., we have the networks representation as follows:

```
   A ╲ s
      ╲
       ⊙ ∨        s, t ∈ [-1, 1].
      ╱
   B ╱ t
```

Fig. 11

In Fig.-11, inputs and outputs in the nodes A and B are same with those in Fig.-10, the inputs in 'node $\vee$' are T(A) and T(B), i.e., $in_1(\vee) = T(A)$, $in_2(\vee) = T(B)$, and the output value out($\vee$) in 'node $\vee$' is defined as

$$out(\vee) \overset{\Delta}{=} bnd\left(\frac{dec(sgn(s))in_1(\vee))}{abs(s)}\right) \oplus bnd\left(\frac{dec(sgn(t))in_2(\vee))}{abs(t)}\right) \quad \text{(Formula-II)}$$

where $\oplus$ could be any one of the bi-operators that correspond to logic operator $\vee$. That is, $\oplus \in \{\max, +, \vee, \text{bounded-sum}, \ldots\}$.

When $t=0$, we define $bnd\left(\frac{dec(sgn(t)T(A))}{abs(t)}\right) = 0$, and use $\Diamond$ to replace $\vee$.

➤ For logic operator $\rightarrow$, we use the logic formula $(A \wedge \neg A) \vee (A \wedge B)$ to replace $A \rightarrow B$. The following is a brief discussion of the reason and the underlying assumptions:

We know that, in a rule "if A can imply B, then the control force is positive big", the really meaning of $A \rightarrow B$ is that A is true and A can imply B. In practice, if we consider T(A) as a measure of information of A we got, and consider T(A) as a confidence degree for A in the same time, then the above rule is actually "if A is true and if A can imply B in the same time, then the control force is positive big". According to this explanation, we can use $A \wedge (A \rightarrow B)$ to replace $A \rightarrow B$ reasonably. Through logic operations, we have $A \wedge (A \rightarrow B) = (A \wedge \neg A) \vee (A \wedge B)$. Therefore we can reasonably use $(A \wedge \neg A) \vee (A \wedge B)$ to replace $A \rightarrow B$ in practice, for example, in financial forecast field, or fuzzy control field.

Through these replacements, those logic formulae containing $\rightarrow$ can be changed into the formulae without $\rightarrow$. Therefore we can represent them by networks according to the previous discussions.

Example 1. In terms of the results we have obtained above, for any logic formula containing $\wedge$, $\vee$, $\neg$, for example,

$$((A \vee B) \wedge \neg(\neg A \wedge C)) \vee \neg((P \wedge \neg Q) \vee (\neg P \wedge \neg R))$$

we can represent it by a network as follows:

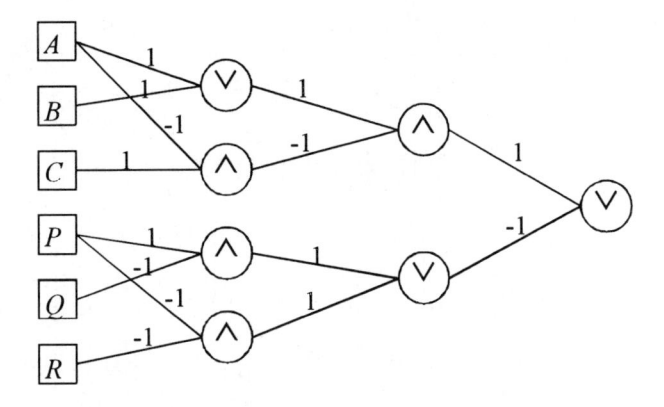

Fig. 12

In this neural network, the transfer function in the nodes '∧' is Formula-I; and the transfer function in the nodes '∨' is Formula-II.

Example 2. Suppose we have a rule as that "among the four propositions A_1, A_2, A_3, and A_4, if there are at least two propositions whose truth values are equal to or larger than 0.8, and in the same time there are at least three propositions whose truth values are equal to or larger than 0.5; or there is no more than one proposition, for its negative proposition, the truth value is equal to or larger than 0.3, then the control force is negative middle", then the logic formula in the antecedence of this rule is that

$$(((A_1 \wedge A_2) \vee (A_1 \wedge A_3) \vee (A_1 \wedge A_4) \vee (A_2 \wedge A_3) \vee (A_2 \wedge A_4) \vee (A_3 \wedge A_4))[0.8] \wedge$$
$$((A_1 \wedge A_2 \wedge A_3) \vee (A_1 \wedge A_2 \wedge A_4) \vee (A_1 \wedge A_3 \wedge A_4) \vee (A_2 \wedge A_3 \wedge A_4))[0.5]) \vee$$
$$(\neg((\neg A_1 \vee \neg A_2 \vee \neg A_3) \wedge (\neg A_1 \vee \neg A_2 \vee \neg A_4) \wedge (\neg A_1 \vee \neg A_3 \vee \neg A_4) \wedge$$
$$(\neg A_2 \vee \neg A_3 \vee \neg A_4)))[0.7]$$

and this logic formula has the representation in Fig.-13.
In the Fig.-13, for those lines there are no weights on them, the dotted lines are supposed to have weights -1, and the real lines are supposed to have weights 1.

Through the logic neural networks, as long as we input the truth values of atomic propositions $A, B, C,...,$ we can get the truth value of any logic formula.

The learning problem of logic neural networks is as follows:
Here we suppose that $W(i:j, i+1:k)$ represents the weight between the j-th node in layer i and the k-th node in layer $i+1$, and use $\Delta W(i: j, i+1:k)$ to denote its difference; and suppose that $I(i, j)$, $O(i, j)$ and $E(i, j)$ to denote the input, output and error of the j-th node in layer i respectively; and use T to denote the target value; then we may define $E(i, j)$ and $\Delta W(i: j, i+1: k)$ as follows:

In the output layer, $E(i,j) = T - O(i,j)$
in other layers,

$$E(i,j) = \sum_k W(i:j, i+1:k)E(i+1,k)$$

$$\Delta W(i:j, i+1:k) = -\alpha E(i+1,k)$$

where α a is proper positive number parameter.

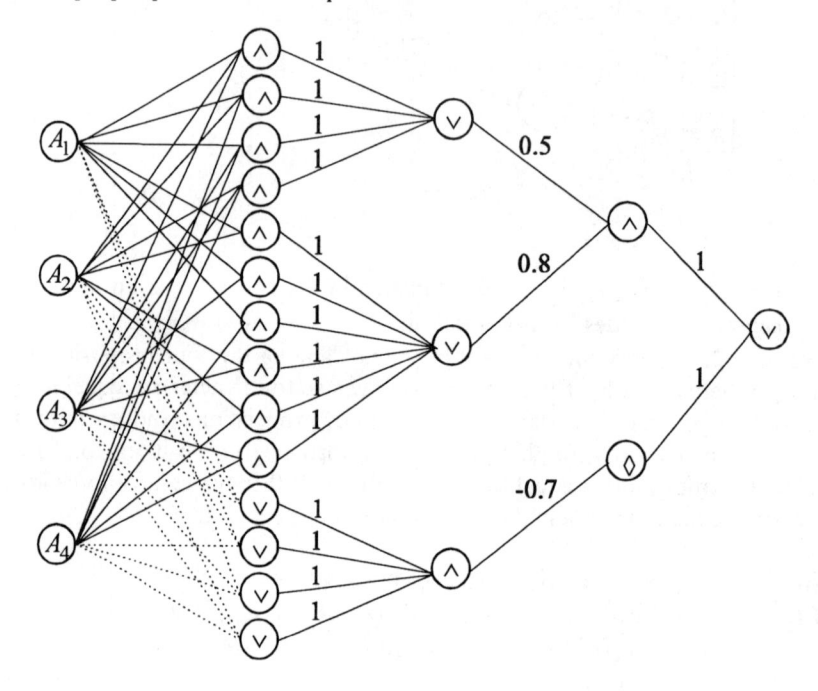

Fig. 13

1.4.2 Learning of Membership Function

One of the essential problems of training FCNN is how to train fuzzy membership functions, and the problem of training fuzzy membership functions is how to set up a reasonable mathematical methods to modify membership function when the membership function is changed at one point. In the following, we will discuss the methods how to modify normal membership functions. A normal fuzzy set has the following definition:

Definition: A normal fuzzy set $\tilde{A}$ is a fuzzy set on the real line that satisfies:

(1) exist at least one point $u_0 \in U$ such that $\tilde{A}(u_0) = 1$;

(2) as a function on the real line, $\tilde{A}$ is non-decreasing on $(-\infty, u_0]$, and is non-increasing on $[u_0, +\infty)$.

The general criteria of modifying membership functions should be:

① Modifying radius is positive proportional with the membership degree and the error δ;

② Considering left wing, right wing and median part (the set which have membership degree 1) separately;

③ For a normal membership function, the modified membership function should also be normal.

In reference to the above general criteria, we can give a mathematical method of modifying a normal membership function. We first give several notations:

$\mathcal{F}_N(U)$ denotes the set of all normal fuzzy sets on the universe U, where U is a subset of the real line.

$$m^- \triangleq \inf\{u \mid \tilde{A}(u) = 1\}, \ m^+ \triangleq \sup\{u \mid \tilde{A}(u) = 1\}.$$

$$A^- \triangleq (-\infty, m^-], \ A^+ \triangleq [m^+, +\infty), \ \overline{A} \triangleq [m^-, m^+].$$

$\tilde{A}_\lambda \triangleq \{x \mid x \in U, \tilde{A}(x) \geq \lambda\}$, and is called λ-cut set of $\tilde{A}$. In the case of $\tilde{A}$ is normal, $\tilde{A}_\lambda$ is an interval, and we denote $\tilde{A}_\lambda = [a_\lambda^-, a_\lambda^+]$, where $\lambda \in [0,1]$.

$\rho: [0,1] \times [0,1] \to [0, +\infty)$ is an increasing function with regard to its two variables respectively, and satisfies $\rho(y, 0) = 0$ for any $y \in [0,1]$.

$M: [0, +\infty) \to Z^+ = \{1, 2, 3, \cdots\}$ is an increasing function.

$\mathcal{U}(u)$ denotes a neighbor of u in U, and $\mathcal{U}(u, r)$ denotes the neighbor of u with radius r, $\mathcal{U}^-(u, r)$ denotes the left half-neighbor of u with radius r, and $\mathcal{U}^+(u, r)$ denotes the right half-neighbor of u with radius r. They can be defined by the following formulae.

$$\mathcal{U}(u, r) \triangleq \{x \mid x \in U, |x - u| \leq r\},$$

$$\mathcal{U}^-(u, r) \triangleq \{x \mid x \in U, 0 \leq (u - x) \leq r\},$$

$$\mathcal{U}^+(u, r) \triangleq \{x \mid x \in U, 0 \leq (x - u) \leq r\}.$$

Using u to denote the adjusted point, we will discuss the modifying membership function problem in the three cases, $u \in A^-$, $u \in A^+$, and $u \in \overline{A}$.

● Case of $u \in A^-$.

Definition: Define mapping Φ as follows:
$$\Phi: \mathcal{F}_N(U) \times U \times [-1, 1] \mapsto \mathcal{F}_N(U)$$
$$(\tilde{A}, u, \delta) \to \tilde{A}_{(u, \delta)}$$

where $\tilde{A}_{(u, \delta)}$ is defined as follows:

$$\tilde{A}_{(u,\delta)}(x) \stackrel{\Delta}{=} \begin{cases} \xi^-(\tilde{A},u,\delta)(x) & x \leq u; \\ \xi^+(\tilde{A},u,\delta)(x) & u < x \leq m^-; \\ \varepsilon(\tilde{A},u,\delta)(x) & m^- < x \leq (m^- + \theta(\tilde{A},\kappa)); \\ \zeta^+(\tilde{A},u,\delta)(x) & (m^- + \theta(\tilde{A},\kappa)) < x \leq \tau; \\ \tilde{A}(x) & \text{else.} \end{cases}$$

where

$$\tau = m^- + \theta(\tilde{A},\kappa) + \rho\left(\tilde{A}(m^- + \theta(\tilde{A},\kappa)), (1 - \tilde{A}(m^- + \theta(\tilde{A},\kappa)))\right)$$

$$\kappa = 1 - \xi^+(\tilde{A},u,\delta)(m^-)$$

and other functions are defined respectively by the following formulae.

$\theta(\tilde{A},\kappa): \mathfrak{F}_N(U) \times [0,1] \mapsto [0,+\infty)$

is an increasing function with respect to κ when $\tilde{A}$ is fixed, and satisfies the condition: $\theta(\tilde{A},0) = 0$ for any $\tilde{A}$.

$\varepsilon(\tilde{A},u,\delta): [m^-, m^- + \theta(\tilde{A},\kappa)] \mapsto [0,1]$

is an increasing function, and satisfies the conditions that

$\varepsilon(\tilde{A},u,\delta)(m^-) = \xi^+(\tilde{A},u,\delta)(m^-)$ and $\varepsilon(\tilde{A},u,\delta)(m^- + \theta(\tilde{A},\kappa)) = 1$.

Suppose for any radius ρ, we divide the two intervals $[u-\rho, u]$ and $[u, u+\rho]$ to be $M(\rho)$ equal segments so that each section has the length $\frac{\rho}{M(\rho)}$, then we have the following function definitions.

Definition: Define function $\xi^-(\tilde{A},u,\delta)$ as follows:

When $x \in \mathcal{U}^-(u,0) = \mathcal{U}^-\left(u, \frac{0 \times \rho}{M(\rho)}\right)$,

$$\xi^-(\tilde{A},u,\delta)(x) \stackrel{\Delta}{=} \alpha_0^- \stackrel{\Delta}{=} \text{bnd}\left(\tilde{A}(x) + \delta\right) = \text{bnd}\left(\tilde{A}(x) + \frac{(M(\rho)-0)}{M(\rho)}\delta\right)$$

When $x \in \mathcal{U}^-\left(u, \frac{1 \times \rho}{M(\rho)}\right) \setminus \mathcal{U}^-\left(u, \frac{0 \times \rho}{M(\rho)}\right)$,

$$\xi^-(\tilde{A},u,\delta)(x) \stackrel{\Delta}{=} \alpha_1^- \stackrel{\Delta}{=} \min\left(\text{bnd}\left(\tilde{A}(x) + \frac{(M(\rho)-1)}{M(\rho)}\delta\right), \alpha_0^-\right)$$

In general situations, when $x \in \mathcal{U}^-\left(u, \frac{i \times \rho}{M(\rho)}\right) \setminus \mathcal{U}^-\left(u, \frac{(i-1) \times \rho}{M(\rho)}\right)$,

$$\xi^-(\tilde{A},u,\delta)(x) \stackrel{\Delta}{=} \alpha_i^- \stackrel{\Delta}{=} \min\left(\text{bnd}\left(\tilde{A}(x) + \frac{(M(\rho)-i)}{M(\rho)}\delta\right), \alpha_{i-1}^-\right)$$

where $i = 1, 2, \ldots, M(\rho)$.

Definition: Define function $\xi^+(\tilde{A}, u, \delta)$ as follows:

When $x \in \mathcal{U}^+(u, 0) \cap A^- = \mathcal{U}^+\left(u, \frac{0 \times \rho}{M(\rho)}\right) \cap A^-$,

$$\xi^+(\tilde{A}, u, \delta)(x) \triangleq \alpha_0^+ \triangleq \text{bnd}(\tilde{A}(x) + \delta) = \text{bnd}\left(\tilde{A}(x) + \frac{(M(\rho) - 0)}{M(\rho)} \delta\right)$$

When $x \in \left(\mathcal{U}^+\left(u, \frac{1 \times \rho}{M(\rho)}\right) \setminus \mathcal{U}^+\left(u, \frac{0 \times \rho}{M(\rho)}\right)\right) \cap A^-$,

$$\xi^+(\tilde{A}, u, \delta)(x) \triangleq \alpha_1^+ \triangleq \max\left(\text{bnd}\left(\tilde{A}(x) + \frac{(M(\rho) - 1)}{M(\rho)} \delta\right), \alpha_0^+\right)$$

In general situations, when $x \in \left(\mathcal{U}^+\left(u, \frac{i \times \rho}{M(\rho)}\right) \setminus \mathcal{U}^+\left(u, \frac{(i-1) \times \rho}{M(\rho)}\right)\right) \cap A^-$,

$$\xi^+(\tilde{A}, u, \delta)(x) \triangleq \alpha_i^+ \triangleq \max\left(\text{bnd}\left(\tilde{A}(x) + \frac{(M(\rho) - i)}{M(\rho)} \delta\right), \alpha_{i-1}^+\right)$$

where $i = 1, 2, \ldots, M(\rho)$.

Denote $\delta' = 1 - \tilde{A}(m^- + \theta(\tilde{A}, \kappa))$, then $\zeta^+(\tilde{A}, u, \delta)$ can be defined as follows:

Definition: Define function $\zeta^+(\tilde{A}, u, \delta)$ as follows:

When $x \in \mathcal{U}^+\left(m^- + \theta(\tilde{A}, \kappa), 0\right) = \mathcal{U}^+\left(m^- + \theta(\tilde{A}, \kappa), \frac{0 \times \rho}{M(\rho)}\right)$,

$$\zeta^+(\tilde{A}, u, \delta)(x) \triangleq \beta_0^+ \triangleq \text{bnd}(\tilde{A}(x) + \delta') = \text{bnd}\left(\tilde{A}(x) + \frac{(M(\rho) - 0)}{M(\rho)} \delta'\right)$$

When $x \in \mathcal{U}^+\left(m^- + \theta(\tilde{A} + \kappa), \frac{1 \times \rho}{M(\rho)}\right) \setminus \mathcal{U}^+\left(m^- + \theta(\tilde{A} + \kappa), \frac{0 \times \rho}{M(\rho)}\right)$,

$$\zeta^+(\tilde{A}, u, \delta)(x) \triangleq \beta_1^+ \triangleq \min\left(\text{bnd}\left(\tilde{A}(x) + \frac{(M(\rho) - 1)}{M(\rho)} \delta'\right), \beta_0^+\right)$$

In general situations, when

$x \in \mathcal{U}^+\left(m^- + \theta(\tilde{A} + \kappa), \frac{i \times \rho}{M(\rho)}\right) \setminus \mathcal{U}^+\left(m^- + \theta(\tilde{A} + \kappa), \frac{(i-1) \times \rho}{M(\rho)}\right)$,

$$\zeta^+(\tilde{A}, u, \delta)(x) \triangleq \beta_i^+ \triangleq \min\left(\text{bnd}\left(\tilde{A}(x) + \frac{(M(\rho) - i)}{M(\rho)} \delta'\right), \beta_{i-1}^+\right)$$

where $i = 1, 2, \ldots, M(\rho)$.

All above procedures can be shown in the following Fig.-14.

Theorem: The mapping Φ is well defined. That is, for given functions ρ, M, θ, and ε, we can get one and only one membership function $\tilde{A}_{(u, \delta)}$ from array $(\tilde{A}, u, \delta)$ according to the above definitions, and this membership function is normal.

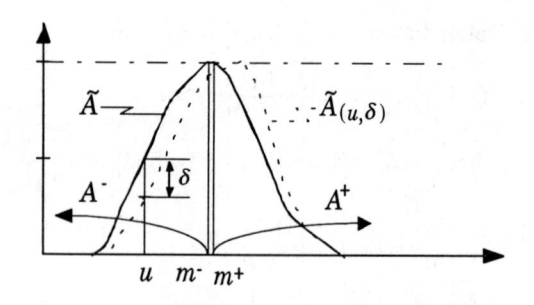

Fig. 14

● Case of $u \in A^+$.

When the adjusted point u is in A^+, we can similarly define a mapping Φ, and have a correspondent result.

● Case of $u \in \overline{A}$.

When the adjusted point u is in $\overline{A}$, we have $u \in \tilde{A}_{1-\delta} = [a^-_{1-\delta}, a^+_{1-\delta}]$, and we may discuss this problem according to the distances between u and $a^-_{1-\delta}$, $a^+_{1-\delta}$.

In the case of $|u - a^-_{1-\delta}| < |a^+_{1-\delta} - u|$, we may define $\Phi(\tilde{A}, u, \delta) \stackrel{\Delta}{=} \tilde{A}_{(u,\delta)}$ to be

$$\tilde{A}_{(u,\delta)}(x) \stackrel{\Delta}{=} \begin{cases} \xi^-(\tilde{A}, u, \delta)(x) & x \leq u; \\ \varepsilon(\tilde{A}, u, \delta)(x) & u < x \leq (u + \theta(\tilde{A}, \kappa)); \\ \zeta^+(\tilde{A}, u, \delta)(x) & (u + \theta(\tilde{A}, \kappa)) < x \leq \tau; \\ \tilde{A}(x) & \text{else.} \end{cases}$$

where $\tau = u + \theta(\tilde{A}, \kappa) + \rho(\tilde{A}(u + \theta(\tilde{A}, \kappa)), (1 - \tilde{A}(u + \theta(\tilde{A}, \kappa))))$,

$\kappa = 1 - \xi^-(\tilde{A}, u, \delta)(u)$,

and other functions have the same meanings as in the previous definitions.

In the case of $|u - a^-_{1-\delta}| > |a^+_{1-\delta} - u|$, we have a corresponding result.

In the case of $|u - a^-_{1-\delta}| = |a^+_{1-\delta} - u|$, we had better take further observing, so we do not take any modification in this moment.

Corresponding to the above theorem, we have that

Theorem: In the cases of $u \in A^+$, or $u \in \overline{A}$, the mapping Φ is well defined.

1.4.3 Learning of FCNN

In control process, it is not necessary to drive object **x** from an initial position to the target position at only one moving, it often needs more than one time to do so. So in our FCNN neural networks, we need not FCNN converging in one loop, but we only need FCNN converging at eventually. In practice, we can give a large enough positive integer L_0, when FCNN converges within L_0 loops, we call it converging; When FCNN does still not converge after L_0 loops, we call it not converging, and we modify the weights and membership functions in FCNN according to these errors. In the following paragraph, we will give the learning rules step by step.

Let's first give the concept of **correlate strength** between factors and control forces as follows:

Definition: The **correlate strength** between factor f_i ($i=1, 2, \ldots, n$) and control force u_j ($j=1, 2, \ldots, m$) is defined as

$$\operatorname{cor}(f_i, u_j) \overset{\Delta}{=} \bigvee_{k_1, k_2, k_3} \{W(f{:}f_i, p{:}k_1) \wedge W(p{:}k_1, r{:}k_2) \wedge W(r{:}k_2, g{:}k_3) \wedge W(g{:}k_3, u{:}u_j)\}$$

where k_1, k_2, and k_3 go through layer p, layer r, and layer g respectively.

Now let's discuss the learning rules of FCNN.

(1) Modifying of layer u

Suppose $\gamma: [0, +\infty) \to [0, +\infty)$ be an increasing function, and satisfy that $\gamma(0)=0$; and denote the target value in node s_i as T_i, and the output value at loop l of node s_i as $O_i(l)$, where $i=1, 2, \ldots, n$. Especially, the output value at loop L_0+1 of node s_i is $O_i(L_0+1)$.

In the case of the output in node s_i converges to a fixed value, then this value is $O_i(L_0+1)$. In this case, the modified value in node u_j caused by s_i is $\Delta_{ji} = \alpha_i (\operatorname{sgn}(u_j)|u_j + \operatorname{sgn}(u_j)\gamma(|T_i - O_i(L_0+1)|)| - u_j)$; In the case of the output in node s_i does not converges, the modified value in node u_j caused by s_i is $\Delta_{ji} = \alpha_i (\operatorname{sgn}(u_j)|u_j - \operatorname{sgn}(u_j)\gamma(|T_i - O_i(L_0+1)|)| - u_j)$. Adding all Δ_{ji}, and considering the *correlate strength* $\operatorname{cor}(f_i, u_j)$ between f_i (i.e., s_i) and u_j, we can get the modification Δ_j in node u_j computed by the following formula:

$$\Delta_j = \sum_{i=1}^{n} \alpha_i \operatorname{cor}(f_i, u_j)(\operatorname{sgn}(u_j)|u_j + \operatorname{sgn}(u_j)\gamma(|T_i - O_i(L_0+1)|)| - u_j)$$

where $j=1, 2, \ldots, m$, and operator $\pm$ means that when s_i converges, it takes *plus* +; when s_i does not converges, it takes *minus* -.

(2) Modifying of layer g

After the Δ-modifications in layer u, we can modify the weights $W(g{:}*, u{:}*)$ and outputs in layer g according to the Δ-rules in Back-propagation neural networks.

(3) Modifying of layer r

From the modifying in layer g, we can modify the weights $W(r:*, g:*)$ and output in layer r according to the Δ-rules in Back-propagation neural networks. Further more, based on the errors of inputs in layer g, we can modify the logic neural networks in layer r, and get the errors of inputs in layer r using the approach given in section 1.4.1.

(4) Modifying of layer p

From the errors of inputs in layer r, we can modify the weights $W(p:*, r:*)$ and outputs in layer p according to the Δ-rules in Back-propagation neural networks, and get the errors of inputs in layer p.

(5) Modifying of membership functions in layer p

In step (4), the errors of inputs can be considered as the δ-changes of the membership degrees. From these δ-changes, we can modify the membership functions in layer p according to the approach given in section 1.4.2.

Theorem: Above learning rules are reasonable and feasible.

This theorem means that the errors in FCNN will become smaller and smaller by using above learning rules to train it. It is not difficult to prove the theorem according to the limit theory in Calculus and neural network theory [11], so here we omit this proof.

1.5 Factor Analysis According to the Learning Results of FCNN

We first give some denotations used in the following.

Define $N_n = \{1, 2, ..., n\}$, and use $N_{n_1} \times N_{n_2} \times \cdots \times N_{n_h}$ to denote the Cartesian Product of $N_{n_1}, N_{n_2}, ..., $ and N_{n_h} as usual. A **passage** through an element i_j of the set N_{n_j} is a point $\sigma_{i_j} \triangleq (i_1, i_2, ..., i_{j-1}, i_j, i_{j+1}, ..., i_h)$ in the Cartesian Space $N_{n_1} \times N_{n_2} \times \cdots \times N_{n_h}$, and we denote the set of all the passages through the element i_j as $\Sigma_{i_j} \triangleq \{\sigma_{i_j} \triangleq (i_1, i_2, ..., i_{j-1}, i_j, i_{j+1}, ..., i_h) \mid 1 \leq i_r \leq n_r; r=1, 2, ..., h; r \neq j\}$.

We also denote $\sigma_{i_j}[r] = i_r, r=1, 2, ..., h$.

Passage-Capacity Method

Suppose net_{ij} is the jth element in the layer i, its effect measure to the neural network with h layers is defined as

$$E(net_{ij}) \triangleq \bigoplus_{\sigma_{i_j} \in \Sigma_{i_j}} \bigotimes_{1 \leq r < h} W\left(r, \sigma_{i_j}[r], \sigma_{i_j}[r+1]\right)$$

where net_{ij} goes through all the nodes of the neural networks, and $W(r, i, j)$ is the weight between node net_{ri} and node $net_{(r+1)j}$. Especially, in the fuzzy control neural networks we are discussing, for factor f_j in the layer f, its effect measure has the following representation:

$$E(f_j,\mathbf{x}) \stackrel{\Delta}{=} E(net_{1_j}) \stackrel{\Delta}{=} \bigoplus_{\sigma_{1_j} \in \Sigma_{1_j}} \bigotimes_{1 \leq r < h} W(r, \sigma_{1_j}[r], \sigma_{1_j}[r+1])$$

where $j=1, 2, \ldots, n$.

In the above formula, when $(\oplus, \otimes)$ takes the dual operators $(\vee, \wedge)$ (*means* max, min), $(+, *)$ (*means* plus, multiply), $(\vee, *)$ (*means* max, multiply), $(P_+, P_\times)$ (*means* probability-plus, probability-multiply), ..., we can take many factor analysis methods of neural networks.

Accumulated-Error Method

The accumulated error method was proposed by F. Wong [12]. In this method, after the neural networks learning, we compute the accumulated errors in the nodes of input layer, according to the accumulated error, we can know that which factor has significant effect on the neural networks, and which factor has insignificant effect.

Differentiation Method

The Differentiation method was proposed by T. H. Goh [10]. In this method, after the neural networks learning, we give a small change in a node of input layer, then we get the effect value of this node on the neural networks according to the change in the output layer caused by the change in the input layer.

Based on the above discussion, the factor analysis process can be stated as follows:
① training a fuzzy control neural network with large enough data,
② finding the factors which have the most insignificant effect on the neural network,
③ eliminating those factors got in step ②, and getting a new neural network,
④ training the new neural network, if it converges, go to step ① to repeat the above process; if it does not converge, go back to step ②, finding the factor which have the second insignificant effect to the original neural network, and going on this process to step ③, and step ④; ....

Through the above training and factor analysis, for a fuzzy control system (FC), we can get the main factors of it and all trained fuzzy rules in the fuzzy control systems. Based on these factors and the fuzzy rules we have got we can implement this fuzzy control system.

2. THE IMPLEMENTATION OF FUZZY CONTROL

In this part, we will discuss the problem how to implement fuzzy control system using the factors and rules got in the first part.

Implementing the Fuzzy Control Systems Using Neural Networks
In the first section, we have used neural networks analyzing factors and training rules for a fuzzy control system. As a result, we have got a trained neural networks which can be used in implementing the fuzzy control system.

Implementing the Fuzzy Control Systems Using Interpolation Methods
Neural Networks, as a special kind of interpolation methods, can be used in fuzzy control system as shown in the above. Besides it, many other interpolation methods, e.g., linear interpolation, polynomial interpolation, etc., can also be used in fuzzy control system. we know that each rule can be represented as a point in the *f*-**u** space (*see* Fig.-15), and each fuzzy rule can be represented as fuzzy point in the *f*-**u** space. Using the linear or polynomial interpolation methods to fit these point, we can get the **fuzzy response curve** in a fuzzy control system. Therefore we can implement the fuzzy control system according to this fuzzy response curve. Fig-15 is an example of linear interpolation method.

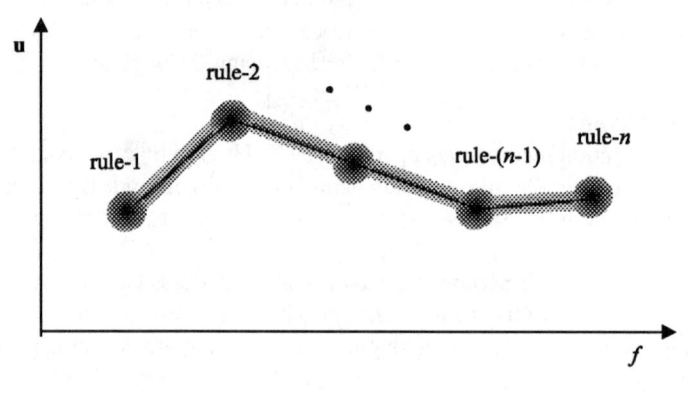

Fig. 15

Implementing the Fuzzy Control Systems Using Matrix Methods
In this method, we list all rules in the matrix form, and then we use the relation inference to implementing the fuzzy control systems.

3. CONTROLLABILITY AND STABILITY OF FUZZY CONTROL

Controllability
Definition: A fuzzy control system FC(**x**, F, **u**, **T**(t)) is said to be **controllable** if it is possible to find a control vector **u** which, in specified finite time L_0, will transfer the system from any initial position to the target position.

By the FCNN's terms, a fuzzy control system is said to be controllable if it is possible to find a control vector **u**, and a big enough factor $f=f_1 \vee f_2 \vee \cdots \vee f_n$ so that FCNN converges.

For a noncontrollable fuzzy control system, it is possible to make it controllable by adding additional control forces. For example, we certainly could make the double-broom system controllable by adding additional forces.

There may be many reasons for a fuzzy control system to be noncontrollable, the most common reasons are the following two cases:
- The control force has no influence on some components of target vector;
- Some components of target vector are not independent.

About the noncontrollable problem, the factor analysis of neural networks is a possible and feasible approach for solving it.

Stability

Definition: A time-invariant fuzzy control system $FC(\mathbf{x}, F, \mathbf{u}, \mathbf{T}(t))$ is said to be **stable** if, *undriven*, its state tends to the target vector from any finite initial state. The system is said to be *locally stable* if when subject to a sudden small perturbation, it tends to remain within a small specified region R surrounding the target vector $\mathbf{T}(t_0)$, and the system is said to be *globally stable* if R includes the entire finite state space.

About the stability, we have the following two conclusions:

▲ A time-invariant fuzzy control system $FC(\mathbf{x}, F, \mathbf{u}, \mathbf{T}(t))$ is likely locally stable if the weights $W(u: u_i, e: e_j)$ ($i=1, 2, \ldots, m; j=1, 2, \ldots, t$) in FCNN may all be nearly equal to zero.

▲ A time-invariant fuzzy control system $FC(\mathbf{x}, F, \mathbf{u}, \mathbf{T}(t))$ is locally stable if its response curve may have the following form (Fig.-16); on the other hand, if any of that form of curves may not be its response curve, then the system is not locally stable.

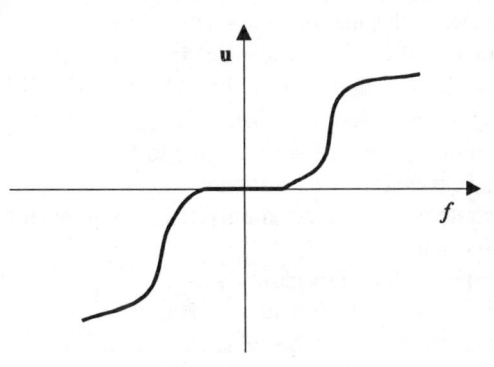

Fig. 16

Robust

Definition: A fuzzy control system FC($\mathbf{x}$, F, $\mathbf{u}$, $\mathbf{T}(t)$) is said to be δ-**robust** if its FCNN neural networks still converges to the target vector in the case of constraining an extra random δ-force on the fuzzy control system.

In order to examine if a fuzzy control system is δ-robust, we can add a bias node in the layer u, connecting this node with the nodes u_1, u_2, ..., u_m in layer u with a group of random weights whose absolute values are not larger than a quite small positive number δ, and then we see if this trained FCNN neural networks still converges to the target vector under the same weights. If yes, then the system is δ-robust. If no, we can decrease the value δ, and re-examine the FCNN neural networks. If for any positive number δ, the FCNN neural networks does not converge after adding a bias force, then the system is not δ-robust. *See* Fig.-2.

4. THE COMPARISONS BETWEEN THE FUZZY CONTROL THEORY AND MODERN CONTROL THEORY

The Comparisons of Methodologies

➻ In modern control theory, the principle approach is **differential equations**. For a deterministic control system, we first derive the differential equations, then linearize them to be the linear ordinary type, the *dimension* of the system depends upon the minimum number of independently chosen variables needed to fully characterize the system. The *state* of a system is the minimum set of numbers or variables, the state variables, which contain sufficient information about the past history of the system to permit us to compute all future states of the system-assuming, of course, that all future inputs (control forces) are known and also the equations (bonds of interactions) describing the system.

In modern control theory, after the state variables are determined, the basic problem facing the control engineer at this juncture is selecting proper *control strategy* or, in other words, specifying a suitable form for the control force $\mathbf{u}$. This is by no means a simple problem, neither does it have a unique solution. The choice will have to be guided by *intuition, experience*, and quite often *luck*, and often guided by choosing a suitable form of force $\mathbf{u}$ so that the differential equations are easily to be solved.

The Principles of Selecting Control Strategy

(1) The control strategy to be selected must result in an overall system for which the primary control objective is met.

(2) The chosen strategy should result in a system which is sound from an engineering viewpoint: i.e., the strategy should be instrumented easily.

(3) The selected strategy should be of such a nature that its actual effect on the systems can be predicated easily. More specifically, this requirement means that when the selected control force $\mathbf{u}$ is substituted back in to Equations, it should be possible to obtain with reasonable efforts an actual solution of the response.

➼ In fuzzy control theory, the principle approach is **fuzzy neural networks**. For a deterministic control system, we first set up the fuzzy neural networks according to the chosen main factors and the chosen control force vector **u**, then train it and analyze the factors to get the minimum number of independently chosen factors needed to fully characterize the system, and get the trained fuzzy control rules used in fuzzy control. The minimum set of factors contains sufficient information about the past history of the system to permit us to compute all future states of the system.

In fuzzy control theory, one of the basic problems facing the control engineer is selecting proper *control strategy* or, in other words, specifying suitable control rules for the control force **u**. This can be done by experts and engineers themselves initially, and then put these rules into the fuzzy control neural networks so that they can be trained. After this training, we can get suitable rules for the control force vector **u**.

The Principles of Selecting Control Rules

(1) The control rules to be selected must result in an overall system for which the primary control objective is met.

(2) The chosen rules should result in a system which is sound from an engineering viewpoint: i.e., the rules should be instrumented easily.

(3) The selected rules should be of such a nature that its actual effect on the systems can be predicated easily. More specifically, this requirement means that when the selected control force **u** is given, it should be possible to obtain with reasonable efforts an actual solution of the response.

Application Fields

Fuzzy control (FC) and modern control (MC) both can be used in various fields. In the cases that differential equations can be easily solved, MC is more feasible and powerful; In the cases that differential equations cannot be easily solved, even not easily to be got, FC is more feasible and powerful, e.g., in social control systems and many other very large, nonlinear and complex systems.

CONCLUSION

In this paper, based on fuzzy neural networks, we have systematically discussed the establishment, implementation, and properties of fuzzy control systems in part 1, part 2, and part 3. In part 4 we have also discussed the comparisons between the fuzzy control theory and modern control theory in the aspects of methodologies and application fields. We have implemented parts of the Fuzzy Control Neural Networks proposed in this paper on Unix platform in C. We hope, through the Fuzzy Control Neural Networks, to give a theoretical foundation of fuzzy control theory.

REFERENCES

[1] Wang P. Z., Zhang H. M., "Truth-valued flow inference and its dynamic analysis". *BEIJING SHIFAN DAXUE XUEBAO*, No. 1, pp. 1-12, 1989.

[2] Wang P. Z., "A factor spaces approach to knowledge representation". *Fuzzy Sets and Systems*, Vol. 36, pp. 113-124, 1990.

[3] Wang P. Z., Sanchez E., "Treating a fuzzy subset as a projectable random set". Gupta M. M., Sanchez E. (eds.): *Fuzzy Information and Decision*, Pergamon Press, 1982, pp. 212-219.

[4] Zhang X. H., Wong F., Lui H. C., Wang P. Z., "Theoretical basis of truth value flow inference and its applications". *The Proceedings of the First Singapore International Conference on Intelligent Systems*, Singapore, Asia Computer Weekly, 1992, pp. 131-136.

[4] Zadeh L. A., "Fuzzy sets as a basis for a theory of possibility". *Fuzzy Sets and Systems*, No. 1, pp. 3-28, 1978.

[5] Elgerd O. I., *Control Systems Theory*. International Student Edition by McGraw-Hill Inc., 1967.

[6] Sugeno M., (eds.): *Industrial Applications of Fuzzy Control*. North-Holland, 1985.

[7] Zhang X. H., Wong F., "Decision-support neural networks and its applications in financial forecast". *ISS technical report*, Singapore, 1992.

[8] Zhang X. H., Lui H. C., Wong F., Shen Z. L., "The coupling of truth value flow neural networks and approximate reasoning based on similarity measure". *ISS technical report*, Singapore, 1992.

[9] Peng X. T., Liu S. M., Yamakawa T., Wang P. Z., "Self-regulating PID controllers and its applications to a temperature controlling process". Gupta M. M., Yamakawa T. (eds.): *Fuzzy Computing*, North-Holland, 1987.

[10] Goh T. H., Wong F., "Semantic extraction using neural network modeling and sensitivity analysis", *ISS technical report*, Singapore, 1992.

[11] Khanna T., *Foundation of Neural Networks*, Addison-Wesley Publishing Company, 1990.

[12] Wong F., "FastProp: a selective training algorithm for fast error propagation". *Proceeding of the International Joint Conference on Neural networks*, Singapore, 1991.

PART C

IMPLEMENTATIONS AND APPLICATIONS

Chapter 21
Human Friendly Fuzzy Transportation System

Introduction, *460*
System Configuration, *461*
Fuzzy Language Understanding and Fuzzy Image Recognition, *462*
Fuzzy Path Planning, *464*
References, *472*

The *human friendly fuzzy transportation system* (HFFTS) described in this chapter is developed as the transportation system for the future factory. The HFFTS understands oral commands issued by the operator and communicates them to a multi-joint handling robot which carries the designated work (servo motors of different sizes and colors) to an autonomous delivery vehicle. The autonomous delivery vehicle then transfers the designated work to its destination while avoiding obstacles.

HUMAN FRIENDLY FUZZY TRANSPORTATION SYSTEM

Tadashi Iokibe   Takashi Kimura
Meidensha Corporation
Tokyo, Japan

1. INTRODUCTION

HEADING: Much research is under way on "language understanding problems", "image recognition problems" and "path planning problems" using fuzzy control theory. In an effort to generalize problems, however, most of them do not have good results from the viewpoint of practical applications. This time, imagining a human friendly transportation system in near future factories by applying the fuzzy control theory to solutions of "language understanding problems", "image recognition problems" and "path planning problems" has been undertaken. Using these algorithms, we proceeded to a computer simulation and also configured a demonstration system by actual articles and presented it at the System Control Fair 1991.

As stated above, a transportation system in a near future factory is assumed. This system understands oral commands by an operator and the autonomous vehicle transfers the designated work to a destination while avoiding obstacles. Further details will be explained below. The work transferred in this system includes servo motors of different sizes and colors.

The oral transfer commands by the operator have a format such as "Carry a large red servo motor." If the corresponding article is present, the text-to-speech synthesizer replies "Yes. I understand. I will carry a large red servo motor." If such an article is not found, the text-to-speech synthesizer replies "There is no large red servo motor. Please give me another command." If the former reply is given, the operator responds with "Yes. Carry it please." to initiate a transfer. It is possible to give continuous commands to transfer several pieces of work.

When a transfer start is instructed, the multi-joint handling robot located at the start position transfers the work to the autonomous vehicle. Thereafter, the autonomous vehicle transfers the work to the target position while avoiding obstacles. When the target position is reached, the multi-joint handling robot located at the target position transfers the work from the autonomous vehicle to the work yard.

The following explains the system configuration, language understanding and image recognition by fuzzy inference.

2. SYSTEM CONFIGURATION

As shown in Fig. 1, the present system consists of a speech recognizer, a text-to-speech synthesizer, an image processor, multi-joint handling robots, an auto-nomous vehicle and a fuzzy controller. All information is input to the fuzzy controller. The Transfer instruction by the operator is input from the speech recognizer, work and transfer area information from the image processor through a CCD camera, and current position of the autonomous vehicle from itself to the fuzzy controller.

The fuzzy controller instructs "from, to" information for work transfer to the multi-joint handling robots, and traveling route and speed to the autonomous vehicle.

The following is a brief explanation of the system component devices.

Speech recognizer Voice recognition equipment for specific speakers. In the present system, the voice of 3 operators is already registered.

Text-to-speech synthesizer An electrical acoustic tube method is used. The word processor text is converted to voice as it is.

Image Processor The system uses 2 image processors. One is for recognizing obstacles in the transfer area and the position of the autonomous vehicle with six CCD cameras connected to it. The other is for recognizing work placed in the work yard installed on the target position and start position with four CCD cameras connected to it.

Multi-joint Handling Robots A robot is placed at each start and target position and transfers work between the work yard and the autonomous vehicle.

Autonomous Vehicle Can store routes. It travels freely without guiderails unlike conventional vehicles. The traveling route is wireless transmitted by the fuzzy controller.

Fuzzy Controller A fuzzy inference package is built in a general-purpose industrial personal computer having real time characteristics. The fuzzy controller can simultaneously drive 4 fuzzy inference motors.

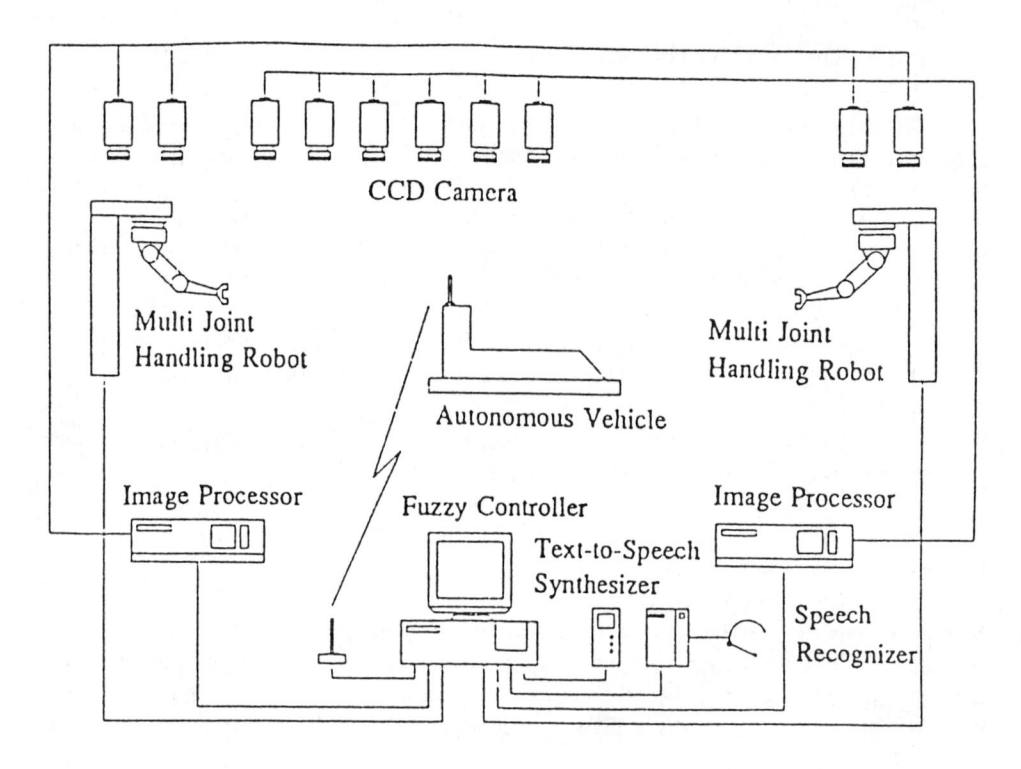

Fig. 1 System Configuration

3. FUZZY LANGUAGE UNDERSTANDING AND FUZZY IMAGE RECOGNITION

3.1 Learning of Work as Language Model

In the system, the operator recognizes the work sizes and colors beforehand, and the fuzzy controller learns the results as combination of work sizes and colors and language as transfer instructions.

(1) Arbitrary work is read by the image processor and the image size and color information is input to the fuzzy controller.

(2) At the same time, the operator watches the work, puts its particulars to language information such as "large and red", and inputs this to the fuzzy controller.

(3) The fuzzy controller stores data input in (1) and (2) and combines language information, image area and RGB intensity.

(4) Steps (1) to (3) are repeated for each of the different types of work. The fuzzy controller automatically generates data given in Table 1 as a work identification file.

Table 1 Work Identification File

Work Type	Language Code	Area	R Intensity	G Intensity	B Intensity	R/G Ratio	G/B Ratio	B/R Ratio
W1	L1	S1	R1	G1	B1	RG1	GB1	BR1
W2	L2	S2	R2	G2	B2	RG2	GB2	BR2
W3	L3	S3	R3	G3	B3	RG3	GB3	BR3
⋮	⋮	⋮	⋮	⋮	⋮	⋮	⋮	⋮
WN	LN	SN	RN	GN	BN	RGN	GBN	BRN

(5) From data generated in (4), the fuzzy controller automatically generates area, R intensity, G intensity, B intensity and other fuzzy control labels and membership functions. Suppose a certain factor (area, R intensity, G intensity, B intensity, etc.) expressing a particularity of work is (C) and that the values indicating that factor are rearranged to ($C_1, C_2, C_3, \cdots, C_i, \cdots, C_N$) in an ascending order, the membership function related to factor (C) is determined as shown in Fig. 2.

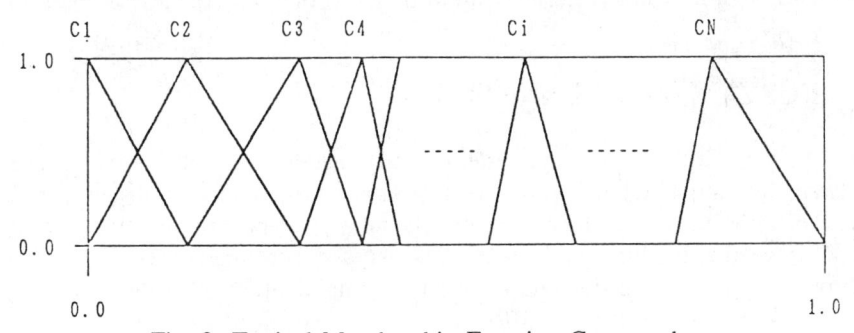

Fig. 2 Typical Membership Function Generated

3.2 Work Identification by Oral Command

In the system, the work is identified in such a manner that the particularities of work for the language are identified by the fuzzy matching technique. The following shows how to identify work.

(1) Generating rules for fuzzy matching

When a transfer instruction by operator voice is input to the fuzzy controller, fuzzy rules corresponding to language (L_i) shown in Table 1 are automatically generated according to the details of instruction.

If S is S_i * R is R_i * G is G_i * B is B_i * RG is RG_i * GB is GB_i * BR is BR_i
Then W is W_i
where,
S   : Variable indicating area
R   : Variable indicating R intensity
G   : Variable indicating G intensity
B   : Variable indicating B intensity
RG : Variable indicating R/G ratio
GB : Variable indicating G/B ratio
BR : Variable indicating B/R ratio
W   : Variable indicating work
*   : AND operator

(2) All particularities (area; R, G, B intensities) of all work are input to the fuzzy controller through the image processor. The fuzzy controller calculates R/G, G/B and B/R ratios given in Table 1. This information and antecedent part of fuzzy rule generated in (1) are subjected to fuzzy matching and the satisfaction degree (μ_i) is obtained. The work where μ_i is highest above a certain threshold is regarded as corresponding to the oral commands. If there is no μ_i exceeding the threshold, no work corresponding to the oral commands is deemed present.

4. FUZZY PATH PLANNING

As stated in Section 2, the system uses an autonomous vehicle which needs no guiderails for transferring the work. At an arbitrary position of the transfer area, several obstacles are installed and their sizes and positions are input from the image processor to the fuzzy controller through the six CCD cameras mounted on the ceiling. Therefore, path planning for the autonomous vehicle is an important factor of the system.

Various proposals are presented for solutions to avoid obstacles regarding automatic traveling of unmanned transfer carriage, autonomous traveling robot, etc. Configuration space method, artificial potential method and heuristic technique are typical ones. These methods are intended to show how to avoid stationary or moving obstacles when traveling on the basically predetermined transfer route or to a near-sighted target position, how to return to the determined route or how to reach a target position. In other words, it is a navigator/supervisor function on how to avoid obstacles [1] - [5].

In the present system, a planner function has been realized resorting to a fuzzy inference. It determines the overall traveling path and speed upon recognizing the transfer area status [6].

The following explains the fuzzy determination technique of the path.

4.1 Traveling Route Planning Technique

Prerequisites: There are the following prerequisites before determining the traveling route for the autonomous vehicle.

(1)   Position and shape

- The area where the autonomous vehicle can travel is a rectangle and dimensions of its sides are known.
- Positions and sizes of obstacles are known.
- Dimensions of autonomous vehicle are known.
- There is only one autonomous vehicle in the transfer system.
- The start and target positions are known.

(2)   Autonomous vehicle

- Forward movement is more controllable than reverse movement.
- Turning a small radius at a high speed is difficult.
- A small steering angle makes the control easy.

Planning the Traveling Route: When determining the traveling route for the autonomous vehicle, we consider how to drive a car. In this case, the direction and speed are controlled while determining subgoals successively. Namely,

- Minimize detours.
- Take a route encountering less obstacles as possible.

The traveling route is determined by successively setting subgoals toward the target position.

(1) Setting a moving target line from position of autonomous vehicle (target line setting)

A straight line interconnecting the current position of autonomous vehicle and final subgoal is an ideal target line. The forwarding direction is determined so that the traveling route will be as near that line as possible and that there will be as less obstacles as possible.

For this purpose, around the forward center point of the autonomous vehicle and on each side of the ideal target line, sectors having angle β and radius r are set up to 90 degrees in α degree steps (when α is a divisor of 90, $(2 \times 90/\alpha) + 1$ sectors are formed) and membership and rules are set for fuzzy inference of advantage U_m of forwarding in the sector direction from area S_m of obstacle included in that sector and angle θ_m (0 or multiple of α) of the ideal target line form the center line of the sector. Fig. 3 shows the target line setting and Fig. 4 indicates the membership functions and Table 2 indicates rules for inferring the advantage.

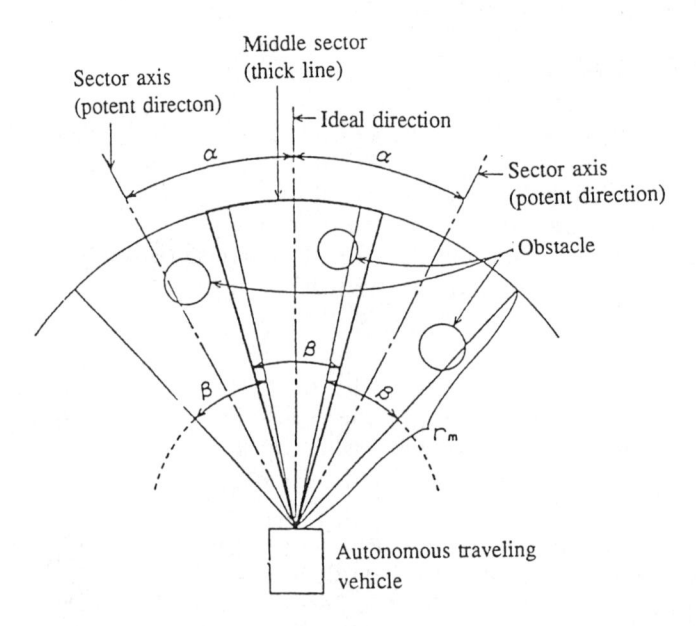

Fig. 3 Setting Target Line

θm MSF

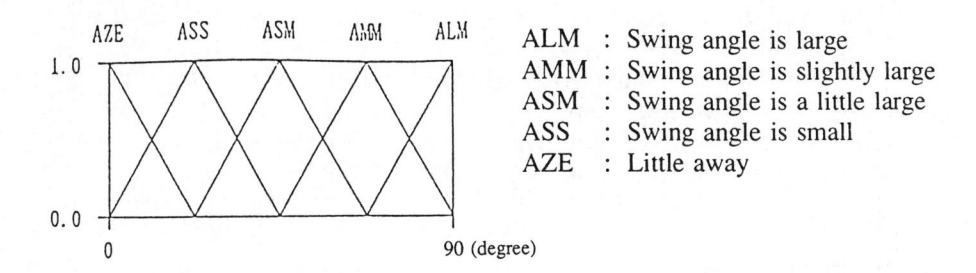

```
        AZE    ASS    ASM    AMM    ALM
   1.0
   
   
   
   0.0
        0                          90 (degree)
```

ALM : Swing angle is large
AMM : Swing angle is slightly large
ASM : Swing angle is a little large
ASS : Swing angle is small
AZE : Little away

Sm MSF

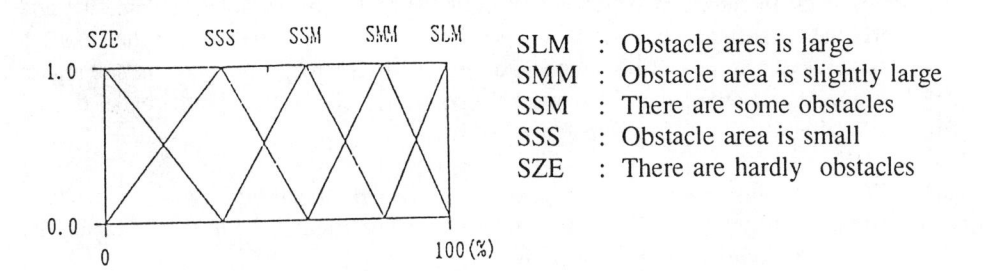

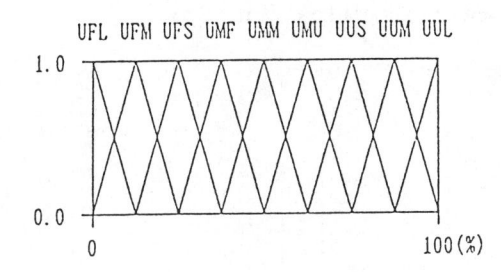

```
        SZE    SSS    SSM    SMM    SLM
   1.0
   
   
   
   0.0
        0                          100 (%)
```

SLM : Obstacle ares is large
SMM : Obstacle area is slightly large
SSM : There are some obstacles
SSS : Obstacle area is small
SZE : There are hardly obstacles

```
       UFL UFM UFS UMF UMM UMU UUS UUM UUL
  1.0
  
  
  
  0.0
       0                              100(%)
```

UFL UFM UFS UMF UMM UMU UUS UUM UUL
Disadvantageous ——————————————→ Advantageous

Fig. 4 Examples of Membership Functions for Inferring Advantage

Table 2 Rules for Inferring Advantage

IF θm is AZE and Sm is SZE THEN Um is UUL	
IF θm is ASS and Sm is SSS THEN Um is UUS	
IF θm is ASM and Sm is SSM THEN Um is UMM	
IF θm is AMM and Sm is SMM THEN Um is UFS	
IF θm is ALM and Sm is SLM THEN Um is UFL	
IF θm is AZE and Sm is SLM THEN Um is UMM	

(2) Determining subgoals according to target line (subgoal determination)

If there is no obstacle up to (sector radius r x b, where $0 < b \leq 1$) from the center of sector on the set target line, that point is set as a subgoal, taking intoaccount the width and minimum turning radius of the autonomous vehicle. If there are obstacles, the next advantageous target line is set by the advantage calculation and whether a subgoal can be or not is calculated. The above is repeated until a subgoal can be set. If no subgoal can be on any sector, the preceding subgoal is resumed and other than the current subgoal is set likewise. If no subgoal can be still, the route is deemed undeterminable.

(3) The first autonomous vehicle position is the start position. Steps (1) and (2) above are repeated until the final subgoal set beforehand is linearly reachable by a subgoal (final subgoal determination).

The route from the final subgoal to the arrival position is stationary according to the turning characteristics and stop accuracy characteristics of the autonomous vehicle. Fig. 5 shows the status where all subgoals are set.

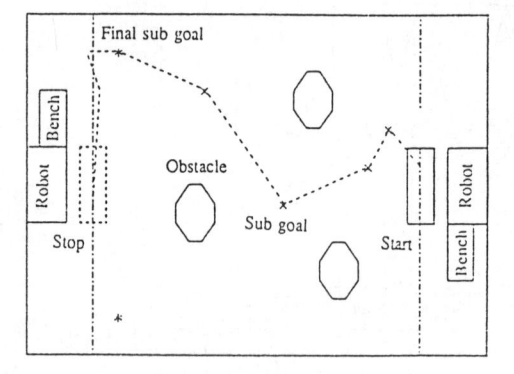

Fig. 5 Status Where All Subgoals Are Set

Setting Speed Changing Point: After determining the traveling route according to the last paragraph, speed changing points are set on the traveling route. They are intended for traveling at as high a speed as possible without deviation from the specified route.

(1)   Determining set position of speed changing point

Set the speed changing points upstream and downstream of the subgoals on the traveling route as a rule.

(2)   Determining operation speed

Determine the traveling speed at speed changing points determined in (1) as follows.

Fig. 6 indicates the membership functions and Table 3 indicates rules for determining speed at speed changing point.

· Upstream subgoal
Generate membership function and fuzzy rules determining the traveling speed according to the curvature at the subgoal.

· Downstream subgoal
Generate membership function and fuzzy rules determining the traveling speed according to the distance up to the next speed changing point.

Table 3 Rules for Determining Speed at Speed Changing Point

IF θ_{sg} is GZE   THEN v is VLL
IF θ_{sg} is GSS   THEN v is VLM
IF θ_{sg} is GSM   THEN v is VMM
IF θ_{sg} is GMM THEN v is VSM
IF θ_{sg} is GLM   THEN v is VSS
IF θ_{sg} is GLL   THEN v is VFL

θsG MSF (θsg: Subgoal bend angle)

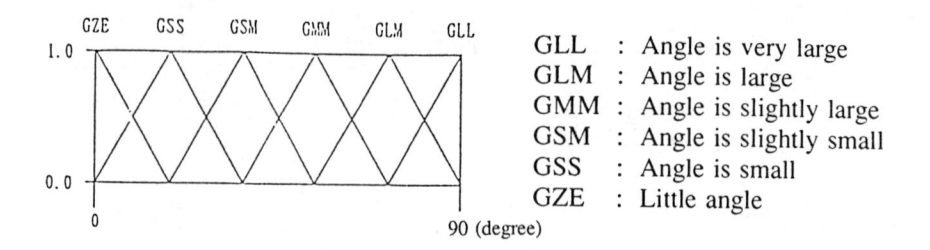

GZE GSS GSM GMM GLM GLL

GLL : Angle is very large
GLM : Angle is large
GMM : Angle is slightly large
GSM : Angle is slightly small
GSS : Angle is small
GZE : Little angle

0 — 90 (degree)

dsg MSF (dsg: Distance between subgoals)

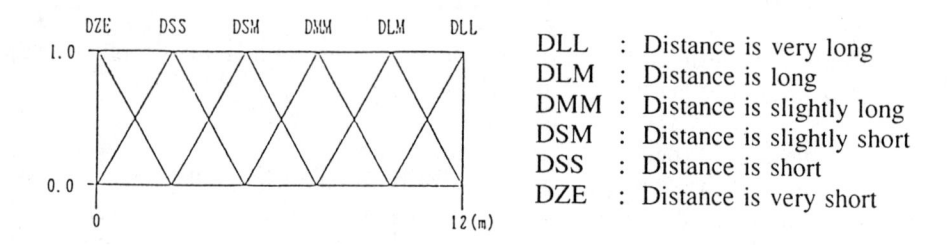

DZE DSS DSM DMM DLM DLL

DLL : Distance is very long
DLM : Distance is long
DMM : Distance is slightly long
DSM : Distance is slightly short
DSS : Distance is short
DZE : Distance is very short

0 — 12 (m)

v MSF (v: Traveling speed)

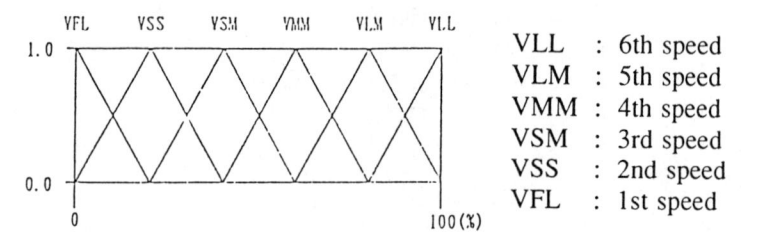

VFL VSS VSM VMM VLM VLL

VLL : 6th speed
VLM : 5th speed
VMM : 4th speed
VSM : 3rd speed
VSS : 2nd speed
VFL : 1st speed

0 — 100 (%)

Fig. 6 Examples of Membership Functions for Determining Speed at Speed Changing Point

4.2 Results

We carried out a test on an actual facility and a simulation on a computer. The results are shown in Fig. 7 and 8.

As for the membership function according to the area of obstacle for determining the traveling route, we adopted an aggregation of membership functions which were bodily shifted right so that judgements for advantage will not be excessive apart from each other when there are not so many obstacles.

Radius r of sector for determining the traveling route and subgoal determination coefficient b considerably affect the route determination ability and route efficiency (how near to the shortened route). Reducing r x b permits fine set subgoals and, at first reflection, a route near the shortest route can be determined. But, increasing subgoals prolongs the computation time and increases speed changing points, thereby lowering the speed and, further, entering cul-de-sacs makes it impossible to determine the route. Thus this is not necessarily the best. Determining to a certain degree may be efficient for route determination resorting to fuzzy inference.

When an autonomous vehicle of 0.8 m wide and 1.6 m long runs on a transfer yard of 6 m wide and 10 m long such as the case here, r x b of 2m is appropriate. When r x b is approximately 2m, the traveling route is almost determined but, as shown in Fig. 8, some are slightly away from the shortest route. However, the route direction determined is near the one determined by human beings.

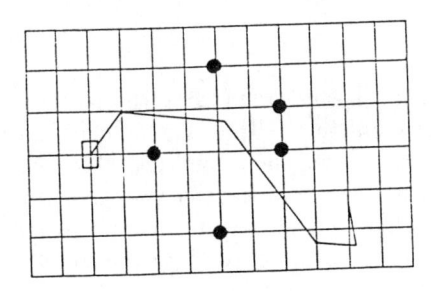

Transfer vehicle
set width = 1.3 m
$\beta = 10°$
$\alpha = 20°$
r = 4.0 m
b = 0.4

Fig. 7 Result 1

Transfer vehicle
set width = 1.3 m
$\beta = 10°$
$\alpha = 20°$
r = 4.0 m
b = 0.4

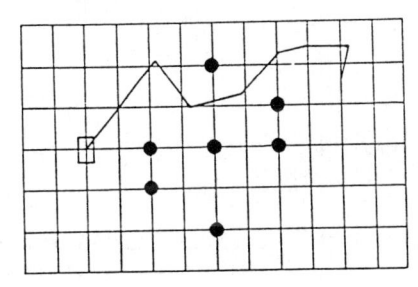

Fig. 8 Result 2

4.3 Future Subjects

The following can be enumerated as future subjects.

(1) Smoothing unnecessary subgoals

As is evident from the example in Fig. 8, it is necessary to add processing which smoothes an unnecessary subgoal (shown by A in Fig. 8) and linearizes upstream and downstream subgoals or rules by which unnecessary subgoals will not be set.

(2) Prevention of collision with obstacles

If the width of the autonomous vehicle is set the same as the measured width when determining a traveling route, the traveling route can be set somehow orother but, when actually traveling around an obstacle, a side of the vehicle might collide with the obstacle on account of the difference in the turning radius between the front and rear inner wheels. In the text of this time, collision was avoided by setting the width of autonomous vehicle to approximately 1.5 times the actual width. Hereafter, it is necessary to set forth fuzzy rules for collision degree attributable to turning radius, thereby determining the route still more appropriately.

(3) Handling outside the transfer yard

If a sector for determining the target line has departed from the transfer yard, the system of this time deems it free from obstacles. Thus, the direction toward the transfer yard periphery will be relatively advantageous and there will be a tendency toward the ends. If the entire outside of transfer yard is handled as an obstacle, on the other hand, there would be a tendency of departing from the ends and an optimum route near an end might be overlooked. It can be considered that the outside of transfer yard is excepted from the sector area but its results have not yet been determined. Handling the outside of transfer yard may have to be combined with other factors.

REFERENCES:

[1] Lozano-Perez, T., *Automatic Planning of Manipulator Transfer Movements*, IEEE Vol.SMC-11, No.10, pp.681-698, October, 1981

[2] Khatib, O., *Real-Time Obstacle Avoidance for Manipulators and Mobile Robots*, The International Journal of Robotics Research, Vol.5, No.1, pp.90-98, 1986

[3] Takeno and Kakikura, *Avoidance of Collision of Mobile Robot with Moving Obstacles*, periodical of Japan Robotics Institute, Vol.4, No.5, pp.33-37, 1986

[4] Maeda, *Fuzzy Control of Avoiding Obstacles Having Operator Control Strategy Based on CMAC Learning*, Japan Fuzzy Control Institute, anthology of lecture treatises at 6th Fuzzy System Symposium, pp.531-534, 1990

[5] Ishikawa, *Study of Guiding Method for Autonomous Moving Robot Resorting to Fuzzy Control*, periodical of Japan Robotics Institute, Vol.9, No.2, pp.149-161, 1991

[6] Iokibe, Kimura and Sasaki, *Fuzzy Transportation System*, Japan Fuzzy Control Institute, anthology of lecture treatises at 8th Fuzzy System Symposium, pp.325-328, 1992

Chapter 22
Control of a Chaotic System using Fuzzy Logic

Introduction, *476*
The Chaotic System, *478*
Surface-Fitting and the Analytic Solution, *480*
Genetic Algorithm-Designed Fuzzy Logic Controller, *482*
Results, *490*
Summary, *494*
References, *495*

Chaotic systems are generally characterized by three tangible qualities: (1) they act irregularly, (2) they are extremely sensitive to initial conditions, and (3) their behavior is described by mathematical equations that may be quite simple. The approach to chaotic control presented in this chapter employs both fuzzy rule-based systems and genetic algorithms. Genetic algorithms are needed to specify the fuzzy membership functions because the fuzzy logic controller must be extremely accurate to manipulate a chaotic system. Unfortunately, using genetic algorithms to define the membership functions is not enough to guarantee the necessary precision. The fuzzy logic controller must utilize a large number of membership function forms, and it must also be provided with great flexibility in defining the consequent for the fuzzy rules. This flexibility is obtained by employing a *real-valued consequent* approach. The results obtained in this chapter demonstrate that fuzzy logic provides a viable alternative to some of the "highly mathematical" techniques currently being investigated, and that "inexact reasoning" can be used to obtain very exact solutions.

CONTROL OF A CHAOTIC SYSTEM USING FUZZY LOGIC

C. L. Karr
U. S. Bureau of Mines
Tuscaloosa Research Center
Tuscaloosa, AL

E. J. Gentry
SEER Technology
Cary, NC 27511

1. INTRODUCTION

Scientific researchers have been aware of chaotic behavior in engineering systems for a number of years [1, 2] — generally they simply termed such systems unstable and avoided operating in chaotic regions. However, in recent years new phenomena have been discovered in nonlinear dynamics, and these discoveries have led to the uncovering of a seemingly underlying order in chaotic systems. This underlying order has produced hope of predicting and controlling chaotic systems.

Chaos, unfortunately, is not easily defined in either nonscientific or scientific terms. In nonscientific terms chaos is generally associated with a system or human behavior that is without apparent pattern or that is "out of control." In scientific terms these systems are no easier to define, but are generally characterized by three tangible qualities: (1) they act irregularly, i.e., they are not repetitive, (2) they are extremely sensitive to initial conditions, and (3) their behavior is described by mathematical equations that are often quite simple. Chaotic systems have been identified in a number of scientific fields including electrical engineering [3], hydrodynamics [4], chemistry [5], and mechanical engineering [6].

The U. S. Bureau of Mines has developed an express interest in chaos because many mining and metallurgical systems have been shown to undergo dramatic changes with small perturbations in boundary conditions. Examples of this chaotic phenomena have been observed in electric arc furnaces [7], the fracture of materials [8], and instabilities of mine structures [9]. These complex systems, when allowed to proceed uncontrolled, can prove to be unduly expensive or quite dangerous. Thus, the ability to successfully predict and control chaotic systems can represent tremendous economic savings and substantial improvements in safety for the minerals industry, as well as for numerous other industries.

Researchers have recently begun to investigate techniques for efficiently manipulating chaotic systems. One of the first and most important efforts in this area was made by Ott and his co-workers [10] who utilized a technique called delay coordinate embedding. This technique is highly mathematical and requires an in-depth understanding of the physical system's response to external stimulus. Shortly thereafter, Romeiras and his co-workers [11] discussed a method whereby motion on a chaotic attractor can be converted to a desired attracting time-periodic motion by applying a small control. However, their results described regions of the control space in which the orbit wandered erratically from the desired setpoint.

Researchers at the U.S. Bureau of Mines have developed a technique for process control in which fuzzy rule-based systems are combined with genetic algorithms (GAs) [12]. Fuzzy rule-based systems allow for the implementation of a human's "rule-of-thumb" approach to decision making [13] by employing linguistic variables, and have been effectively used for control purposes [14]. GAs are search algorithms based on the mechanics of natural genetics [15]. They are capable of rapidly locating near-optimum solutions to difficult search problems, and have been shown to be quite effective in locating efficient fuzzy membership functions [12]. This technique for designing fuzzy logic controllers (FLCs) has been successfully implemented in a number of control problems including an inverted pendulum [16], a chemical reactor [17], and a pH titration [18]. All of these problem environments are highly nonlinear and are characterized by changing process dynamics. Therefore, gaining suitable control of these environments is not a trivial task. Since GA-FLCs can be used to control such highly nonlinear and time-varying systems, it seems like a natural extension to utilize them for controlling chaotic systems. In this chapter, a particular chaotic system, a ball bouncing on an oscillating table is studied. The control surface for this system has been obtained using an analytical method as will be discussed.

Virtually any control problem with two input variables can be viewed as a surface-fitting problem; an appropriate control action is simply a function of the state of the two variables in the problem environment. Both fuzzy logic [19] and GAs [20] have been used to solve curve-fitting problems. Thus, the precedence exists for utilizing a GA-FLC for the task of defining a control surface based on data from a problem environment. Control surfaces for chaotic systems are generally very difficult to obtain, and once obtained, they are quite complex. In this chapter, a particular chaotic system is studied for which the control surface has been obtained. Specifically, a ball bouncing on an oscillating table is considered. The frequency of oscillation of the table is adjusted so that the ball bounces to a constant height. The control variable, the desired frequency of oscillation, ω, is a function of two state variables: (1) the velocity of the ball when it leaves the table, **v**, and (2) the phase angle of the table at impact, θ.

The control surface of the chaotic bouncing ball system is very detailed. However, despite its complexity, this surface can be defined with *inexact reasoning*. A GA is needed to specify the fuzzy membership functions because the FLC must be extremely accurate to suitably manipulate the chaotic system. Unfortunately, using GAs to define the membership functions is not enough to guarantee the necessary precision. The FLC must utilize a large number of membership function forms, and it must also be provided with great flexibility in defining the *consequent* (the {action} portion of the fuzzy rules of the form IF {condition} THEN {action}) for the fuzzy rules. Triangular, trapezoidal, sinusoidal, and exponential membership functions are necessary. The flexibility in defining the membership functions for the consequent of the rules is obtained by employing a *real-valued consequent* approach.

In this chapter, details are provided for a FLC used to control a computer simulation of the chaotic bouncing ball system. The results that are presented demonstrate the effectiveness of using a FLC to manipulate a chaotic system. These results are important for two reasons. First, controlling chaotic systems is very difficult, and fuzzy logic provides a viable alternative to some of the highly mathematical techniques currently being investigated. Second, the successful use of fuzzy logic to control a chaotic system demonstrates that inexact reasoning can be used to obtain very precise control.

2. THE CHAOTIC SYSTEM

The chaotic system considered here is a computer-simulated mechanical system studied by Tufillaro and Albano [6]. It is a member of a class of chaotic systems often termed *impact-force* systems, and consists of a ball bouncing on an oscillating table. The frequency of oscillation of the table is adjustable. Certainly, the bouncing ball system does not appear in the minerals industry (nor is it a staple of many other industries), but that was not the point behind its selection. The bouncing ball system was selected simply to demonstrate the effectiveness of using a particular control strategy to manipulate chaotic systems. It was chosen, of the many potential chaotic systems, because of the nature of its chaotic characteristics, because it is easy for most people to gain a "feel" for the response of this system, and because it is relatively simple.

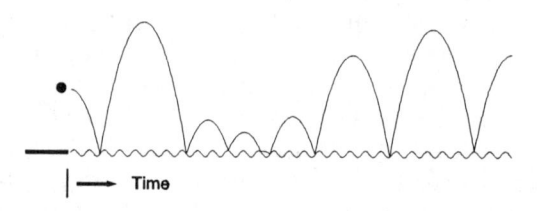

Figure 1. The Chaotic Bouncing Ball System

The bouncing ball system consists of a ball and a table. The ball is initially dropped from an arbitrary height, and is free to bounce on a table which moves sinusoidally up and down as shown in Figure 1. When the gravitational force of the system and the coefficient of restitution of the ball are constant, the height to which the ball bounces is determined strictly by the velocity of the ball relative to the table at impact, and by the phase angle of the table when an impact occurs. If the ball impacts the table when the table is moving upward, the relative velocity is large and the ball bounces higher than it would if the table were stationary. If the ball impacts the table when the table is moving downward, the relative velocity is small and the ball bounces lower than it would if the table were stationary. The net effect is that the height to which the ball bounces is deterministic, yet it has no apparent long-term pattern. Furthermore, the time-history of the ball (its position at a particular time) is extremely sensitive to the initial height from which it is dropped — the system is chaotic.

Poincaré maps are tools that are particularly useful in determining whether or not a system is chaotic. Poincaré maps are basically plots of the phase space, or state of the system, at particular times. A Poincaré map for the bouncing ball system appears in Figure 2. In this plot, the polar coordinate **r**, the distance from the origin, represents the velocity of the ball at impact. The angular coordinate θ represents the phase angle of the table when an impact occurs. Thus, a point on the Poincaré map completely describes the state of the bouncing ball system at the time of each collision, the time of the next impact, and the state of the bouncing ball at the time of the next impact can be computed from this information. Since the Poincaré map does not consist of either a finite set of points (indicating the system is in some way repetitive) or a closed orbit (indicating the system is *almost periodic* or *quasiperiodic*), the motion is chaotic.

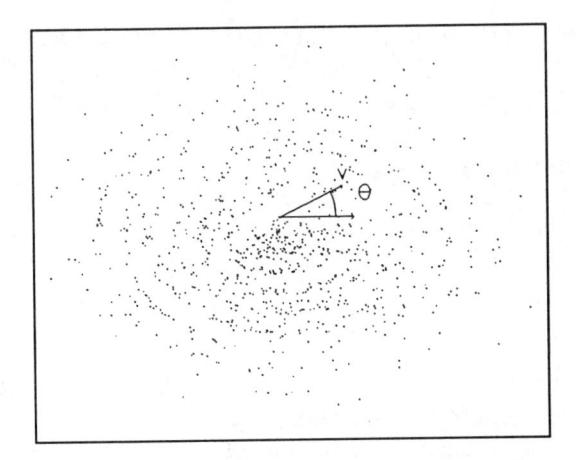

Figure 2. A Poincaré Map of the Bouncing Ball System

The bouncing ball system demonstrates chaotic behavior; the ball bounces to a seemingly random height each time it contacts the table. However, this does not have to be the case. The frequency at which the table oscillates is adjustable, and changing this frequency can dramatically affect the height to which the ball bounces. In fact, the frequency of oscillation can be altered to effectively force the ball to bounce to a constant height, i.e., to force the ball to always strike the table with the same relative velocity at the same phase angle. Thus, this becomes the objective of a control problem: to adjust the table's frequency of oscillation in such a way as to cause the ball to rise to the same maximum height after each bounce. Moreover, a Poincaré map provides a convenient mechanism for evaluating the effectiveness of a controller. An effective control system should produce a Poincaré map that consists of a single point, because the response of the system will be wholly repetitive.

The development of a GA-FLC for the bouncing ball system is discussed in the remainder of this chapter. First, the control problem is cast in the light of a surface-fitting problem. The bouncing ball system presents an especially appealing problem environment in this realm because a control surface can be produced analytically using the equations of motion of the system, and thus there is an exact solution to which the GA-FLC can be compared. Second, the procedure for developing a GA-FLC is described. This description includes the necessary information for utilizing a GA to tune a FLC. Third, results are presented that portray the effectiveness of using a FLC to manipulate the chaotic bouncing ball system. These results are important because they demonstrate the effectiveness of using inexact reasoning (fuzzy logic) to obtain a precise solution.

3. SURFACE-FITTING AND THE ANALYTICAL SOLUTION

As stated above, the problem of controlling the bouncing ball system is a surface-fitting problem: for all possible values of the ball's velocity at impact, **v**, and all possible values of the phase angle of the table, $0 \leq \theta \leq 2\pi$, the appropriate value to which the frequency of oscillation of the table, ω, must be adjusted to ensure that the ball bounces to the desired height, $y_{desired}$.

The equations of motion are used to compute the control surface. The equations of motion for the system are rather simple, although obtaining the solution to these equations can be quite tedious. First, consider the equation of motion for the table:

$$y_{table} = A \sin(\omega t) \qquad (1)$$

where y_{table} is the height of the table, A is a constant that indicates the magnitude of the oscillation, ω is the frequency of oscillation, and t is time. Thus, for given values of A and ω, the position of the table is known at any time. Second, consider the equation of motion for the ball:

$$y_{ball} = y_o + v_o t + \frac{1}{2}gt^2 \qquad (2)$$

where y_o is the height of the ball at an impact, v_o is the velocity of the ball immediately after impact, g is the acceleration of gravity, and t is time. Thus, as with the table, the height of the ball can be computed at any time if the appropriate initial conditions are provided. Furthermore, equations (1) and (2) can be differentiated to compute the velocities of the ball and table, respectively, at any given time. Finally, an equation is needed that relates the velocity of the ball immediately before an impact and the velocity of the ball immediately after an impact. Fortunately, this relationship exists in the form of the definition of the coefficient of restitution for the ball:

$$e = \frac{v^a_{ball} - v^a_{table}}{v^b_{table} - v^b_{ball}} \qquad (3)$$

where e is the coefficient of restitution, v^a_{ball} is the velocity of the ball immediately after impact, v^a_{table} is the velocity of the table immediately after impact, v^b_{ball} is the velocity of the ball immediately before impact, and v^b_{table} is the velocity of the table immediately before impact. It is important to note that when using equation (3), the mass of the table is considered to be substantially greater than the mass of the ball. Therefore, the velocity of the table is unchanged when the ball collides with the table, i.e., $v^a_{table} = v^b_{table}$.

The above three equations can be combined to compute the time at which a collision occurs, t_c, i.e., the time at which $y_{ball} = y_{table}$. This value can then be used to compute both the height and the velocity of the ball at impact. These values serve as the new values of y_o and v_o, respectively, when computing the next collision. Given values of y_o and v_o, the maximum height to which the ball bounces can easily be computed by solving for the time at which the ball reaches zero velocity:

$$t_n = -\frac{v^a_{ball}}{g} \qquad (4)$$

where t_n is the time at which the ball reaches its apex. Keep in mind that the objective of the control problem is to adjust ω so that the ball bounces to a constant height, which will be called $y_{desired}$. The following equation which includes both $y_{desired}$ and the value of ω necessary to ensure the ball bounces to a maximum height of $y_{desired}$, must be solved:

$$\sqrt{2g[A\sin(\omega t) - y_{desired}]} = ev_o + egt_c + (1 - e)A\omega\cos(\theta) \qquad (5)$$

where $\theta = \omega t_c$.

Equation (5) provides a means for computing the appropriate value of ω for any given values of v_o, θ, e, A, and $y_{desired}$. In this study, the following values were used: $e = 0.9$, $A = 1.0$, and $y_{desired} = 20.0$. With these values of the parameters, an appropriate control space consisted of $10.0 \leq v_{ball} \leq 40.0$ and $0 \leq \theta \leq 2\pi$. For values of v_{ball} and θ in this range, equation (5) must be solved numerically. The process of solving this equation, although not difficult using a traditional root-finding method, is quite tedious. There are multiple roots to the equation, and great care must be taken to ensure that the correct root is found and used.

Equation (5) was solved using a combination of a Newton method and a bisection method. This approach allowed for the location and selection of the appropriate roots. The results yield the analytical control surface shown in Figure 3. Note the complex nature of this surface. This complexity must be fully described with fuzzy logic, because the bouncing ball system is very sensitive to alterations in ω.

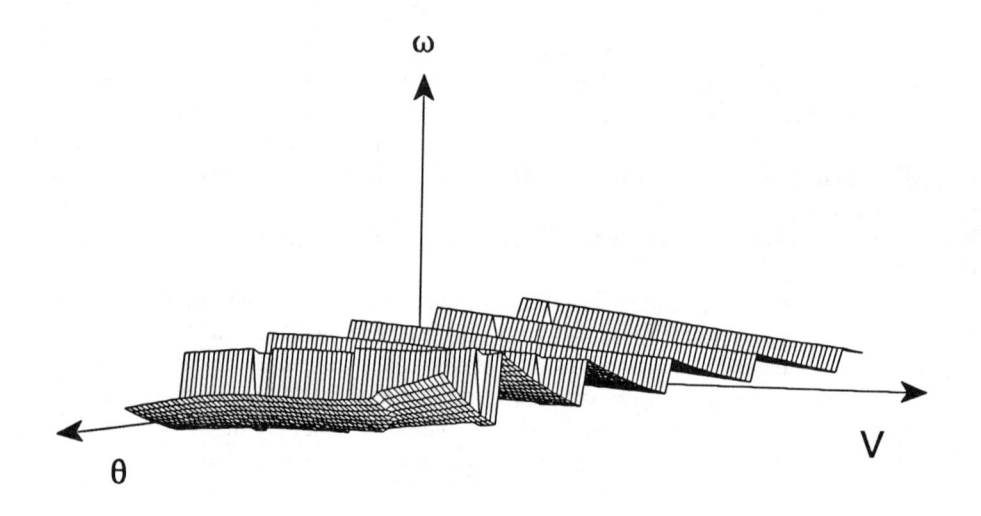

Figure 3. Analytical Control Surface

4. GENETIC ALGORITHM-DESIGNED FUZZY LOGIC CONTROLLER

The procedure for developing a GA-FLC for the bouncing ball system is presented in this section. First, the basic procedure for designing the FLC is discussed. The details of both the membership function forms and the real-valued consequent approach are described. Second, the use of a GA to tune the FLC is examined. The main focus of this examination is on the application of a GA to the tuning of the FLC, not on the details of the particular GA employed.

4.1 The Fuzzy Logic Controller

A large number of the approaches to developing FLCs that are currently popular utilize cumbersome fuzzy mathematics that can be rather confusing to the uninitiated. This fact is rather unfortunate since it has, to some extent, limited the use of the powerful and straightforward implementation of approximate reasoning or fuzzy logic. Researchers at the U.S. Bureau of Mines have adopted an approach to the development of FLCs that is easily applied to complex systems. This approach is outlined in a step-by-step procedure.

The first step in developing a FLC for the bouncing ball system is to determine which variables are important for effective control. Two such *controlled variables* were identified in the last section. The state of the bouncing ball system at any time can be completely described with the determination of the velocity of the ball immediately after impact, **v**, and the phase angle of the table at which a collision occurs, θ. Recall that these are the variables that are used to plot a Poincaré map for the bouncing ball system. They are also the variables that will appear on the left hand side of fuzzy rules that are of the form: IF {condition} THEN {action}.

The second step in FLC development is to select the *manipulated variables*, those that will in effect be used to control the system. In the bouncing ball system, as set forth in the previous section, there is only one manipulated variable: the frequency of oscillation of the table, ω. The manipulated variable appears on the right hand side of the fuzzy rules. Certainly, there are numerous variables that can be altered in the bouncing ball system that could be used as manipulated variables. For instance, changes in either the acceleration of gravity or in the coefficient of restitution of the ball could have been selected. The choice of an appropriate manipulated variable is generally limited by restrictions on the physical system, or as in this case, by the problem definition.

Once the important controlled and manipulated variables have been identified, the linguistic terms that will be used to describe these variables must be defined (the *fuzzy sets*). These linguistic variables will allow the FLC developer to look at the analytical surface (or to experiment with controlling the system in those cases in which the analytical control surface is not available) and rapidly write some fuzzy rules that begin to describe the analytical control surface. Since the bouncing ball system is chaotic, and therefore very sensitive to slight changes in ω, the analytical control surface shown in Figure 3 must be represented very accurately. Because of the detailed nature of this terrain, numerous linguistic terms are required to adequately represent each of the pertinent controlled variables, **v** and θ. Twenty-nine linguistic terms were used to represent both **v** and θ. The manipulated variable, ω, was represented using triangles only, and an infinite number of linguistic terms were used as will be explained shortly with the presentation of the real-valued consequent approach. It is important to point out that although the use of a large number of linguistic terms to describe the controlled variables allows for a detailed description of the analytical control surface, there is a price to be paid for this accuracy. As the number of linguistic terms used to describe the controlled variables increases (thereby increasing the accuracy with which the surface can be described), so too does the size of the FLC rule set. This increase in the size of the rule base translates directly to slower controller response.

The driving force behind a FLC is the idea that some uncertainty exists in categorizing the values of the system variables. The linguistic variables mean different things to different people. As a result, there must exist some mechanism for interpreting the fuzzy sets. This mechanism is the fuzzy membership function. The fuzzy membership functions traditionally used in FLCs are either triangular or trapezoidal. Triangular membership function forms are used in the bouncing ball FLC. However, triangular membership function forms alone are not adequate to allow for an acceptable representation of the control surface. The analytical control surface shown in Figure 3 depicts the intricate detail of the terrain that must be represented with a fuzzy rule base. Notice that the contours of the control surface seem to have a linear nature in some places, a sinusoidal shape in some places, and an exponential form in other places. Those sections of the surface that appear to be linear are accurately represented with triangular membership functions (trapezoidal functions could have also been used but were not necessary in this example). However, to accurately represent the other regions, more complex membership function forms were needed. Intuitively, one might try to represent the sinusoidal regions with sinusoidal membership functions, and the exponential regions with exponential membership functions. Both sinusoidal and exponential membership functions can be represented with general equations. The sinusoidal membership functions are described with an equation of the following form:

$$f(x) = \frac{1}{2} [\sin(\frac{\pi x}{w} + \frac{\pi}{2} - \frac{c\pi}{w}) + 1.0] \qquad (6)$$

where c is the center (location of maximum height) and w is the width. The exponential membership functions are described with an equation of the following form:

$$f(x) = 1.0 + \frac{(e^{k*abs(x-c)} - 1.0)}{(1.0 - e^{kw})} \qquad (7)$$

where c is the center (location of maximum height), w is the width, and k is a constant defining the curvature of the exponential function. This happens to be a case where intuition is well served. It turns out that the above three membership function forms (triangular, sinusoidal, and exponential) are adequate, yet also necessary, to represent the control surface. These membership function forms are shown in Figure 4.

Now that the crisp conditions (definite values of v and θ) existing in the bouncing ball system at the time of a collision can be categorized in a fuzzy set with a degree of membership, a process for determining a crisp action to take on the bouncing ball system, an adjustment to ω, must be developed. This process involves a set of fuzzy production rules. The set of fuzzy production rules provides a fuzzy action for any possible condition that could possibly exist in the problem environment, as that problem environment has been discretized with fuzzy sets. With the 29 fuzzy sets used to describe v and θ, there exist 841 possible conditions in the bouncing ball system, and a human expert provides a desirable action for each condition. In most traditional FLCs the manipulated variable, like the controlled variables, are prescribed using one of a discrete number of fuzzy sets that have been set aside to describe the action to be taken on the table. However, in this study, again because of the great degree of accuracy that is required, a real-valued consequent approach has been adopted. In this approach, the value of the consequent, i.e., the value of the frequency of oscillation of the table, is prescribed using a real number. The developer is actually free to choose an action from an infinite number of possible choices for each combination of the controlled variables. Some fuzzy logic proponents will frown on this informal use of fuzzy logic in the bouncing ball FLC, however, such an approach was deemed necessary, and is arguably a special case of traditional fuzzy logic. In actuality, the real number that is used to prescribe the fuzzy action can be thought of as representing a triangular membership function. When this view of the real-valued consequent approach is adopted, the real number simply indicates the center of an isosceles triangle that has a base width of 2.0. The only difference is that the developer is not limited by a finite choice of fuzzy actions, and therefore is capable of far greater accuracy than with a more traditional fuzzy logic approach. When the stance is that the developer is simply provided with a choice of an infinite number of fuzzy sets for the consequent, or manipulated variable, the only thing that is lost from traditional FLCs is the familiarity with the language. However, this slight loss in familiarity is more than compensated for with the greater accuracy that can be attained.

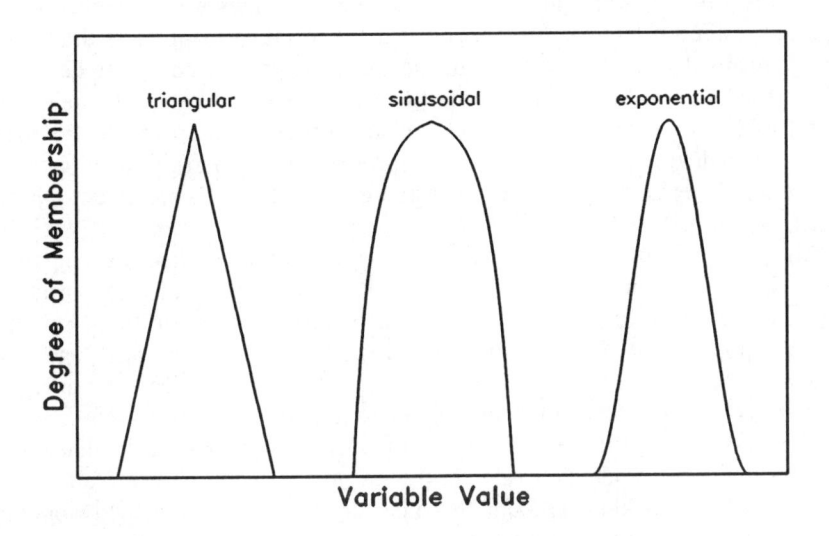

Figure 4. Membership Function Forms

It has been stated that both a large number of membership function forms and a real-valued consequent approach are necessary in the development of the bouncing ball FLC. The need for these two aspects of the controller can be seen when the following brief example is considered. This simple example will demonstrate how a real-valued consequent approach can be combined with a number of membership function forms to produce very fine control of system, and thus the production of a detailed curve or surface.

Consider a one-dimensional curve-fitting problem that has one control variable, $0.0 \leq x \leq 5.0$, and one manipulated variable, $0.0 \leq y \leq 10.0$. The following linguistic terms will be used to describe x: LOW and HIGH. In classical fuzzy logic, two rules that could be used to describe y would be of the form: IF {x is LOW} THEN {y is SMALL} and IF {x is HIGH} THEN {y is LARGE}. However, using a real-valued consequent approach these same two rules would be: IF {x is LOW} THEN {y is 0.0} and IF {x is HIGH} THEN {y is 10.0}. When these two rules are combined with the triangular membership functions shown in Figure 5(a), the curve that is shown in Figure 5(b) results. Clearly if $x=0.0$, then $y=0.0$ because the confidence that x is LARGE is 0.0. Also, if $x=5.0$, $y=10.0$. The value of y that results when x is between 0.0 and 5.0 depends exclusively on the shape of the two membership functions in Figure 5(a). With triangular membership functions, y maps (not surprisingly) to a line. However, the mapping that is obtained does not have to be linear. The use of alternative membership functions produces different mappings. Moreover, when exponential and sinusoidal membership function forms are used, virtually any mapping of y is attainable. Figures 6(a) and 7(a) show two combinations of exponential membership functions for the variable x. Figures 6(b) and 7(b) show the resulting mappings for y. Notice that the mappings that result are far from linear, and that the resulting mapping is dramatically altered with relatively slight alterations in the membership functions that describe x. With the tremendous flexibility afforded the FLC developer using multiple membership function forms in conjunction with the real-valued consequence approach, the natural question is how to determine the most efficient form of the membership functions to use. In a later section, an effective procedure for making these choices is presented. Namely, the use of a GA to make these decisions is presented.

At this point a means for converting a crisp set of conditions existing in the bouncing ball level system to a set of fuzzy conditions, and a set of fuzzy production rules prescribing a real-valued action associated with a particular set of fuzzy conditions have been developed. Unlike in conventional expert systems where only one rule is eligible to take effect, all of the rules in a FLC take effect to some degree at every time step. (Certainly, many of the rules take effect to a degree of 0.0 due to confidence levels.) Therefore, there still remains the task of converting the 841 real-valued actions prescribed by the fuzzy production rules into a single, crisp action to be taken on the bouncing ball system. Larkin [21] found that a center of area (COA) method (sometimes called the centroid method) provides an efficient means for determining this crisp action. In the COA method, the fuzzy sets and the degrees of membership associated with the fuzzy sets are used in a weighted averaging technique to compute a single value of ω.

In the case of the real-valued consequence approach, the COA method simply yields the average of the action values (real-values) weighted by the product of the confidence levels. It is important to note here that the product of the confidence levels for **v** and θ were used, not the minimum values. The use of the multiplication operator provided smooth mappings that simply were not possible using the minimum operator.

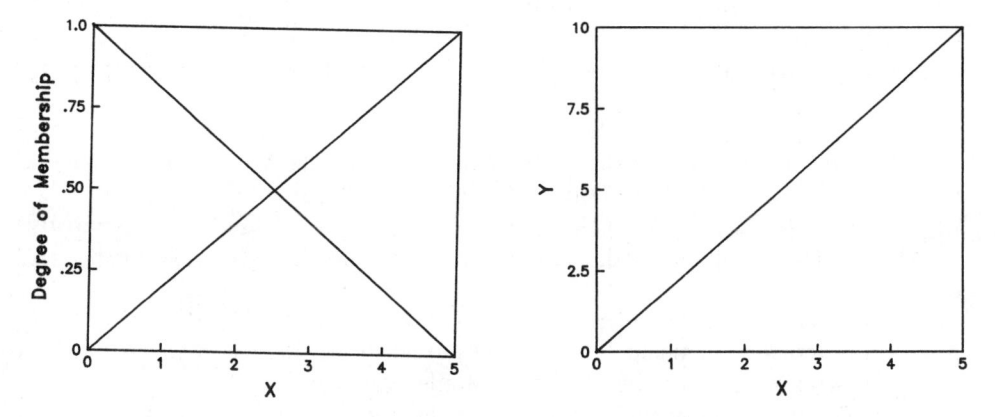

Figure 5. (a) Membership Function Forms for the Variable x
(b) The Resulting Mapping for the Variable y

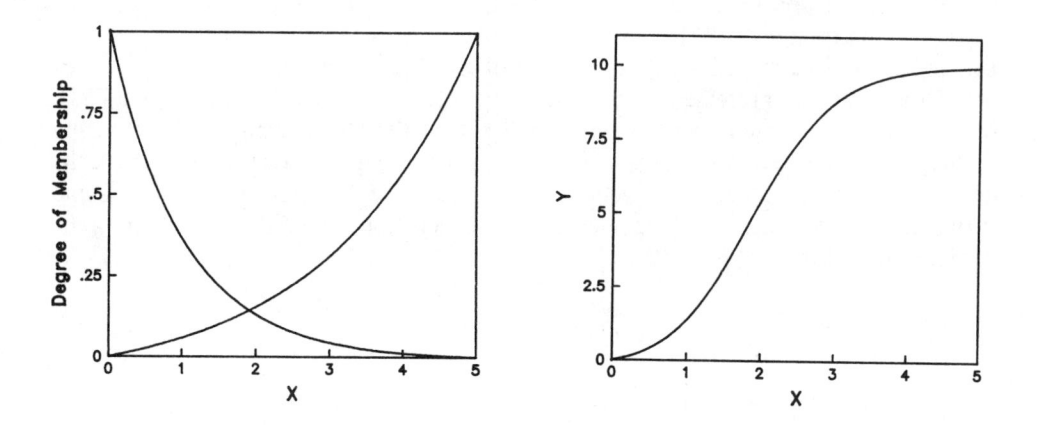

Figure 6. (a) Membership Function Forms for the Variable x
(b) The Resulting Mapping for the Variable y

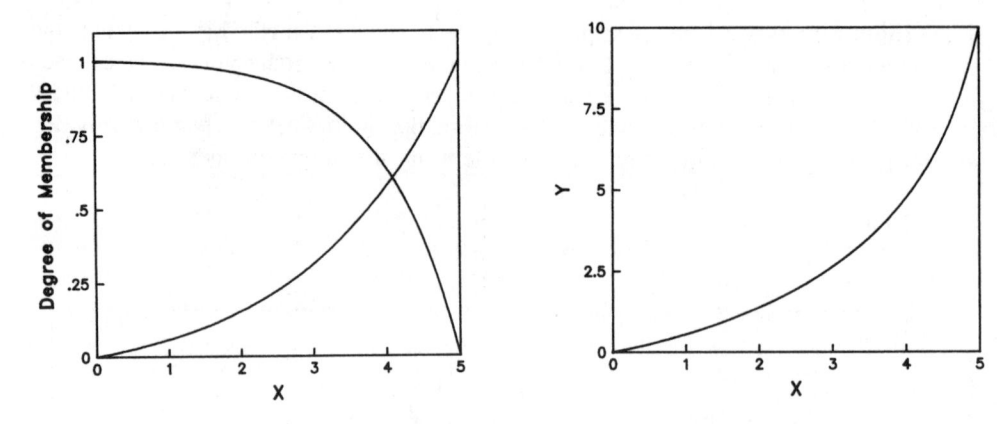

Figure 7. (a) Membership Function Forms for the Variable x
(b) The Resulting Mapping for the Variable y

The step-by-step procedure described above for developing an FLC is summarized below:
1) Determine the controlled variables to be considered;
2) Determine the manipulated variables to be considered;
3) Describe the fuzzy sets for the controlled variables;
4) Establish a set of fuzzy production rules that cover all of the possible conditions that could exist in the problem environment;
5) Define the fuzzy membership functions (using triangular, sinusoidal, and/or exponential forms);
6) Compare the set of conditions existing in the problem environment to the production rules, and compute the desirable value of ω using the COA approach.

This procedure can be easily put in the form of a computer program.

This section has presented a straightforward approach to the development of a FLC. Although the steps in this development for the bouncing ball system are conceptually simple, two of the necessary steps can indeed be frustrating. The selection of both the rule set and the membership functions can present the FLC developer with some intriguing decisions. In the remainder of this section, the use of a GA for tuning the membership functions is presented.

4.2 Tuning The Fuzzy Logic Controller With A Genetic Algorithm

A FLC that allows for reasonably effective control of the bouncing ball system can be designed simply by utilizing the analytic control surface depicted in Figure 3. However, further steps must be taken if optimal FLC performance is to be attained. It has been shown in the past that tuning the membership functions associated with a FLC can dramatically improve controller performance [12]. GAs are search algorithms based on the mechanics of natural genetics. These efficient search algorithms rapidly locate near-optimum solutions to large and difficult search problems by representing possible solutions as coded strings of characters (most frequently as bit strings). GAs require only information concerning the effectiveness of the potential solutions; they require no derivative information, nor do they require information concerning the continuity of the search space. These facts make them very inviting search techniques for the problem of tuning fuzzy membership functions.

There are numerous "flavors" of GAs: several genetic operators and variations of the basic scheme have been developed and implemented [15]. The discussion in this section is kept intentionally generic with respect to the particular GA employed. This is due to the fact that virtually any GA will provide better FLC performance, although in some problem domains one particular GA scheme may out-perform the other. Once the details of the particular GA to be employed have been determined, there are basically two decisions to be made when utilizing a GA to select FLC membership functions: (1) how to code the possible choices of membership functions as finite bit-strings and (2) how to evaluate the performance of the FLC composed of the chosen membership functions.

Consider the selection of a coding scheme. To define an entire set of membership functions (functions for v, θ, and ω), several parameters must be selected. However, before the representation of an entire set of membership functions is described, consider the representation of a single membership function. First, some representation of the form of the membership function must be made. This can be accomplished with two bits. For instance, the bit combination 00 can represent a triangular membership function, 01 can represent a trapezoidal membership function (even though no trapezoidal membership functions were necessary, and the GA bore this out, the membership function form was offered as a possibility), 10 can represent a sinusoidal membership function, and 11 can represent an exponential membership function. Second, the exact location and size of the membership function must be represented. Fortunately, each of the four membership function forms chosen can be completely defined using only two parameters. The triangular membership functions are limited to isosceles triangles, in which case only the center and the width of the base must be defined. The trapezoidal membership functions are forced to be symmetric, in which case only the location of two points must be defined. Both the sinusoidal and exponential membership functions are described by equations that require only two parameters for complete definition. Thus, for each membership function, two parameters, C_1 and C_2, must be defined. These parameters can be coded using a conventional coding scheme called *concatenated, mapped, unsigned binary coding*. In this coding scheme each individual parameter is discretized by mapping linearly from a minimum value (C_{min}) to a maximum value (C_{max}) using an *m*-bit, unsigned binary integer according to the equation:

$$C = C_{min} + \frac{b}{(2^m - 1)} * (C_{max} - C_{min}) \qquad (8)$$

where C is the value of the parameter of interest, and *b* is the decimal value represented by an *m*-bit string.

The above approach to coding provides a mechanism by which a single membership function can be represented using a bit string of length $2+m$ (two bits for the membership function form and *m* bits for the exact definition of the membership function). Now, *k* membership functions can be represented simply by concatenating *k* of the $(2+m)$ length bit strings. In this way, the GA can operate on bit strings that completely define a set of membership functions. In the case of the bouncing ball FLC, it is important to note that not all of the membership functions required tuning with a GA; many of the membership functions were adequately defined by the developer. This is an important point, because the size of the search space could be reduced to a manageable size.

Now for the second decision: how are the strings, or the potential membership functions, evaluated? In judging the performance of the bouncing ball FLC, there is an analytical solution for comparison. Recall that the development of the bouncing ball controller is being viewed as a surface-fitting problem. Therefore, the task is to use inexact reasoning to accurately reproduce the analytical control surface. Thus, each string (each set of potential membership functions) can be judged strictly on how well the surface they produce matches the analytical surface. A least squares criterion was used, and the surface was evaluated at 1000 points.

This section has provided a quick glimpse at the approach developed at the U.S. Bureau of Mines for developing GA-FLCs. The details of the GA have been kept intentionally generic because the art of GAs is growing at a rapid pace and there are a number of good references now available. Basically, there are two decisions that must be made when applying a GA, and each of these have been discussed in light of the bouncing ball control problem. In the next section, the performance of the GA-FLC will be evaluated.

5. RESULTS

A GA-FLC has been developed that is capable of efficiently manipulating the bouncing ball system. This controller alters the frequency of oscillation of the table, based solely on the velocity with which the ball strikes the table and on the phase angle of the table at the time of impact, in such a manner to ensure the ball bounces to a constant height. As pointed out earlier, the effectiveness of such a controller can be judged from a Poincaré map. Figure 8 shows a Poincaré map for the bouncing ball system wherein the GA-FLC has been put into operation. Notice that there are only 5 points in the space. There are such few points because the controller is put into operation after 4 bounces. Figure 8 is far different than Figure 2 which shows a Poincaré map for the bouncing ball system in which no attempt has been made at control. The GA-FLC has instilled order in an otherwise chaotic system.

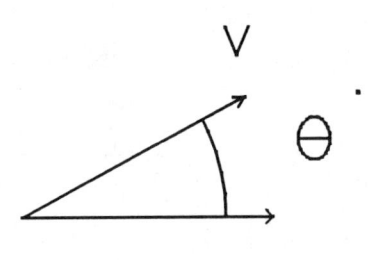

Figure 8. Poincaré Map with GA-FLC

For many interested observers, a Poincaré map is a natural vehicle for judging the performance of a controller in the realm of chaos. For others, however, it is difficult to disseminate the information portrayed by this type of map. To assist any such observers, Figure 9 shows the maximum height to which the ball bounces as a function of time. Recall that the objective of the controller was to adjust the frequency of oscillation of the table in such a manner as to force the ball to always bounce to a constant height. Figure 9 leaves no doubt as to the effectiveness of the GA-FLC. Figure 9 shows the same simulation as does the Poincaré map depicted in Figure 8. In both, the controller has been put into operation after 4 bounces.

For the simulation depicted in Figures 8 and 9, the controller is able to adjust the oscillatory frequency of the table so as to immediately cause the ball to bounce to the target height. However, this is certainly not always the case. There were limits placed on the range of acceptable values for ω. This limitation restricts the controller by restricting the amount of energy that can be imparted to or dissipated from the bouncing ball. As an example, if the ball is bouncing well above the target height, the maximum energy dissipation will occur when the ball strikes the table when the table has its maximum downward velocity. Since the value of ω is restricted, so to is the energy that can be dissipated. Likewise, if the ball is bouncing to a height well below the target height, the maximum energy that can be imparted to the ball will occur when the ball strikes the table when the table has its maximum upward velocity. Since the value of ω is restricted, so to is the energy that can be imparted to the ball.

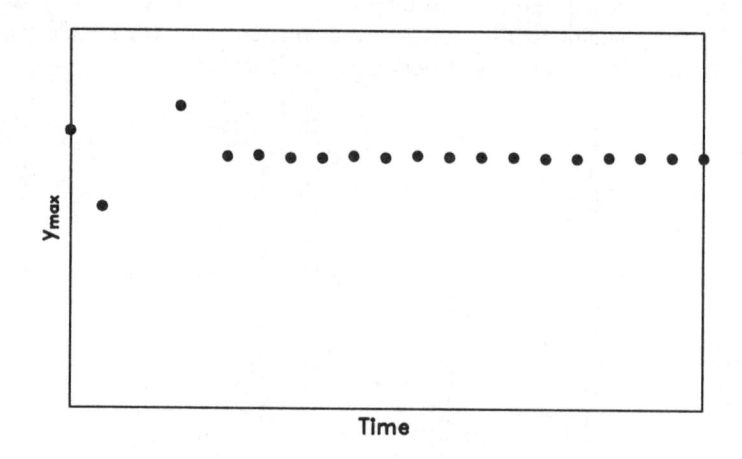

Figure 9. Maximum Ball Height as a Function of Time

The GA-FLC produces a control surface that can be compared to the analytical control surface shown in Figure 3. The surface produced by the GA-FLC is shown in Figure 10. Notice that the GA-FLC control surface is characterized by all of the complexities contained in the analytical control surface. However, simply comparing these two surfaces in such an inexact fashion is by no means adequate. Therefore, Figure 11 shows an error surface, which is simply the absolute value of the difference between the control actions prescribed by the analytical controller and the by the GA-FLC. Note that there is excellent agreement between the two controllers for the majority of the space. However, there are a few places in which the difference between the two controllers is substantial. These regions are generally characterized by discontinuities in the control surface. The GA-FLC is sometimes unable to represent these discontinuities adequately because of the fact that fuzzy logic requires a discretization of the search space. Certainly this discretization is more flexible than the discretization possible with conventional set theory. At this point, the authors yield to the classical trade-off made in virtually all computational work; that of accuracy versus computational time. The magnitude of the error surface could well be reduced with greater discretization (the definition of more fuzzy sets), but this adds to the complexity of the controller and increases the time the controller needs to prescribe a control action.

For most people, the information depicted by Figures 8, 9, and 11 should be adequate to demonstrate the effectiveness of the GA-FLC. However, there are some who prefer statistical analysis over graphical comparisons. For these people, the following analysis is included. For the purposes of this analysis, error, E, is defined as:

$$E = \frac{\omega_{analytical} - \omega_{GA-FLC}}{\omega_{analytical}} * 100\% \qquad (9)$$

where $\omega_{analytical}$ is the oscillatory frequency of the table prescribed by the analytical controller and ω_{GA-FLC} is the oscillatory frequency of the table prescribed by the GA-FLC. One hundred thousand points were sampled to compute the average error, $E_{avg}=0.4673\%$, and the standard deviation, $\sigma=3.141$. Moreover, 98.54% of the points sampled had an average error of less than 1.0%. The GA-FLC produces an accurate depiction of the analytical control surface.

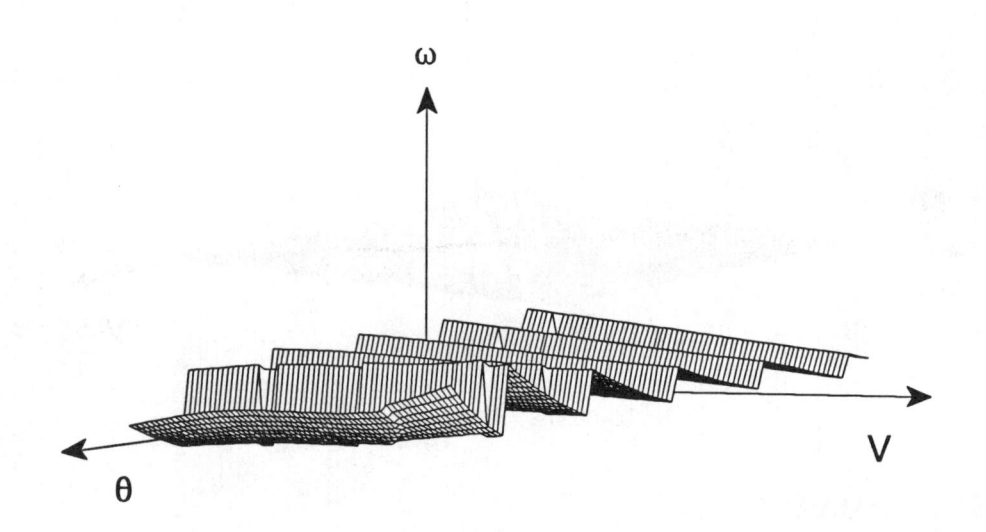

Figure 10. GA-FLC Control Surface

The above discussion demonstrates the accuracy of the GA-FLC for the bouncing ball system. As with most chaotic systems, a controller for the bouncing ball system must be more than just accurate; it must also be fast. It is quite important to note that the analytical controller described earlier in this chapter takes on the order of one second to compute its control action, whereas the GA-FLC takes on the order of milliseconds to compute its control action. This requirement for fast and accurate controllers becomes even more important in some other chaotic systems. An example of a chaotic system in which speed is at a premium appears in the steel industry in the form of an electric arc furnace [7].

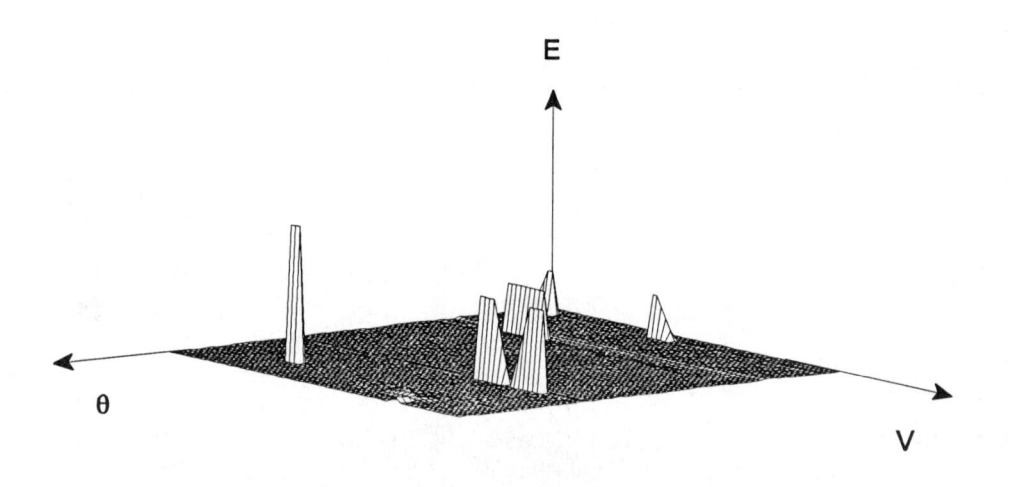

Figure 11. Error Surface

6. SUMMARY

In this chapter, a control system has been described that effectively manipulates a chaotic system. The control algorithm is based on a technique developed at the U. S. Bureau of Mines in which a GA is used to design a FLC. The use of fuzzy logic enables a person with limited experience working with a system to establish a set of rules for manipulating that system, and a set of membership functions the define terms used in the rule-set. GAs are search algorithms based on natural genetics. They are used in the control algorithm to tune the membership functions so that the inexact reasoning characteristic of the FLC is sufficient to control a system that requires precise control actions.

The chaotic system considered consists of a ball bouncing on an oscillating table. The objective of the control problem is to adjust the table's frequency of oscillation so that the ball always bounces to a constant height. Results demonstrate the effectiveness of the control system by tracing the position of the ball through time, and by considering the state space of the chaotic system at every impact via a Poincaré map. Furthermore, a statistical analysis in which 100,000 points were sampled demonstrates the accuracy of the GA-FLC in reproducing a control surface developed using an analytical approach. Thus, the GA-FLC is able to solve a complex surface-fitting problem.

The GA-FLC required two special attributes to effectively manage the complexities associated with controlling the bouncing ball system. First, the FLC required a large number of membership functions, and four forms of membership function. Specifically, triangular, trapezoidal, sinusoidal, and exponential membership functions were all used. Second, a real-valued consequence approach was employed to account for the extreme precision required in reproducing the analytical control surface. This approach allowed for the use of fuzzy numbers on the consequent side of the rules, as opposed to the conventional approach in which traditional fuzzy sets are used as the consequence. This approach, in effect, provides the developer with an infinite number of choices for each consequence.

The results presented are important because they demonstrate the effectiveness of manipulating a chaotic system using a FLC. FLCs are generally more easily written than other rule-based systems because FLCs employ fuzzy, linguistic terms with which humans are comfortable using. Thus, the use of fuzzy logic in a rule-based control system allows for the production of a "common-sense", yet a very effective control strategy. Furthermore, the results presented yield hope for effectively manipulating the numerous chaotic systems occurring in industry that have historically been costly due to a lack of control.

REFERENCES

1. Poincaré, H., *The Foundation of Science: Science and Method*, English Translation. New York, NY, The Science Press, 1921.

2. Van der Pol, B., & Van der Mark, J., "Frequency Demultiplication." *Nature*, Vol. 120, pp. 363-364, 1927.

3. Moon, F. C., *Chaotic Vibrations*. New York, NY, John Wiley & Sons, 1987.

4. Berge, P., Pomeau, Y., & Vidal, C., *Order Within Chaos*. New York, NY, John Wiley & Sons, 1984.

5. Borman, S., "Researchers find order, beauty in chaotic chemical systems." *Chemical and Engineering News*, January, pp. 18-29, 1991.

6. Tufillaro, N. B., & Albano, A. M., "Chaotic Dynamics of a Bouncing Ball." *American Journal of Physics*, Vol. 54, No. 10, pp. 939-944, 1986.

7. Ochs, T. L., & Hartman, A. D., "Improved arc stability in electric arc furnace steel making." In The U.S. Bureau of Mines' (eds): *New Steel-Making Technology from the Bureau of Mines*, Bureau of Mines Information Circular number 9195, 1988, pp. 2-11.

8. Brady, B. T., "Prediction of failures in mines — An overview." *Bureau of Mines Report of Investigations* (number 8285), 1978.

9. Coughlin, J., & Kranz, R., "New approaches to studying rock burst associated seismicity in mines." *Proceedings of the 32nd U.S. Rock Mechanics Symposium*, 1991, pp. 105-109.

10. Ott, E., Grebogi, C., & Yorke, J. A., "Controlling chaos." *Physical Review Letters*, Vol. 64, No. 11, pp. 1196-1199, 1990.

11. Romeiras, F. J., Ott, E. Grebogi, C., & Dayawansa, W. P., "Controlling Chaotic Dynamical Systems." *Proceedings of the American Control Conference*, pp. 121-130, 1991.

12. Karr, C. L., "Genetic algorithms for fuzzy logic controllers." *AI Expert*, Vol. 6, No. 2, pp. 26-33, 1991.

13. Zadeh, L. A., "Outline of a New Approach to the Analysis of Complex Systems and Decision Processes." *IEEE Transactions on Systems, Man, and Cybernetics*, Vol. SMC-3, No. 1, pp. 28-44, 1975.

14. Sugeno, M., (ed), *Industrial Applications of Fuzzy Control*. Amsterdam, Elsevier Science Publishers, 1985.

15. Goldberg, D. E., *Genetic Algorithms in Search, Optimization, and Machine Learning*. Reading, MA, Addison-Wesley, 1989.

16. Karr, C. L., "Design of an adaptive fuzzy logic controller using a genetic algorithm." In Belew, R. K., & Booker, L. B. (eds): *Proceedings of the Fourth International Conference on Genetic Algorithms*. San Mateo, CA, Morgan Kaufmann Publishers, 1991, pp. 450-457.

17. Karr, C. L., Sharma, S. K., Hatcher, W. J., & Harper, T. R., "Control of an exothermic chemical reaction using fuzzy logic and genetic algorithms." In Ralston, P. A. S., Stoll, K. E., & Ward, T. L. (eds): *Proceedings of International Symposium on Artificial Intelligence in Real-Time Control*. 1992, pp. 246-254.

18. Karr, C. L., & Gentry, E. J. "Real-time pH control using fuzzy logic and genetic algorithms." Paper presented at the Annual Meeting of the Society for Mining, Metallurgy, and Exploration (preprint number 92-49), Phoenix, AZ, 1992.

19. Cao, Z., Kandel, A., & Li, L., "A New Model of Fuzzy Reasoning." In *Fuzzy Sets and Systems*. 1991, pp. 311-325.

20. Karr, C. L., Stanley, D. A., & Scheiner, B. J., "A genetic algorithm applied to least squares curve fitting." *Bureau of Mines Report of Investigations* (number 9339), 1991.

21. Larkin, L. I., "A fuzzy logic controller for aircraft flight control." In Sugeno, M. (ed): *Industrial Applications of fuzzy Control*. 1985, pp. 87-100.

Chapter 23
Applications of a Fuzzy Control Technique to Superconducting Actuators using High-Tc Superconductors

Introduction, *500*
Superconducting Levitation Mechanism, *500*
Superconducting Radial Bearing, *506*
Superconducting Linear Actuator, *510*
Superconducting Pump Actuator, *515*
Conclusions, *519*
References, *520*

A levitation mechanism of high-Tc superconductors is created with the use of alternating-polarity magnets of the same size. The basic characteristics of the superconducting levitation mechanism are described in this chapter. Coupling the characteristics of the levitation mechanism makes it possible to design a variety of superconducting actuators including a superconducting linear bearing and a superconducting journal bearing. Fuzzy control techniques are useful for controlling the position of superconducting actuators because of some reasons depending on superconductivity. The fuzzy control technique in this chapter is applied to two different types of actuators such as a superconducting linear slider and a superconducting pump actuator.

APPLICATIONS OF A FUZZY CONTROL TECHNIQUE TO SUPERCONDUCTING ACTUATORS USING HIGH-Tc SUPERCONDUCTORS

M. Komori T. Kitamura
Department of Mechanical System Engineering
College of Computer Science and System Engineering
Kyushu Institute of Technology
Iizuka, Fukuoka 820, JAPAN

1. INTRODUCTION

The Meissner effect and the pinning effect using high-Tc superconductor are very promising for stable levitation of superconductor in magnetic field with neither contacts nor additional control mechanisms[1,2]. Many efforts using high-Tc superconductor have been made to apply the Meissner effect and the pinning effect to a variety of levitation systems[3-7]. Our group has investigated a levitation mechanism using a high-Tc superconductor and a set of alternating-polarity magnets[8-11]. The static experiments about the levitation mechanism give larger levitation pressure and uniform levitation pressure.

By using the levitation mechanism a superconducting linear actuator has been developed[12]. Fuzzy theory is applied to control the actuator. Conventional control techniques with the use of a kinetic model of the slider in the liquid nitrogen are less practical due to several uncertain factors in its modeling.

2. SUPERCONDUCTING LEVITATION MECHANISM

Two different types of superconducting levitation mechanisms are shown in Fig. 1 with magnetic flux lines. Both levitation mechanisms consist of a superconductor and a set of alternating polarity magnets. One type has a magnetization placement perpendicular to the magnet plane as shown in Fig. 1(a). The other has a magnetization placement parallel to the magnet plane as shown in Fig. 1(b). In the figures the magnetic flux going out of each N pole goes into the adjacent S poles to reduce the magnetic energy of the alternating-polarity magnets. Some flux lines are trapped or pinned in the superconductor by the pinning effect and other

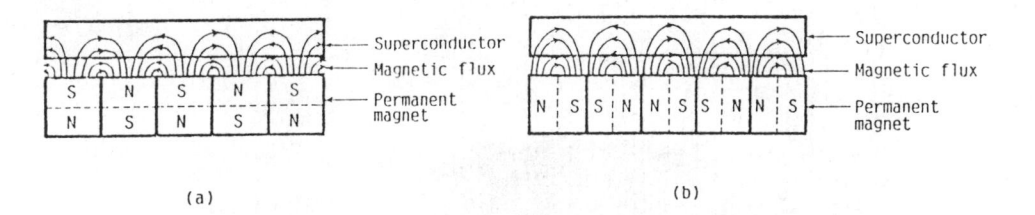

Fig.1. Superconducting Levitation Mechanism Consisting of a Superconductor and a Set of Magnets

flux lines are expelled from the superconductor by the Meissner effect. The magnetic flux penetrates easily in the superconductor as long as type-II superconductors such as high-Tc superconductors are used. The flux pinning effect is much more important than the Meissner effect to the levitation force of the superconductor [3]. The magnetic flux works to support the superconductor like mechanical springs. These two types of the levitation mechanisms are applicable to a variety of actuators.

Fig. 2(a) shows a photograph of the developed superconducting linear slide. The magnetic guide includes ten pieces of bar magnets ($2 \times 20 \times 5$ mm^3) at the bottom for levitation and three pieces of the same dimensions for its each side. The magnets are samarium cobalt rare-earth with a residual magnetization of $Br \simeq 1.0$ T. The superconducting slider has disk-shaped nine superconductors ($7\phi \times 1$ mm) on the base for levitation and square eight superconductors ($5 \times 5 \times 1$ mm^3) on its each side for guide. The superconductors adhere to a Bakelite board ($1 \times 20 \times 80$ mm^3). The superconductors used are made of $Y_1 Ba_2 Cu_3 O_x$ ($Tc \simeq 90$ K) in a standard fashion of sintering in the authors' laboratory. It is observed that the slider is levitated with no mechanical contacts with any side of the guide in the liquid nitrogen. A longitudinal view of this observation is shown in Fig. 2(b).

Levitation pressures on the superconducting levitation mechanism were measured by a beam type load cell containing strain gauges[10]. The levitation pressure is defined as the levitation force measured by the load cell divided by the bottom pole-face area of the magnet assembly. In each experiment the superconductor tile was first warmed to the normal state and then zero-field cooled to the superconducting state. The high-Tc superconductor used is $Bi_{1.85} Pb_{0.35} Sr_{1.90} Ca_{2.05} Cu_{3.05} O_x$ with the critical temperature of $Tc \simeq 100$ K, made by the process of free-sintering method. The superconductor tile is $55 \times 28 \times 5$ mm^3 in size and weighs 37.5 g. The set of alternating-polarity magnets consists of several magnet bars of the same size ($W \times 5 \times 20$ mm^3; $W = 2 \sim 16$ mm). W is the width of a bar

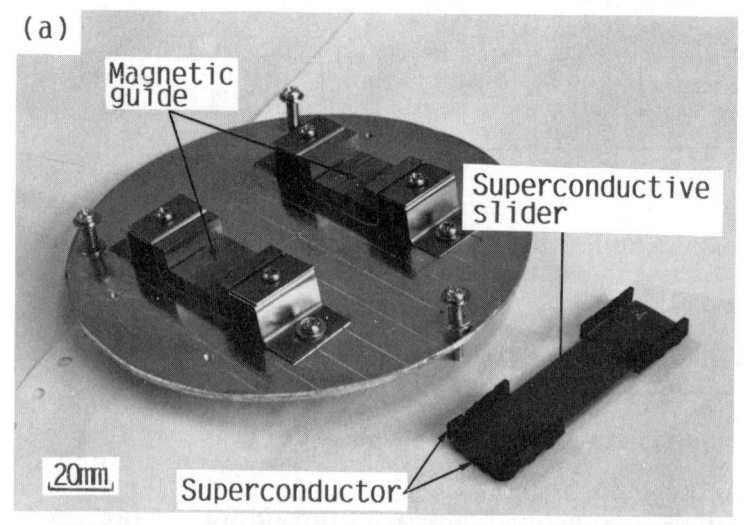

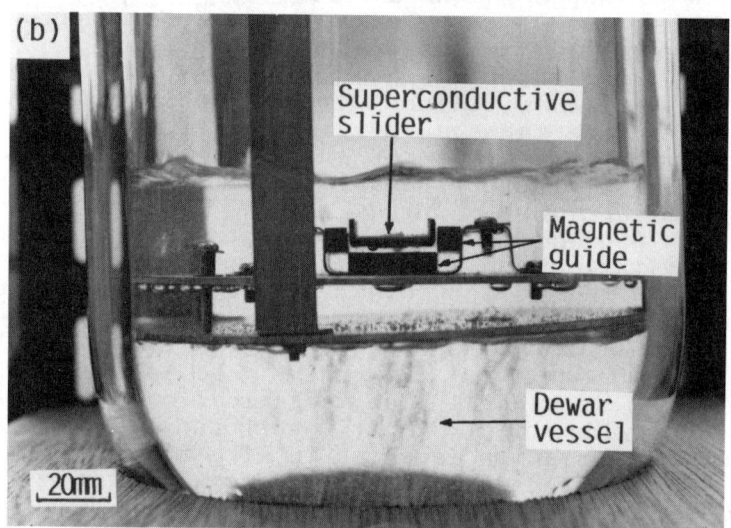

Fig.2. Superconducting Linear Slide. (a) Structure of the Linear Slide. (b) Longitudinal Observation of the Slide Working in the Liquid Nitrogen

magnet. The magnets are samarium cobalt rare-earth with a residual magnetization of $Br \simeq 1.0$ T.
Fig. 3 shows levitation pressure P of the superconductor tile versus distance D between the

superconductor tile and the magnets. In the experiment the superconductor tile was continuously moved at a speed of less than 0.5 mm/sec toward the magnets from the distance D=30 to 0.5 mm (providing an approach curve) and then moved away (providing a retreat curve). The measurements were carried out in the width range from 2 to 16 mm. The results of the widths W = 2, 4, 8, and 16 mm are shown in Figs. 3(a), (b), (c) and (d), respectively. There exist various hysteresis loops corresponding to their widths. These loops were repeatable when the experiment was carried out in the same distance range. The hysteresis loops are

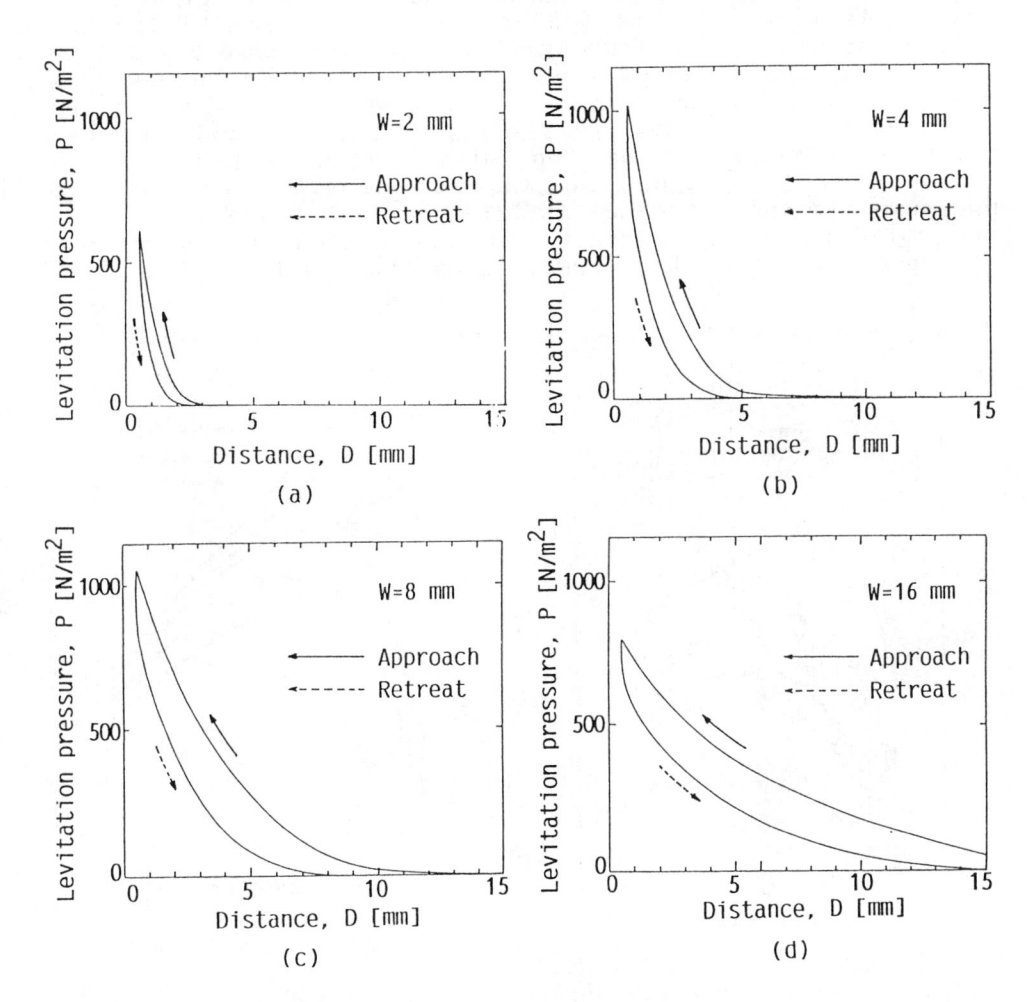

Fig.3. Relationship Between the Levitation Pressure and the Superconductor-magnet Distance for Various Widths

believed to result from the flux pinning effect as already reported by other investigations[3,4]. By using these levitational mechanisms with a set of alternating-polarity magnets the levitation pressure can be obtained over a wide area of the magnetic pattern. Choosing the optimal width of a bar magnet is important for designing levitation mechanisms.

The relationship between the levitation pressure P and the width W of a bar magnet for various distances D is shown in Fig. 4. For $D = 1.0$, 2.0, and 3.0 mm the approach curves (solid lines) and retreat curves (dotted lines) of the P-D hysteresis loops are obtained respectively. From Fig. 4 the maximum pressure ($P \simeq 1100$ N/m^2) for the curve of $D = 0.5$ mm exists near the width of $W = 6$ mm. For the other curves from $D = 1.0$ to 3.0 mm the maximum pressure exists near the width of $W = 6 \sim 10$ mm. This suggests that the alternating-polarity magnets can be optimized with respect to the width of a bar magnet so that it provides the maximum pressure if the distance is fixed. For each fixed distance the solid curves are larger than the dotted curves because of the existence of the pressures by the pinning effect.

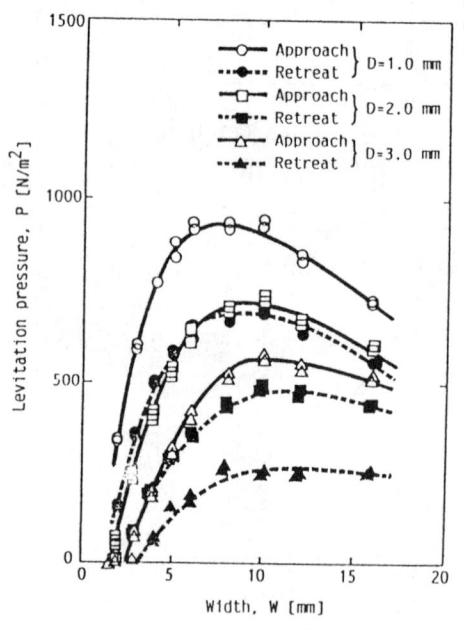

Fig.4. Levitation Pressure vs. Width of a Bar Magnet

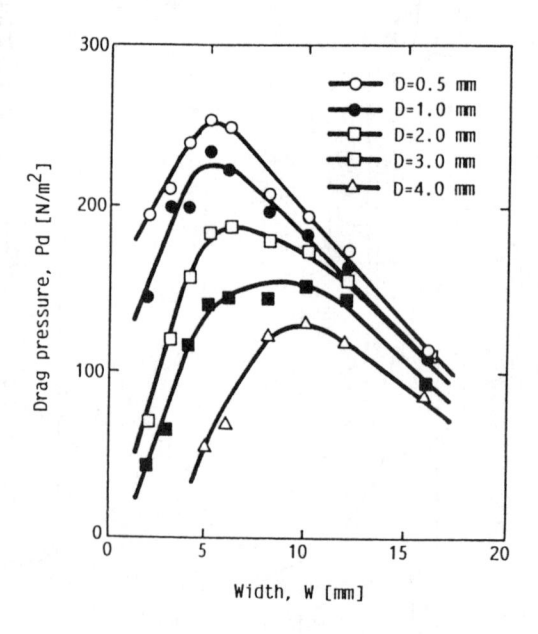

Fig.5. Pressure by the Pinning Effect vs. Width of a Bar Magnet

The pressures by the flux pinning effect are given by subtracting the dotted curve from the solid curve of the same distance. The magnetic flux trapped in the superconductor is larger with decreasing distance[10]. It is stable for the superconductor tile to levitate on the set of alternating-polarity magnets which gives a maximum pressure by the pinning effect. Because the superconductor tile can not move easily due to the trapped flux penetrating in itself. The force by the pinning effect is important for designing a variety of actuators because of the stability.

Drag pressures were measured by a beam type load cell containing strain gauges[11]. The drag pressure is defined as the drag force measured by the load cell divided by the bottom pole-face area of the magnet assembly. In the experiment the magnets were slowly moved over the superconductor tile keeping parallel to the superconductor tile by the mechanical stage. In each experiment the superconductor tile was first warmed to the normal state and then zero-field cooled to the superconducting state in the liquid nitrogen.

The drag pressure of the superconducting levitation mechanism shows various hysteresis loops corresponding to the distances from 0.5 to 4.0 mm. The drag pressures do not depend on the moving speed of the magnets[3]. The relationship between the drag pressure and the width of a bar magnet for various distances is shown in Fig. 5. As the distance becomes small the drag pressure becomes large. This is because the flux pinning effect contributing to the pressure becomes large with decreasing distance. There exist maximum drag pressures near the width from 4 to 10 mm for each curve. The pinning effect depends on the magnetic flux trapped in the superconductor. The existence of the maximum pressures is due to the magnetic flux distributions over the magnets which are different from each other due to the different widths of bar magnets.

The relationship between the drag pressure and the angle between the alternating-polarity and the moving direction is shown in Fig. 6 for various distances. In this case the width is fixed at 6 mm. The drag pressure becomes large with decreasing because of the pinning effect depending on the distance. From this figure the angle dependence of the drag pressure in the levitation mechanism is remarkable. In each curve there exists maximum drag pressure near the angle of 0 degree. The drag pressures near ±90 degree are negligible compared with those near 0 degree. As long as the magnet does not move the drag pressure is not generated by the magnetic flux trapped in the superconductor tile. The number of the magnetic flux unpinned and jumping to a new stable position by moving the magnets depends on the angle as shown in Fig. 6. For designing superconducting actuators using the levitation mechanism with alternating-polarity magnets it is important to reduce the pressure. It is also necessary to use the drag pressure to stop the undesirable motion like vibration.

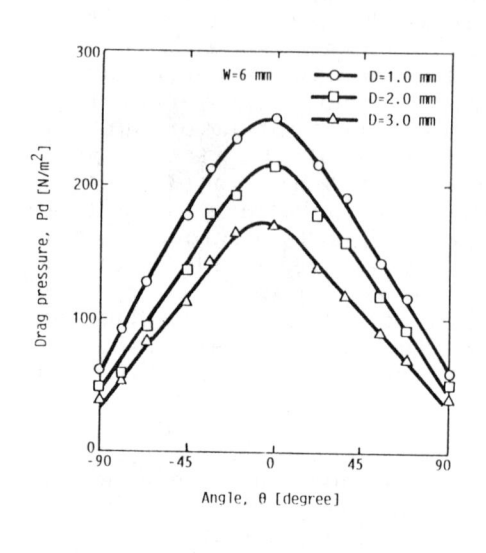

Fig.6. Drag Pressure vs. the Angle for Various Distances

Fig.7. Superconducting Bearing. (a) Linear Bearing. (b) Journal Bbearing.

3. SUPERCONDUCTING RADIAL BEARING

As applications of this frictionless mechanism a linear bearing (see Fig. 7(a)) and a journal bearing (see Fig. 7(b)) are realized. In the linear bearing as shown in Fig. 7(a) the pinning effect stops the rotary motion of the superconductor shaft. In the journal bearing as shown in Fig. 7(b) the superconductor shaft can rotate with no friction. But the pinning force stops the linear motion of the shaft.

Fig. 8 shows photographs of a superconducting linear bearing. The magnet housing (ID10ϕ×OD22ϕ×20 mm) includes 10 pieces of bar magnets of the same dimensions (3×5×20 mm^3). The radius and length of the columnar shaft of $Y_1Ba_2Cu_3O_x$ are 7 mm and 16 mm respectively, weighing 2.5 g. Fig. 8(b) shows a photograph of the longitudinal view of the bearing working in the liquid nitrogen.

The relationship between the levitation force and the displacement for the superconducting linear bearing is shown

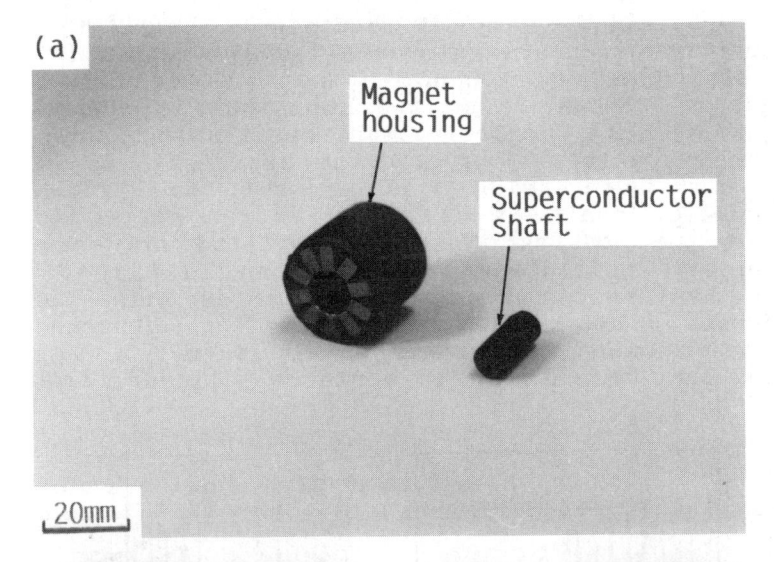

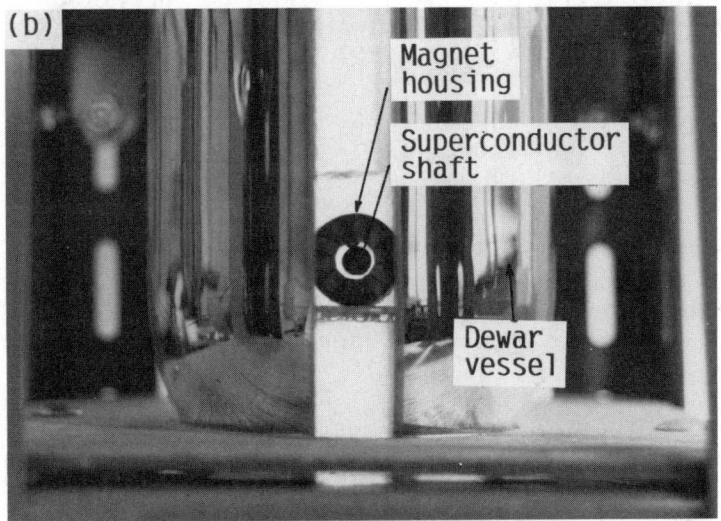

Fig.8. Superconducting Linear Bearing. (a) Structure of the Bearing. (b) Longitudinal Observation of the Bearing Working in the Liquid Nitrogen.

in Fig. 9. The measurements were carried out within the displacement range from -0.5 to 0.5 mm. In the origin the displacement is the center of the magnet housing. The shaft was moved manually by a mechanical stage. The result indicates the circular hysteresis loops as shown in Fig. 9. In Fig.9(a) the superconductor shaft is at first in the amplitude of 0.5 mm. Then the amplitude is approached to zero (end point) following an oscillating and damping displacement pattern. The circular hysteresis loops become smaller with decreasing displacement. In Fig. 9(b) the superconductor shaft is at first in the center (starting point) of the housing. Next the amplitude is made larger following an oscillating pattern shown in Fig. 9(b). The circular hysteresis loops become larger with increasing displacement. The levitation force of the superconductor shaft works toward the center of the magnet housing. This indicates that this mechanism works as a linear bearing.

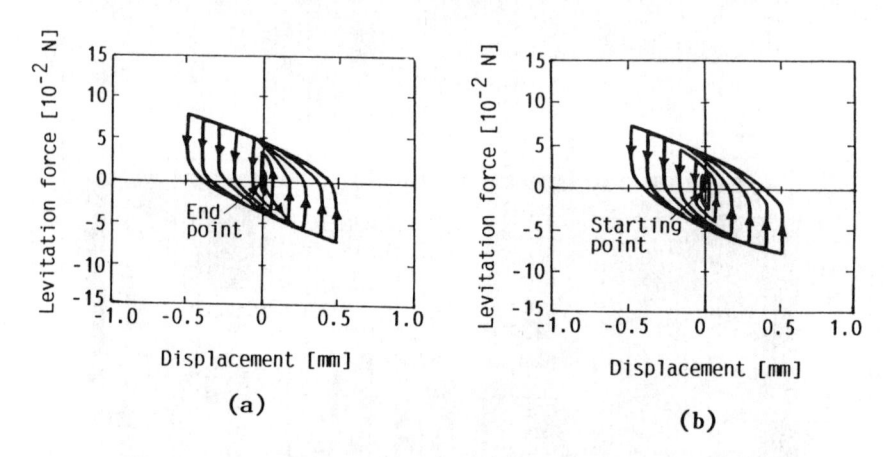

Fig.9. Relationship Between the Levitation Force and the Displacement of the Superconducting Linear Bearing

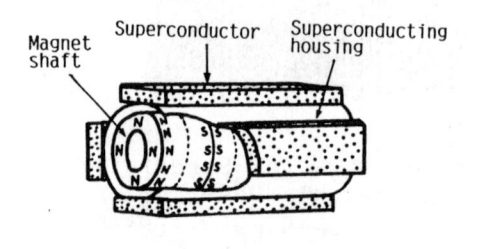

Fig.10. Superconducting Journal Bearing With Four Superconductors

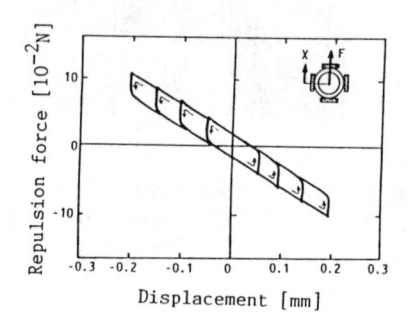

Fig.11. Repulsion Force vs. Displacement of the Bearing

As another application of the levitation mechanism shown in Fig. 7(b), a superconducting journal bearing is made. The schematic picture of the journal bearing is shown in Fig. 10. The journal bearing consists of a superconducting housing and a magnet shaft. The superconducting housing is covered with superconductors (YBa$_2$Cu$_3$O$_x$, $\simeq 1.5 \times 3.5 \times 10$ mm^3). The magnet shaft has some ring magnets (OD6ϕ ×ID3ϕ ×3 mm^3, $Br \simeq 1.0$ T) whose magnetization is parallel to the axial direction.

The restoring force in the radial direction was measured in the superconducting journal bearing. The superconductors used in the housing was four pieces. The magnet shaft has three ring magnets. Fig. 11 shows the relationship between the repulsion force and the displacement. The result shows the hysteresis loops. In the displacement range shown, the repulsion force is linear with respect to the displacement. In the case of eight pieces of the superconductors, the repulsion force is twice as large as that for four pieces. And the repulsion-displacement relationship is also linear in the small displacement range shown in Fig. 11.

In order to investigate the rotation characteristics of the new superconducting journal bearing, a rotation test of the bearing by air drive was performed. An air turbine (aluminium, 10 mmϕ ×5 mm, 1.55g) was attached to the bearing with a non-magnetic yoke in the center. The rotation speed was measured using a tachometer. After the air was cut, the speed decay was measured as a function of time. Fig. 12 shows the relationship between the rotation speed and the time for various initial rotation speeds. The rotation speed decreases monotonically with increasing time. In each cases the shaft continues rotating for more than 60 sec. Each curve has an exponential decay in the experiment. This shows that a constant drag torque on the rotating shaft due to magnet's asymmetry is negligible. The shaft rotated stably at rotation speeds less than 26000 rpm.

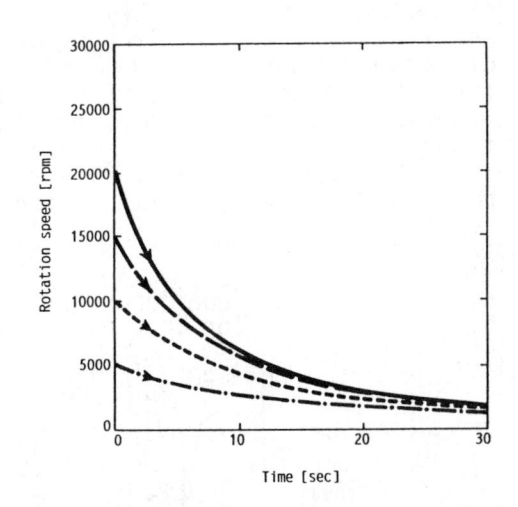

Fig.12. Rotation Characteristics of the Bearing

4. SUPERCONDUCTING LINEAR ACTUATOR

Actuator system Fig. 13 shows a schematic picture of the superconducting linear actuator. The superconducting slider is 2.8 g in weight with disk-shaped two superconductors ($Y_1Ba_2Cu_3O_x$, $7\phi \times 1$ mm) on the base for levitation on its each side and 81.7 mm in length. A solenoid (0.3 mmϕ, $\simeq$ 500 Turn) is placed facing the superconductor at each end of the slider. The resistances of the solenoids are $\simeq$ 4 Ω at room temperature and 0.5 Ω at liquid nitrogen temperature (77K). The resistances are also changed by the driving current applied to the solenoids. The displacement of the slider is detected by a Hall sensor and a magnet ($6 \times 5 \times 3$ mm^3, $Br \simeq 0.4$ T). The magnetic guide includes ten pieces of samarium cobalt bar magnets ($2 \times 20 \times 5$ mm^3, $Br \simeq 1.0$ T) for levitation. Two iron magnetic screenings ($53 \times 120 \times 0.5$ mm^3) are placed between the slider and the solenoids to protect the Hall sensor from magnetic disturbance produced by the solenoids. The slider is stably levitated at about 1.5 mm heigh with no mechanical contacts with any side of the guide in the liquid nitrogen as shown in Fig. 14. A longitudinal view of the superconducting linear actuator working in the liquid nitrogen is shown in Fig. 14. The superconducting slider is levitated over the magnetic guide with alternating-polarity magnets. The slider has no contact with any side of the guide.

The relationship between the Hall voltage V_H and the position X of the superconducting slider is shown in

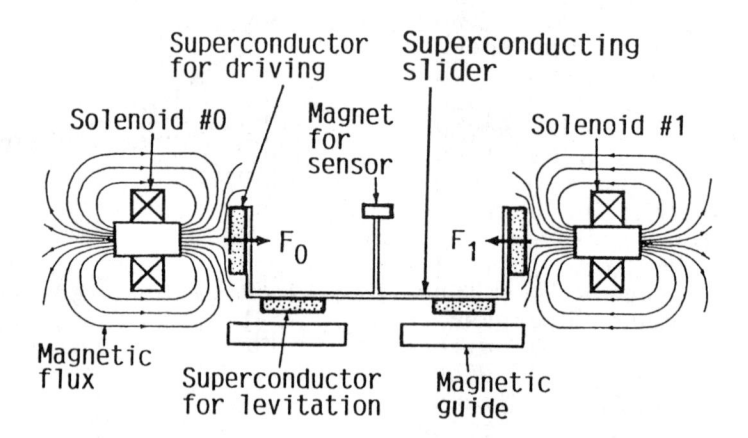

Fig.13. Schematic Picture of the Superconducting Linear Actuator

Fig.14. Longitudinal Observation of the Superconducting Linear Actuator Working in Liquid Nitrogen

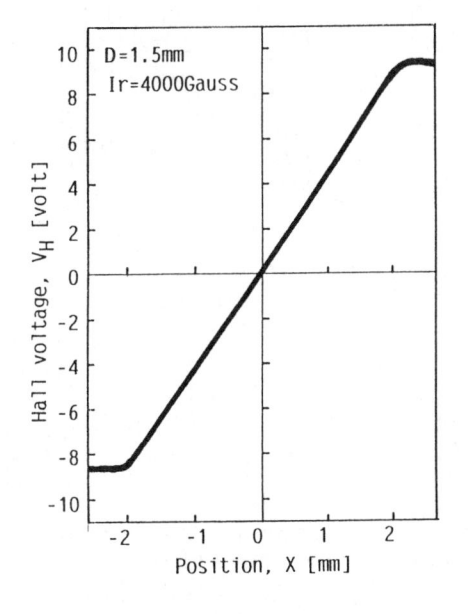

Fig.15. Hall Voltage vs. Position of the Slider

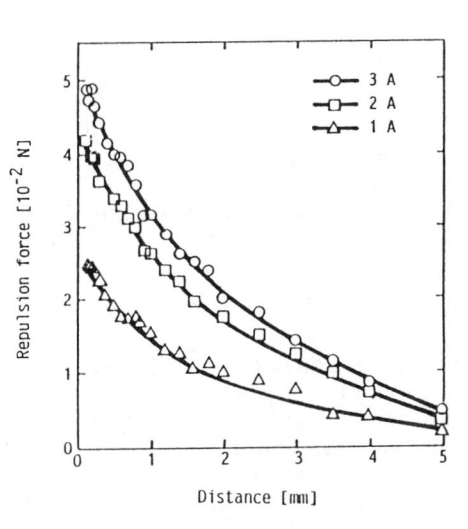

Fig.16. Repulsion Force vs. Distance

Fig. 15. This relationship between them is linear between -2 and 2mm.

The relationship between the repulsion force and the distance for various applied currents of 1, 2 and 3 A is shown in Fig. 16. The repulsion force decreases rapidly with increasing distance. To drive the superconducting slider the force curve with rapid decrease is applied to the slider.

Total system consisting the actuator, controller, and amplifier is shown in Fig. 17. The slider is controlled by using the fuzzy logic. The fuzzy control law is generated using 16 bit personal computer with 12 bit A/D and D/A converters and two amplifiers for D/A output and the Hall voltage as shown in Fig. 17. The fuzzy logic is programed by an assembler software. Trapezoid referential curves V_R are given by a oscillator over the frequency range from 0.1 to 1.0 Hz.

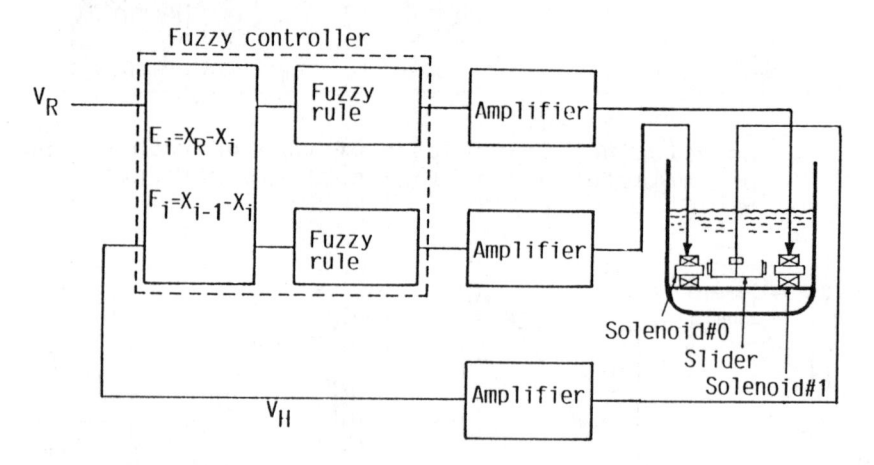

Fig.17. Schematic Picture of the Total Actuator System Consisting of the Controller

Fuzzy control There exist several difficulties in controlling the superconducting actuator by conventional control techniques such as PID control technique for the following reasons:
(1) nonlinear inverse relationship between the repulsion force and the position (distance) of the slider.
(2) hard hysteresis relationship between the repulsion force and the position to deal with.
(3) only use of repulsion forces to actuate the slider.
(4) unknown frictions caused by the pinning effect in the superconductors.

(5) unknown resistances of the solenoids in controlling the slider.
(6) unknown viscosity of liquid nitrogen in driving the slider.
(7) disturbance of nitrogen bubble.

The fuzzy logic is designed based only on the qualitative knowledge (1). The fuzzy controller is composed of linguistic control rules as shown in Fig. 18. PB (positive big), PM (positive medium), PS (positive small), ZO (approximately zero), NS (negative small), NM (negative medium), and NB (negative small) are labels of fuzzy sets. Error Ei and difference Fi are given by $Ei = X_R - Xi$ and $Fi = Ei_{-1} - Ei$ where Xi is the present position, X_R is the referential position [13]. To simplify the assembler program membership functions of the fuzzy sets are expressed by the shape of Λ-type with 31 elements (-15~15) and a grade (0~10) in the staircase. The membership functions for condition part are the same ones as those for operation part. The center of gravity method is used for defuzzification to obtain a certain value of control.

Variable value control result for trapezoid referential curves over the frequency range from 0.2 to 1.0 Hz is shown in Fig. 19. The trapezoid referential curve V_R, the position of the slider expressed by the Hall voltage V_H, applied voltages to the solenoid #0 (V_{S0}) and #1 (V_{S1}) are shown in Fig. 19. The Hall voltage of $V_R = 5V$ corresponds to the slider position of $X = 1.2$ mm. The slider is well controlled over the frequency range less than 0.4 Hz. The frequency dependence of this result is due to the small driving force and the viscosity of liquid nitrogen. The discontinuous curves of V_{S0} and V_{S1} result from the fuzzy membership functions in the staircase.

		F_i						
		PB	PM	PS	ZO	NS	NM	NB
E_i	PB				PB			
	PM				PM			
	PS				PS			
	ZO	PB	PM	PS	ZO	NS	NM	NB
	NS				NS			
	NM				NM			
	NB				NB			

Fig.18. Control Rules for the Actuator

Fig. 20 shows the variable value control result at a frequency of 0.2 Hz for dynamic disturbances. The disturbances are given to the actuator by swinging it in the liquid nitrogen. The positive angle means that the actuator is rotated clockwise. The angle of +1 degree corresponds to the disturbance of 0.5×10^{-3} N. When the disturbances are

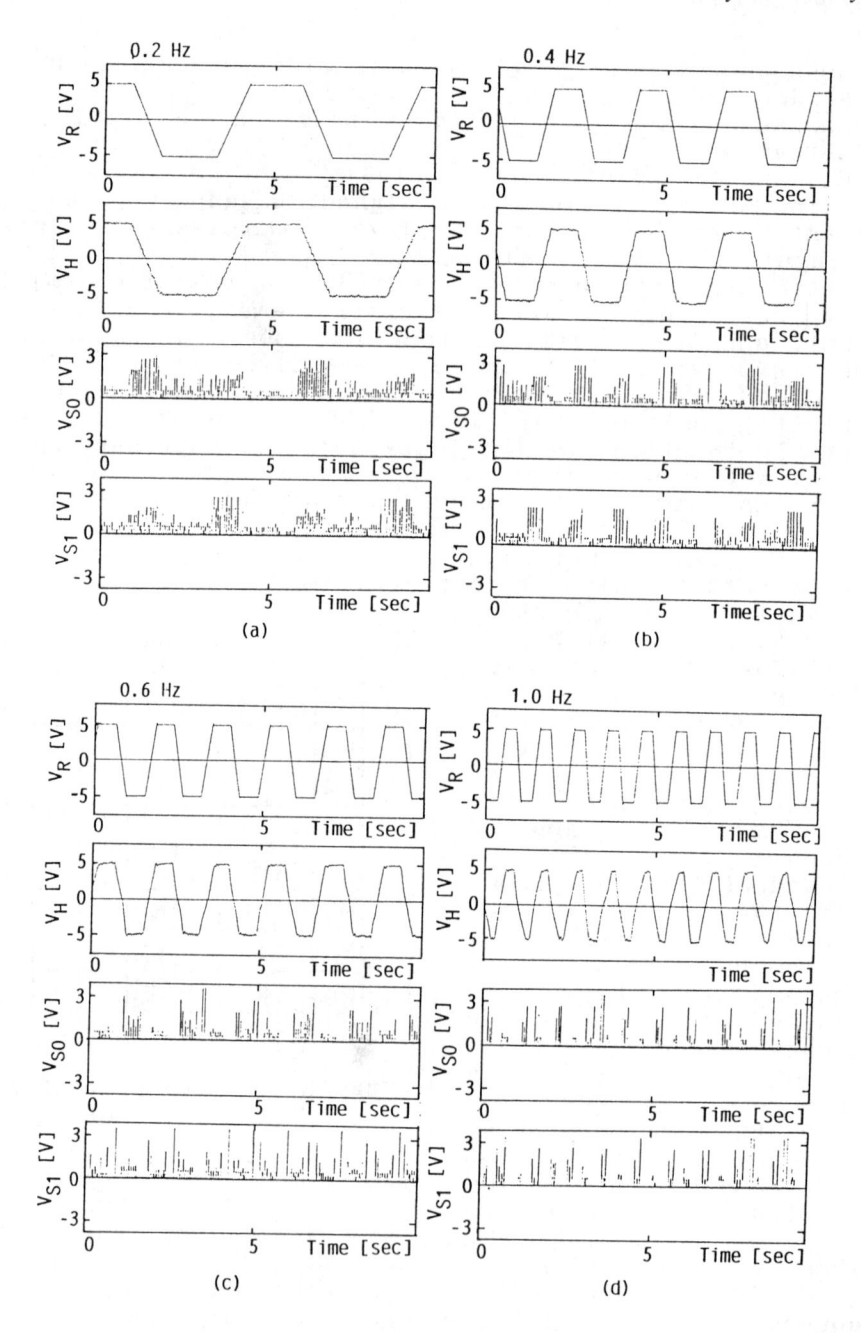

Fig.19. Variable Value Control Result for the Trapezoid Referential Curve

applied to the actuator small position errors are generated over the time range from 0 to 13 sec. However the error disappear when the disturbances are removed.

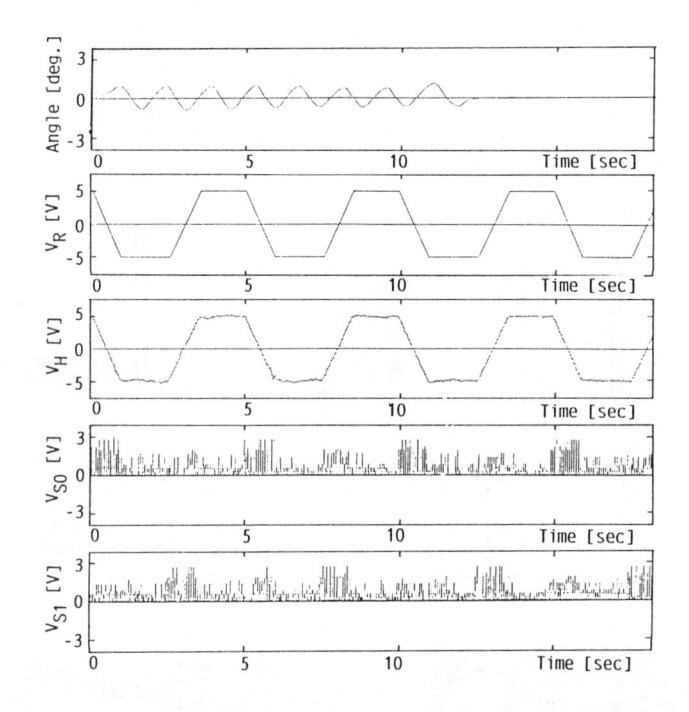

Fig.20. Variable Value Control Result for Dynamic Disturbances

5. SUPERCONDUCTING PUMP ACTUATOR

Fig. 21 shows a schematic picture of a superconducting pump actuator for extremely low temperature of 4.2 K (liquid helium) or 77 K (nitrogen temperature). This system consists of a superconducting piston, two solenoids to drive the piston, and a magnet cylinder for levitation. This system is one of the applications of the linear bearing mentioned in Fig. 7(a). This system is being advanced using the superconducting linear actuator and the control technique as previously mentioned. The picture of the superconducting pump actuator is shown in Fig. 22.

The bearing consists of a superconductor disk-shaped piston (YBa$_2$Cu$_3$O$_x$, 44.5ϕ×11.5mm, 99.1 g) supported by two

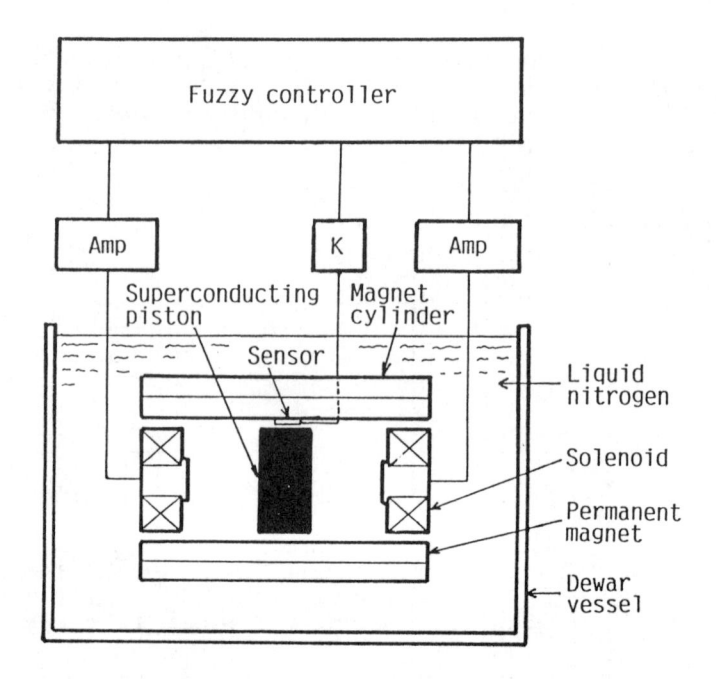

Fig.21. Schematic Picture of the Superconducting Pump Actuator

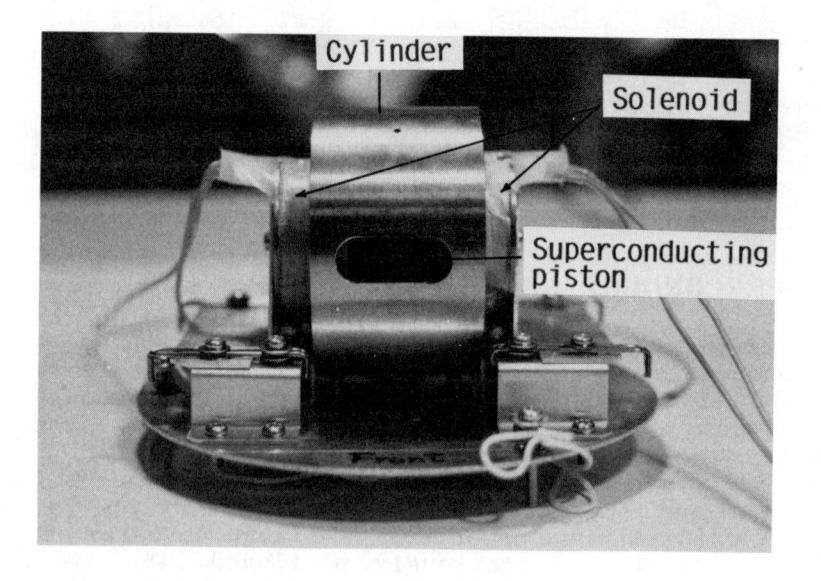

Fig.22. Picture of the Superconducting Pump Actuator

sets of magnet arrangements. A bar magnet of the sets is $4\times5\times40$ mm^3 with $Br \simeq 1.0$ T. The superconductor piston is prepared by QMG process with a critical current $Jc \simeq 1\times10^4$ A/cm^2 at 1.0 T.

Repulsion force about the linear bearing was measured. The measurements were carried out within the displacement range from -0.5 to 0.5 mm. The origin for the displacement is the center of the magnet housing. The disk-shaped piston was moved manually by a mechanical stage.

Fig. 23 shows the relationship between the repulsion force and the displacement. In Fig 23(a) the direction of the displacement is perpendicular to the magnet plane. The result shows the hysteresis loops for various displacements. The hysteresis loop becomes small with decreasing displacement. The levitation force of the superconductor shaft acts toward the center of the magnet housing as a restoring force. It is considered not so good that the hysteresis loop exists in the repulsion force. The pinning effect stops the rotary motion of the superconductor piston. However there is no friction and drag force in the axial motion of the superconductor piston.

In Fig. 23(b) the direction of the displacement is parallel to the magnet plane different from that in Fig. 23(a). The repulsion force is almost equal to that shown in Fig. 23(a) in spite of the different directions. From these experiments the repulsion force is almost the same as long as the displacements are in the radial directions. This indicates that two sets of magnet placed in the housing produce enough levitation force to support the superconductor piston. These are attributed to the flux pinning effect.

The restoring torque was measured as shown in Fig. 24. θ is an angle which indicates the rotation of the superconductor piston. The result shows hysteresis loops which resemble those shown in Fig. 23. The restoring torque for the rotation angle of 10 degree is about 3×10^{-2} Nm. It is expected that this restoring force becomes larger with increasing thickness of the superconductor piston.

The relationship between the repulsion force and the applied current to the solenoid for various distances is shown in Fig. 25. Large hysteresis loops are observed in Fig. 25. These loops are related to the magnetic flux pinning effect. The pinning force is difficult to deal with. This is the reason why fuzzy control technique is adopted in controlling the superconducting actuator.

A sufficient condition of asymptotic stability of the fuzzy control system is obtained with the use of an extended circular criterion [14]. Under this condition it is shown that the system is asymptotically stable for a conventional PD controller.

Fig. 26 shows the variable value control result. The

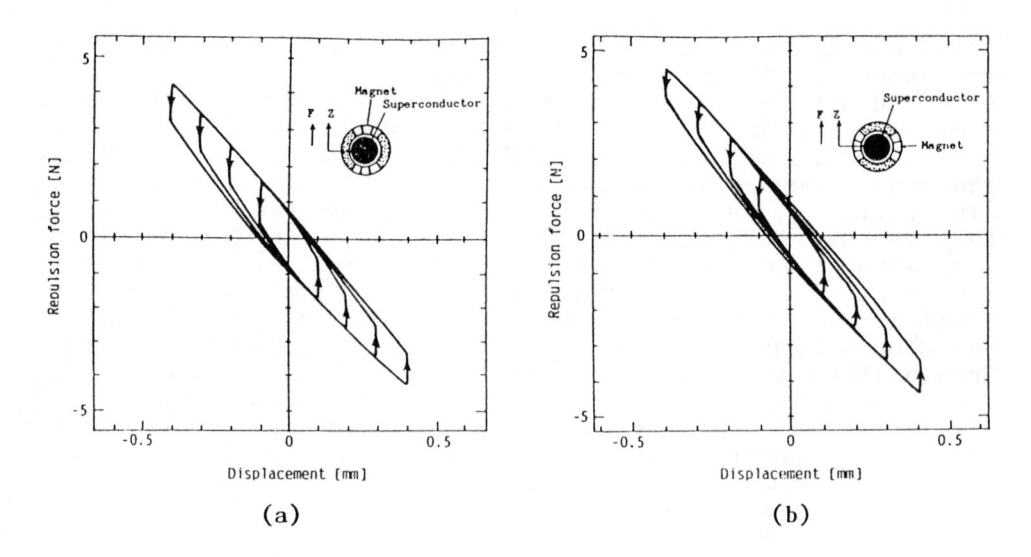

Fig.23. Repulsion Force vs. Displacement. (a) The displacement is perpendicular to the magnet plane. (b) The displacement is parallel to the magnet plane.

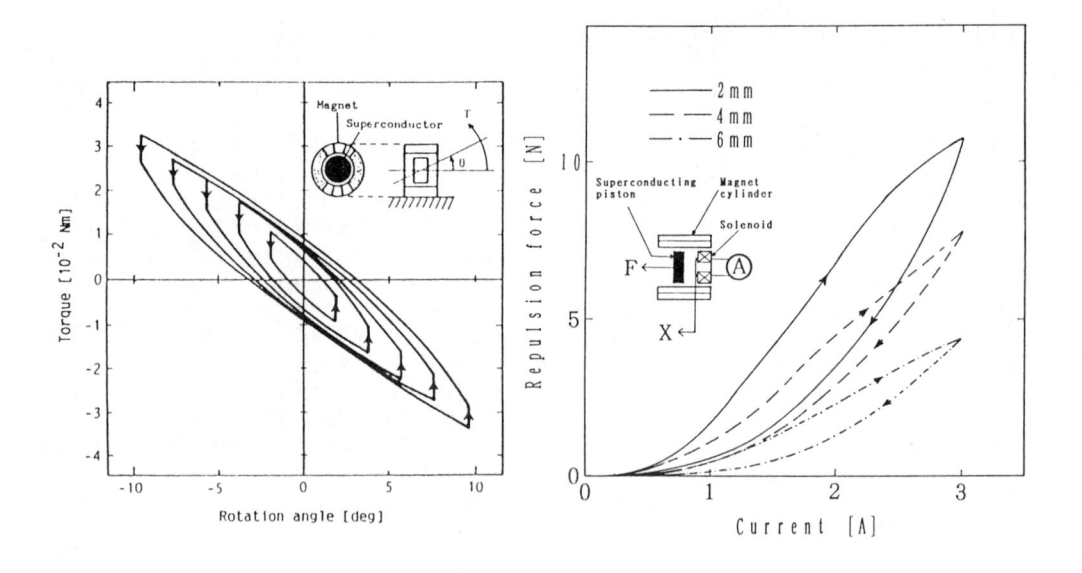

Fig.24. Torque vs. Rotation Angle

Fig.25. Repulsion Force vs. Applied Current

trapezoid referential curve V_R, the position of the piston expressed by the Hall voltage V_H, applied voltages to the solenoid #0 (V_{S0}) and #1 (V_{S1}) are shown in Fig. 26. From the result the position of the superconductor piston is almost under control. Small oscillations are observed in the response curve. This is believed related to the magnetic flux pinning effect in the superconductor. The detail about this is now studied.

The fuzzy theory is suitable for controlling the superconducting actuator with high-Tc superconductor for driving it. It is possible that better controllability is obtained by using better superconductors by Quench and Melt Growth (QMG) process with high critical current density and other better membership functions. Using more powerful superconductors thrust force more than 100N can be generated from the pump actuator.

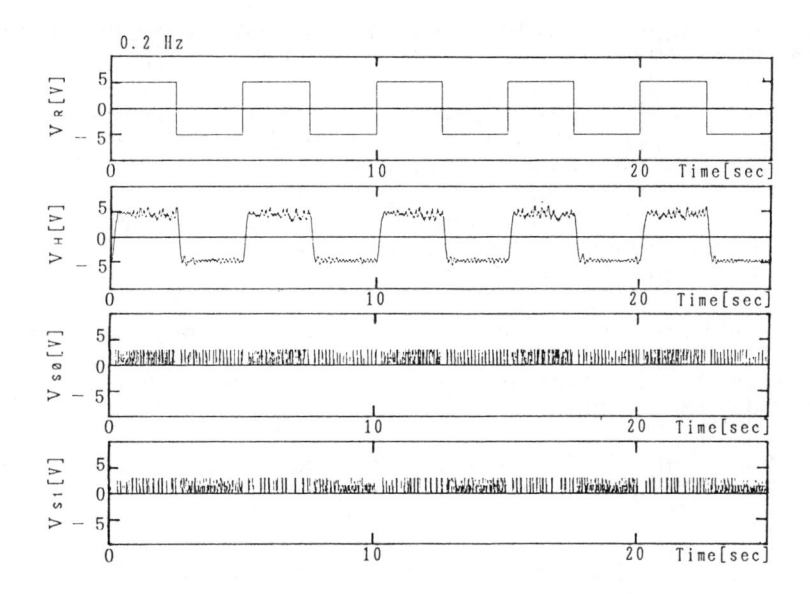

Fig.26. Variable Valuecontrol Result for the Trapezoid Referential Curve

6. CONCLUSIONS

Characteristics of the levitation mechanism are studied. The results show that the levitation pressure and the drag pressure depend greatly upon the width of a bar magnet of the mechanism. The drag pressure is also dependent on the angle between the alternating-polarity and the moving direction. These characteristics are available for designing the frictionless actuator by using the superconducting levitation mechanism.

Superconducting actuators using levitation mechanism have been developed by our group. The actuators are controlled by the fuzzy control technique with no contacts. The fuzzy control technique is found to be useful for controlling the superconducting actuator working in the extremely low temperature.

REFERENCES

[1] F.Hellman, E.M. Gyorgy, D. W. Johnson, Jr., H.M.O'Bryan, and R.C.Sherwood, "Levitation of a magnet over a flat type II superconductor", *J. Appl.Phys.*, Vol.63, pp.447-450, 1988.

[2] E.H.Brandt, "Friction in levitated superconductors", *Appl.Phys.Lett.*, Vol. 53, pp.1554-1556, 1988.

[3] F.C.Moon, M.M.Yanoviak, and R.Ware, "Hysteretic levitation forces in superconducting ceramics", *Appl. Phys. Lett.*, Vol.52, pp.1534-1536, 1988.

[4] F.C.Moon, K.-C.Weng, and P.-Z.Chang, "Dynamic magnetic forces in superconducting ceramics", *J. Appl. Phys.*, Vol.66 pp.5643-5645, 1989.

[5] P.-Z.Chang, F.C.Moon, J.R.Hull, and T.M.Mulcahy, "Levitation force and magnetic stiffness in bulk high-temperature superconductors", *J. Appl. Phys.*, Vol.67, pp.4358-4360, 1990.

[6] F.C.Moon and P.-Z.Chang, "High-speed rotation of magnets on high Tc superconducting bearings", *Appl. Phys.Lett.*, Vol.56 pp.397-399, 1990.

[7] M.Murakami, T.Oyama, H.Fujimoto, T.Taguchi, S.Gotoh, Y.Shiohara, N.Kosizuka, and S.Tanaka, "Large levitation force due to flux pinning in YBaCuO superconductors fabricated by melt-powder-melt-growth process", *Jpn. J. Appl. Phys.*, Vol.29, pp.L1991-L1994, 1990.

[8] T.Kitamura and M.Komori, "Levitational bearing systems by Meissner effect", *Micro System Technology '90*, Berlin, Springer-Verlag, 1990, pp.826-831.

[9] M.Komori and T.Kitamura, "Development of a superconductive levitational mechanism and its application to a superconductive radial bearing", *Cryogenic Engineering*, Vol. 25, pp.411-413, 1990.

[10] M.Komori and T.Kitamura, "Static levitation in a high-Tc superconductor tile on magnet arrangements", *J.Appl.Phys.*, Vol.69, pp.7306-7309, 1991.

[11] M.Komori and T.Kitamura, "Drag pressures of a set of alternating-polarity magnets over a superconductor tile", *J.Appl.Phys.*, Vol.70, pp.2226-2229, 1991.

[12] M.Komori and T.Kitamura, "Fuzzy control of a superconductive linear sider", *International Coference on Fuzzy Logic & Neural Networks*, Iizuka, Japan, 1990, pp.87-90.

[13] M.Sugeno, *Fuzzy Control*, Nikkan Kogyo Shinbun Corp., Tokyo, 1988, pp.67-136.

[14] T.Kitamura and M.Komori, "Stability analysis of a fuzzy control system of a superconducting actuator", *International Conference on Fuzzy Logic & Neural Networks*, Iizuka, Japan, 1992, pp. 461-464.

Chapter 24
A Fuzzy Logic Based Approach to Machine-Tool Control Optimization

Introduction, *524*
Position Control, *525*
Direct Rule-Based Fuzzy Controller, *526*
Self-Organizing Fuzzy Controller, *530*
The SPB System, *532*
Results Resume, *534*
Fuzzy Position Control of Machine Tools: Conclusions, *536*
A More General Approach: Towards the Concept of Intelligent Machining, *537*
References, *538*

This chapter presents a new approach to machine-tool control optimization based on fuzzy logic. First, research results pertaining to position control of a machine-tool are presented. Fuzzy algorithms for direct and self-organizing fuzzy controllers are described and results of their application to a microprocessor-based servomotor control in real-time are presented. A system is developed to easily test and compare the described algorithms on real-time servomotor control. Finally, a more general approach is suggested, directed at the concept of intelligent machining, and a hierarchical intelligent control system, working under multiple objectives, is proposed. This is accomplished by setting some control rules under a merit index and others under a different index (multi-index optimization criteria).

A FUZZY LOGIC BASED APPROACH TO MACHINE-TOOL CONTROL OPTIMIZATION

José Ramón Alique López
Instituto de Automática - CSIC
28500 Arganda del Rey
Madrid, Spain

Elena Agüero Gutierrez
Instituto de Automática
Univ. Nacional de San Juan
Argentina

Leonardo C. Rosa
Grupo de Pesquisa e Treinamento em Comando Numérico
Universidade Federal de Santa Catarina
Brasil

1. INTRODUCTION

With the appearance in 1970 of CNC Systems (Computer Numerical Control), a step of great importance towards automation of machine-tool was made. However, numerical control, whether DNC (direct numerical control) or CNC, is based on old techniques and it is time to consider new ideas of machine-tool control and to develop more powerful techniques.

The limitations of present numerical control are well-known. The uncertainty of machine-tool control depends on the use of pre-process developed NC programs based on unreliable pre-process data. In addition to this, the integration of CNC systems in flexible manufacturing cells (FMS) demands improvements in machine-tool automation. By improving automation of machine-tool control the following could be achieved: higher production rates, better product quality, unmanned machining, etc. Each of these improvements can reduce manufacturing costs.

Nowadays, although there has been considerable research on the development of new machine-tool control systems, few, if any, of these systems are used in practice. The reasons are found in the characteristics of the cutting process:

> A. The cutting process is very complex. The knowledge about the machining process is limited. There exists lack of reliable data from sensors. The uncertainty and incomplete knowledge are inherent in machine-tool control. There are no models, in the process domain, dealing with material hardness, tool wear, chatter, thermal effects, deformations, surface finish, etc.

B. There is uncertainty in the measures from sensors (noise) and in the relation between this measures and the process state.

C. The cutting process is subject to large disturbances because of variations in machines, raw material and the machining conditions.

2. POSITION CONTROL

When trying to improve machine-tool automation with higher production rates or better surface finish, the first problem related to todays CNCs is the bad dynamic performance of the position control loop. In general, systems with a closed-loop in its position control use a PI algorithm as controller. Some PI controller characteristics that justify the option are simplicity, reliability and small geometrical errors in the workpieces when interpolated axis gains are well compensated. By contrast, there are some cases where the performance is not so good:

▶ High-speed milling. A great following error, that is the lag between the actual position values and the reference values can generate geometrical errors in the workpiece. Gains adjustments can not simply be increased because oscillations would occur. State-of-the-art machines solve the problem with status controllers and observers.

▶ High-accuracy workpieces with high-productivity. To minimize geometrical errors, and also for surface finishing considerations, the feedrate used, in many cases, is much smaller than the machine-tool is able to admit.

▶ High-speed drilling. Positioning time between successive holes is important when there are hundreds or thousands of these in the same workpiece.

To improve the CNC performance, industry has made a considerable effort in hardware and software. Some examples are:

- Multiprocessor CNC, with CPU per axis [1].
- Fast 32-bit microprocessors [2],[3].
- Stick motion compensation [2].
- Load compensation [1],[2].
- Feed-forward control [2].
- Adaptive feedrate control [3].
- Compensation of relative position error in interpolation[1].

However, all these advances do not change the fact that PI controllers are still in use. Despite that, they do not consider the different requirements of the position control when it is positioning, dressing roughly or finishing, doing linear or circular interpolation. The gains are set by the machine builder (commissioning data) and do not change during operation. They are set considering the worst operation case, that will probably occur a few times in the machine life time.

The main objective in this stage of our work has been applying Fuzzy Logic to the position controller design, to obtain better performance than the actual PI controller.

3. DIRECT RULE-BASED FUZZY CONTROLLER

The objective of the direct fuzzy controller is the angular position control of a DC motor subject to changes of inertial load with step or ramp input signals. Fig.1 shows physical system structure and Fig.2 the block system diagram.

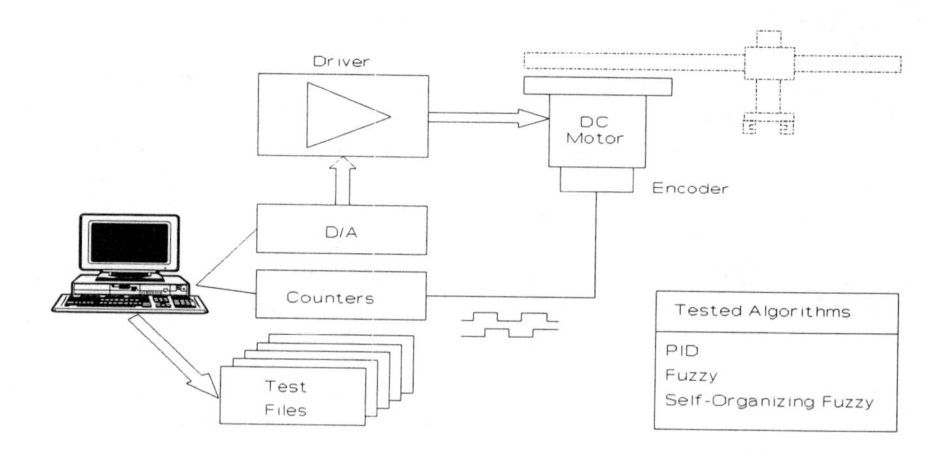

Figure 1: Physical System Structure

From the setpoint and the instant output process value, both inputs to the fuzzy algorithm can be calculated: position error (e) and change in position error (ce). The universe of discourse of change in position error or motor speed ranges between (-10.000, +10.000) mm/min. The universe of discourse of position error has the same range as setpoint: (-3000, +3000) mm. The output variable is a voltage reference which is sent to the DAC. Its universe of discourse is included in (-10, +10) Volts.

An interesting feature of a fuzzy controller is that it works with normalized input and output variables, in order to obtain a Knowledge Base independent of changes in the universes of discourse of these variables. In our controller, input variables are continuously normalized into the interval [-1, +1], by multiplying them by GE and GCE, respectively (see Fig.2).

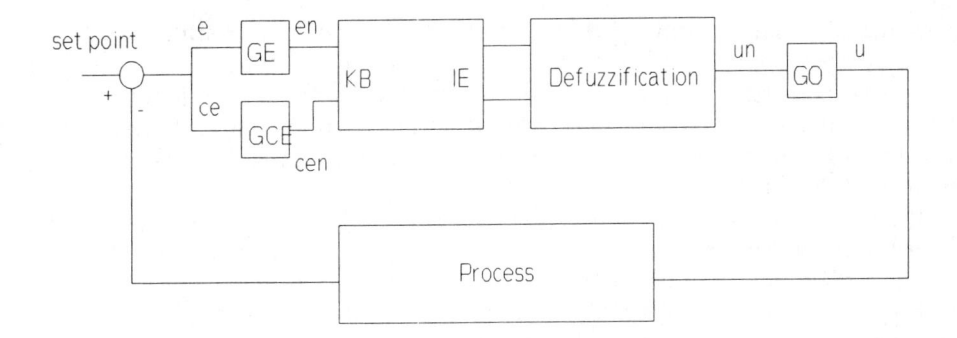

Figure 2: Fuzzy Controller Diagram

It is usual in fuzzy controllers to transform the continuous universe of discourse of the input variables into a discrete one [4],[5]. In this way computations are reduced but problems with the convergence of the system arise. Since our system has strong requirements about accuracy, and the computation time obtained was acceptable, working with continuous variables was preferred. The membership functions used for the input and output variables are of triangular type, as in Figure 3.

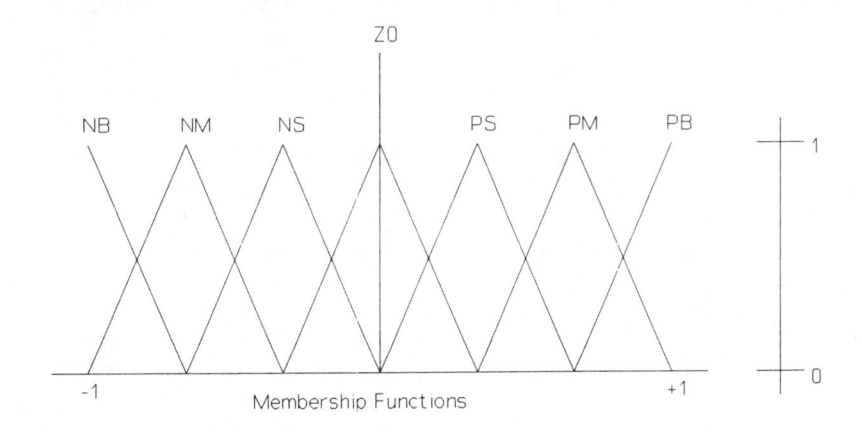

Figure 3: Membership Functions Used

This type of function is preferred because of the facility for computation associated with it. There are 7 linguistic levels: small, medium and big, both positive or negative, with a zero central level.

Following the schematic diagram of Fig.2, the fuzzy algorithm is arrived at. It is composed of a Knowledge Base (KB) and an Inference Engine (IE). In rule based fuzzy controllers, KB and IE are not always completely separated, but simply form a block [6].

The KB is composed of a Rule Base (RB), and a Data Base (DB), which have the same functions as the RB and the DB of a classical expert system. The only difference is that a fuzzy KB contains fuzzy variable values. The RB is composed of a set of 49 rules, as shown in Figure 4. However, not all the 49 rules are inferred because a simple set of metarules selects 4 rules each sample time to be inferred.

The IE uses the data driven (forward chaining) strategy to make the inferences. All chains of reasoning are only one inference long, because in process control there is no need to generate and test hypothesis [6].

Our IE can apply two types of implication functions: Max-min rule (proposed by Mamdani) and Max-product rule (applied by Yamazaki). Both algorithms use the center of gravity as defuzzification method:

$$u(t) = \frac{b\,[E_i * Y_i]}{b\,E_i}$$

where E_i represents the fulfilment grade of the premises, as the result of applying the minimum or product implication function. Y_i represents the linguistic output variables.

RULES SET V6 R0		VELOCITY						
		NB	NM	NS	ZO	PS	PM	PB
POSITION ERROR	PB	PB	PB	PB	PB	PB	PB	PB
	PM	PB	PB	PB	PB	PB	PB	PB
	PS	PB	PB	PB	PM	PM	PM	PM
	ZO	NS	NS	ZO	ZO	ZO	PS	PS
	NS	NM	NM	NM	NM	NB	NB	NB
	NM	NB	NB	NB	NB	NB	NB	NB
	NB	NB	NB	NB	NB	NB	NB	NB

Figure 4: Fuzzy Controller Rule Base

The normalized output (un) is also restricted to the interval [-1, +1]. Thus, it is necessary to multiply it by the output gain GO in order to obtain a control signal included in (-10, +10) Volt. This control signal u(t) or velocity reference, is sent to the velocity controller (driver).

3.1 Metarules

First versions of our fuzzy algorithms were more time consuming than later ones, because all the rules were executed, each sample time, consuming about 8 ms. New versions use metarules to decrease the time consumed by the IE. They actuate over the rule base decreasing the number of rules which is needed to infer to obtain the control output.

Metarules select, first according to the error value and then, according to the velocity value, which rules are convenient to infer and which are not.

Error metarules are of type:

"IF error is bigger than 0.666
THEN only consider rules 1 to 14"

Next, inside this subset, metarules of similar form to the error metarules are introduced, now asking for the velocity value:

"IF 0.333 < velocity < 0.666,
THEN only infer rules 5,6,12 and 13"

Figure 5 shows an explanatory diagram. In this way, the rule number to infer is finally, and in all the cases, reduced to 4, and the computing time to 2 msec, approximately.

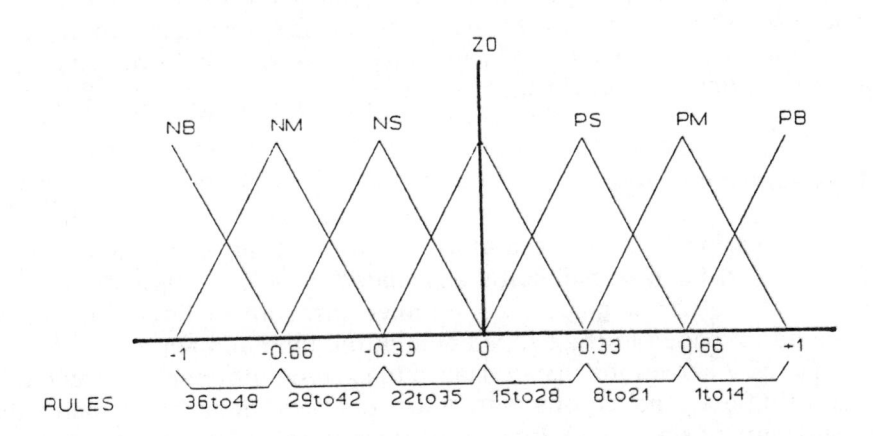

Figure 5: Rule Subsets Generated by Error Metarules

4. SELF-ORGANIZING FUZZY CONTROLLER

One of the main problems of rule-based controllers is to obtain the rules [7]. Even knowing them in advance, there is always the difficulty of achieving the optimum control.

It would be desirable for the controller to be able to modify or add new rules while comparing its performance with a pattern. Such a controller is called "Self-Organizing Controller" (see Fig. 6).

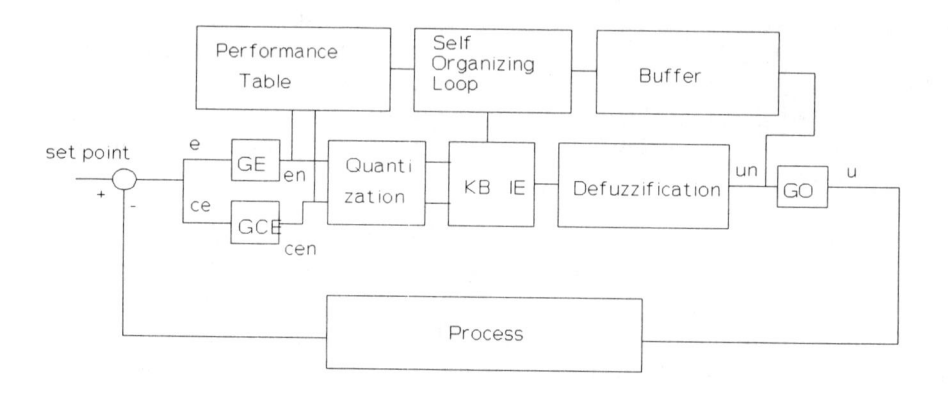

Figure 6: Self-organizing Fuzzy Controller Diagram

The self-organizing controller is a hierarchical controller with two levels. In the lower level there is a rule-base in charge of the control task, as a usual direct fuzzy controller. In the higher level, the algorithm compares at every sample time if the system behavior is the desired, based on a "Performance Table". If the system time response does not match the given by the table, then an algorithm modifies rules of the lower level until reaching the desired performance.

4.1 Performance Table

The Performance Table is a matrix with error (e) and change in error (ce) as inputs. Each pair (e,ce) is associated with a matrix element. The value of that element gives a quantitative measure of how close is the system performance to the desired one, in the time domain.

Figure 7 shows the matrix used in the servomotor control. The matrix has three different areas: one with zeros, one with negative elements and another with positive elements. The zeros area shows the path to be followed by the system in order to achieve the desired time response. As the elements move further from the zeros path they will take bigger values, positive or negative. As greater is the element value (positive or negative), the system will be further away from the desired behavior. The positive and negative signs indicate the acceleration and deceleration effect in the system, by modifying the rules.

MATRIX PT		VELOCITY										
		-1.0	-0.8	-0.6	-0.4	-0.2	0	+0.2	+0.4	+0.6	+0.8	+1.0
ERROR	+1.0	1	1	1	1	1	0	0	0	0	0	0
	+0.8	1	1	1	0.62	0.62	0.25	0.25	0.25	0.25	0.25	0
	+0.6	1	1	1	0.62	0.62	1	1	1	0.62	0.25	0
	+0.4	1	1	1	0.62	0.62	1	1	0.62	0.25	0	-0.25
	+0.2	1	1	1	0.25	0.62	0.25	0.25	0.25	0	-0.25	-0.62
	0	1	0.25	0.25	0	0	0	0	0	-0.25	-0.62	-1
	-0.2	0.62	0.25	0	-0.25	-0.25	-0.25	-0.62	-0.25	-1	-1	-1
	-0.4	0.25	0	-0.25	-0.62	-1	-1	-0.62	-0.62	-1	-1	-1
	-0.6	0	-0.25	-0.62	-1	-1	-1	-0.62	-0.62	-1	-1	-1
	-0.8	0	-0.25	-0.25	-0.25	-0.25	-0.25	-0.62	-0.62	-1	-1	-1
	-1.0	0	0	0	0	0	0	-1	-1	-1	-1	-1

Figure 7: Performance Table

The criteria followed to construct the Performance Table are:
a. The system must accelerate until it reaches the motor nominal speed, if the displacement is enough to reach it.
b. It must keep that speed as much as possible.
c. If the displacement is very little, it must go at the maximum possible speed until reaching the reference position.
d. Fast deceleration and stop in the desired position, with a maximum error of 0.02 mm (1 encoder pulse) and without overshoot.

4.2 Algorithm to Modify Rules

Each sample time, the controller looks for the performance table element, corresponding to (e) and (ce) values. With this value, it modifies the lower level rules according to:

$$R(t) = R(t-nT) + PT(t)$$

where:
R(t): modified rule at (t) instant.
R(t-nT): rule corresponding to (t-nT) instant.
T: sample time.
PT(t): performance table element chose by the controller, according to e and ce at t instant.

If PT(t) is different from zero, the rules inferred n sample times before are considered responsible for the bad controller performance. Then, their consequents are modified by PT(t) value, according to the rule weight in the final output, starting at a 0.2 threshold.

The n factor, called delay in reward, was selected as 4. With this value the best results were achieved.

5. THE SPB SYSTEM

The SPB system (Sistema Posicionador Borroso, in Spanish) was developed to easily test and compare the described algorithms on real-time servomotor control. Its characteristics are the following:

- Possibility of working with simulated or real processes.
- Possibility of selecting the control algorithm, between PID, BOR_POND (max-min rule direct fuzzy controller) BOR_PROD (max-product rule direct fuzzy controller) or SOC.
- Calculation of performance indices for each test.
- Graphic representation of both control signal and velocity response in time domain. In addition, graphic representation of the response in the performance table.
- Possibility of easily changing rules, input and output gains, and so on.

5.1 Hardware

The SPB system consists of the following components:

- IBM-PC AT computer with 80287 coprocessor. Counter card with one 9513 counter for each rotation direction. This card can receive information from 5 axes.
- 12 bits DAC card. It is used to send the speed reference to the driver.
- Driver INFRANOR SOCS 150.30.
- DC servomotor model MOVINOR MM4040, with tachometer and encoder.

The counter card also generates an interrupt (IRQ3) every 10 ms (sample time), to call the position control routines.

5.2 Software

The languages used were PASCAL (Turbo Pascal 5.0 compiler) and MACRO ASSEMBLER. The Pascal routines include a file to define the system variables, a unit to initialize them, a unit to implement the user-friendly interface, a unit to manipulate PC interrupts, and a unit to calculate the performance indices.

The time critical routines are in Assembler, making intensive use of the mathematical coprocessor. The first includes control of the counters and DAC cards, and calls the routines containing the control algorithms: PID, BORPOND (fuzzy direct control with max-min inference), BORPROD (fuzzy direct control with max-product inference), and SOC (self-organizing algorithm).

5.3 Performance Indices

Criteria and method are necessary to compare the performance of two or more controllers. In the case of a machine-tool, the kind of operation that it is carrying out is important. For example, if an overshoot can be admitted in a positioning, it can not be present in a finishing operation. The criteria that in principle looks intuitive, is perhaps not so simple.

Studies on controlling position with fuzzy techniques are starting at IAI and the performance indices proposed here are for two types of tests: step and ramp setpoint.

1. Overshoot (SO): This value measures damping characteristics.
2. Stationary State Error Average (PEEE): Absolute error average in the last 20 samples. This index measures the magnitude of the control error. It is important principally to the SOC, because large control error is one of its weak points.
3. Settling time (TE): The time spent for the output to reach up the strip of 5% error. This index measures the convergence characteristics of the response to the setpoint.
4. Rise time (TS): From 10% to 90% of final value. This value measures the speed of the response.
5. Absolute Integral Error (IAE): Integral of time x absolute error. This measures the quality of overall response.
6. Velocity Rise time (TSV): The same as (4) for ramp setpoint.
7. Velocity Settling time (TEV): The same as (3) for ramp setpoint.
8. Velocity Overshoot (SOV): The same as (1) for ramp setpoint.
9. Maximum following error (EMS): The maximum difference between ramp setpoint and real position.
10. Integral of Velocity Absolute Error (IAEV): The same as (5) for ramp setpoint.

The first test, or the classical test, for a controller is the response to a step input signal. The machine-tool does not use that function, it uses ramp (G00) and (G01), sinusoidal (G03) or polynomial functions (G06), but this test gives a good information about the algorithm performance.

The system must fulfil the following conditions to obtain the best performance:

- SO and PEEE: smaller or equal to system resolution.
- TE - TS and IAE : minimum

The encoder divisions (1250) and the transmission relation define, in our case, the controller resolution: 0.02 mm (1 encoder pulse).

The second test, which gets closer to the machine-tool world, is the response to a ramp input signal. This function is used with the instructions G00 (positioning) and G01 (linear interpolation). Now the most important thing is that the following error, the acceleration and deceleration time be small. The order of priority is:

- SOV smaller than 6%.
- Minimum EMS.
- Minimum IAEV.

The limit of 6% to SOV is more or less arbitrary. The most important idea is that there should be a definitive limit.

6. RESULTS RESUME

The Direct Fuzzy Controller was tested under step and ramp input signals. The results obtained are similar to the PID ones. In some cases they

Table 1: Some Results Obtained with Direct Fuzzy Controller with Step Input

POS. REF. (mm)	FUZZY		PID	TS		TE		SO		PEEE		IAE	
	GE	GCE	Kp	PID	FUZZY	PID	FUZZY	PID	FUZZY	PID	FUZZY	PID	FUZZY
1	100	40000	0.6	0.02	0.03	0.03	0.04	0.02	0.02	0	0	0.02	0.03
10	120	40000	0.5	0.02	0.03	0.05	0.06	0	0	0	0	0.31	0.36
100	138	41000	0.24	0.09	0.09	0.16	0.15	0.02	0.02	0	0	8.58	8.55
G0= 1 Ki= 0.03 Kd= 0													

are better, in others worse. The disadvantage is that adequate normalizing factors are necessary to different step or ramp input signals. Functions or tables with that data must be introduced in the algorithm for all the operational range of the variables.

The new algorithms versions reduced the calculation time from the order of 10 msec to about 2 msec. It was achieved with:

- Metarules. They select, first according to the error value and then according to the speed value, which rules are convenient to be inferred and which are not.
- Intensive use of a mathematical coprocessor.

The time reduction achieved allowed us to see that the calculation time, one of the highest difficulties of fuzzy controllers, can be significantly lessened with conventional techniques (metarules, coprocessor, etc.). Despite that, the actual time values made the use of these algorithms in a simple CNC (with just one processor for all functions) difficult.

Nevertheless, the use of special hardware [8], [9] must be studied. Fuzzy inference and defuzzification at a speed of 15 Ms per rule are reported.

Table 2: Some Results Obtained with the Direct Fuzzy Controller with Ramp Input

POS. REF. (mm)	VELOC. (mm/sg)	FUZZY GE	FUZZY GCE	FUZZY GO	PID Kp	TS PID	TS FUZZY	TE PID	TE FUZZY	SO PID	SO FUZZY	PFEE PID	PFEE FUZZY	IAE PID	IAE FUZZY	ESM PID	ESM FUZZY	SOV PID	SOV FUZZY	IAEV PID	IAEV FUZZY
50	10000	20	18000	0.4	1	0.24	0.24	0.3	0.3	0	0.02	0.04	0	0.97	1.11	3.29	3.64	4.4	3.2	0.7	0.7
50	5000	28	18000	0.38	1	0.48	0.48	0.58	0.58	0.02	0.02	0	0	0.88	0.98	1.6	1.6	10.0	0	0.2	0.2
50	1000	15	20000	0.2	1.25	2.4	2.4	2.86	2.86	0.02	0.02	0	0	0.82	0.81	0.33	0.3	0	1.0	0.3	0.1
100	10000	39	35000	0.52	0.4	0.49	0.48	0.59	0.59	0.02	0.02	0	0	2.02	2.17	3.85	3.87	1.99	2.0	0.4	0.4
100	5000	50	35000	0.45	0.4	0.96	0.96	1.16	1.16	0.02	0.02	0	0	2.84	2.57	2.47	2.31	0	0	0.9	0.9
100	1000	30	12000	0.39	0.6	4.8	4.8	5.72	5.72	0	0	0	0	2.13	2.01	0.42	0.4	0	0	0.1	0.1
500	10000	28	70000	0.48	0.6	2.4	2.4	2.87	2.87	0.01	0.02	0	0	10.35	10.39	3.57	3.7	2.0	0	10	5
500	5000	28	70000	0.35	0.49	4.8	4.8	5.72	5.72	0.02	0.02	0	0	12.75	12.27	2.19	2.18	0	0	2.5	2.5
500	1000	25	70000	0.23	0.59	24.0	24.1	28.5	28.54	0.01	0	0	0	10.69	10.72	1.75	1.7	3.0	2.0	0.2	0.7

Ki= 0.03   Kd= 0

SOC algorithms look to be more promising. They have been used with good results in controlling robot arms in tracking [10]. In the case of SOC, the objective is not just to achieve good performance indices, but also that the rules converge. When the work conditions return to a past situation, the rules also return to their corresponding set.

To test the convergence we used successive identical step changes in setpoint. Table 3 shows the obtained results. Between the first and the 12th test (all steps of 100 mm) the SOC automatically changed the rules adapting the response to suit the zero path in the phase plane. Then the load was introduced. In successive tests the rules were changed and the performance indices converged to a new value. The SO in the 12th run was reduced in the following runs. In the 24th run the load was removed and the system converged again to the original values (run 11 similar to 32).

It must be noted that a stronger self-organization action is easily obtained if the PI matrix is multiplied by a number greater than 1.

Table 3: Some Results Obtained with the SOC Controller

Run	IAE	SO	Run	IAE	SO	Run	IAE	SO	Run	IAE	SO
1	10.15	0.02	9	11.08	0.02	*17	16.30	0.02	25	11.62	0.0
2	11.14	0.02	10	10.99	0.04	*18	16.30	0.02	26	11.30	0.02
3	11.13	0.02	11	11.01	0.02	*19	16.31	0.02	27	11.02	0.0
4	11.08	0.02	*12	15.94	6.32	*20	16.35	0.02	28	10.30	0.02
5	10.99	0.02	*13	16.75	0.98	*21	16.36	0.02	29	11.02	0.02
6	11.16	0.02	*14	16.34	0.24	*22	16.31	0.0	30	11.04	0.02
7	11.05	0.02	*15	16.40	0.14	*23	16.35	0.0	31	11.16	0.02
8	10.96	0.02	*16	16.30	0.02	*24	16.36	0.0	32	11.18	0.02

System resolution=0.02 mm.   Run 1 to 11 and 25 to 32: NO LOAD   * LOAD = 100%

7. FUZZY POSITION CONTROL OF MACHINE-TOOLS: CONCLUSIONS

Starting from the premise that "intelligence" will still be introduced into machine-tools, we can present the two basic situations in which fuzzy control will and will not be applicable in the position control of machine-tool, from our point of view:

IT WILL BE APPLICABLE IF:

- The evolution of sensor technology creates efficient sensors in the high level control variables.
- The fuzzy multicriterion machine-tool controllers prove to work well in the lower control levels. In this case, the system block diagram should be integrated by three independent blocks: machine-tool, sensorial system and an integrated CNC and optimization system.

IT WILL NOT BE APPLICABLE IF:

- Good sensors were not technically or economically viable for the high level control variables.
- Fuzzy multi-criterion machine-tool controller work better using the lower level just as command level. In this case the position controllers (and the CNC) will continue being of the classical type.

8. A MORE GENERAL APPROACH: TOWARDS THE CONCEPT OF INTELLIGENT MACHINING

Up to this point we have studied the possibility of applying Fuzzy Logic as a way to avoid the problems which appear in todays CNC systems, specially at high speed. However, even developing the ideal position control we will still be far from reaching the optimization of the machining process. The main reason has been mentioned previously: the uncertainty of machine-tool control depends on the use of pre-process developed NC programs, based on unreliable pre-process data. Besides, in practice the merit index will not always be the productivity increase. Other merit indices exist, such as an economic function which in many circumstances may be more important than the previous one. Other possible merit indices which could be used would be, for example, surface finish or dimensional precision.

If we look deeper the machining process, so called "intelligent", it seems evident that it should be possible to use non exclusive merit indices. In this way, depending on each particular situation, different optimization global criteria could be applied as a first step toward the concept of "intelligent machining".

Adaptive numerical control has been the first approach to achieve the system optimization. It has been shown that with these techniques it is not possible to optimize the machining process due to the many existing restrictions, such as: the control information comes from only one source, the operation sequence is fixed, the control process is numerical (it processes numbers instead of machining procedures) and process data are not obtained in real time.

Nowadays, uncertainty and incomplete knowledge of machine-tool are the main problems. Manufacturing systems and particularly machine-tool control are fields in which fuzzy logic control techniques could be applied to solve many of these problems: the knowledge can be expressed imprecisely and an incomplete set of rules is available. A very important reason to choose fuzzy control is the simplicity of obtaining a multi-index control by using a set of fuzzy control rules. By applying fuzzy control we can obtain an "intelligent" machine controller working under multiple objectives. This is accomplished by setting some control rules under a merit index and others under a different index.

All this leads us to design a hierarchical intelligent control system for a machine-tool as shown in Fig.8. The lower level contains the position-speed loops. The higher level is composed of different controllers, one per each merit index. These would be:

- Fuzzy controller for maximum productivity.
- Fuzzy controller for minimum cost.
- Fuzzy controller for maximum surface finish.
- Fuzzy controller for minimum geometrical errors.

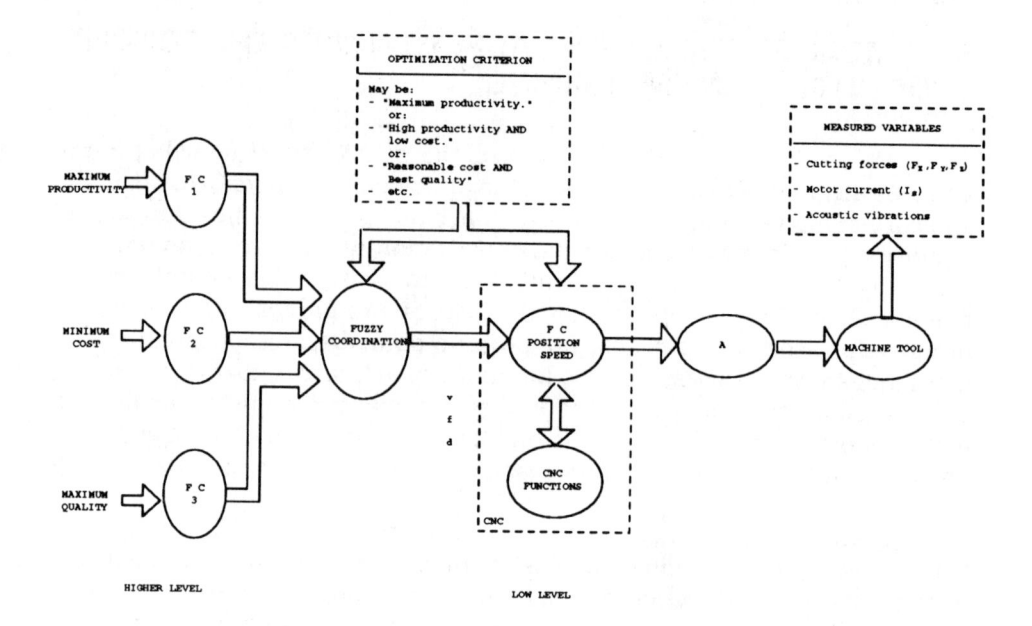

Figure 8: The Proposed Control System

The system could work under multi-index optimization criteria. The coordination of different objectives will be made by fuzzy reasoning. One example of optimization criteria could be: "Reasonable cost and best quality".

Note that the system requires the existence of a multi-sensorial system, which has to provide information about process state in real time.

Finally, at this moment we are designing the fuzzy controller for a maximum productivity. As input variables we are using cutting forces, motor current and acoustic vibrations. Once implemented we will begin the design of the fuzzy controller for minimum cost.

ACKNOWLEDMENTS

This research have been supported by the National Research Agency (CICYT) under proyects Pb 87-0260 and ROB 91-0357.

REFERENCES

1. Kirchner, G. and Pfatschbacher, "Electronic gearbox means new machine-tool functions". *Energy & Automation XI*, pp. 37-38, 1989.

2. *Mitsubishi CNC - M300 - V AI - CNC Series Catalog*. Mitsubishi Electric Corp., pp. 10-11, 1989.

3. *Mazatrol M-32 Catalog*. Yamazaki Mazak Corp., pp 10-11, 16-17, 1989.

4. Mamdani, E.H. et al., "Use of fuzzy logic for implementing rule-based control of industrial processes". In Paul Wang (Ed.): *Advances In Fuzzy Sets, Possibility Theory, and Applications*, Plenum Press, 1983.

5. Li, Y.F. and Lau, C.C., "Development of fuzzy algorithms for servo systems", 1988 *IEEE International Conference on Robotics and Automation*. Philadelphia, Pennsylvania, April 24-29, 1988.

6. Efstathiou, J., "Expert systems, fuzzy logic and rule-based control explained at last". *Trans. Inst. MC, Vol. 10, No. 4*, July-September 1988.

7. Sugeno, M., "An introductory survey of fuzzy control". *Information Sciences*, Vol. 36, pp. 59-83, 1985.

8. Mc Cusker, T., "Neural networks and fuzzy logic, tools of promise for control". *Control Engeneering*, pp 84-85, May, 1990.

9. Yamakawa, T., "High speed fuzzy controller hardware system". *Proc. 2nd Fuzzy Systems Symposium*, pp. 122-130, 1986.

10. Tanscheit, R. and Scharf, L.M., "Experiments with the use of rule-based self-organizing controller for robotics applications". *Fuzzy Sets and Systems*, Vol. 26, pp 195-214, 1988.

Chapter 25
Fuzzy Management of Cache Memories

Introduction, *542*
Organization of Cache Memory, *543*
Fuzzy Cache Management, *544*
Simulation Results, *545*
Conclusion, *546*
References, *549*

This chapter investigates the performance of cache systems in which swapping between the cache memory and the main memory is based on fuzzy logic. The rationale behind this approach is the fact that management of scarce resources by resorting to a clever scheduling strategy is a part and parcel of human endeavor. Therefore, the proposed scheduling strategy or replacement algorithm uses fuzzy logic to mimic a human expert. The simulation results indicate that the fuzzy replacement algorithm performs as well as the other traditional replacement algorithms such as the *first-in first-out* and the *least recently used*. However, the fuzzy algorithm has good potential because changing the rules may improve the performance furthermore.

FUZZY MANAGEMENT OF CACHE MEMORIES

Mahmoud A. Manzoul
Department of Electrical Engineering
Southern Illinois University
Carbondale, Illinois 62901

1. INTRODUCTION

As the size of the main memory (MM) of a computer system is increased, the access time of the main memory also increases. Consequently, the speed disparity between the central processing unit (CPU) and the main memory could severely degrade the performance of the computer system. An elegant solution to this problem is to place a smaller and faster buffer memory, called cache memory (CM), between the CPU and MM. The primary objective of the cache memory is to hold the information that is immediately needed by the CPU [1]. Today almost all processors, including Intel 80386 and 80486 and Motorola 68020, 68030 and 68040, employ cache memories. This success may be attributed to a property called locality of reference which is exhibited by the vast majority of programs [2]. According to this property, within a short period of time, the addresses generated by a CPU have the tendency to get clustered around a small region of the address space. This implies that the memory locations that are being referenced now, have the strong likelihood to get referenced in the future. Therefore, a significant improvement in performance can be realized if the locality cluster is held in the cache memory in a dynamic manner.

In a typical cache based system, as shown in Figure 1, the CPU generates requests for data. Each request is first presented to the CM. If the required data is found in the CM then it is called cache hit and the data is transferred to the CPU right away. On the other hand, a cache miss occurs if the requested data is not found in the CM. In such cases, the slower MM is accessed and a block that contains the required data is first transferred to the CM, and then the CPU request is satisfied by routing the data from the cache to the CPU. This block transfer is the manifestation of the locality property mentioned earlier. Also, situations such as tight loops, repeated subroutine calls clearly indicate a strong sequential correlation in the address streams generated by a CPU [2]. In a system that entails a cache, it is mandatory to swap information between MM and CM because the size of the CM is 25 to 40 times smaller than the MM. This process is carried out by using a replacement algorithm (RA). Basically, if there is a cache miss and if the cache is full then the RA selects one of the existing cache blocks for replacement so as to provide room for an incoming block. Some of the developed replacement algorithms [3] include:

1. Least Recently Used (LRU): This procedure selects the least recently used cache block as the victim.
2. First In First Out (FIFO) : This algorithm selects the oldest cache block as the victim.

3. RANDOM: As the name implies, this policy randomly selects a resident block as the victim.

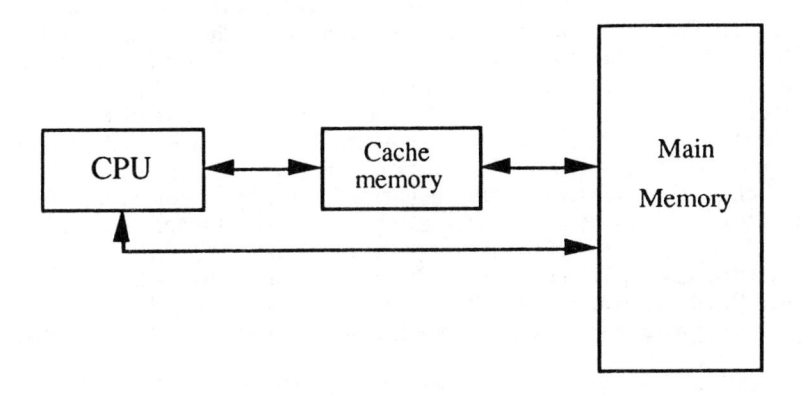

Figure 1 - Cache Memory

In this work, a replacement algorithm that is based on fuzzy logic [4] is presented. Our motivation is the fact that management of scarce resources by resorting to a clever scheduling strategy is a part and parcel human endeavor. Therefore, any scheduling strategy that mimics a human expert using fuzzy logic has a great likelihood to excel. Furthermore, the success of any cache replacement algorithm depends on how well it can lend itself to a simple, structured, fast and cost effective silicon realization [5]. The current trends in VLSI as well as fuzzy hardware architecture [6][7] strongly suggest a new dimension for further explorations.

For convenience, section 2 gives a brief background on cache memories. The next section, section 3, describes the fuzzy replacement algorithm and its operational characteristics such as the input, output and the knowledge base (KB). Section 4 is devoted to the simulation results while section 5 is the conclusion.

2. ORGANIZATION OF CACHE MEMORY

The purpose of including a fast CM, Figure 1, is to increase the overall system throughput by reducing the time the CPU must spend waiting for data to arrive from the slower MM. The success of CM is largely dependent on the Locality of Reference criterion exhibited by the program being executed [8]. This Criterion elucidates two distinct phenomena that the program entails -- first, a particular memory location that has been presently addressed by the CPU stands a high possibility of being referenced in near future (Locality in Time), second, a neighbor location also stands a high possibility of being referenced in near future (Locality in Space). The purpose of CM is to hold these Space Localities over Time for faster response to the CPU request. The idea of having a CM has been widely accepted with respect to a wide range of processors.

Since the CM is 20 to 40 times smaller than the MM, a user-transparent design scheme has to be employed for mapping the entire MM into the CM. For simplifying hardware complexities, both CM and MM are divided into equal sized units, called Blocks in MM and Block-Frames (BF) in CM. At present, the following three mapping schemes are employed [9][10]: (a) Direct Mapping (DM) (b) Fully Associative Mapping (FAM) (c) Set Associative Mapping (SAM). DM and FAM schemes can be viewed as special cases of the SAM scheme and a brief description of these schemes are in order.

Consider a CM having N BFs as shown in Figure 2. In the SAM scheme, the CM is divided into S sets with each set having N/S BFs. A block i of MM can be mapped in any BF belonging to the set (i mod (N/S)). The SAM scheme reduces to DM scheme when N/S equals 1 and to the FAM scheme when S is 1. The SAM scheme is an elegant compromise between DM and FAM schemes. The FAM scheme is the most expensive because it calls for associative search. In contrast, the DM scheme suffers from the risk of severe contention among several MM blocks which are all mapped into a single BF in the CM. The inherent lack of flexibility of the DM scheme results in such a contention even when the CM is largely empty [9]. SAM scheme has therefore gained wide acceptance (Intel 80486 and Motorola MC 68040 both use 4-way SAM scheme) and the effect of varying set-associativity from one extreme to another has been reported [10].

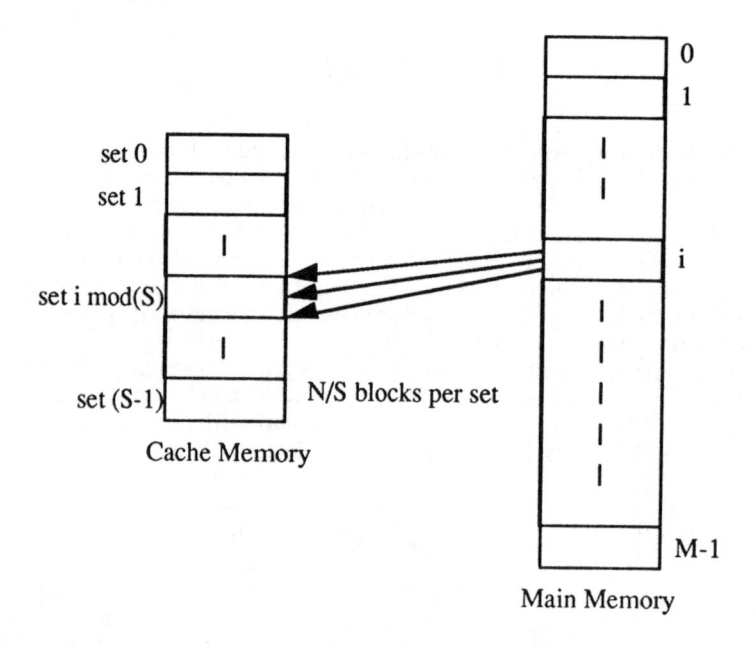

Figure 2 - Set Associative Mapping

3. FUZZY CACHE MANAGEMENT

Our cache management approach is based on a 2-input single-output fuzzy controller. The fuzzy controller implements a fuzzy replacement algorithm (FRA). The two inputs of the fuzzy controller are:

(1) Block Age (Block Age). Each MM block that is currently mapped to CM has an age associated with it. This is an integer number that indicates how long the block in question has been residing in the CM. Typically, this parameter is updated after every CPU cycle.
(2) Block Reference Count (BlockRefCount). It indicates the number of times a given block in the cache has been referenced by the CPU. This value is updated every time a new reference causes a cache hit.

The output of the fuzzy system is the Block Replacement Index (BlockRepIndex). The fuzzy controller is invoked only when the present CPU reference has caused a cache miss and the cache is full [11]. Due to the invocation, the fuzzy controller calculates a BlockRepIndex for each cache block. The block with the high BlockRepIndex is chosen for replacement. The BlockRepIndex for each block is computed by using a set of well defined 34 rules that represents the knowledge base of the fuzzy controller. The 34 rules in the knowledge base represent a collective attempt to exploit the temporal and spatial locality exhibited by the different programs.

Each of the two inputs as well as the output is described by 7 fuzzy variables. For the inputs BlockRefCount and BlockAge *very low, low, low medium, medium, high medium, high, and very high* are used. On the other hand BlockRepIndex is described by *very young, young, teen, old, older, very old, or very very old*. The complete set of rules is given in Table 1. As a sample, the first rule is:

IF the BlockRefCount is *very low* AND the BlockAge is *very young*
THEN the BlockRepIndex is *very low*.

Table 1. The set of rules: BlockRepIndex

BlockAge	BlockRefCount						
	v. low	low	l. med.	med	h. med.	high	v.high
very young	*v. low*	*v. low*	--	--	--	--	--
young	*med.*	*low*	*v. low*	--	--	--	--
teen	*med.*	*l. med.*	*v. low*	*v. low*	--	--	--
old	*h. med.*	*l. med.*	*low*	*v. low*	*v. low*	--	--
older	*high*	*h. med.*	*med.*	*low*	*v. low*	*v. low*	--
very old	*high*	*h. med.*	*med.*	*l. med.*	*low*	*v. low*	*v. low*
v. very old	*v. high*	*v. high*	*h. med.*	*med.*	*l. med.*	*low*	*v. low*

4. SIMULATION RESULTS

Our results are obtained by extensive trace driven simulations. Trace driven simulation is an effective method of evaluating the performance of a cache memory. This involves driving a simulation model of a cache system with an external trace usually generated by interpretively executing a program and recording every memory referenced by the program during its execution [12]. Besides being easy to use, a simulator is very flexible and faster to build than the construction of a prototype design. These features of trace driven simulators make them very attractive for the generation of accurate performance statistics for a given cache organization [13][14].

In this study, we have used Dinero-formatted [15] benchmark program trace including 1,000,000 references (SPICE trace). The benchmarks and traces are gathered with a technique called Address Tracing Using Microcode (ATUM) on a VAX-11 system [10][12]. ATUM provides more realistic traces by capturing complete information of the workload including the operating system kernel and multiprogramming. The details of the traces used in this study are listed below.

SPICE: benchmark trace of SPICE program (VMS) (1,000,000 references).

DEC0.000: a behavioral simulator at DEC, DECSIM, simulating some cache hardware(VMS) (361,982 references).

Using the above discussed traces, the performance of the Fuzzy Replacement Algorithm (FRA) is analyzed and compared with the traditional replacement algorithms LRU and FIFO. For each of these traces, the cache system is partitioned into instructions and data sections. The miss ratio of the two sections were added so as to get the overall miss ratio of the system. For this cache system, the simulations were run for associatively of 2, 4 and 8. These simulations were repeated for line sizes of 8 and 16. Due to space limitation, not all the results are shown. The results shown in Figures 3 and 4 represent the SPICE trace while Figures 5 and 6 for the DEC0.000 trace. From these graphs it is evident that the FRA performs as well as FIFO and LRU for a wide configuration of cache such as different cache size, associativity and line size. In fact, it is noticed that FRA out-performs FIFO and performs at-par with LRU. This is very promising because it indicates that by tuning the rules performance could be increased. With these observations, we find that fuzzy logic is not only well suited for cache replacement algorithm but shows the sign of emerging as a serious contender for traditional replacement algorithms like LRU and FIFO.

5. CONCLUSION

This paper has presented a new replacement algorithm, which is based on a fuzzy controller, for cache memories. The performance of the introduced Fuzzy Replacement Algorithm (FRA) is compared with the existing LRU and FIFO algorithms. This may be due to the fact that the salient features of both LRU and FIFO algorithms have been amalgamated into the fuzzy rules. Simulation results indicate that the FRA performs at more or less the same level as the existing algorithms. However, FRA is a simple and a structured approach with an ample room for improvement by tailoring the set of rules.

The BlockRepIndices for the resident cache blocks give a gradient of priority with which they can be replaced by the incoming MM blocks. So, by setting a threshold index value, one can evict all blocks whose BlockRepIndices are larger than the threshold. Thus, this idea can pave way for further performance improvement because when it is the time for replacement, one gets rid of several useless blocks as opposed to just one block.

The implementation of any cache RA requires a very fast hardware because time is a very crucial parameter. At present, several fast hardware architecture for fuzzy logic systems [6][7] have been reported.

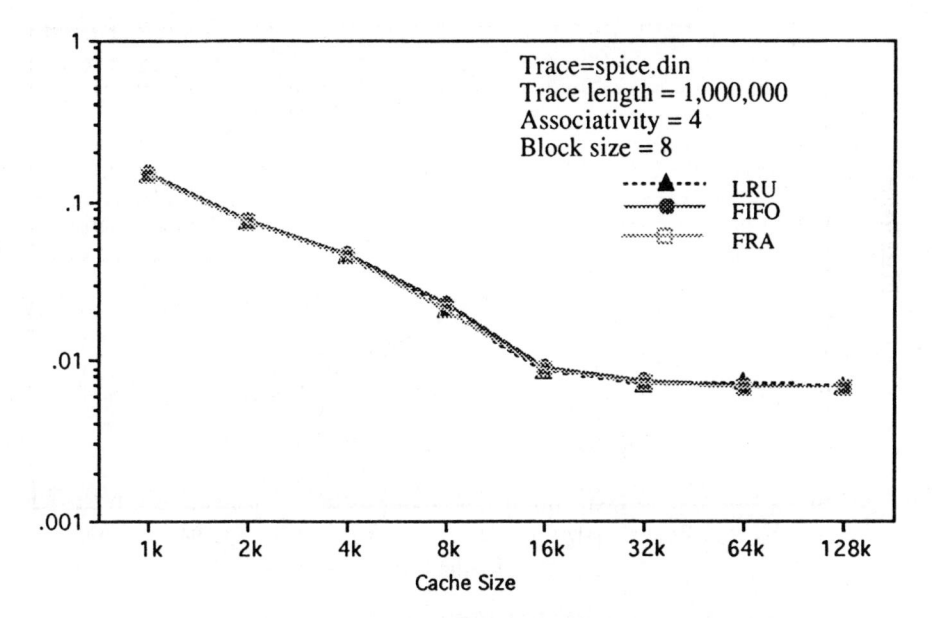

Figure 3 - Comparison of LRU, FIFO and FRA with SPICE trace
Associativity = 4, Block size = 8

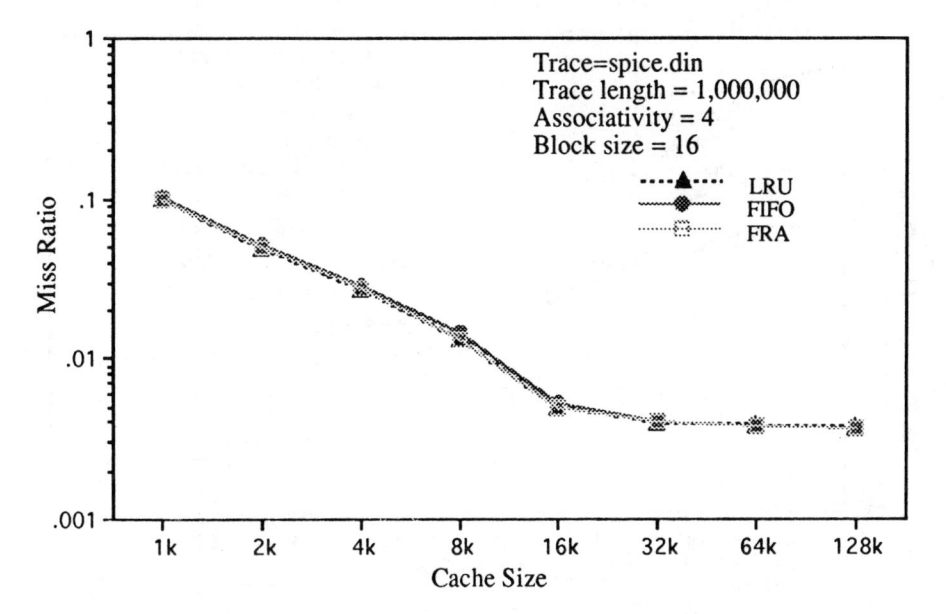

Figure 4 - Comparison of LRU, FIFO and FRA with SPICE trace
Associativity = 4, Block size = 16

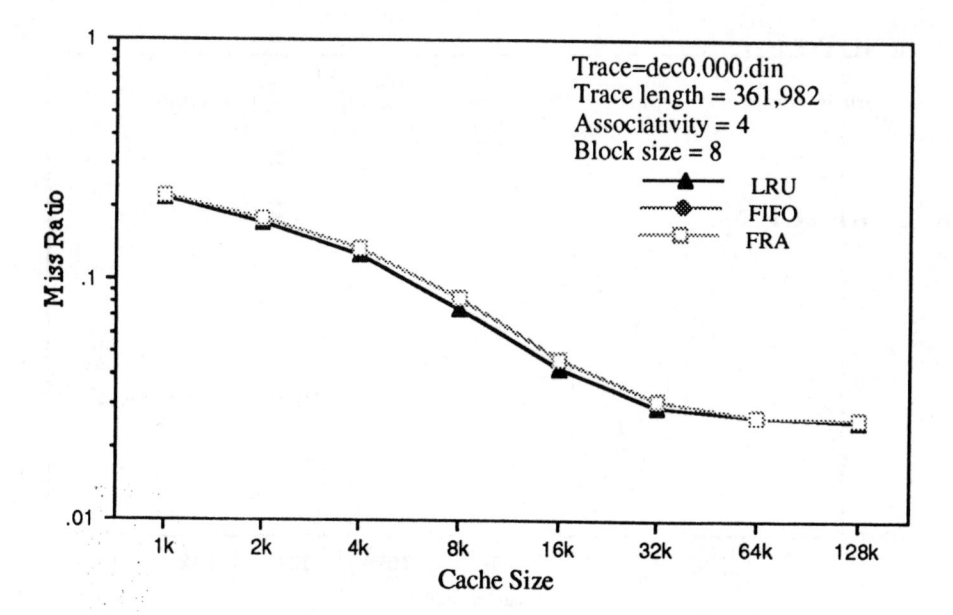

Figure 5 - Comparison of LRU, FIFO and FRA with DEC0.000 trace
Associativity = 4, Block size = 8

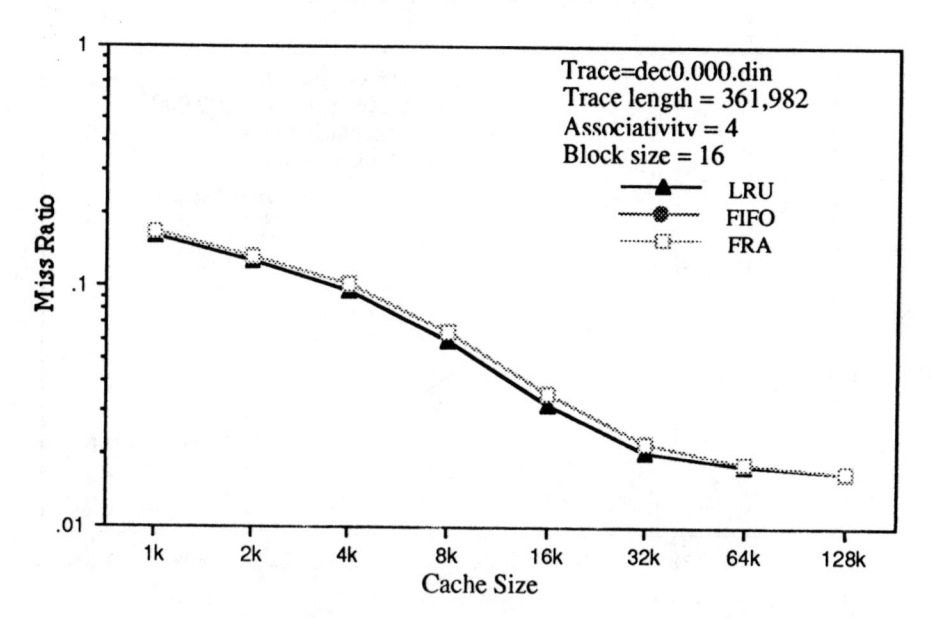

Figure 6 - Comparison of LRU, FIFO and FRA with DEC0.000 trace
Associativity = 4, Block size = 16

ACKNOWLEDGMENT

This work is supported in part by the National Science Foundation under grant number MIP 9011478. Thanks to A. Hossain, A.R. Marudrajan, and S. Hiremath for their work.

REFERENCES

[1] Hamacher V.C., Vranesic Z.G., and Zaky S.G., Computer Organization, New York, NY: McGraw-Hill, 1984.

[2] Stone H.S., High Performance Computer Architecture, Reading, MA: Addison_Wesley, 1990.

[3] Smith A.J., "Cache Memory Design: An Evolving Art," IEEE Spectrum, pp 40-44, December 1987.

[4] Zadeh L.A., "Fuzzy Logic," IEEE Computer, pp. 83-93, April 1988.

[5] Kassimidis S., " Use FPGAs to Match a CPU to Its Memory Subsystem," Electronic Design, pp. 77 - 84, February 1990.

[6] Manzoul M.A. and Tayal S., "Systolic VLSI Array for Multi- Variable Fuzzy Control Systems," International Journal of Cybernetics and Systems, Volume 21, Number 1, pp. 27- 42, 1990.

[7] Manzoul M.A. and Jayabharathi D., "Fuzzy Controller on FPGA Chip", IEEE International Conference on Fuzzy Systems, pp. 1309-1316, March , 1992.

[8] Madnick, S.E. and Donovan, J.J., Operating Systems, McGraw-Hill Book Company, New York, 1974.

[9] Hill M.D., "A Case for Direct-Mapped Caches", Computer, pp. 25-40, December 1988.

[10] Hill, M.D. and Smith, A.J., "Evaluating Associativity in CPU Caches", IEEE Transactions on Computers, Vol. 38, No. 12, pp. 1612-1630, December 1989.

[12] Anant Agarwal, Mark Horowitz and John Hennessy, "An Analytical Cache Model", ACM Trans. Comput. Syst. Vol 7, pp. 184-215, May 1989.

[11] Hossain A., Marudrajan A.R., Manzoul M.A., "Fuzzy Replacement Algorithm for Cache Memory", Cybernetic and Systems: an International Journal, Volume 22, pp 733-746, 1991.

[13] Smith A.J., "Cache Memories", "Comput.. Surveys, vol. 14, pp 473-530, September 1982.

[14] Smith A.J., "Line(Block) Size Choice for CPU Cache Memories", IEEE Trans. Comput. , vol C-36, pp. 1063-1075, September 1987.

[15] Hill M.D., DineroIII Documentation, Unpublished Unix-style Man Page, University of California, Berkeley, October 1985.

Chapter 26
Fuzzy Controllers on Semi-Custom VLSI Chips

Introduction, *552*
Logic Synthesis of Fuzzy Controllers, *553*
FPGA Implementation, *558*
Conclusions, *559*
References, *560*

This chapter describes the implementation of fuzzy controllers on field programmable gate arrays (FPGAs). FPGAs are inexpensive semi-custom VLSI circuits whose development tools can run on personal computers. They combine the advantages of gate arrays (high density, flexible architecture, and advanced software) with those of programmable logic devices (short design time, user programmable, and standard product). The presented approach reduces the fuzzy controller to a look-up table or a set of Boolean equations. The development system of the FPGA takes the set of Boolean equations and generates the necessary code for programming the FPGA chip. It turns out that the size of the FPGA is independent of the number of rules in the knowledge base. Consequently, the speed of the fuzzy controller is independent of the number of rules, and a speed over 50M fuzzy logical inferences per second (FLIPS) can be achieved.

FUZZY CONTROLLERS ON SEMI-CUSTOM VLSI CHIPS

Mahmoud A. Manzoul
Department of Electrical Engineering
Southern Illinois University
Carbondale, Illinois 62901

1. INTRODUCTION

Fuzzy controllers have been applied in diverse fields [1]. A typical fuzzy controller consists of a knowledge base, which contains the set of rules, and an inference mechanism. The inference mechanism is responsible for making the fuzzy decisions based on the inputs and the knowledge base. Fuzzy encoder (fuzzifier) and fuzzy decoder (defuzzifier) interface the fuzzy controller with the outside world. Most of the current applications employ software as the medium for the implementation of the fuzzy controller. It would be cheaper and faster to use hardware implementation. Furthermore, the lack of fast hardware has limited and sometimes prevented the actual implementation of some applications. The following three factors indicate that the speed of fuzzy controllers is of prime importance.

1. Fuzzy controllers are real-time systems. The success of any real-time system relies heavily on how fast the necessary computations are executed.
2. When human operators are modeled using fuzzy logic, a large set of rules is usually developed. These rules constitute the knowledge base. The larger the knowledge base the slower the system; because all data in the knowledge base must be processed.
3. Most of the actual processes that are suitable for fuzzy logic have more than one variable to be monitored. Consequently, multi-variable fuzzy controllers must be used. The amount of information contained in each rule is approximately proportional to the number of variables. Therefore, multi-variable controllers have specially large knowledge bases.

Since 1985, great deal of work have been done in the hardware implementation of fuzzy controllers. M. Togai and H. Watanabe [2] built a special VLSI chip with a speed of 80K Fuzzy Logic Inferences Per Second (FLIPS). A 580K-FLIPS fuzzy controller with re-configurable and cascadable architecture was developed by H. Watanabe, W.D. Dettloff, and K.E. Yount [3]. A Mega FLIPS machine, which can perform more than 1,000,000 FLIPS, has been introduced by T. Yamakawa [4]. A systolic array for fuzzy controllers was reported by Manzoul and Tayal [5].

This paper describes a method for implementing fuzzy controllers on semi-custom VLSI chips, in particular Field Programmable Gate Arrays (FPGAs). FPGAs are inexpensive and their development systems run on personal computers [6]. FPGAs combine the advantages of both Gate Arrays [7] and Programmable Logic Devices (PLDs) [8]. They are similar to the Gate Arrays in their high density structure, but can be field-programmed [9][10][11]. The basic idea behind the presented method is to logically synthesize the fuzzy controller into a set of Boolean equations. A closer look at the fuzzy controller reveals that it can be reduced to a look-up table. Once the knowledge base and the set of all possible inputs are known, the look-up table for the fuzzy controller can be formed. The size of the look-up table is reasonably small. In fact, the number of rows is equal to the product of the dimensions of the inputs universes of discourse [12]. The look-up table is described by a set of Boolean equations which is an acceptable input to the Development System of the FPGA.

This paper is organized as follows. The representation of fuzzy controllers by Boolean equations, logic synthesis, is discussed in section 2. It contains the synthesis of single-input single-output controllers as well as multi-input single-output controllers. Also, a simple numerical example is included in section 2. Section 3 describes the implementation of the Boolean equations on FPGAs. The conclusion is section 4.

2. LOGIC SYNTHESIS OF FUZZY CONTROLLERS

The first step in realizing fuzzy controllers on FPGA chips is to reduce the fuzzy controller to a set of Boolean equations. The set of Boolean equations is then implemented on an FPGA chip. In this section the process of reducing the fuzzy controller to a set of Boolean equations is introduced. For illustration, the logic synthesis of single-input single-output fuzzy controllers is considered first, then a numerical example is presented. Finally the general case of fuzzy controllers is considered. However, only multi-input single-output fuzzy controllers are considered because any multi-input multi-output fuzzy controller can be decomposed into k, where k is equal to the number of outputs, multi-input single-output fuzzy controllers [5].

2.1 Single-input Single-output Fuzzy Controllers

Consider a single-input single-output fuzzy controller with n rules where rule i is:

$$\text{IF } \alpha_1 \text{ is } A_{1i} \text{ THEN } \beta \text{ is } B_i$$

Here α_1 are A_{1i} are defined in the input universe of discourse A^1; while β and B_i are defined in the output universe of discourse B^1. Let the dimensions of the two universes of discourse be

$$\dim [A^1] = q_1 \tag{1}$$
$$\dim [B^1] = p \tag{2}$$

The fuzzy relation of the system is computed as [13]

$$R = V (A_{1i} B_i) \tag{3}$$

where V is the max operator and

$$\dim [R^1] = q_1 \times p \tag{4}$$

Given any input α_1, a fuzzy output β may be computed using the compositional rule of inference [13]; that is

$$\beta = \alpha_1 \text{ O } R \tag{5}$$

where O denotes the max-min composition of the fuzzy relation.

Our objective is to represent the fuzzy controller by a look-up table that can be described by a set of Boolean equations. In order to do that, we have to determine all the unique fuzzy inputs to the controller. For every unique fuzzy input, an output must be computed. Initially one would think that the number of unique fuzzy inputs is large; but a closer look reveals that this number is equal to q_1, the dimension of the input universe of discourse [12]. The reason is the fuzzification process. In the fuzzification process, the range of the input is mapped on to the input universe of discourse. Therefore, each fuzzified input has a full membership grade at one element in the input universe of discourse and zero membership grade at all other elements, see Figure 1. The element with full membership corresponds to the actual input presented to the controller. Since there are q_1 elements in the input universe of discourse, the number of unique fuzzy inputs (ufi) is equal to the dimension of the input universe of discourse. That is,

$$\text{ufi} = \dim[A^1] = q_1 \tag{6}$$

Since each unique fuzzy input has only one output value, the maximum number of distinct output values is also equal to q_1. Therefore, the number of rows in the look-up table is also equal to q_1. The size of the look-up table depends largely upon the way the information is presented. For each fuzzified input value, it is observed that all elements have zero membership grades except one which has full membership grade. Therefore, all we need to know or store is the element whose membership grade is equal to one. There are q_1 elements in the input

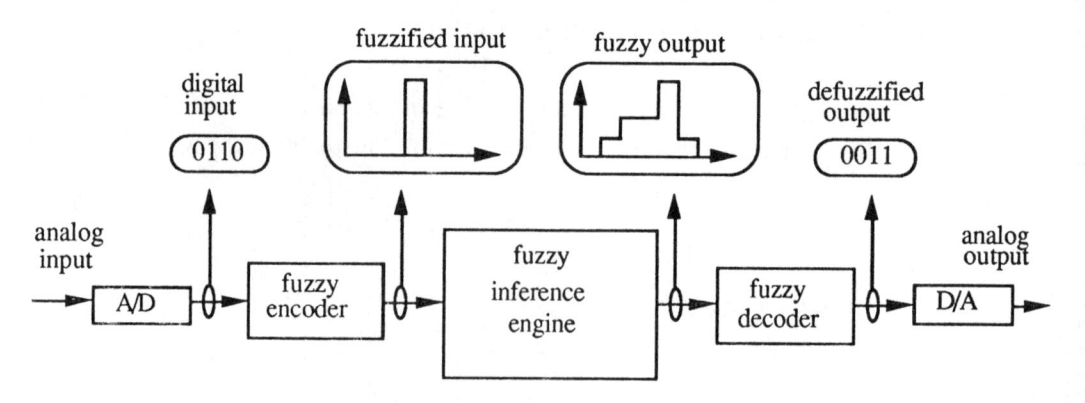

Figure 1 - Fuzzy Controller

universe of discourse; therefore, the number of binary bits needed to represent the different input values of the look-up table is equal to the least integer containing $\log_2 q_1$. As for the output, it is easier and economical to use the defuzzified outputs in the look-up table. Using the center of gravity method, the defuzzification process is basically the selection of one element, in the output universe of discourse, which corresponds to the non-fuzzy output. There are p elements in the output universe of discourse; therefore, the number of binary bits needed to represent the outputs of the look-up table is equal to the least integer containing $\log_2 p$.

2.2 Numerical Example

Consider a single-input single-output controller which operates a tank valve. The valve adjusts the supply rate of a fluid into the tank. The input variable is the pressure and the output variable is the supply rate. The system may be quite complex in real situations, but for convenience the system is described by the following three rules.

1. IF (*HIGH*) pressure   THEN (*DECREASE*) supply rate
2. IF (*OK*) pressure   THEN (*ZERO INCREASE/DECREASE*) supply rate
3. IF (*LOW*) pressure   THEN (*INCREASE*) supply rate

The original ranges for the input and output are (0 to 1400) psi and (0.03 to .21) kg/s, respectively. The membership values are in the interval [0,1], where 0 denotes no membership and 1 denotes full membership. Assume dim $[A^1]$ = dim $[B^1]$ = 7, and the three rules of the system are expressed numerically as:

1. IF [.0 .1 .2 .3 .6 .9 1.0]   THEN [.8 1.0 .8 .6 .3 .1 .0]
2. IF [.0 .5 1.0 1.0 1.0 .5 .0]   THEN [.0 .2 .8 1.0 .8 .2 .0]
3. IF [1.0 .9 .6 .3 .2 .1 .0]   THEN [.0 .1 .2 .3 .6 1.0 .3]

Using equation (3), the fuzzy relation R is computed.

$$R = \begin{bmatrix} .0 & .1 & .2 & .3 & .6 & 1.0 & .3 \\ .1 & .2 & .5 & .5 & .6 & .9 & .3 \\ .2 & .2 & .8 & 1.0 & .8 & .6 & .3 \\ .3 & .3 & .8 & 1.0 & .8 & .3 & .3 \\ .6 & .6 & .8 & 1.0 & .8 & .2 & .2 \\ .8 & .9 & .8 & .6 & .5 & .2 & .1 \\ .8 & 1.0 & .8 & .6 & .3 & .1 & .0 \end{bmatrix}$$

Now, if the input pressure is equal to 1200 psi, that corresponds to the fuzzy value

1200 psi = [.0 .0 .0 .0 .0 1.0 .0]

To determine the fuzzy output, the compositional rule of inference, given by equation (5), is employed. The fuzzy output is found to be

Fuzzy output = [.8 .9 .8 .6 .5 .2 .1]

After defuzzification, the output is found to be 0.09 kg/s. The complete computations of the controller is shown in Table 1. The look-up table representing the fuzzy controller is given in Table 2. The inputs of the table are the non-zero elements of the fuzzified input. The outputs of the table are the actual outputs after defuzzification.

Table 1 - Summary of the controller computations

input psi	fuzzified input							fuzzy output							output 0.03kg/s
000 to 200	1.0	.0	.0	.0	.0	.0	.0	.0	.1	.2	.3	.6	1.0	.3	6
201 to 400	.0	1.0	.0	.0	.0	.0	.0	.1	.2	.5	.5	.6	.9	.3	5
401 to 600	.0	.0	1.0	.0	.0	.0	.0	.2	.2	.8	1.0	.8	.6	.3	4
601 to 800	.0	.0	.0	1.0	.0	.0	.0	.3	.3	.8	1.0	.8	.3	.3	4
801 to 1000	.0	.0	.0	.0	1.0	.0	.0	.6	.6	.8	1.0	.8	.2	.2	4
1001 to 1200	.0	.0	.0	.0	.0	1.0	.0	.8	.9	.8	.6	.5	.2	.1	3
1201 to 1400	.0	.0	.0	.0	.0	.0	1.0	.8	1.0	.8	.6	.3	.1	.0	1

Table 2 - Look-up Table for the Fuzzy Controller

input			output f_2 f_1 f_0		
0	0	1	1	1	0
0	1	0	1	0	1
0	1	1	1	0	0
1	0	0	1	0	0
1	0	1	1	0	0
1	1	0	0	1	1
1	1	1	0	0	1

Since the input universe of discourse as well as the output universe of discourse have dimensions equal to seven, the look-up table can be described by three 3-variable Boolean equations, that is

$f_0 = \Sigma\ (2,6,7)$
$f_1 = \Sigma\ (1,6)$
$f_2 = \Sigma\ (1,2,3,4,5)$

2.2 Multi-input Single-output Fuzzy Controllers

Consider an m-input single-output fuzzy controller with n rules. Rule i of the system can be described as:

IF α_1 is A_{1i} AND α_2 is A_{2i} AND ... AND α_m is A_{mi} THEN β is B_i

where α_j and A_{ji} are defined in the input universe of discourse A^j; while β and B_i are defined in the output universe of discourse B^1, that is

$$\dim [A^j] = q_j \quad \text{for } 1 \leq j \leq m; \tag{7}$$
$$\dim [B^1] = p \tag{8}$$

The fuzzy relation of the system is given by [13]:

$$R = V(A_{1i}\ A_{2i}\ ...\ A_{mi}\ B_i) \tag{9}$$

where V is the max operator and the dimension of the fuzzy relation R is,

$$\dim [R^m] = q_1 \times q_2 \times ... \times q_m \times p \tag{10}$$

Given any input combination $\alpha_1\ \alpha_2\ ...\ \alpha_m$, the fuzzy output β is computed using the compositional rule of inference [13],

$$\beta = (\alpha_1\ \alpha_2\ ...\ \alpha_m)\ O\ R \tag{11}$$

where O denotes the max-min composition of the fuzzy relation.

In this case, the number of unique fuzzy inputs (ufi) is equal to the product of all the dimensions of the input universes of discourse, that is:

$$\text{ufi} = \prod_{j=1}^{m} q_j \tag{12}$$

Once the fuzzy outputs are calculated, the outputs are defuzzified using the center of gravity approach. Let Y be the number of binary bits needed to represent the different output values. Therefore, Y is equal to the least integer containing $\log_2 p$, where p is the dimension of the output universe of discourse. Now, the fuzzy controller can be described by Y Boolean equations. Each equation is a function of X variables, where X is equal to the sum of the least integers containing $\log_2 q_j$ for $j = 1$ to m.

$$X = \sum_{j=1}^{m} \log_2 q_j \tag{13}$$

3. FPGA IMPLEMENTATION

A field programmable gate array chip is used to implement the Boolean equations obtained as a result of the logic synthesis of the fuzzy controller. The FPGA architecture [14] consists of a configuration program store, and three types of configurable elements: a perimeter of I/O blocks, a core of Configurable Logic Blocks (CLBs), and interconnection area. Figure 2 shows the general features of the FPGA chip. The configurable program is loaded automatically from an external PROM, on power-up. The FPGA Development System [14] takes the Boolean equations and generates the necessary configurable program. Therefore, the file containing the Boolean equations is the input file to the Development System. The input file is prepared in PALASM format. The Boolean equations are realized using PAL structures in the FPGA.

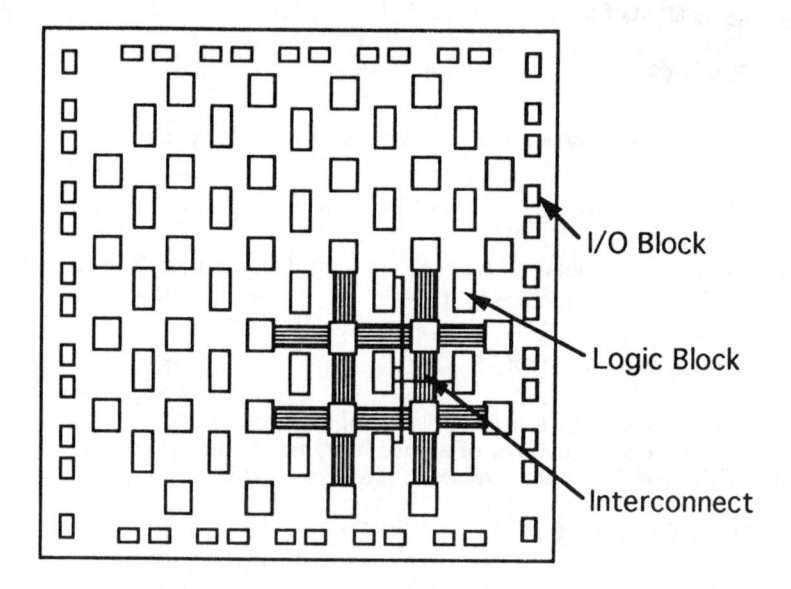

Figure 2 - Architecture of the FPGA

Several fuzzy controllers were implemented and timing simulations were performed to calculate the worst delay. Speed over 50M FLIPS was obtained for simpler designs. For a Single-Input Single-Output System, speed up to 60M FLIPS were obtained using a 70MHz FPGA chip, Table 3. For Two-Input Single-Output System speeds close to 50M FLIPS were obtained using a 100MHz FPGA chip, Table 4. It is observed that the speed of the fuzzy controller is inversely proportional to the dimensions of the universes of discourse.

Table 3 - Speed of Single-input Single-output Fuzzy Controllers
(Based on XC3020PC68-70MHz FPGA chip)

dimensions of the universes of discourse		speed (FLIPS)
input	output	
4	4	60M
7	7	58M
12	12	55M
32	32	22M

Table 4 - Speed of Two-input Single-output Fuzzy Controllers
(Based on XC3030PC84-100MHz FPGA chip)

dimensions of the universes of discourse		speed (FLIPS)
input	output	
4	4	50M
7	7	35M
12	12	12.5M
32	32	9M

4. CONCLUSIONS

A method for implementing fuzzy controllers on FPGAs, which are semi-custom VLSI chips, has been presented. In order to enhance the speed of the fuzzy controller, most of the computations are done off-line. The size and speed of the fuzzy controller on the FPGA are independent of the number of rules. In addition, different types of fuzzy relations can be used.

ACKNOWLEDGMENT

This work is supported in part by the National Science Foundation under grant number MIP 9011478. We would like to thank Xilinx Inc. for making their products available at reduced cost. Thanks as well are due to Dinesh Jayabharathi for his work on this project.

REFERENCES

[1] Maiers J. and Sherif Y.S., "Application of Fuzzy Set Theory," IEEE Transactions on System, Man, and Cybernetics, Vol. SMC-15, No. 1, pp. 175-189, Jan/Feb 1985.

[2] Togai M. and Watanabe H., "Expert System on a chip: An Engine for Real-Time Approximate Reasoning," IEEE Expert, pp. 55-62, Fall 1986.

[3] Watanabe H., Dettloff W.D. and Yount K.E., "A VLSI Fuzzy Logic Controller with Re-configurable, Cascadable Architecture," IEEE J. of Solid-State Circuits, Vol. 25, No. 2, pp. 376-382, April 1990.

[4] Yamakawa T., "High-Speed Fuzzy Controller Hardware System: The Mega FLIPS Machine," Information Sciences, Elsevier Science Publishing Company, Inc., 1988, pp. 113-128.

[5] Manzoul M.A., and Tayal S., "Systolic Arrays for Multi-Variable Fuzzy Control Systems," International Journal of Cybernetics and Systems, Vol. 21, No. 1, pp. 27-42, 1990.

[6] Freeman R., "User-Programmable Gate Array," IEEE Spectrum, pp. 32-35, December 1988.

[7] Di Giacomo J., "Design Methodology," in *VLSI Handbook*, J. Di Giacomo, Ed., New York: McGraw-Hill, 1989, pp. 1.3-1.9.

[8] Coli V.J., "VLSI Programmable Devices," in *VLSI Handbook*, J. Di Giacomo, Ed., New York: McGraw-Hill, 1989, pp. 4.1-4.39.

[9] Rose J., Francis R.J., Lewis D., and Chow P., "Architecture of Field-Programmable Gate Arrays: The Effect of Logic Block Functionality on Area Efficiency," IEEE Journal of Solid-State Circuits, Vol. 25, No. 5, pp. 1217-1225, Oct 1990.

[10] Gamal A.E., et al., "An Architecture for Electrically Configurable Gate Arrays," IEEE Journal of Solid-State Circuits, Vol. 24, No. 2, pp. 394-398, April 1989.

[11] Rose J., and Brown S., "Flexibility of Interconnection Structures for Field-Programmable Gate Arrays," IEEE Journal of Solid-State Circuits, Vol. 26, No. 3, pp. 277-282, March 1991,.

[12] Manzoul M.A., and Jayabharathi D., "Implementation of Fuzzy Controllers Using Combinational Circuits," Proceedings of NAFIPS '91 Workshop, pp. 163-167, May 1991.

[13] Zadeh L.A, "Outline of a New Approach to the Analysis of Complex Systems and Decision Processes," IEEE Transactions on System, Man and Cybernetics, Vol. SMC-28, No. 1, pp. 28-44, 1973.

[14] The Programmable Gate Array Data Book, 1991, Xilinx Inc.

Chapter 27
General Analysis of Fuzzy-Controlled Phase-Locked Loop

Introduction, *563*
Possibilities of Fuzzy Control in PLLs, *564*
PLL Parameters and Adaptability, *568*
The Case of Synchronous and Coherent Phase-Locked Synchronous Oscillators, *569*
Performance Analysis of an Analog, Fuzzy-Controlled PLL, *571*
The Stability of F-PLL and Chaotic F-PLL, *574*
Advantages and Limits of F-PLL, *575*
References, *576*

This chapter deals with aspects of design and analysis of fuzzy-controlled phase-locked loop (PLL). It considers two methods of PLL control, the control of the loop gain and the control of the phase comparator, and introduces the basic concepts of applying fuzzy control to PLL circuits. Two important PLL parameters, the capture range and the lock range, are analyzed in terms of the fuzzy-controlled PLL and compared with the classic PLL. The behavior of a class of analog fuzzy-controlled PLL circuits is analyzed, including stability and possible chaotic behavior. Also described is a new class of oscillators, synchronous and coherent phase-locked synchronous oscillators, and the fuzzy control of these devices is introduced.

GENERAL ANALYSIS OF FUZZY-CONTROLLED PHASE-LOCKED LOOP

H. N. Teodorescu and A. Brezulianu
Center for Fuzzy Systems and Artificial Intelligence
Iasi, Romania[*]

List of abbreviations

AM	– amplitude modulation
A-PLL	– analog PLL
CMOS	– complementary MOS (technology)
CPSO	– coherent phase synchronous oscillators
CR	– capture range
D-PLL	– digital PLLs
FCS	– fuzzy controlled system
F-PLL	– fuzzy controlled PLL
LPF	– low-pass filter
LR	– lock range
NB	– negative big
NM	– negative medium
NS	– negative small
PhC	– phase controller
PhD	– phase detector
PLL	– phase-locked loop
OA	– operational amplifier
PB	– positive big
PM	– positive medium
PS	– positive small
SO	– synchronous oscillator
S/N	– signal-to-noise (power) ratio
VCO	– voltage controlled oscillator
ZR	– zero

[*] Part of this work was written during the stay of the first author as a Visiting Professor in FLSI, Iizuka, Japan

1. INTRODUCTION

The phase-locked loop (PPL) concept dates to the early days of radio technology. Phase-locked loops (PPLs) devices are systems primarily aimed to generate signals in phase with the input (control) signal phase, while the input signal is (slowly) changing. If the input signal is noisy, the output signal should follow the carrier (basic signal) phase.

Thus, the PLL can act as a nonlinear bandpass filter tuned by the incoming signal. In fact, the PLL re-creates the original signal rather than just filter the input signal. Following, the PLL basically consists of two circuits (see **Figure 1**): a controlled

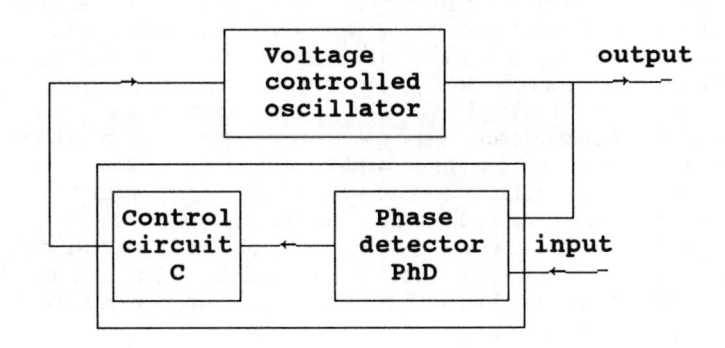

Figure 1. PLL Basic Diagram

signal generator (voltage controlled oscillator - VCO), and a phase detector and control circuit (PhD-C). As the control signal is an estimate of the phase (or a function of it), the PLL can be used for demodulation purposes (frequency or phase demodulation). The PLL can also be used in AM demodulation since it generates a constant level output signal, as required by AM demodulator. Moreover, PLLs are used in frequency synthesizers. In this application, a fixed precise generator provides the input signal and the control loop includes a frequency divider to allow for frequency changes. Industrial applications such as motor-speed control [1] were also announced. Other applications include signal synthesis for broadcasting, telemetry, control and instrumentation, and whenever it is necessary to extract frequency or phase information from signals embedded in noise or other interference.

In many such applications, the dynamical characteristics of the PLL play an important part, mainly the acquisition time and the noise immunity. The time needed to reach the quasi - stationary regime, for a given hop in frequency/phase is most usually determined in terms of equivalent number of periods. This characteristic is important in MF/FSK demodulator and in fast switching frequency synthesizers that must often change the output frequency. (Such devices are used for example in frequency hopping systems.) Noise output spurious signal suppression power versus noise input power is important in (tele)communications applications such as carrier recovery.

In the last two decades, PLLs turned from the analog technology to the digital one, due to some important advantages [1]: high frequency range (up to 30 MHz in monolithic IC), insensitivity to changes in temperature and power-supply voltage, programmable bandwidth and center frequencies.

Moreover, in the digital technology, very high (i.e., narrow-bandwidth) loops can be achieved, and high-order loops are easy to construct by simple cascading [1]. Unlike the analog PLLs, where the error signal provided by the phase detector PhD corrects the (analog) VCO frequency, in digital PLLs the error signal controls the direction of a up-down counter [1].

Much used is a class of integrated (monolithic) hybrid PLLs, including an analog VCO, an input signal amplifier, a low-pass filter (LPF), a digital PhD, and dividers. These devices are usually manufactured in CMOS or TTL technology (MSI) and a classical example is the 4046 circuit. (Such devices are often named "digital PLLs" although they are hybrid, while the true digital PLLs are named "all - digital PLLs".)

Further details on basic analog PLL circuits and a complete bibliography until 1970 can be found in [2]. A basic study on PLL acquisition was published by [3]. An introduction to digital PLLs is presented in [1].

2. POSSIBILITIES OF FUZZY CONTROL IN PLLs

Both analog and (all-)digital PLLs can be controlled by fuzzy phase controllers (FPhC). Moreover, the control may act at various actual stages of the loop, according to the application in hand. In what follows, a brief analysis of different ways of control is performed.

2.1 Control Of The Loop Gain

For analog PLLs, probably the simplest manner to control the loop is that of changing the loop gain and thus the input (control) voltage to the VCO. This may be simply performed using an automatic gain control (AGC) amplifier in the loop. Such AGC - amplifiers can be implemented either by using a controlled resistance in the input (or output) attenuator (**Figure 2**), or in the feedback loop of an amplifier (**Figure 3**).

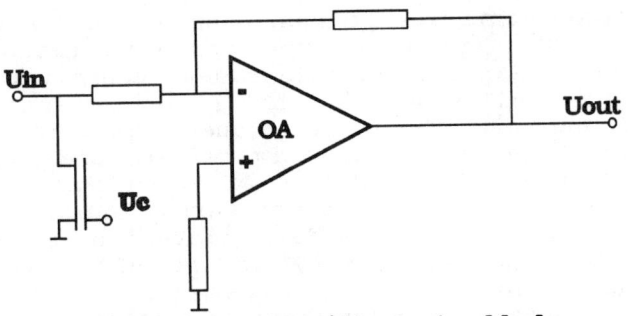

Figure 2. AGC with Controlled Resistance in the Input Attenuator

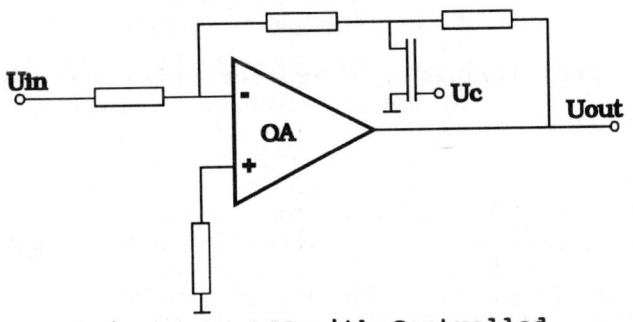

Figure 3. AGC with Controlled Resistance in the Feedback Loop

For example, consider the (simplest) case of the second order analog PLL (**Figure 4**).

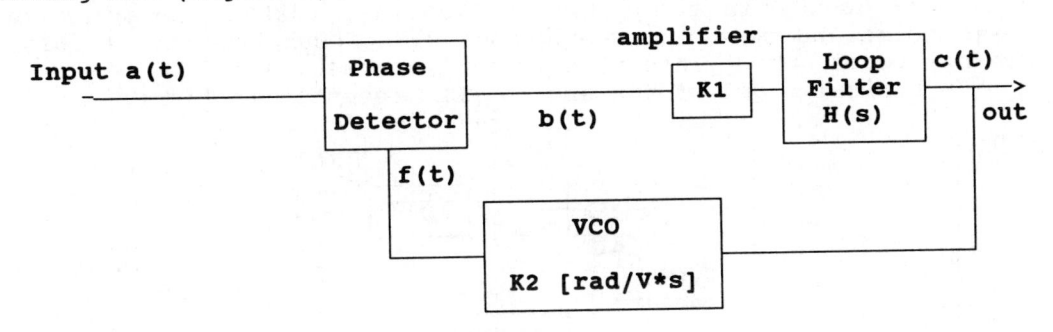

Figure 4

where
$a(t) = A*\cos[\omega t + \varnothing_i(t)] + N(t)$,
$b(t) = \sin[\varnothing_i(t) - \varnothing_r(t)] + n(t)$,
$f(t) = -2/A*\sin[\varnothing_i(t) + \varnothing_r(t)]$.

The gain loop control also has another important advantage, namely it can both speed-up the loop acquisition time and compensate for the change of static lock-in characteristics.

Indeed, it is well known that the lock-in characteristics of an analog PLL change with the input frequency [4]. It is relatively simple to achieve the desired performances of the PLL at a fixed frequency by design, but the change of the frequency causes the variation of certain internal parameters. Thus, the transient behavior of the loop as well as the sideband noise are degraded [4]. For example, the sensitivity of the VCO - which in fact consists in an integrator - decreases when the frequency is increased. The decrease is a nonlinear function of frequency. In applications such as synthesizer-based hopping spread-spectrum ratio system, fast-switching, low-noise synthesis is needed, thus asking for non-degrading properties versus frequency, in a wide range of frequencies. Thus, in such applications, an automatic, nonlinear compensation of the loop gain, according to the frequency change, is needed.

The loop gain frequency characteristics can be altered by varying either the phase detector characteristics, or the loop filter characteristics. (The VCO characteristic is fixed and very difficult to alter.) On the other hand, the loop filter characteristic is rather difficult to alter as this operation requires switching of R and/or C components. Switching capacitors or resistors in the filter is undesirable since changing DC voltage on the switched capacitors can introduce severe transient inputs into the PLL [4]. In general, the R and C components of the filter are fixed values components. Although it is possible to use FET transistors as variable resistors, or active filters to get a controlled filter, technological reasons limit the use of this alternative.

Following, only changes of the PhD gain or of the gain of the amplifier introduced in the loop are technically recommended.

A reasonable control will provide a high loop gain in the acquisition phase to achieve fast acquisition, and a constant gain versus frequency in the almost locked-in situation to minimize phase noise and to maximize spurious signal suppression. A fuzzy control seems appropriate to perform this task.

A typical model of the analog PLL presented in Figure 5.

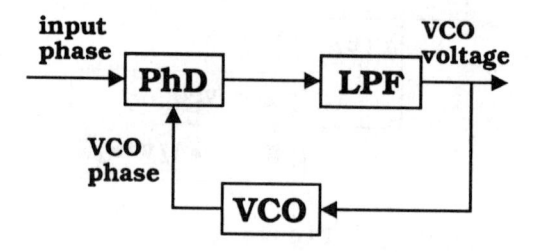

Figure 5. Analog PLL Circuit

The block diagram of a fuzzy controlled PLL circuit is presented in Figure 6.

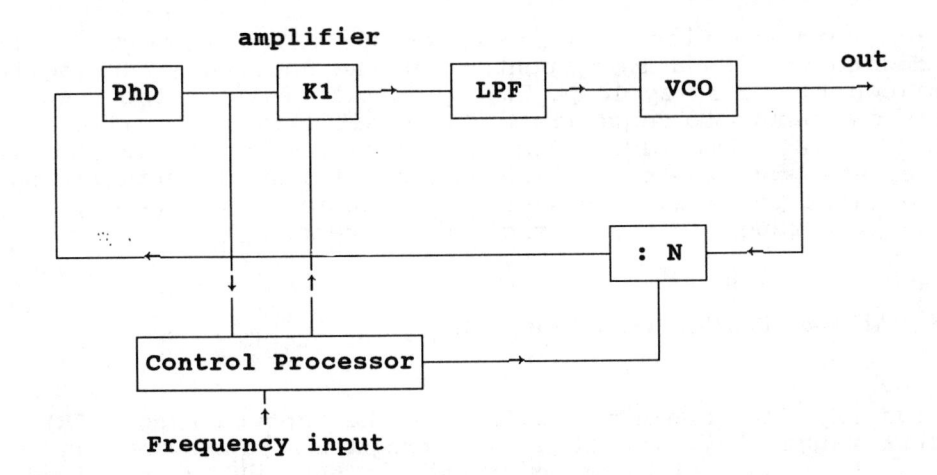

Figure 6. Basic Diagram of the Fuzzy Controlled Analog PLL

The controller receives information on the desired frequency and on the phase difference. It controls both the programming divider (: N) and the loop gain (K1).

2.2 Control Of The Phase Comparator

Most hybrid PLLs, including the common 4046 and the SP 8850 type 1.5 GHz professional synthesizer (manufactured by Plessey Semiconductors), include two phase comparators. The use of two different PhC is justified for obtaining best performance results in both phase locked and unlocked conditions.

In the 4046 IC, the two comparators are digital. The first is used on the frequency locked condition thus obtaining the highest achievable noise suppression.

The SP 8850 IC uses a digital and an analog PhC. Because the digital comparator is sensitive to both frequency and phase error over a wide linear range, it is used to rapidly bring the loop close to phase lock condition.

When this condition is fulfilled, the digital PhC is automatically disabled and the PLL switches on the analog PC. This occurs when the loop phase error is within the linear range of the analog comparator. In this condition, minimum sideband generation (i.e., pure output signal) occurs. Moreover, the SP 8850 offers the possibility of using an external phase comparator. Following, this circuit is suitable for use in conjunction with a fuzzy PhC. On the other hand, the 4046 IC can also be used in conjunction with an external PhC, but then only the VCO is used from this IC. Moreover, both comparators in the 4046 IC can be used simultaneously, in conjunction with a single external circuit as described in [5], to make the PLL adaptive.

In all-digital PLLs, the standard method to make the loop adaptive is to change the PhC characteristic (see for example [6]).

In conclusion, the use of a controlled PhC in an adaptive PLL is a standard solution even in the crisp PLL, and can be easily extended to provide a fuzzy control of the PhC, thus turning the crisp loop into a fuzzy loop. This possibility is in fact used in our first attempts to demonstrate the feasibility of an all-digital F-PLL [7], [8]. The advantages of fuzzy control of the PhD is, among others, the ease of continuous control over the entire range, while in most implementations of crisp adaptive PLLs (e.g., [6]), at least one range of the control is discrete.

3. PLL PARAMETERS AND ADAPTABILITY

Two major parameters of a PLL are the **capture range (CR)** and the **lock range (LR)**. The CR is the frequency range over which a loop can lock into an incoming signal. In the F-PLLs described in this volume and in [7], [8], the CR is given by the frequency (phase) range covered by ϕ_n (and ϕ_{n-1}) membership functions. Obviously, there is a compromise between the number of membership functions (mf) - i.e., of computation complexity - the setting time, and the CR.

The lock range is the range of frequency over which a loop will remain locked on when the input signal frequency (slowly) shifts from the central frequency of the loop. As in crisp PLLs, in F-PLL the lock range is determined by the capabilities of the blocks in the loop (e.g., by the LPF and by VCO parameters/maximum control range in A-PLLs, or by the counter in D-PLLs).

Usually, the center frequency f_o of the loop is the same as the free - running frequency, i.e., the frequency of the output signal when no input signal is applied. The "tracking range" or "hold-in range" is defined as the maximum deviation from f_o for the loop remains in lock and is therefor one-half of the lock range if the lock range is centered on f_o.

In analog PLLs, in general the capture range is lower than the lock range. The CR is also usually centered on f_o. The (equal) right and left maximal deviations from f_o for the loop captures the frequency of the incoming signal are named **"lock-in"** or **"pull-in"** ranges [13].

In a F-PLL as described above, that the condition the CR and LR ranges are centered on f_o, the free-running frequency, is that the truth table of the controller is centered on ($\emptyset_n=0$, $\emptyset_{n-1}=0$).

The settling time of a complex digital PLL is determined by both the control algorithm and the computation time. Indeed, if the rate of change of the phase correction is chosen very high (a few number of periods of the output signal between two successive corrections) than the time of computing t_c, the correction may become larger than the time between two such corrections - a situation which must be avoided. Thus, for such PLLs, t_c is an important parameter, as well as the number of necessary corrections to get the desired frequency. In F-PLLs, t_c can be large enough when using classical processors and processing algorithms. Hence the need for fast fuzzy inference algorithms, or for using specialized fuzzy processors. (For example, in an application developed in Iasi, a 1/1000 division was used.)

The F-PLL can be reconfigured easily to make it adapt to large capture range by simply rescaling the input membership functions. The overshoot, i.e., the damping of the loop, and the acquisition time are related parameters. Both are dependent on the PhC characteristics. In analog PLLs, only three PhC characteristics are easy to obtain: the saw-tooth, the triangular, and the sinusoidal ones [9] - see **Figure 7**.

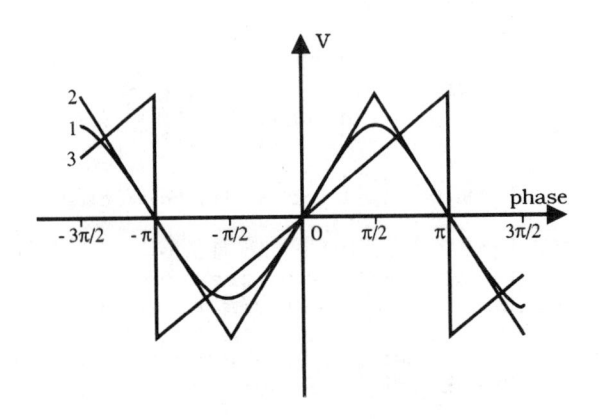

Figure 7. Usual Characteristic of PhD

More elaborate PhC characteristics can be achieved in D-PLLs, using microprocessors to control the loop. Best results are obtained by using adaptive D-PLLs.

The major benefit of F-PLLs is related to its controlled system structure: the overall characteristics of the device using PLLs can be changed according to the application in hand by re-resetting the PLL loop response. Thus, the need for locking the loop over a wide range of acquisition/setting frequencies in a low setting time determined an important research effort in making the loop adaptive.

The adaptation can be as simple as the dynamical change of the loop time constant as in [1], can involve the variation of the loop gain [4], or can involve elaborate adaptive techniques as in [10].

4. THE CASE OF SYNCHRONOUS AND COHERENT PHASE-LOCKED SYNCHRONOUS OSCILLATORS

Recently, a new class of oscillators which synchronize, track, filter, amplify and, if necessary, divide in a single process was introduced [11]. These devices are named **synchronous oscillators** (SO), and were reported to work in the frequency range 100 Hz - 1.2 GHz [11].

The SO is built as a Colpitts oscillator to which a phase modulator transistor is added. The phase modulator acts as a current injection circuit in the oscillating transistor and contributes to the regenerating process in the oscillator transistor. The block diagram of the SO is shown in **Figure 8**.

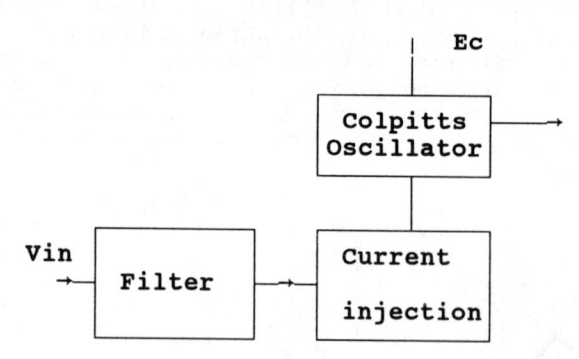

Figure 8. Block Diagram of a Synchronous Oscillator

Based on SOs, and by adding two loops, coherent phase locked synchronous oscillators (CPSO) are created.

The CPSO is simply a SO with a phase correction added. Compared to the block diagram of a PLL, the CPSO has the same blocks, but they are connected in another order and they fulfill different tasks. The block diagram of a CPSO is sketched in **Figure 9**.

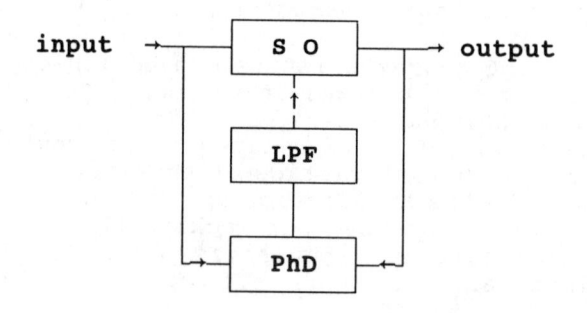

Figure 9. Block Diagram of Coherent Phase-locked Synchronous Oscillator

The SO and CPSO have a number of advantages over the PLLs loops, mainly:

i) The input signal sensitivity of a SO is several orders of sensitivity better than a PLL;

ii) SOs acquire much faster than second-order PLLs (since SOs act as first-order loops for small phase offsets);

iii) The noise bandwidth of a SO can be designed to be almost independent of its tracking range;

iv) For the same input S/N ratio, the tracking range is several orders of magnitude wider in a CPSO than a PLL (because noise bandwidth and the tracking range are designed independently).

v) The circuit of a SO is a very simple one;

vi) SO works well in a very wide band (from 100 Hz to over

1 GHz) [11];

The existence of these circuits was mentioned here, because the designer has to choose among PLLs - both analog and digital - with crisp control, CPSOs, and PLLs with fuzzy control. It seems that CPSOs can share a part of the special applications palette with fuzzy controlled PLLs, and with all-digital PLLs.

Moreover, to apply some fuzzy control to CPSO could be also a challenging research project. What would be necessary to design and manufacture a sound fuzzy controlled CPSO is a very fast fuzzy controller. This could be achieved with fast bipolar MIN/MAX gates, and using a very fast defuzzifier (maybe, not based on the center-of-gravity method). A nonlinear control law could much accelerate the synchronization.

5. PERFORMANCE ANALYSIS OF AN ANALOG, FUZZY-CONTROLLED PLL

In accordance with [12], Figure 10 shows how the fuzzy controller (realized with two inputs/one output fuzzy system) is inserted between the phase detector (PD) and the low pass filter (LPF), based on a classical diagram of the PLL devices (see Figure 5). In Figure 10, E1 represents the first input in the fuzzy controller, and stands for the phase error $\Delta\emptyset(n)$ on the current moment $t = t_n$, and E2 represents the second input in the fuzzy controller and stands for the phase error $\Delta\emptyset(n-1)$ on the antecedent moment $t = t_{n-1}$.

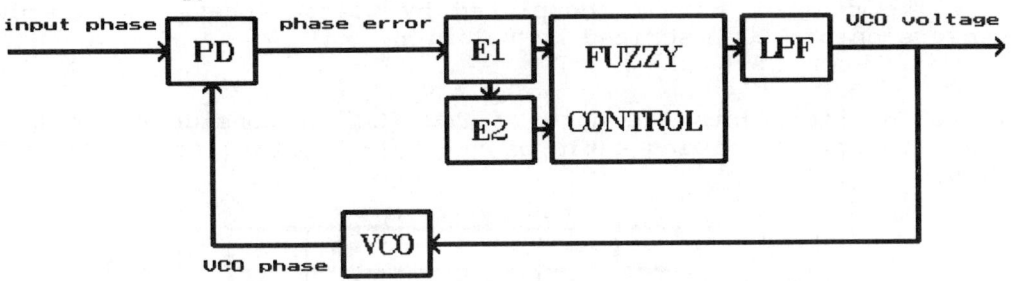

Figure 10. Fuzzy Controlled Analog PLL Circuit

The controller input signals are:

$$\Delta\emptyset(n-1) = \emptyset_{input}(n-1) - \emptyset_{vco}(n-2)$$
$$\Delta\emptyset(n) = \emptyset_{input}(n) - \emptyset_{vco}(n-1)$$

where $\emptyset_{input}$ is the phase of the input signal, $\emptyset_{vco}$ is the phase of VCO signal, and the symbols (n-1), respectively (n) represent the values of the variables at successive moments.

The fuzzy control is determined by five membership functions in antecedent on each of the inputs. The membership functions are sketched in Figure 11. This number of input membership functions is a compromise between the quality of control and the rules table dimension. Seven triangular membership functions with equal bases, overlapping two by two are used for both inputs. Seven membership functions, with equal bases, overlapping three by three, as sketched in Figure 12, are used for the consequent (i.e., output).

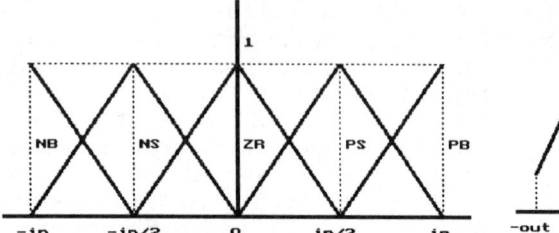

Figure 11. Input Membership Functions

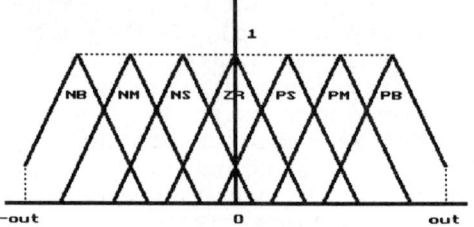

Figure 12. Output Membership Functions

The fuzzy controller yields a control voltage VCO_n, applied to the VCO input.

The center of gravity method is used for defuzzification. The value Cg(n) represents the current correction applied to the antecedent value of VCO input voltage, noted $V_{vco}(n-1)$, by:

$$V_{vco}(n) = V_{vco}(n-1) + Cg(n)$$

Usualy, the anterior relation is realized by the fuzzy controller, but in special conditions the LPF can achieved that adding.

The control loop is completed by voltage-phase (frequency) conversion of VCO described by relation:

$$\emptyset_{vco} = G_{vco} * V_{vco},$$

where G_{vco} represents the gain of VCO (G_{cvo} is considered to be a fixed value). The rules table of the fuzzy controller is shown in Figure 13.

		phase error n-1				
		NB	NS	ZR	PS	PB
phase error n	NB	PB	PB	PM	PM	PS
	NS	PB	PM	PM	PS	PS
	ZR	PM	PS	ZR	NS	NM
	PS	NS	NS	NM	NM	NB
	PB	NS	NM	NM	NB	NB

Figure 13. Rules Table of the Fuzzy Controller

In the first place, the fuzzy controlled PLL circuit must be analyzed for phase tracking behavior.

Figure 14 presents the tracking curve for a 3 radian step phase signal applied to the PLL circuit input, for three different values of the controller output range **[-OUT, +OUT]**. If the controller output range is too small or too large (see the curves for **OUT** = 0.5 and for **OUT** = 1.9 of Figure 14) the phase tracking performance decreases.

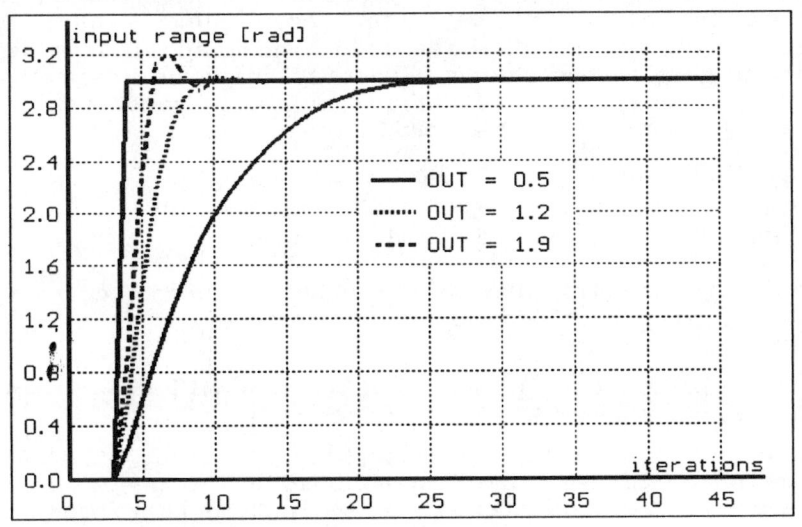

Figure 14. Phase Tracking Curves

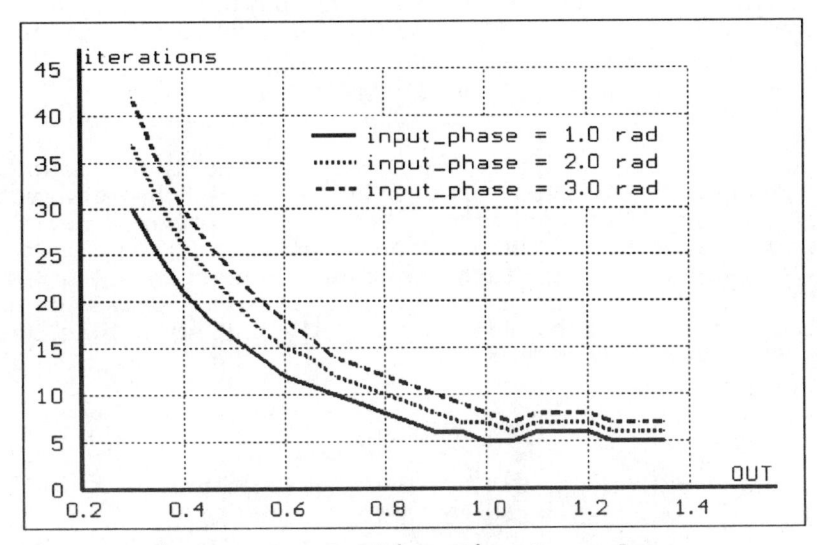

Figure 15. Locking Time Dependence

For analysis of the F-PLL behavior the "phase locked" condition was adopted as follows: for three succeeding iterations the absolute phase error must be less than 0.01 radian. An important parameter is the number N_o of iterations until the "locked phase" condition is fulfilled.

Figure 15 presents the dependence of the number N_o with respect to the output discourse universe range **[-OUT; +OUT]** for three different values of phase step signal applied to input.

The optimum of the output discourse universe range (see Figure 15) is **[-OUT; +OUT]** = [-1.10; +1.10]. For this range the parameters of the fuzzy controlled PLL are:

     phase locking time   = 8   number of iterations
     phase overshoot value   = $8*10^{-2}$ radian
     phase undershoot value = $6*10^{-3}$ radian

An important parameter of the PLL loop is the error when time tends to infinity (with fixed input). This error is $e = 5*10^{-4}$ rad.

6. THE STABILITY OF F-PLL AND CHAOTIC F-PLL

It was noted by Endo and Chua [14], [15] that PLLs can present dramatic instabilities, leading to chaos. The question arises if the same danger is encountered when using the fuzzy control of the PLLs.

Briefly, in [15], the PLL-based FM (frequency modulation) demodulator is addressed, with the input $\varnothing i$ given by eq.(1), where Δw denotes the detuning between the carrier frequency and the VCO free running frequency, M denotes the maximum angular frequency deviation, and w_m denotes the modulation angular frequency:

$$d\frac{\phi_i}{dt} = \Delta\omega + M \sin\omega_m t \quad (1)$$

Let $\varnothing$ denotes the difference between the input phase and the output phase of the VCO. Let $h(\varnothing)$ denotes the characteristic function of the phase detector (PhD) -- remember that $h(t)$ is a periodic function of period 2π. Consider, as in [15], that $h(.)$ and $h'(.)$ can be expanded into Fourier convergent series. (This assumption is valid too for the usual fuzzy phase control, as presented before.) Then, following [15], the equations describing the PLL FM demodulator are:

$$\phi' = y \quad (2)$$
$$y' = [-\beta + (2\zeta - \beta)h'(\phi)]y - h(\phi) + \beta\sigma + m\beta\sin\Omega t + m\Omega\cos\Omega t$$

where:

Ω is the normalized modulation frequency;
β - the normalized natural frequency of the PLL;
m - the normalized maximum frequency deviation;
ζ - the dumping factor of the loop; σ - normalized frequency detuning.

The above equations are derived considering that the PhD is followed by a low-pass filter (LPF) with linear transfer function KoF(s), and that the VCO acts as an integrator (with the transfer function 1/s).

The fact that a linear LPF is used does not change the basic assumption of this discussion, as a LPF can also be used after the fuzzy controller (and, in fact, it was used by us in one simulation [16]).

As the fuzzy controller, represented for a F-PLL by $h(\emptyset)$, satisfies the conditions imposed in [15], it follows that the F-PLL can also behave in a chaotic manner.

7. ADVANTAGES AND LIMITS OF F-PLL

As was recently proved [17], [18], any fuzzy system with triangular membership functions is equivalent to a crisp rational system. Thus, there is a membership function which can not be replaced by a crisp one and, also proved, the converse is also true. Following, any F-PLL could be implemented as a crisp (adaptive) PLL, and a natural question arises: When should we prefer the F-PLL to a crisp PLL ?

The answer is according to the application in hand and it must take into account the advantages and limits of the F-PLL, and of the **fuzzy control systems (FCS)** in general.

The first major advantage of F-PLL -- and in general of FCS -- is that it can be easily changed. The change can be easily performed in a manner that almost surely improves -- or at least do not dramatically degrades -- the overall performances of the system (supposing an expert in PLLs implements his knowledge in the system). Moreover, it can be easily re-configured, for example by extending the ranges of control. In contrast, changing a crisp PLL by means of a set of numerical coefficients is not at all transparent to the designer (in general, there is no obvious relation between the coefficients of the loop and the desired characteristics of complex PLLs).

The advantages of F-PLL (and of FCSs in general) are as follows:

i) The system can be implemented as the linguistic description of a fuzzy system much easier than the description of a (high order) control system;

ii) The output of the F-PLL is linguistically settled in the design (description) phase, while the output of the crisp PLL (FCS) is not always transparent to the designer while writing the equations of the system. Thus, the expert can determine -- at least partly -- the general behavior of the system from the beginning;

iii) The parameters of the system are easily changed and, moreover, the designer can easy determine the way of changing (because he includes both linguistic conditions of change and the results of change);

iv) The system can be easily reconfigured or rescaled, even without the help of the expert in PLLs. For example, the widening or the shrinking of the input or output range can be obtained by simply re-scaling the base of the input/output membership function, or by adding linguistic degrees and mapping the input/output relations in the system of the previously defined one.

v) Instability regions for the FCSs are determined more easily then for crisp control systems;

vi) The behavior of the F-PLL can be easily altered without changing the overall behavior of the PLL. This is due to the description on segments of the input/output sets for the F-PLL.

The main drawbacks of F-PLL refers to the cost and complexity of the hardware required (a fuzzy inference machine with two inputs is required if the CS is based on fuzzy processors). If one adopts a usual processor, the computation time can be too high for PLLs with free frequencies above 1KHz or, alternatively, the acquisition time can be degraded by the computation time if higher f_o are used.

Further details on the principles, performances and behavior optimization of F-PLLs can be found in [7], [8].

REFERENCES

[1]. Greer, W.T.; Kean, B. - *Digital phase-locked loops move into analog territory* - Electronic Design, March 31, 1982, pp.95-99

[2]. Klapper, Jacob; Frankle, John T. - *Phase-Locked and frequency-feed-back systems* - Academic Press, 1972, N.Y., San Francisco, London

[3]. Frazier, J.P.; Page, J. - *Phase-Locked Loop frequency acquisition Study* - IRE Trans. on Space Electronics and Telemetry, Sept. 1962, pp. 210-227

[4]. Yeager, Richard - *Loop Gain Compensation in Phase-Locked Loops* - RCA Review, vol. 47 (March 1986), pp. 78-87

[5]. Marz, Daniel - *Simple technique Speeds PLL Lock-in time* - RF Design, June, 1986, pp. 53

[6]. Makajimo, O.; Hikawa, H.; Mori, S. - *Performance improvement of all digital phase-locked loop with adaptive multiband-quantized phase comparator* - Proc. ISCAS '88, IEEE Rev., 1988, pp. 603-606

[7]. Teodorescu H.N., Bogdan I., D. Galea, E. Sofron - *Adaptive fuzzy phase locked loops;* Proc. Int. Conference on Fuzzy Systems & Neural Networks, Iizuka '90, vol.1, pp. 165 - 168

[8]. Bogdan, I., Teodorescu H.N. - *Dynamical behavior of fuzzy controlled PLLs* - In: Fuzzy Systems and Signals H.N.Teodorescu, Editor), 1990, AMSE Press, France

[9]. Vatasescu A., et al. - *Circuite Integrate liniare. Manual de utilizare. Vol.1, Cap. 1* - Editura Tehnica, Bucuresti 1979

[10]. Endo, Tetsuro, Chua, L.O, Marito, T. - *Chaos from Phase-Locked Loops - Part II, High-Dissipation Case* - IEEE Trans. on Circuits & Systems, vol 35, Feb. 1989, pp. 225-263

[11]. Uzunoglu, V.; White, M.H. - *Synchronous and coherent phase-locked synchronous Oscillators; New Techniques in synchronization and Tracking* - IEEE Trans. on Circuits and Systems, vol.36, no 7, July 1989, pp. 997-1004

[12]. Teodorescu, H.N., Brezulianu, A., Bogdan, I, Sofron, E., - *Analog PLL fuzzy controlled circuit* - 2nd BUFSA Conference Proceedings, Turkey, 1992, pp. 88-92

[13]. Phillips - Linear LSI Catalog IC 11M/1985, page. 562

[14]. Endo, T., Chua, L.O. - *Chaos from phase locked-loops* - IEEE Trans. Cicuits and Syst., 1988, vol.35, pp. 987-1003

[15]. Endo T., Chua L.O., Narita T. - *Chaos from phase-locked loops -Part III: High dissipation Case* - IEEE Trans. Circuits and Syst., vol.35, no.2, pp. 255-263 (Feb. 1989)

[16]. Brezulianu A., Teodorescu H.N. - *Fuzzy control of an analog PLL* - (Report, Workshop on Fuzzy Systems and Applications, Iasi, March 1992).

[17]. E.P. Klement, H.N. Teodorescu (Editors) - *An introduction to Fuzzy Systems in Engineering (chapters 4,5)* - Iasi Polytehnic Publishing House, Iasi, Romania, 1991.

[18]. H.N. Teodorescu - *Equivalence fuzzy and crisp systems* - Fuzzy signals and systems (AMSE proceedings), AMSE Conference, Cetinje, Iugoslavia, 1991.

Chapter 28
A Fuzzy Logic Controller for a Rigid Disk Drive

Introduction, *580*
Seek Control Method, *581*
Seek Employing Fuzzy Logic, *583*
Trial Seek Employing Fuzzy Logic, *584*
Seek Table Reference, *588*
Reversed Seek Table reference, *590*
Correcting Force Unevenness, *595*
Conclusion, *598*
References, *598*

This chapter considers the application of fuzzy inference to the head-access control system of a hard disk drive to improve access performance. The fuzzy algorithm is applied to a bang-bang controller of the disk drive. The algorithm has two functions, one of *trial seek* and the other of *correction for uneven force*. Fuzzy logic is used to estimate the desired switching time for the bang-bang control, and to correct for changes in actuator coil resistance, due to temperature, and actuator force unevenness. Also discussed is another approach to head seek control in which a quasi-fuzzy inference rule is established based on the relation between the moving distance and the acceleration command time. From it, the acceleration command time for minimum deviation of the moving distance is inferred. Changes in the actuator coil resistance, due to temperature, can be compensated for only by varying the rule.

A FUZZY LOGIC CONTROLLER FOR A RIGID DISK DRIVE

Shuichi Yoshida
Information Equipment Research Laboratory
Matsushita Electric Industrial Co., Ltd.
Osaka, 571 Japan

1. INTRODUCTION

With the recent trends toward more powerful personal computers and workstations has emerged a demand for magnetic rigid disk drives (RDD) and other peripheral storage devices which are smaller in size and provide greater storage capacities with increased rates of data transfer to host computers. (See Fig. 1.)

The time for data transfer is determined by the seek time required by the head reading the data. This is the time to move from one data cylinder to the target cylinder. The seek time is limited by the performance of the actuator moving the head as well as by the control method.

This chapter shows how to reduce seek time, through a bang-bang controller employing fuzzy logic together with a method for correcting for changes in actuator coil resistance and actuator force unevenness[1,2,3].

Fig. 1 External View of a Rigid Disk Drive

2. SEEK CONTROL METHOD

2.1 Seek Control Method Of RDD

The seek control used in conventional disk drives relies on closed-loop velocity profile control. In other words, the controller command tries to match a velocity profile when the head moves from one data cylinder to another. In this control, first the actuator is accelerated to the maximum acceleration, and the reference velocity corresponding to the remaining distance is given as a deceleration profile so as to decelerate sufficiently. Then, it is changed over to the position control just before the target position is reached, in order to follow up the target cylinder. In this system, when the voice coil motor (VCM) is used as the actuator, the transfer function of the control object will be approximated by a simple second order integral system. The deceleration profile is governed by the time-optimal, bang-bang control, expressed below [4] :

$$Ve = Kv \cdot Xe^{1/2} \qquad (1)$$

where, Ve: reference velocity; Xe: remaining distance; Kv: constant.

In practice, however, the deceleration profile is set lower than when the actuator exhibits its intrinsic performance; or the tail edge of the profile is at a moderate setting, in order to absorb characteristic changes due to actuator fluctuations, temperature variations, or external disturbance of the actuator [4,5]. The limits on the deceleration profile constrain the seek time and it is difficult to exploit the full capabilities of the actuator. (See Fig. 2.)

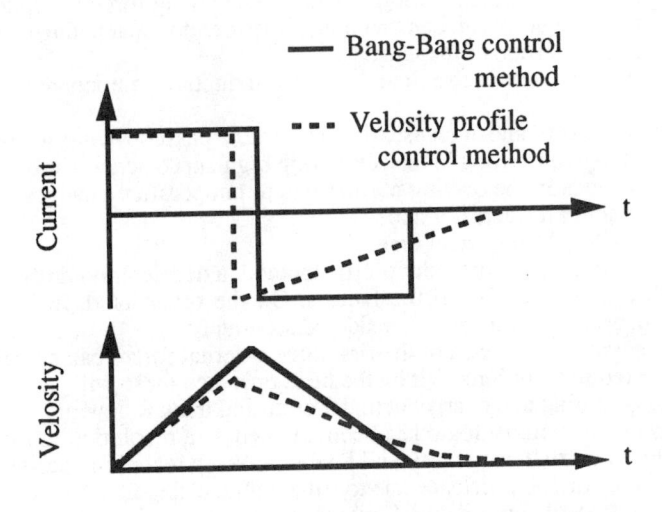

Fig. 2 Comparison of Seek Control Methods

2.2 Open-loop, Bang-bang Seek Control Method

To replace the conventional seek control method mentioned in Section 2.1, we propose an open-loop bang-bang seek control method [1,2,3]. Maximum acceleration is used not only for acceleration, but for deceleration as well. One of the problems attending this bang-bang method is the timing for switching between acceleration and deceleration given the distance to the target. (See Fig. 2.)

Procedures for timing the switch can be classified into two types. In the first, switching is performed when the midpoint of the range of motion is detected; in the second, switching relies on time management. The first case requires highly accurate and expensive position sensor in order to detect the position of the actuator moving assembly in real time. So the switching procedure adopted in this section relies on time management.

When time management is used with bang-bang control, driving commands must be given such that the durations of acceleration and deceleration result in zero velocity when the head has moved through the target distance. As is well known, the relation between the distance of motion and the bang-bang acceleration time resulting in zero velocity at the target position is uniquely determined, so the driving command need only specify the acceleration time. Execution of the deceleration command is judged to be complete when the velocity reaches zero.

There are a number of factors which complicate the timing of the switch.
(1) External disturbing forces are exerted on the actuator moving assembly by flexible wires passing current to the actuator and transmitting signals to the head.
(2) External disturbing forces are exerted on the head slider flexure by the air flow generated between disk platters by the disk revolution.
(3) External disturbing forces are caused by friction in the bearings supporting the rotating assembly of the actuator.
(4) There is unevenness in the position at which the driving force is applied owing to unbalance in the magnetic circuit construction of the actuator, machining tolerances, and unevenness in magnetization of the magnet.
(5) There is lag in the current rise time due to the inductive component of the actuator coil.
(6) There are decreases and increases in the current passed owing to the back emf (electro-motive force) appearing in the coil during high-speed actuator movement.
(7) There are changes in the driving current due to temperature changes in the purely resistive component of the actuator coil.
(8) All of these factors change with time.

Thus, when using the velocity profile method, a deceleration profile is designed which includes tolerances for all the factors. On the other hand, in the bang-bang driving method, each factor must be considered separately.

If the magnitudes of each of the first three external forces can be held to within one gram force (equivalent force felt by the head tip), then these will pose no problems when the head is settling at the target cylinder or during track following.

In this chapter, fuzzy logic has been adopted as a resolution to the remaining factors relating to current sensitivity and a table look-up technique has been used for correcting for actuator coil resistance due to temperature change.

In the dedicated servo method, which is the method for detection of the head position adopted in this work, linear position detection is only possible within one cylinder in the disk radial direction. Hence, when positioning the magnetic head at the target cylinder during the seek process, the velocity of head motion must be accurately controlled right before it reaches the target track. The method presented here allows positioning at the target track under position control, without the occurrence of seek errors, even when there is considerable scattering in the velocity of head motion immediately before reaching the target track.[1]

3. SEEK EMPLOYING FUZZY LOGIC

Figure 3 shows the flow of operations in the production of an RDD adopting bang-bang driving by a manufacturer and in use of the RDD by a user.

In production line of disk drives adopting bang-bang seek, the manufacturer determines bang-bang seek commands appropriate to different target distances through trial seek operations and stores the data in memory prior to shipment of the unit. In other word, the relation between the target distance and the bang-bang acceleration time (when starting from a certain cylinder on the disk) has to be determined in advance, and the results stored as a table look-up. However, the unevenness in the driving force of the actuator must be omitted from the table.

In use of the disk drive by a user, the bang-bang acceleration time corresponding to any given target distance is determined by interpolation based on the table.

In disk drive seek control, the temperature of the actuator coil increases when the frequency of seek increases, and the driving current decreases. Therefore the acceleration time must be increased accordingly.

In seek operations the acceleration must be corrected for unevenness in the driving force of the actuator over the range of acceleration. The acceleration time is corrected in accordance with the start position and the range of unevenness which overlaps the range of acceleration.

The trial seek operation is done off-line iteratively during initial setup to obtain the nominal acceleration times, and the temperature and unevenness corrections are done on-line to adjust these nominal values in real-time.

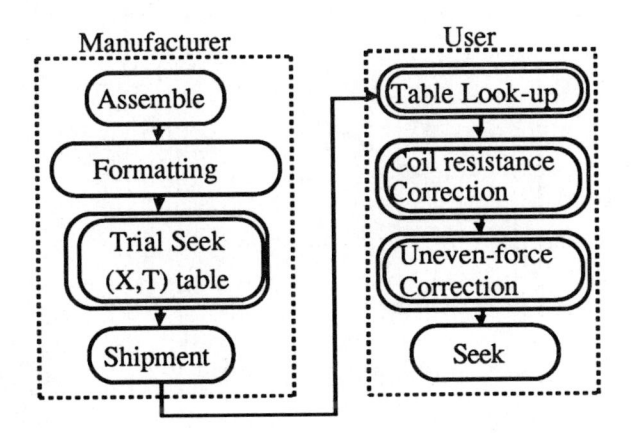

Fig. 3 Flow of Processes from Manufacture to Use of an RDD

4. TRIAL SEEK EMPLOYING FUZZY LOGIC

4.1 Simulation

Figure 4 is a block diagram of an actuator transfer function, including a current driving circuit which employs a voice coil motor (VCM) as the actuator.

As shown in the model in Fig. 4, the response was simulated for the input of a bang-bang driving command signal. The relation between the target distance X and the

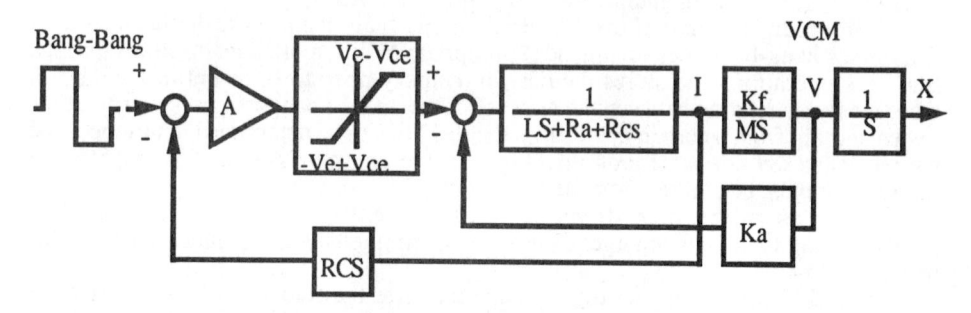

Ra : coil resistance   L : coil inductance
Kf : force factor   Rcs : current sense
Ka : back e.m.f. constant

Fig. 4 Block Diagram of Actuator Driving Circuit for RDD

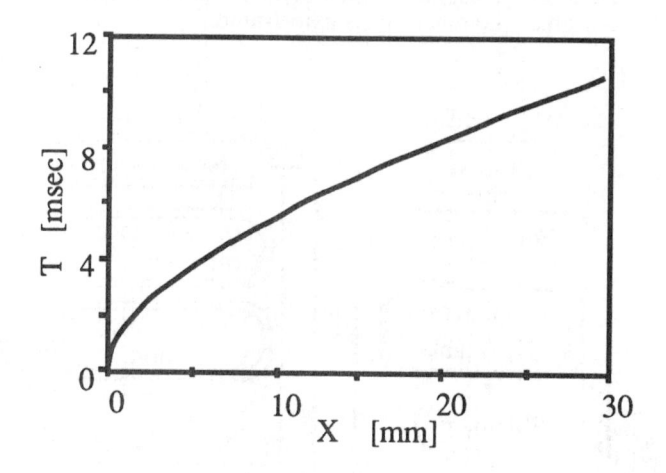

Fig. 5 Target Distance Versus Acceleration Time (Simulation)

corresponding acceleration time T was obtained as shown in Fig. 5. This relation may be expressed approximately as follows:

$$X = Ka \cdot T^2 \qquad (2)$$

where Ka is a constant.

Because of the various factors discussed previously there are differences in disk drive units and the seek direction, making it difficult to quantify this relation unambiguously. In order to obtain the relation (2) in a form which includes these factors, we have developed a fuzzy logic trial seek method.

4.2 Trial seek

In the fuzzy logic trial seek method, operations are performed for a certain target distance X given an assumed bang-bang acceleration time. Based on the deviation dX of the distance actually moved from the target distance X and the general tendency of eq. (2), the most appropriate bang-bang acceleration time is determined.

The target distance X and the deviation dX from this distance X when using bang-bang seeking with a certain acceleration time T were taken as the input fuzzy logic variable, and the correction dT to the acceleration time was the output fuzzy logic variable.

The specific procedure follows. After the first trial seek operation, fuzzy logic is implemented based on the rules of Table 1 and the membership functions of Fig. 6 to calculate dT. Trial seeking is performed with the acceleration time T+dT. This procedure is repeated to obtain the bang-bang acceleration time T most appropriate for the target distance X.

In Fig. 6, the linguistic description PB (positive big) is represented by the membership function PB(dX), where the abscissa is an input value and the ordinate is the degree to which the input value can be classified as PB. In this figure, one cylinder is equal to one track pitch, and is equal to 14 micro-meter.

Table 1 Fuzzy Logic Rules for Trial Seeking

		dX						
		NB	NM	NS	ZR	PS	PM	PB
X	S	PB	PM/PB	PS	ZR	NS	NM/NB	NB
	M	PB	PM	PS	ZR	NS	NM	NB
	B	PB	PM/PS	PS	ZR	NS	NM/NS	NB

Because the trial seek routine interrupts other production processes, it should require as little time as possible. Other methods like gradient-based identification technique need more repetitions of trial seek in short distances than in long distances caused by the non-linearity of the X-T curve. Fuzzy trial seek could overcome this problem by tuning the membership function of dX, so that the peaks gather in small dX as shown in Fig. 6, and total repetitions become smaller than by the other methods.

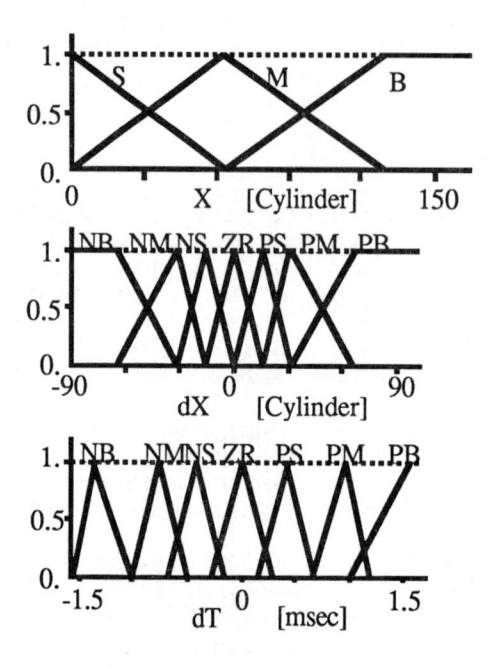

Fig. 6 Membership Functions for Fuzzy Logic in Trial Seeking

4.3 Experimental results

Figure 7 is a block diagram of the experimental system. The overall sequence control was performed with an 80C196 microcomputer, the fuzzy logic rules were designed and tuned on a personal computer, and data was downloaded to a fuzzy controller for operation.

Results of Trial Seek Figure 8 shows the relation between the target distance X and the acceleration time T as determined using fuzzy logic. The rules and membership functions of Table 1 and Fig. 6 were employed. Rule tuning was performed by trial and error, such that the repeated logic routine would converge after three or four executions, regardless of the target distance.

Average seek time was 9.1 msec including settling. This was a reduction of about 20 to 30% compared to the velocity profile method (when 60 to 75% deceleration profile was used, see Fig. 9).

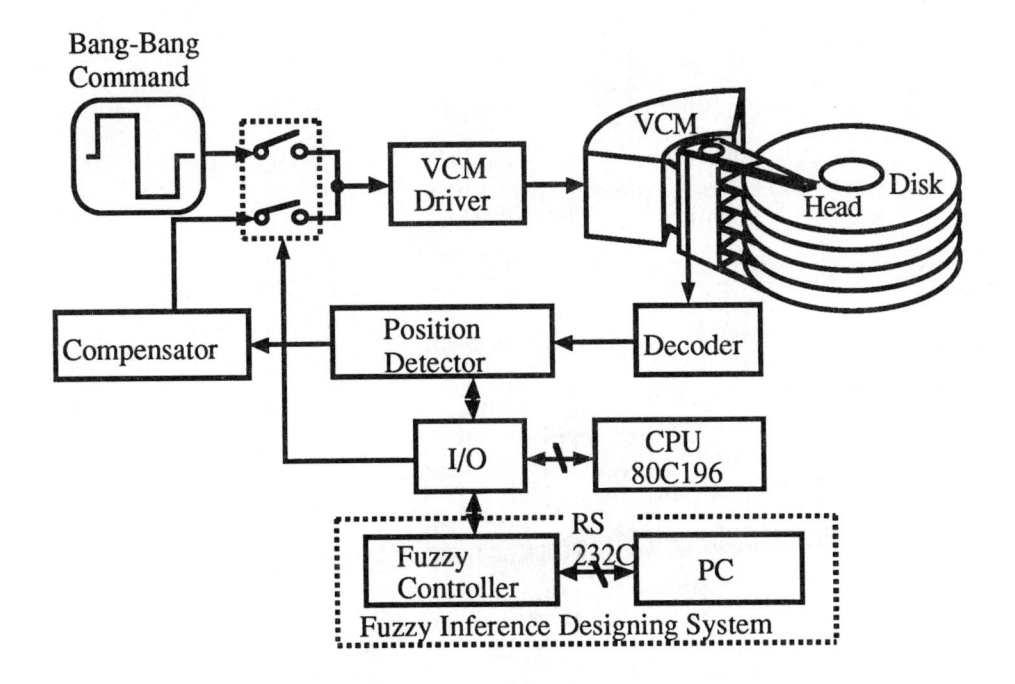

Fig. 7 Block Diagram of the Experimental System

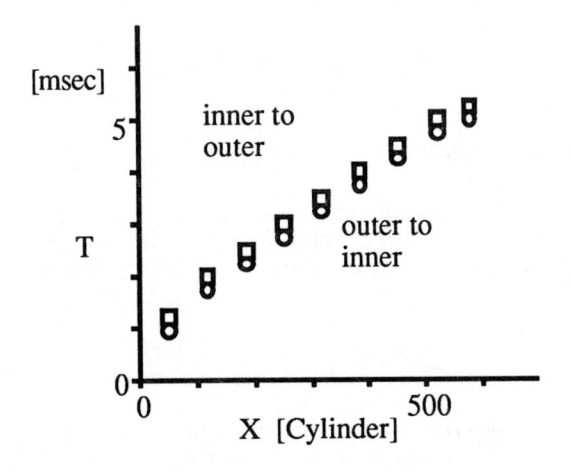

Fig. 8 Results of Trial Seeking

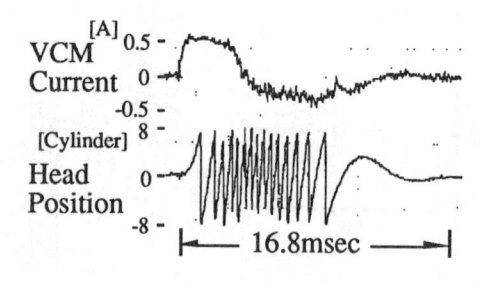

(a) Profile Seek

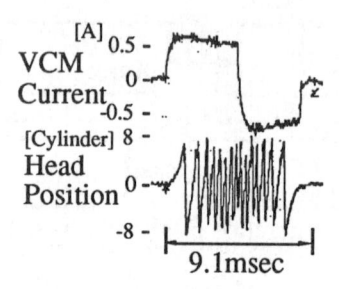

(b) Bang-Bang Seek

Fig. 9   Seek Wave Forms

5. SEEK TABLE REFERENCE

The bang-bang acceleration time corresponding to any given target distance is determined by interpolation based on the Figure 8, which is the relation between the target distance X and the acceleration time T, as determined using fuzzy logic in Section 4.

5.1 Correcting For Actuator Coil Resistance Due To Temperature Change [2]

In disk drive seek control, the temperature of the actuator coil increases when the frequency of seek increases, and the driving current decreases. Therefore the acceleration time must be increased accordingly.

Figure 10 shows the relation between bang-bang acceleration time T and actuator coil temperature T_c with the parameter of seek distance. Figure 11 shows the relation between seek distance and "acceleration time/coil temperature", which is the inclination of the straight line in Fig. 10 . Fig. 10 and Fig. 11 are both obtained from computer simulation.

If the relation between a seek distance and the most appropriate acceleration time is known in advance at the normal temperature (25°C), one can obtain an optimum acceleration for any coil temperature due to the relations of Fig. 10 and Fig. 11.

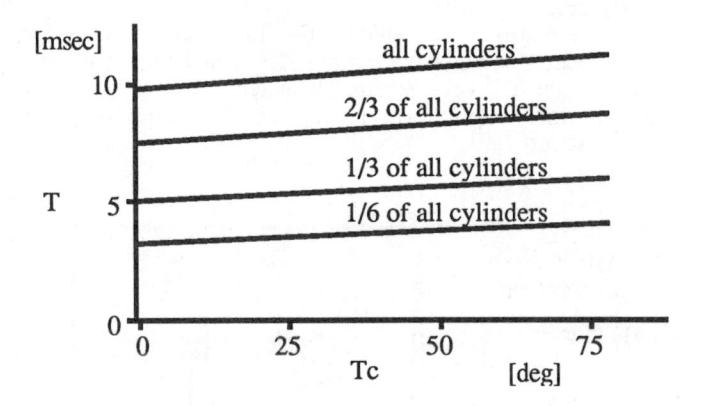

Fig. 10 Seeking Period and Coil Temperature

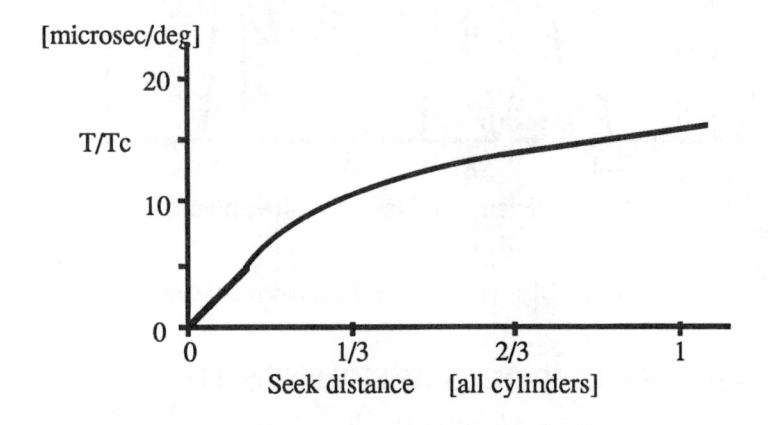

Fig. 11 Seeking Period/Coil Temperature and Seek Distance

5.2 Results Of Correction For Coil Resistance

In the experiment, continuous seek was performed for more than three minutes so that coil temperature became in a state of equilibrium, and the relation between temperature increase and optimum acceleration time was recorded , which is equivalent to the relation of Fig. 10. Based on this relation, the temperature increase was corrected. Coil temperature was measured by a thermo-couple.

Figure 12 shows distribution of seek distance deviation. In this figure, (a) shows the distribution before correction and (b) after correction for coil resistance. In both cases, seeks over 1/3 cylinders were performed 200 times each with an interval of 30msec. In the last half of these seeks, environmental temperature was gradually increased by five degrees.

The white-painted area of (a), which is distributed near -8 cylinder, is gone to that of (b), near 0 cylinder, due to the correction. The distance deviation is improved to within ±0.5 cylinders at the half value width. The hatched area of (a) shows that before correction the distribution becomes worse caused by temperature change, but after correction it is greatly improved.

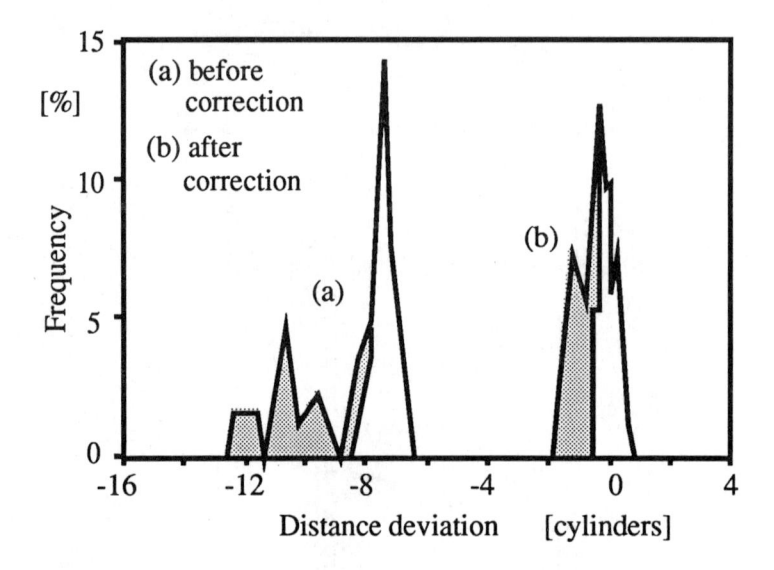

Fig. 12 Distribution of Seek Distance Deviation

6. REVISED SEEK TABLE REFERENCE [3]

The method of Section 5 requires software for the nonlinear interpolation operation used to determine the precise acceleration time corresponding to a desired target distance taken from the table of target distance versus acceleration time. In addition to this table, it also requires a thermal sensor for detecting the ambient temperature, a table corresponding to the temperature changes and the operation software to correct them.

It therefore creates a demand for an algorithm with smaller hardware resources, such as the thermal sensing unit and memory devices to store the software and the tables.

6.1 Acceleration Time Reference Method Employing Quasi-Fuzzy Inference

This technique, which differs from the method of directly interpolating the table of acceleration time, is intended to determine the acceleration time with the smallest deviation of moving distance, depending on the target distance, by employing an original method based on fuzzy logic. More specifically, by making up the rules describing the acceleration time corresponding to several representative target distances, and a membership function expressing the grade of the target distance corresponding to each rule; the acceleration time corresponding to an given target distance is determined by calculating the weighted mean. This is described in detail below.

In fuzzy control, it is a general practice to employ the "min-max composition method" proposed by Mamdani [6]. In our case, however, since the nonlinearity of the control object is very intense, and in order to simplify the calculation, the fuzzy variable in the action part is made a crisp value so as to render unnecessary the membership function in the action part. In this method of calculation, as representative values of target distance X, six points

$$X = 0, 1, 2, 4, 6, 9 \text{ [mm]}$$

are selected, and the acceleration times T corresponding to these points are preliminarily determined as

$$T = T_0, T_1, T_2, T_4, T_6, T_9 \text{ [msec]}$$

and the following inference forms are introduced.

$$\text{Rule 0 : If } X = X_0, \text{ then } T = T_0$$
$$\vdots$$
$$\text{Rule i : If } X = X_i, \text{ then } T = T_i$$
$$\vdots$$
$$\text{Rule 9 : If } X = X_9, \text{ then } T = T_9$$

$$\underline{\text{Fact} \quad : \quad X = X_a}$$

$$\text{Conclusion : } \quad T = T_a$$

Here, the membership function of the situation part with a representative value X (ex. X = 1 [mm]) is denoted as X_i (ex. $X_1 = 1$ [mm]), which is shown in Fig. 13.

The conclusion is determined as follows:

$$\text{Acceleration time } T(X_a) = \frac{m_0(X_a)T_0 + \cdots + m_i(X_a)T_i + \cdots + m_9(X_a)T_9}{m_0(X_a) + \cdots + m_i(X_a) + \cdots + m_9(X_a)} \quad (3)$$

where, $m_i(X_a)$: membership value of rule i with respect to X_a, (i = 0, 1, 2, 4, 6, 9).

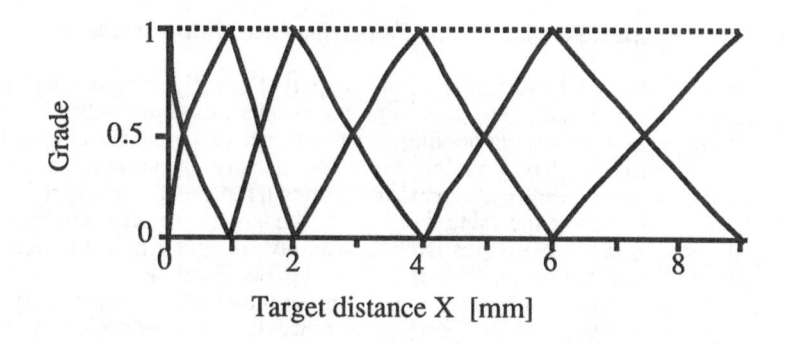

Fig. 13 Membership Functions

6.2 Temperature Correction Method For Coil Resistance Variation

Along with the temperature increase of the coil due to ambient temperature changes or seeking, the pure resistance of the coil varies, and the rule obtained in Section 6.1 cannot be applied.

There are minute temperature differences between the surface and interior of the coil, and it is therefore difficult to know the true value. Accordingly, for the following discussion we consider changes of magnitude of coil resistance as changes of ambient temperature.

According to the simulation, the relation between the acceleration time and coil resistance is as shown in Fig. 14. At each moving distance Xd, the acceleration time varies approximately linearly with respect to coil resistance.

(A) In Fig. 14, taking note of the discrete coil resistance as indicated by dots, each acceleration time in this case is assumed to be the value of Ti in the action part of the quasi-fuzzy inference rule in the case of each coil resistance, and the membership function shown in Section 6.1 is used.

(B) If the ambient temperature varies, using the same membership function as in the case of ordinary temperature (25°C), a quasi-fuzzy inference rule is composed. Without varying the profile of the membership function; only varying each Ti value depending on the resistance, a sufficiently small deviation of moving distance is obtained.

(C) As for an arbitrary coil resistance r, the Ti values for coil resistance values of r_1 and r_2, indicated by dots in Fig. 14, are linearly interpolated, and a rule corresponding to the case is made up.

$$T_i(r) = \frac{T_i(r_1) - T_i(r_2)}{r_1 - r_2}(r - r_2) + T_i(r_2) \tag{4}$$

where, $T_i(r)$: acceleration time supposing the coil resistance to be r   (i = 0, 1, 2, 4, 6, 9).

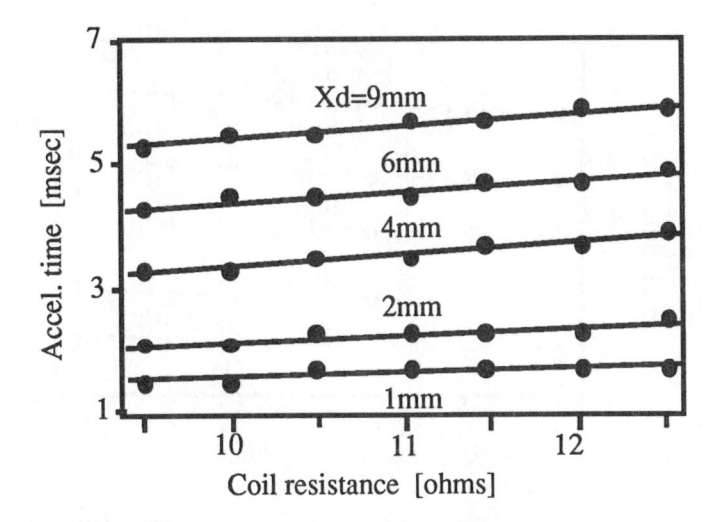

Fig. 14 Coil Resistance Versus Acceleration Time (Simulation)

6.3 Experimental Results

The experiment was carried out on the RDD using the head position detecting method with the dedicated servo system; however, it is essentially the same using the embedded servo system. (See Fig. 7.)

Results of quasi-fuzzy inference Figure 15 shows the experimental results of the Ti values in the action part of the quasi-fuzzy logic rules. The membership function was nearly equal to that obtained by simulation of Fig. 13.

Results of temperature correction Using the Ti values shown in Fig. 15, the temperature correction method in Section 6.2 is experimentally verified. For example, applying

$$r = 12.25 \text{ [ohms]}, r_1 = 12.0 \text{ [ohms]}, r_2 = 12.5 \text{ [ohms]}$$

into equation (4), the interpolation value of Ti is determined, and accordingly the quasi-fuzzy logic rule is composed. The target distance is set at every 0.1 mm in a range of $X_d = (0, 9)$ [mm], and the deviation of moving distance is determined by seeking. Fig. 16 is a histogram of the deviation of moving distance resulting from using the values. The distribution is a Gaussian-like profile with a mean of the deviation of moving distance of 0.4 tracks and a dispersion of not more than ±0.5 tracks.

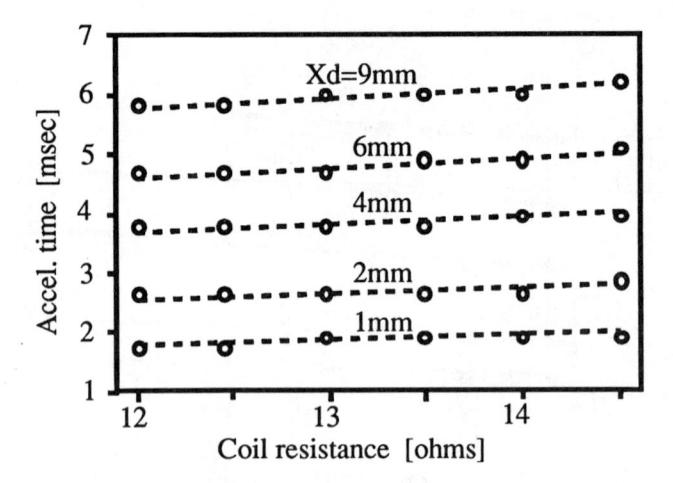

Fig. 15 Coil Resistance Versus Acceleration Time (Experiment)

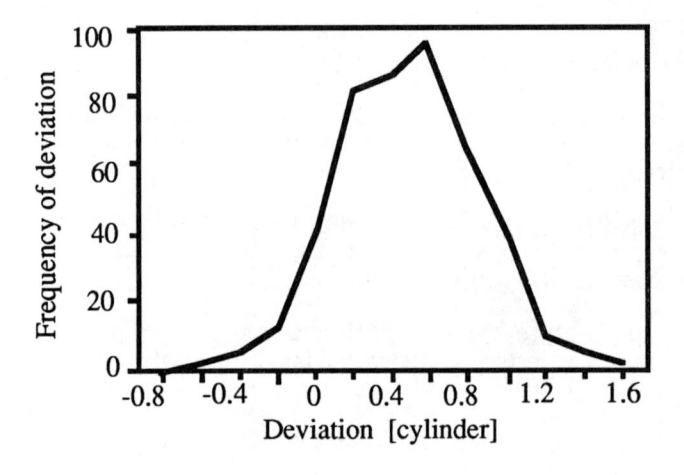

Fig. 16 Histogram of Deviation of Moving Distance

7. CORRECTING FORCE UNEVENNESS

Unevenness in the force near the limits of the actuator shaft rotation scatters considerably from one unit to another.(See Fig. 17.) Moreover, even if scattering can be confined to within a given range, the relative positional relationship between the moving assembly of the actuator and the disk must be controlled rigorously, if the flat part of the force characteristic and the range of motion of the actuator are to coincide. Therefore, the range of force unevenness is determined during trial seek operations, and when this range is contained in the region of bang-bang acceleration, the bang-bang acceleration time is corrected accordingly. In doing so, table look-up method needs massive memory as one has to memorize all the acceleration time for every start cylinder and every moving distance.

Fuzzy logic is used to determine the amount of correction corresponding to the extent to which the region of acceleration contains ranges of force unevenness.

The shape of the force unevenness is estimated in advance at different radial positions, and the result is recorded. Before seeking, fuzzy logic is used to determine the optimum bang-bang acceleration time based on these stored shapes of the uneven force characteristic. Here the unevenness correction rules are used to achieve a nonlinear correction function for the bang-bang acceleration command.

The fuzzy logic parameters for input are the starting position Xs and the acceleration time T ; the output fuzzy logic variable is the acceleration time correction dT . Fuzzy logic is executed based on the rules of Table 2 and the membership functions of Fig. 18 to determine dT, which is added to the acceleration time T .

Unevenness characteristics changes with drive units. But its shape can be classified into a few patterns. One can set rules for each pattern in designing of the actuator and select the nearest ones to a drive unit and re-tune them before shipment of the drive unit.

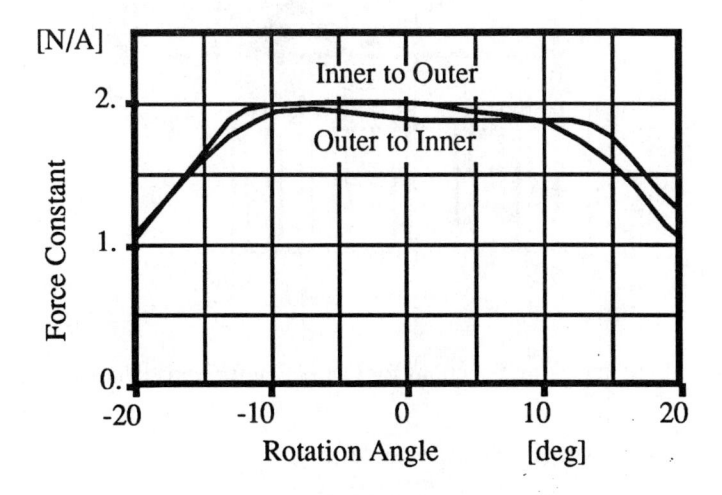

Fig. 17 Unevenness in the Force Constant

Table 2 Fuzzy Logic Rules for Correction for Force Unevenness

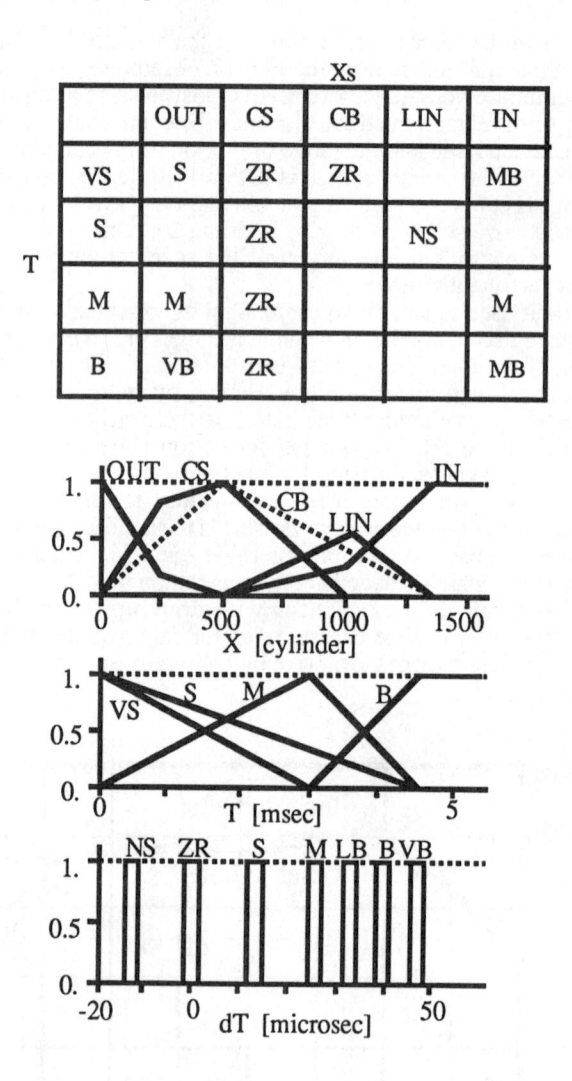

		\	Xs			
		OUT	CS	CB	LIN	IN
	VS	S	ZR	ZR		MB
T	S			ZR	NS	
	M	M	M	ZR		M
	B	B	VB	ZR		MB

Fig. 18 Membership Functions for Correction for Force Unevenness

7.1 Results Of Force Unevenness Correction

Figure 19 shows the relation between the correction in the bang-bang acceleration time dT for force unevenness and the starting position X_s, for the cases where the target distance is 1/6 the total number of cylinders and 1/3 the total number of cylinders. In the figure, the solid lines approximate the optimum acceleration time as determined in advance by experiment, while the discrete values are results obtained using fuzzy logic. There is good agreement between the experimental results and the results obtained by fuzzy logic.

The rules and membership functions of Table 2 and Figure 18 were used. Rule-tuning was performed by trial and error, with reference to the shape of the force unevenness.

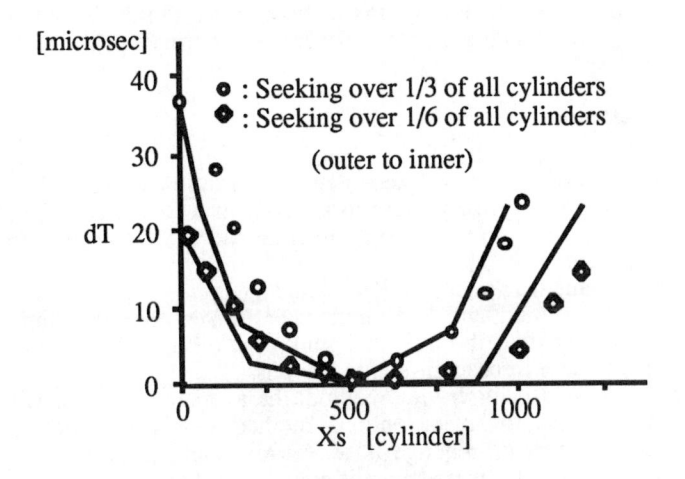

Fig. 19 Correction for Force Unevenness

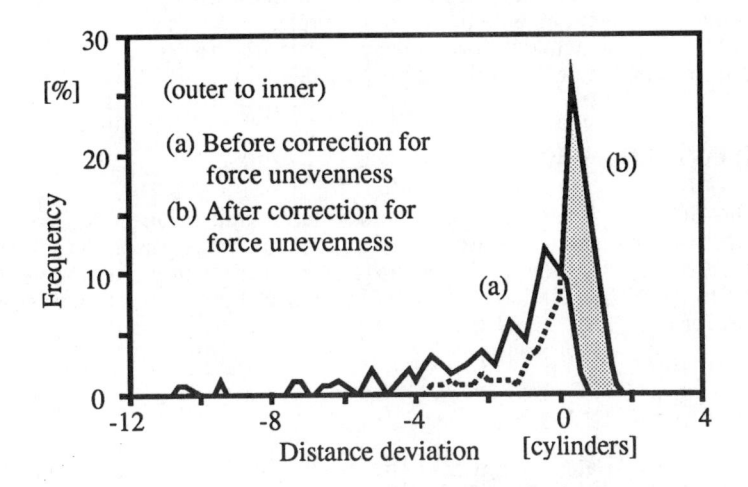

Fig. 20 Distribution of Seek Distance Deviation

Figure 20 shows the frequency distribution of the deviation of the seek distance from the target distance. In this figure, (a) shows the distribution before correction, and (b) after correction for unevenness in the force ; seeks over 1/24, 1/12, 1/6, 1/4 and 1/3 the total cylinders were performed 200 times total for the entire seek span of a disk. The deviations distributed between -10 and -2 cylinders are gone, there is a greater distribution at and near zero cylinders, and the scattering deviation is improved to within ±0.6 cylinders at the half value width. The distance deviations left between -3 and 0 cylinders are caused by miss-tuning of fuzzy rules (Table 2) and membership functions (Fig. 18). So re-tuning is needed for further improvement.

8. CONCLUSION

This chapter presents a novel seek method in a rigid disk drive which consists of bang-bang driving using fuzzy logic together with a method for correcting for actuator coil resistance due to temperature change and a method for correcting for actuator force unevenness.

In the bang-bang driving method, time management is used to switch between acceleration and deceleration, and fuzzy logic is employed both for estimation of the bang-bang acceleration time through trial seek operations, and for correction for unevenness in the actuator force at different positions.

By adopting bang-bang driving method, the average seek time is reduced by about 20 to 30% compared to the conventional method. In addition, correction for coil resistance due to temperature change improves distance deviation greatly and is effective even if environmental temperature is gradually changed.

A method of estimating an acceleration time is proposed by employing the quasi-fuzzy inference rule. The validity of this method is proven, and it has been shown experimentally that it is possible to cope with the changes of the actuator coil resistance due to temperature only by varying the rule.

Correction for actuator force unevenness using fuzzy logic also enables a marked improvement in the scattering of position deviations.

ACKNOWLEDGEMENT

The author wishes to express his sincere appreciation to Mr. Noriaki Wakabayashi (Matsushita Electric Ind. Co., Ltd.) and Professor Seiji Kobayashi (Waseda University) for their kind guidance and instruction in promoting the present study, and to Dr. Hideo Ishihara (Senshu University) for his enthusiastic assistance and suggestions.

REFERENCES

[1] Yoshida S. et al, "Bang-Bang seek control for HDD with fuzzy algorithm", *IEEE Translation Journal on Mag. in Japan*, Vol.6, No.3, pp. 227-239, Mar. 1991.

[2] Yoshida S. et al., "A fuzzy logic controller for a rigid disk drive",*IEEE Control Syst. Mag.*, vol.12, pp.65-70, June 1992.

[3] Yoshida S. et al., "Quasi-fuzzy inference bang-bang access servo", *Proc. of IECON '91*, vol.2, pp.1639-1644, Oct. 1991.

[4] Oswald R.K., "Design of a disk file head-positioning servo", *IBM Journal of Research and Development*, No.18, pp.506-512, 1974.

[5] Workman M.L. et al., "Adaptive proximate time-optimal servomechanisms: Continuous time case", *Proceedings of Automatic Control Conference*, pp.589-594, 1987.

[6] Mamdani E.H., "Applications of Fuzzy Algorithms for Control of Simple Dynamic Plant", *Proceedings of IEE*, Vol.121, No.12, 1585-1588, 1974.

AUTHOR'S BIOGRAPHICAL INFORMATION

Javier Aracil: Received the Ing. Industrial and the Dr. Industrial degrees from the University Politecnica de Madrid, Spain, in 1965 and 1969, respectively. He is also Licenciado en Informatica by the same university. Since 1969 he has been at the Escuela Tecnica Superior de Ingenieros Industriales de Sevilla. Dr. Aracil is council member of IFAC's System Dynamic Society, member of the Comite Scientifique of the *Revue Internationale de Systemique*, and the editorial board of the *International Journal of Systems and Policy Modeling*. In addition, he is an Associate Editor of *Automatica e Instrumentacion* (Journal of the IFAC Spanish Committee). Dr. Aracil received the 1986 Jay W. Forrester Award for his contributions to the qualitative analysis of system dynamic models, and the Research Award of Andalucia, Spain, in 1991. He is also a member of the Expert Committee of Expo'92 in Seville, and is organizing the Automation and Robotics Research Institute of Andalucia, Spain. Dr. Aracil is the author of numerous papers and the books: *Introduccion a la Dinamica de Sistemas* (1978) and *Maquinas, Modelos y Sistemas* (1988), and is co-author of *Practice of Integrated Automation* (1975). His research interests are in the areas of the theory and philosophy of systems modeling and control, with emphasis on the application of qualitative methods (stability, bifurcations, qualitative change, chaos, etc.).
Current Address: Dpto. Ingenieria Sistemas y Automatica, ETS Ingenieros Industriales, Avenida Reina Mercedes, 41012 Sevilla, Spain.

Karl Astrom: Was educated at the Royal Institute of Technology (KTH), Stockholm, where he has held various teaching appointments. He received Docteur Honoris Causa from l'Institut National Polytechnique de Grenoble in 1987. He has been Professor of Automatic Control at Lund Institute of Technology since 1965. Before that he worked for IBM and the Research Institute of National Defense in Stockholm. Professor Astrom's interests cover broad aspects of automatic control, stochastic control, system identification, adaptive control, computer control, and computer-aided control engineering. He has published five books, *Reglerteori* (in Swedish), *Introduction to Stochastic Control Theory, Computer Controlled Systems - Theory and Design* (co-author, B. Wittenmark), *Automatic Tuning of PID Controllers* (co-author, T. Hagglund), and *Adaptive Control* (co-author, B. Wittenmark). He has contributed to several other books and has written many papers. The paper "System Identification," *Automatica,* **7**, pp. 123-162, 1971 (co-authored with P. Eykhoff) is a Citation Classics. The paper "Theory and Application of Adaptive Control," *Automatica,* **19**, pp. 471-486, 1983, was given the Automatica Prize Paper Award, and he was recipient of the Donald G. Fink Prize Paper Award from the IEEE for his paper "Adaptive Feedback Control," *Proc. IEEE,* **75**, pp. 185-217, 1987. Professor Astrom has supervised 36 Ph.D. students and numerous M.Sc. students. He holds three patents. He is a Fellow of the IEEE, a member of the Royal Swedish Academy of Sciences, the Swedish Academy of Engineering Sciences (IVA), and the Royal Physiographical Society. He has received several awards among them the Rufus Oldenburger Medal, the Quazza medal, the IEEE Control Systems Science and Engineering Award, and the IEEE Medal of Honor.
Current Address: Department of Automatic Control, Lund Institute of Technology, Lund, Sweden.

Adrian Brezulianu: Graduated from the Faculty of Electronics and Telecommunications, Isai, Romania, in 1992, with the diploma thesis on fuzzy control of analog phase-locked loop. He is an Assistant Professor in the Faculty of Electronics and Telecommunications, Technical University of Isai. Mr. Brezulianu is a member of the Balkanic Union for Fuzzy Systems and Artificial Intelligence. In 1992 he was awarded the Grigore and George Moisil Foundation grant.
Current Address: Faculty of Electronics and Telecommunications, Technical University of Isai, Av. Copou, No. 11, Isai 6000, Romania.

David J. Comer: Received the B.S. degree from San Jose State University in 1961, the M.S. degree from the University of California (Berkeley) in 1962, and the Ph.D. degree from Washington State University in 1966, all in Electrical Engineering. From 1959 to 1964 he was also employed by IBM in San Jose, California. While at IBM, Dr. Comer was involved in research in the areas of automatic speech recognition and automated guided vehicles for warehousing. He received two patents and had three other inventions published for protection during his work at IBM. In 1964, Dr. Comer became an Assistant Professor at the University of Idaho. He moved to the University of Calgary as an

associate Professor in 1966. In his three years at the University of Calgary, Dr. Comer received research funds from the National Research Council and the Defense Research Board. He published two textbooks, one with Prentice-Hall and one with Addison-Wesley, and published a dozen research articles. Dr. Comer accepted a position as Professor and Chairman (Dean), Division of Engineering, California State University, Chico, in 1969. While in this administrative position, he published three research papers and three textbooks. He assumed the position of Professor of Electrical and Computer Engineering at Brigham Young University in 1981, and was named, in 1990, Chairman of the Electrical and Computer Engineering Department. Returning to a research environment, he established research programs in electronic circuits and robotics. He has published a dozen research articles and four more textbooks while at Brigham Young University. In 1984, Dr. Comer was awarded the Research Chair for the College of Engineering and Technology at Brigham Young University. During this year, his research was concerned with the application of fuzzy logic to the control of a mobile robot. Dr. Comer has published nine textbooks and over 30 articles in the fields of circuit synthesis, computer-aided circuit design, digital system design, and fuzzy logic control theory. He is credited with the development of the loop-gain modulator and the high-frequency, narrowband active filter based on all-pass networks. His present areas of research are circuit synthesis and fuzzy logic control. Dr. Comer is a Senior Member of The IEEE.
Current Address: Electrical and Computer Engineering Department, 435 Clyde Building, Brigham Young University, Provo, UT 84602.

Alfonso García-Cerezo: Received the Ind. Elec. Eng. and the Doctoral Eng. degrees from the Escuela Tecnica Superior de Ingenieros Industriales of Vigo, University of Santiago, Spain, in 1983 and 1987, respectively. From 1983 to 1988 he was Associate Professor in the Department of Electrical Engineering, Computer and Systems at the University of Santiago de Compostela, and from 1988 to 1991 he was Assistant Professor at the same university. Since 1992 he has been a Professor of System Engineering and Automation at the University of Malaga, Spain, and Head of the Department of System Engineering and Automation. Dr. Garcia-Cerezo is a member of the Automation and Robotics Research Institute of Andalucia. He has authored or co-authored about 40 journal articles, conference papers, book chapters, and technical reports. His current research interests include the applications of artificial intelligence techniques to real-time control and supervision of industrial processes and robotics.
Current Address: Dpto. Ingenieria Sistemas y Automatica, Universidade de Malaga, Plaza El Ejido, 29013 Malaga, Spain.

Edward J. Gentry: Received the B.S. degree in Computer Science from the University of Alabama in 1992. Prior to graduation, he worked as a computer assistant at the U.S. Bureau of Mines, Tuscaloosa Research Center, where he developed inventive approaches to combining various artificial intelligence techniques. Currently he works for SEER Technology as a software developer. His interests include genetic algorithms, fuzzy logic, and event driven object oriented programming.

Current Address: SEER Technology, 8000 Regency Parkway, Cary, NC 27511.

Pierre Yves Glorrenec: Received the Ph.D. in Mathematics, probability theory, in 1977. From 1967 to date, he has been teacher in Mathematics and Computer Sciences at the Institut National des Sciences Appliquees, an engineering school, in Rennes, France. His research interests are fuzzy logic, neural networks, and genetic algorithms, with applications to building energy management and adaptive control.
Current Address: Department d'Informatique, Institut National des Sciences Appliquees, 35043 Rennes Cedex, France.

Janos Grantner: Received his M.S. and Ph.D. in Computer Engineering in Electrical Engineering from the Technical University of Budapest, Hungary, in 1971 and 1979, respectively. From 1990 to 1992 he was Visiting Associate Professor at the Department of Electrical Engineering of the University of Minnesota, Twin Cities. He also held a Visiting Assistant Professorship at the Department of Electrical and Computer Engineering of Syracuse University. Previously, he served as Research Fellow, later progressing to the position of Lecturer, and finally as Senior Lecturer and Head of the Hardware Section at the Department of Process Control of the Technical University of Budapest. Dr. Grantner's main research areas include not only parallel algorithms and architectures for designing fuzzy logic hardware accelerators, but also the design of finite state machines based on fuzzy logic, computer-controlled microcomputer development systems, and multi-microprocessor systems for real-time process control.
Current Address: Department of Electrical Engineering, University of Minnesota, Minneapolis, MN 55455.

Elena Agüero Gutierrez: Received her degree in Electronic Engineering from the National University of San Juan (UNSJ), Argentine, in 1984, and worked as Research Assistant at the Automation Institute, UNSJ, in the field of automatic control. In 1988 she joined the Industrial Automation Institute where she is working on the application of artificial intelligence techniques to manufacturing systems. Her current work is related to fuzzy control applied to machining processes.
Current Address: Instituto de Automatica, Univ. Nacional de San Juan, Argentina.

Lawrence O. Hall: Received a B.S. in Applied Mathematics from Florida Institute of Technology in 1980, and the Ph.D. in Computer Science from Florida State University in 1986. He is an Associate Professor of Computer Science and Engineering at the University of South Florida. His current research in artificial intelligence is in parallel algorithms, hybrid connectionist, symbolic learning models, expert systems, and the use of fuzzy sets and logic for uncertainty handling. Dr. Hall has written over 50 research papers and co-authored one book.
Current Address: Department of Computer Science and Engineering, University of South Florida, Tampa, FL 33620.

Isao Hayashi: Received the Dr.Eng. degree in Industrial Engineering from the University of Osaka Prefecture, Osaka, Japan. In 1987 he joined the Matsushita Electric Industrial Co., Osaka, where he is currently a Senior Researcher in the Central Research Laboratories. His primary research interests are fuzzy logic and neural networks, and their fusion technology, and fuzzy retrieval. Dr. Hayashi is a member of the International Fuzzy Systems Association, the Japan Society for Fuzzy Theory and Systems, and the Institute of Electrical Engineers of Japan.
Current Address: Central Research Laboratories, Matsushita Electric Industrial Co. Ltd., 3-1-1 Yagumo-Nakamachi, Moriguchi, Osaka 570, Japan.

Masaaki Ida: Received his B.S. and M.S. degrees in Systems Engineering from Kyoto University, Japan, in 1988 and 1990, respectively. Currently, he is an Instructor at the Department of Precision Mechanics, Kyoto University, engaged in research on systems engineering and artificial intelligence, particularly in the field of fuzzy logic and decision theory.
Current Address: Department of Precision Mechanics, Kyoto University Sakyo-ku, Kyoto 606-01, Japan.

Tadashi Iokibe: Received his B.S. degree from Osaka Institute of Technology. He is the Manager of the System Development Division of Meidensha Corporation. He received the 39th OHM prize. His areas of interest are control and prediction by fuzzy and chaotic logic. Mr. Iokibe is a member of the Japan Society for Fuzzy Theory and Systems, the Society of Instrument and Control Engineers of Japan, and the Institute of Electrical Engineers of Japan.
Current Address: Computer Systems Division, Engineering Operations, Meidensha Corporation, 5-5 Ohsaki 5-Chome, Shinagawa-ku, Tokyo 141, Japan.

Sosuke Iwai: Received his B.S., M.S., and Ph.D. degrees in Electrical Engineering from Kyoto University, Japan. In 1956 he was with Mitsubishi Heavy Industries Ltd. He returned, in 1957, as a Lecturer to the Department of Electrical and Electronics Engineering, Kyoto University. In 1965 he moved to the Department of Precision Mechanics at Kyoto University as an Associate Professor. Currently, Dr. Iwai is Professor at the Department of Precision Mechanics. He has been engaged in research and teaching on automatic control, medical electronics, systems engineering, and knowledge information processing.
Current Address: Department of Precision Mechanics, Kyoto University Sakyo-ku, Kyoto 606-01, Japan.

Yuji Kajitani: Received the B.E. degree in Electrical Engineering from Tottori University, Tottori, Japan, in 1983. In 1983, he joined the Central Research Center of Sanyo Electric Co. Ltd., and was engaged in research on signal processing, expert systems, fuzzy theory, and neural networks. He is now Chief Researcher at Sanyo's

Information and Communication Systems Research Center. Mr. Katjitani is a member of the Information Processing Society of Japan. His current research interests are fuzzy modeling algorithms, and neuro and fuzzy models.
Current Address: Sanyo Electric Co., Ltd., Information and Communication Research Center, Computer Department, AI and Systems Laboratory, 1-18-13 Hashiridani, Hirakata, Osaka 573, Japan.

Hiroaki Kaminaga: Received the B.S., M.S., and D.E. degrees from Tohoku University, Sendai, Japan, in 1979, 1983, and 1989, respectively. He did research on LOTOS specification as an assistant in the Computer Center of Tohoku University. Currently he is a Lecturer at Yamagata University, Yonezawa, Japan. He is a member of the IEEE. Dr. Kaminaga's current research interests are fuzzy theory and image processing.
Current Address: Department of Electronic Engineering, Yamagata University, 4-3-16 Jonan, Yonezawa 992, Japan.

Abraham Kandel: Received the B.Sc. in Electrical Engineering from the Technion, Israel Institute of Technology, the M.Sc. from the University of California, and the Ph.D. in Electrical Engineering and Computer Science from the University of New Mexico. Dr. Kandel is the Chair and Endowed Eminent Scholar of the Computer Science and Engineering Department at the University of South Florida. Previously, he was Professor and Chairman of the Computer Science Department at Florida State University as well as the Director of the Institute of Expert Systems and Robotics at FSU and the Director of the State University System Center for Artificial Intelligence. He is a Fellow of the IEEE, a member of the ACM, and an Advisory Editor to the international journals *Fuzzy Sets and Systems, Information Sciences, Expert Systems,* and *Engineering Applications of Artificial Intelligence.* Dr. Kandel has published over 250 research papers for numerous professional journals in Computer Science and Engineering. He is co-author of *Fuzzy Switching and Automata: Theory and Applications* (1979); author of *Fuzzy Techniques in Pattern Recognition* (1982); co-author of *Discrete Mathematics for Computer Scientists* (1983), and *Fuzzy Relational Databases - A Key to Expert Systems* (1985); author of *Fuzzy Mathematical Techniques with Applications* (1986); co-author of *Designing Fuzzy Expert Systems* (1986), and *Digital Logic Design* (1988); co-editor of *Engineering Risk and Hazard assessment* (1988), and *Hybrid Architectures for Intelligent Systems* (1992); co-author of *Elements of Computer Organization* (1989), and *Real-Time Expert Systems Computer Architecture* (1991).
Current Address: Department of Computer Science and Engineering, University of South Florida, Tampa, FL 33620.

Charles L. Karr: Received the Ph.D. degree in Engineering Mechanics from the University of Alabama in 1989. Since that time, he has been investigating the use of artificial intelligence techniques for process control in the minerals industry. Currently, Dr. Karr is a mechanical engineer with the U.S. Bureau of Mines, Tuscaloosa Research Center. He leads a team of researchers who are implementing innovative process control

strategies in industrial settings, specifically in mineral processing plants. His interest include genetic algorithms, fuzzy logic, equipment design, and fluid dynamics.
Current Address: U.S. Bureau of Mines, Tuscaloosa Research Center, Tuscaloosa, AL 35486-9777.

Osamu Katai: Received the B.S., M.S., and Ph.D. degrees in Precision Mechanics from Kyoto University, Japan. He was an Instructor at the Department of Precision Mechanics, Kyoto University, in 1974. From 1980 to 1981 he was a Visiting Scholar at INRIA (National Institute for Research on Information and Automation Technologies), France, where he was engaged in research on stochastic control and logics of time. Currently, Dr. Katai is an Associate Professor at the Department of Precision Mechanics, Kyoto University. His research interests are in self-organizing mechanisms in natural and artificial systems.
Current Address: Department of Precision Mechanics, Kyoto University Sakyo-ku, Kyoto 606-01, Japan.

Tetsuji Kataoka: Received the B.S. degree in Precision Mechanics from Kyoto University, Japan, in 1992. Presently he is a graduate student in the Department of Precision Mechanics at Kyoto University. His current field of interest is in nonlinear conceptual design optimization using genetic algorithms.
Current Address: Department of Precision Mechanics, Kyoto University Sakyo-ku, Kyoto 606-01, Japan.

Ryu Katayama: Received the B.E. and M.E. degrees in Instrumentation Engineering from Keio University, Yokohama, Japan, in 1979 and 1981, respectively. He was engaged in the study of optimal control, mathematical programming, and optimization theory. In 1981, he joined the Central Research Center of Sanyo Electric Co. Ltd., and was engaged in research on signal processing, expert systems, fuzzy theory, and neural networks. He is now a Chief Researcher at Sanyo's Information and Communication Systems Research Center. Mr. Katayama is a member of the Information Processing Society of Japan, the Institute of Electronics, Information and Communication Engineers, the Society of Instrumentation and Control Engineers, and Japan Society for Fuzzy Theory and Systems. His current research interests are fuzzy modeling algorithms, neuro and fuzzy models, identification of dynamical systems such as chaos, and adaptive control.
Current Address: Sanyo Electric Co., Ltd., Information and Communication Research Center, Computer Department, AI and Systems Laboratory, 1-18-13 Hashiridani, Hirakata, Osaka 573, Japan.

Shigeyasu Kawaji: Received the M.E. degree in Electrical Engineering from Kumamoto University, Kumamoto, in 1969, and the Dr. Eng. degree from Tokyo Institute of Technology in 1980. He joined the Department of Electrical Engineering, Kumamoto University in 1969 as a Research Associate. From 1975 to 1976 he was a Visiting Researcher of JSPS in the Department of Control Engineering at Tokyo Institute of

Technology. Since 1988 he has been a Professor at the Department of Electrical Engineering and Computer Science of Kumamoto University. Dr. Kawaji's research interests are linear control theory, descriptor systems, robust control with applications to robotics. Currently, his investigation areas are concerned with intelligent control, fuzzy control, and neural networks.
Current Address: Department of Electrical Engineering and Computer Science, Kumamoto University, Kurokami 2-39-1, Kumamoto 860, Japan.

Takashi Kimura: Received the B.S. degree from Nagoya Institute of Technology. He is pursuing computer and fuzzy engineering in the Computer System Division of Meidensha Corporation. He is a member of Japan Society for Fuzzy Theory and Systems.
Current Address: Meidensha Corporation, 5-5 Ohsaki 5-Chome Shinagawa-ku, Tokyo 141, Japan.

T. Kitamura: Received the B.S. degree in Mechanical Engineering from Waseda University, Tokyo, Japan, in 1973, and the M.S. and Ph.D. degrees from Kyoto University in 1975 and 1981, respectively. From 1984 to 1986 he was an Assistant Professor at the University of Houston. Currently, he is a Professor at the Kyushu Institute of Technology. Dr. Kitamura is a member of the American Society of Mechanical Engineering, the Japan Society of Mechanical Engineering, the International Society for Artificial Organs, and the Japanese Society of Life Support Technology. His main fields of interest are applications of system control theory, and intelligent control and estimation for artificial hearts.
Current Address: Department of Mechanical System Engineering, Kyushu Institute of Technology, 680-4 Kawazu, Iizuka-City, Fukuoka 820, Japan.

Shinichi Kohno: Received the B.S. degree in Precision Mechanics from Kyoto University, Japan, in 1992. Presently he is a graduate student in the Department of Aeronautic Engineering at the University of Tokyo. His current field of interest is in the analysis support system for FEM using knowledge information processing techniques.
Current Address: Department of Aeronautic Engineering, University of Tokyo Minato-ku, Tokyo 153, Japan.

Ladislav J. Kohout: Graduated in Computing and Electronic Engineering at the Czech Technological University, Prague. He also studied theoretical physics in a postdoctoral course (part-time) at Charles University, Prague, and obtained a Ph.D. degree in Man-Computer Studies from the University of Essex, United Kingdom. Currently, Dr. Kohout is Professor of Computer Science at the Florida State University, where he is also a member of the Institute for Cognitive Sciences. Until 1968 he was Deputy Head of Computing Department, Institute of Astronomy, Czechoslovak Academy of Science. In 1974-1979 he was a lecturer in Medical Computing and Bio-Engineering at University College (Medical school), University of London. While there, he was also a technical director of a large Nuffield Foundation sponsored project "Psychopharmacology of the Elderly," a member of the Psychopharmacology Working Party, and a honorary lecturer in

Geriatric Medicine. From 1979 to 1988, he was a Reader in Computer Science at Brunel University. Dr. Kohout is the US editor of the *Journal of Intelligent Systems*. He has published over 150 papers. His recent books include *A Perspective on Intelligent Systems: Framework for Analysis and Design* (1990), for which he received a prize for "the best book of the year in AI" from the International Institute for Advanced Studies in Systems Research and Cybernetics. His latest book (jointly with J. Anderson and W. Bandler) is a monograph entitled *Knowledge-Based Systems for Multiple Environments* (1992). Dr. Kohout's research interests include fuzzy sets and systems, many-valued and non-classical logics, artificial intelligence, knowledge engineering, and design of knowledge-based systems. He is also interested in parallel relational computer architectures, computer protection, general systems methodology, clinical computing, neuro-computational theories, and the history, conceptual foundation and formalization of medieval logics and semiotics. He is a member of several professional societies, a Charted Engineer registered with the British Council of Engineering Institutions, member of the Board of Directors of the International Institute for Advanced Studies in Systems Research and Cybernetics, and a Fellow of the Cybernetics Society. In addition to the previously received awards, Dr. Kohout recently received an international prize from the Systems Research Foundation, the *Outstanding Scholarly Contribution Award* "in recognition of his high quality research and life long scholarly contribution to the development of Systemic Constructs and Methodology of Activity Structures."
Current Address: Department of Computer Science, Florida State University, Tallahassee, FL 32306-4019.

M. Komori: Received the B.S. and M.S. degrees in Control Engineering from Osaka University, Toyonaka, Japan, in 1982 and 1984, respectively. He was with the Center for MAGLEV, Railway Technical Research Institute, Japanese National Railways. Currently he is with the Department of Mechanical System Engineering, Kyushu Institute of Technology. He is a member of the Japan Society of Mechanical Engineering, the Magnetic Society of Japan, and the Cryogenic Association of Japan. At present, his main fields of interests are control engineering and cryogenic engineering.
Current Address: Department of Mechanical System Engineering, Kyushu Institute of Technology, 680-4 Kawazu, Iizuka-City, Fukuoka 820, Japan.

Vladik Kreinovich: Received the M.S. degree in Mathematics and Computer Science from Leningrad University in 1974, and the Ph.D. degree from Novosibirsk Institute of Mathematics in 1979. He joined the Institute of Mathematics, USSR Academy of Sciences, in 1975. He worked on computational problems of radio astronomy and SETI for the Special Astrophysical Observatory in 1978-1980, followed by research on mathematical and computational aspects in automated education for the Laboratory for Psychological Systems in 1980-1982. In 1982-1989 he worked on error estimation and intelligent information processing for the Institute of Electromeasuring Devices. Dr. Kreinovich was a Visiting Scholar at Stanford University in 1989, and joined the University of Texas at El Paso as an Associate Professor in Computer Science. His current

research is in artificial intelligence, dealing with representation and processing of uncertainty.
Current Address: Computer Science Department, University of Texas, El Paso, TX 79968.

Benjamin Kuipers: Received the B.A. in Mathematics from Swarthmore College in 1970, and the Ph.D. in Mathematics from MIT in 1977. He has held research or faculty appointments at MIT, Tufts University, and the University of Texas. Dr. Kuipers is currently David Bruton Centennial Professor in Computer Sciences at the University of Texas at Austin. He investigates the representation of commonsense and expert knowledge, with particular emphasis on the effective use of incomplete knowledge and limited computational resources. His research accomplishments include developing the TOUR model of spatial knowledge in the cognitive map, the QSIM algorithm for qualitative simulation, Access-Limited Logic for knowledge representation, and a robot exploration and mapping strategy based on qualitative recognition of distinctive places. Dr. Kuipers is a Fellow of AAAI and is currently a member of its Executive Council.
Current Address: Department of Computer Sciences, University of Texas at Austin, Austin, TX 78712-1188.

Reza Langari: Received the Ph.D. degree from the University of California at Berkeley where he worked closely with Professor Lotfi Zadeh. He has held consulting positions with companies active in the manufacturing process control area such as Measurex Corporation and Firestone Tire & Rubber Company. Currently, Dr. Langari is an Assistant Professor in the Department of Mechanical Engineering and Deputy Director of the Center for Fuzzy Logic and Intelligent Systems Research at Texas A&M. His expertise is in the area of fuzzy information processing and control, nonlinear and adaptive control systems, and computing architecture for real-time control. He has published a number of articles on analysis, design, and implementation of fuzzy logic control systems.
Current Address: Department of Mechanical Engineering, Texas A&M University, College Station, TX 77843-3123.

Gideon Langholz: Received the B.Sc. in Electrical Engineering from the Technion, Israel Institute of Technology, and the Ph.D. degree from the University of London, both in Electrical Engineering. He is Professor of Electrical Engineering at Florida State University and at Tel-Aviv University. He is a Senior Member of the IEEE and a member of the editorial board of the international journal *Engineering Applications of Artificial Intelligence*. Dr. Langholz has authored or co-authored over 60 research papers in Electrical Engineering for various professional journals. He is co-author of *Digital Logic Design* (1988) and *Elements of Computer Organization* (1989), and is co-editor of *Hybrid Architectures for Intelligent Systems* (1992).
Current Address: Department of Electrical Engineering - Systems, Tel-Aviv University, Tel-Aviv 69978, Israel.

José Ramón Alique López: Received his B.S. degree in Physics (Electronics) from Madrid University, Spain, and then worked for three years for Philips (Eindhoven, Holland) and as teacher at Madrid University. He received his B.S. degree in Informatics from Polytechnic University of Madrid, and his Ph.D. degree in 1973 from Bilbao University with specialization in switching theory. His dissertation topic was digital circuits optimization. In 1973 he joined the Industrial Automation Institute at the National Research Council (CSIC). His research interests include control applications to manufacturing processes and mechanical systems, particularly computer control of manufacturing systems. He is currently working on the application of artificial intelligence to optimization of the machining process (intelligent machining).
Current Address: Instituto de Automatica Industial, CSIC, 28500 Arganda del Rey, Madrid, Spain.

Mahmoud A. Manzoul: Received the B.Sc. degree in Electrical Engineering from the University of Khartoum, Sudan, in 1977. He received the M.S. and Ph.D. degrees in Electrical Engineering from West Virginia University, Morgantown, in 1981 and 1985, respectively. Since 1985, he has been with the Electrical Engineering Department of Southern Illinois University, Carbondale. At present, he is an Associate Professor of Electrical Engineering. His current research interests include fuzzy logic, array processors, semi-custom VLSI, and applications of microprocessors. Dr. Manzoul is a member of the Institute of Electrical and Electronics Engineers (IEEE), the Association for Computing Machinery (ACM), the International Neural Network Society (INNS), the North American Fuzzy Information Processing Society (NAFIPS), and the American Society for Engineering Education (ASEE).
Current Address: Department of Electrical Engineering, Southern Illinois University, Carbondale, Illinois 62901-6603.

Nobutomo Matsunaga: Received the B.E., M.E., and Ph.D. degrees from Kumamoto University, Japan, in 1985, 1987, and 1993, respectively. From 1987 he has been employed by Omron Corporation, Japan. His current research area is design of intelligent control systems with applications to factory automation. In particular, Dr. Matsunaga is interested in hierarchically structured systems, using fuzzy control theory and neural networks.
Current Address: Department of Electrical Engineering and Computer Science, Kumamoto University, Kurokami 2-39-1, Kumamoto 860, Japan.

Masaharu Mizumoto: Received the B.Eng., M.Eng., and Dr.Eng. degrees in Electrical Engineering from Osaka University, Japan, in 1966, 1968, and 1971, respectively. From 1980 to 1981 he was an Alexander von Humboldt Foundation Fellow at the Technical University of Aachen, Germany. From 1989 to 1991 he was Vice President of the International Fuzzy Systems Association (IFSA). Currently, Dr. Mizumoto is a Professor at the Division of Information and Computer Sciences, Graduate School of Engineering,

Osaka Electro-Communication University, Osaka, Japan. He is a Director of Japan Society for Fuzzy Theory and Systems (SOFT) and an Editor-in-Chief of the *Journal of SOFT*. He is an Advisory Editor of the international journal *Fuzzy Sets and Systems*. His current research interests include approximate reasoning and its applications to fuzzy control methods, neural networks, and fuzzy expert systems.

Current Address: Division of Information and Computer Sciences, Osaka Electro-Communication University, Neyagawa, Osaka 572, Japan.

Yoshiteru Nakamori: Received the B.S., M.S., and Ph.D. degrees, all in Applied Mathematics and Physics, from Kyoto University, Japan, in 1974, 1976, and 1980, respectively. He has served in the Department of Applied Mathematics, Konan University, as an assistant Professor from 1981 to 1986, an associate Professor from 1986 to 1991, and a Professor since 1991. From September 1984 to December 1985 he stayed at the International Institute for Applied Systems Analysis, Laxenburg, Austria, where he joined the Regional Water Policies Project. Dr. Nakamori is a member of the Society of Instrument and Control Engineers of Japan, the Institute of Systems, Control and Information Engineers, the Japan Society for Fuzzy Theory and Systems, The Japan Association of Simulation and Gaming, the Society of Environmental Science of Japan, and the IEEE. His fields of research interest cover identification and measurement optimization of distributed parameter systems, modeling and control of environmental systems, and methodology and software of decision support systems.

Current Address: Department of Applied Mathematics, Konan University, 8-9-1 Okamoto, Higashinada-ku, Kobe 658, Japan.

Mikio Nakatsuyama: Received the B.S., M.S., and Ph.D. degrees in Electrical Engineering from Tokyo Institute of Technology, Tokyo, Japan, in 1955, 1957, and 1965, respectively. He is currently a Professor of Electronic Engineering at Yamagata University, Yonezawa, Japan. He is a member of the IEEE and the ACM. Dr. Nakatsuyama is engaged in research of chopper circuits for the precise operational amplifier, wave transformation from high frequency to very low frequency, processors for fast Walsh-Hadamard transformation, and fuzzy approximate reasoning. His current interest is in fuzzy control.

Current Address: Department of Electronic Engineering, Yamagata University, 4-3-16 Jonan, Yonezawa 992, Japan.

Yukiteru Nishida: Received the B.E. degree in Electrical Engineering from Kyoto University, Kyoto, Japan, in 1971. In 1971, he joined the Central Research Center of Sanyo Electric Co. Ltd., and was engaged in research on natural language processing, machine translation, etc. He is now Manager of the AI and Systems Laboratory at Sanyo's Information and Communication Systems Research Center. He is a member of the Information Processing Society of Japan and the Japanese Society for Artificial Intelligence.

Current Address: Sanyo Electric Co., Ltd., Information and Communication Research Center, Computer Department, AI and Systems Laboratory, 1-18-13 Hashiridani, Hirakata, Osaka 573, Japan.

Brent R. Nokleby: Received the B.S. and M.S. degrees in Electrical Engineering in 1987 from Brigham Young University. He is presently completing his Ph.D. degree in Electrical Engineering at Brigham Young University with a research emphasis in fuzzy logic control. Other research interests include autonomous robot navigation. During the summer of 1985, he was employed by Hughes Aircraft in Fullerton, California.
Current Address: Electrical and Computer Engineering Department, 435 Clyde Building, Brigham Young University, Provo, UT 84602.

Hiroyoshi Nomura: Received the B.S. degree in Control Engineering from Kyushu Institute of Technology, Fukuoka, Japan. In 1986 he joined the Matsushita Electric Industrial Co., Osaka, where he is currently a Researcher in the Central Research Laboratories. He is engaged in research and development in fuzzy logic, neural networks, and genetic algorithms. Mr. Nomura is a member of the Institute of Systems, Control and Information Engineering.
Current Address: Central Research Laboratories, Matsushita Electric Industrial Co. Ltd., 3-1-1 Yagumo-Nakamachi, Moriguchi, Osaka 570, Japan.

Anibal Ollero: Received the "Ingeniero Industrial Electrico" degree (in 1977) and the "Doctor Ingeniero" degree with honors (in 1980) from the University of Seville where he was an Assistant Professor. He was at the University of Santiago in Vigo, Spain, where he became Associate Professor in 1982, Professor in 1986, and Head of Department. Since 1988 he is Professor and Head of Department at the University of Malaga, as well as Director of the Escuela Tecnica Superior de Ingenieros Industriales. In 1979 he visited the Laboratoire d'Automatique et d'Analyse des Systemes (LAAS-CNRS), Toulouse, France. From May 1990 to May 1991 he was Visiting Scientist at the Robotics Institute, Carnegie Mellon University. Dr. Ollero is the author of a book on computer control ("Premio Mundo Electronico" Spanish award), and of about 80 contributions including book chapters, and papers in journals and conference proceedings. He participated and lead several research and development projects on robotics, process control, and computer integrated manufacturing funded by various research agencies and industrial companies. He also developed industrial applications and consulting activities for Spanish companies and agencies. He is currently Head of IFAC's working group on Intelligent Components and Instruments for Control, member of the IEEE, and member of several scientific societies on Robotics, Automatic Control, and Artificial Intelligence. Dr. Ollero is also organizing the Automation and Robotics Research Institute of Andalucia, Spain. His current research interests are in robotics and intelligent control, including fuzzy control.
Current Address: Dpto. Ingenieria Sistemas y Automatica, Universidade de Malaga, Plaza El Ejido, 29013 Malaga, Spain.

Dahee Park: Received the B.S. and M.S. degrees in Mathematics from Korea University, Seoul, Korea, in 1982 and 1984, respectively. He also received the M.S. and Ph.D. degrees in Computer Science at Florida State University in 1989 and 1992, respectively. Currently, Dr. Park is an Assistant Professor in the Department of Computer Science, Korea University at Chochiwon. His research interests include neural networks, fuzzy logic, genetic algorithms, and machine learning.
Current Address: Seocho Gu Banpo Dong 32-5, Hanyang APT 5 Dong 1008 Ho, Seoul, Korea.

Marek J. Patyra: Received his M.S. and Ph.D. degrees in Electrical Engineering from Warsaw University of Technology, Poland, in 1978 and 1986, respectively. From 1978 to 1989 he was an assistant Professor at the Institute of Electron Technology, Warsaw, Poland. From 1989 to 1991 he was Visiting Researcher at the Department of Electrical and Computer Engineering, Carnegie Mellon University, where he researched VLSI circuit design for manufacturability and testability. Since 1991 he has been an Assistant Professor at the Department of Computer Engineering at the University of Minnesota, Duluth. During the last ten years he has also held several visiting professor positions, including those in the Department of Microelectronics and Optoelectronics, and the Department of Information Sciences, both at Warsaw University of Technology, and also in the Department of Electrical Engineering at the Swiss Federal Institute of Technology, Lausanne, Switzerland. Dr. Patyra's main research interests include fault tolerant VLSI circuit design and analog/digital implementations of both fuzzy logic circuits and neural systems.
Current Address: Department of Computer Engineering, University of Minnesota, Duluth, MN 55812.

Witold Pedrycz: Received his M.Sc., Ph.D., and D.Sc. in 1977, 1980, and 1984, respectively. He is a Professor and Head (Computer Engineering) at the Department of Electrical and Computer Engineering, University of Manitoba, Winnipeg, Canada. Dr. Pedrycz has published numerous papers on fuzzy sets, pattern recognition, fuzzy control and fuzzy controllers, and system modeling. He authored or co-authored two research monographs on fuzzy control and fuzzy relational equations. He serves as an Associate Editor of *IEEE Transactions on Fuzzy Systems* and is a member of the editorial boards of *Fuzzy Sets and Systems* (Editor of section of Technological Developments), *Pattern Recognition Letters,* and *Journal of Intelligent Manufacturing.* His current research interest is in knowledge-based neurocomputations. He is interested in their theoretical foundations as well as the applications of this paradigm of computations in intelligent control, pattern recognition, diagnosis, identification, and fuzzy simulations.
Current Address: Department of Electrical and Computer Engineering, University of Manitoba, Winnipeg, Manitoba, Canada R3T 2N2.

Xiantu Peng: Received the B.S. degree in Mathematics from Jishou University (Jishou, China) in 1981, the M.S. degree in Mathematics from Beijing Normal University in 1986, and the M.S. and Ph.D. degrees in Computer Science from the Florida State University in 1990 and 1991, respectively. From 1981 to 1984 he was a Lecturer at the Mathematics Department of Jishou University. From 1988 to 1989 he was a Visiting Scholar at the Florida State University. Currently, he is a Senior Research scientist at Aptronix Inc., San Jose, CA, a company that specializes in fuzzy logic technology. Dr. Peng has published over 20 papers in mathematics and computer science. His current research interests include fuzzy systems, neural networks, knowledge engineering, category theory in computer science, and mathematical problems in cognitive science.
Current Address: Aptronix Inc., 2150 North First Street, San Jose, CA 95131.

Arthur Ramer: Holds a Ph.D. in Computer Science from SUNY. He worked as a mathematician in 1963-1972, an actuary in 1972-1980, and a computer scientist since 1981. He is currently a Senior Lecturer at the University of New South Wales. Dr. Ramer authored nearly 30 journal papers and over 40 international conference papers. His current research interests are in models of uncertainty and approximate reasoning, and theory of information systems. He has also worked in complexity of algorithms, and in software testing.
Current Address: School of Computer Science, University of New South Wales, Kensington, N.S.W. 2033, Australia.

Steve G. Romaniuk: Received the B.S., M.S., and Ph.D. degrees in Computer Science from the University of South Florida in 1988, 1989, and 1991, respectively. His current research is in hybrid connectionist, symbolic learning models, expert systems, and neural networks. Dr. Romaniuk has written over a dozen research papers.
Current Address: Department of Computer Science and Engineering, University of South Florida, Tampa, FL 33620.

Leonardo Cunha Rosa: Received his Mechanical Engineer degree in 1984 from the Fundacao Universidade de Rio Grande, and the Masters degree in 1988 from the Universidade Federal de Santa Catarina (UFSC), Brazil. Since 1985 he has been working as Senior Researcher at the R&D group in Numerical Control and Industrial Automation (GRUCON) of the Mechanical Engineering Department at UFSC. From 1989 to 1990, he worked at the Industrial Automation Institute, Madrid, Spain, as a visiting researcher. His main research interests are in the fields of numerical control, machine automation, and fuzzy control techniques.
Current Address: Grupo de Pesquisa e Treinamento em Comando Numerico, Universidade Federal de Santa Catarina, Brasil.

Tetsuo Sawaragi: Received his B.S., M.S., and Ph.D. degrees in Systems Engineering from Kyoto University, Japan, in 1981, 1983, and 1988, respectively. From 1991 to 1992

he was a Visiting Scholar at the Department of Engineering Economic Systems, Stanford University. Currently, he is an Instructor at the Department of Precision Mechanics, Kyoto University. Dr. Sawaragi has been engaged in research on systems engineering, cognitive science and artificial intelligence, particularly in the analysis of learning systems and in the development of intelligent control systems.
Current Address: Department of Precision Mechanics, Kyoto University Sakyo-ku, Kyoto 606-01, Japan.

Zuliang Shen: Received the B.S. degree in Electrical Engineering from Shaghai Mechanical Institute, China, in 1978, and the Ph.D. degree in Computer Science from Meiji University, Japan, in 1988. From 1978 to 1986 he was a Lecturer and Vice-Chairman of Computer Engineering Department at Shanghai University of Technology. From 1986 to 1990 he was a Visiting scholar at Meiji University in Japan. From 1990 to 1991 he was working in the R&D Department of PFU (Fujutsu Group), Japan. In 1991 he joined the Institute of Systems Science, National University of Singapore, as a Research Staff. Dr. Shen has published 10 books and about 50 papers in computer science and fuzzy systems. His current main research interests include fuzzy theory, approximate reasoning, neural networks, knowledge bases, mathematical linguistics, software engineering, compiler theory, and cybernetics.
Current Address: Institute of Systems Science, National University of Singapore, Kent Ridge, Singapore 0511.

Samuel M. Smith: Received the B.S. and M.S. degrees in 1987 and the Ph.D. degree in 1991, all in Electrical and Computer Engineering at Brigham Young University. He was a graduate research assistant in the Electrical Engineering Department, Robotics Group, at Brigham Young University. Currently, Dr. Smith is an Assistant Professor of Ocean Engineering at Florida Atlantic University and is part of the Advanced Marine Systems Center. His major research interest is the application of fuzzy logic to intelligent control, autonomous robotics, and signal processing. Other research interests include neural networks and expert systems. He is developing command and control systems for autonomous underwater vehicles using fuzzy logic and fuzzy decision-making techniques. His publications have primarily been in the area of automated fuzzy logic controller design and analysis. Dr. Smith is currently a member of IEEE, NAFIPS, and Sigma Xi.
Current Address: Ocean Engineering Department, Florida Atlantic University, 500 NW 20th St., P.O. Box 3091, Boca Raton, FL 33431.

Marian S. Stachowicz: Received his M.S. degree in Control and Computer Engineering from the Leningrad Electrotechnical Institute, and both his Ph.D. and D.Sc. degrees in Digital Electronics and Computer Control Systems from the University of Cracow, Poland. He held a professorship at the University of Cracow, Poland, as well as a number of appointments as Visiting Professor at various institutions, including the University of Minnesota, Minneapolis, the Moscow Power Institute, Leipzig Technical University, the University of Arizona, and Arizona State University. Since 1991, Dr. Stachowicz has been

at the University of Minnesota, Duluth, as Jack Rowe Chair. Currently, he heads the Laboratory for Intelligent Systems at the UMD Computer Engineering Department. His work centers on system theory and fuzzy set theory and their applications. Dr. Stachowicz received two awards from the Polish Ministry of Higher Education and Science for the introduction of digital fuzzy sets into fuzzy set theory which has facilitated applications of fuzzy set theory to computer engineering. He is also a consultant to 3M.
Current Address: Department of Computer Engineering, University of Minnesota, Duluth, MN 55812.

Horia-Nicolai Teodorescu: Graduated from the Faculty of Electronics, Technical University of Bucharest, Romania, in 1975. He received the Doctor in Technical Physics from the Polytechnic Institute of Isai, Romania, in 1981. He is an Associate Professor, Faculty of Electronics and Telecommunications, Technical University of Isai, and Honorary Director of the Center for Fuzzy Systems and Artificial Intelligence. Dr. Teodorescu is General Secretary of the Balkanic Union for Fuzzy Systems and Artificial Intelligence, Vice-President of the Romanian Society for Fuzzy Systems, member on the International Committee of FLSI, Iizuka, Japan, and Senior Member of the IEEE.
Current Address: Faculty of Electronics and Telecommunications, Technical University of Isai, Av. Copou, No. 11, Isai 6000, Romania.

Noboru Wakami: Received the B.S. and Dr.Eng. degrees from Osaka University, Japan. Currently, he is the General Manager of the Central Research Laboratories of Matsushita Electric Industrial Co. He joined the VTR project in the Central Research Laboratories from 1971 to 1978, and worked on a standard alignment tape recorder for VHS VTR. His current research interests are control systems using fuzzy logic, neural networks, and artificial intelligence. Dr. Wakami was an editor of the Institute of Systems, Control and Information Engineers from 1987 to 1988.
Current Address: Central Research Laboratories, Matsushita Electric Industrial Co. Ltd., 3-1-1 Yagumo-Nakamachi, Moriguchi, Osaka 570, Japan.

Peizhuang Wang: After graduating from the Mathematics Department of Beijing Normal University in 1957, he has been a Lecturer, Associate Professor, and Professor at Beijing Normal University. Currently, he is a Professor at the National University of Singapore, a Professor at Beijing Normal University, and the Head of the National Fuzzy Information Processing and Fuzzy Computing Laboratory in Beijing, China. Professor Wang is a Vice-President of the International Fuzzy Systems Association, Secretary of the Chinese Chapter of the International Fuzzy Systems Association, Chairman of Apt. Instruments, Inc. (Santa Clara), the Vice-President of Guangzhou University (Guangzhou, China), a Standing Council Member of the Chinese Systems Engineering association, and a Standing Council Member of the Chinese Artificial Intelligence association. He is also the Co-Chief Editor of *Fuzzy Systems and Mathematics* (Wuhan, China), the Co-Chief Editor of *Applied Mathematics Journal* (Beijing), and a Member of the Editorial Board of *Mathematics Research and Exposition* (Wuhan, China). He has been member of the

Program Committees of more than 20 international conferences. Professor Wang is the author of four books and co-author of three books. He has published about 120 papers in mathematics and fuzzy logic. His current research interests include knowledge engineering, fuzzy logic, uncertainty theory, and neural networks.
Current Address: Institute of Systems Science, National University of Singapore, Kent Ridge, Singapore 0511.

Ronald R. Yager: Received his undergraduate degree from the City College of New York, and his Ph.D. from the Polytechnic Institute of Brooklyn. Dr. Yager is Director of the Machine Intelligence Institute and Professor of Information Systems at Iona College. He has served at the National Science Foundation as program director in the Information Sciences program. He was a Visiting Scholar at Hunter College of the City University of New York. He was a NASA/Stanford Fellow, a Research Associate at the University of California, Berkeley, and has served as a lecturer at the NATO Advanced Study Institute. Dr. Yager is a Research Fellow of the Knowledge Engineering Institute, Guangzhou University, China, and is on the Scientific Committee of the Fuzzy Logic Systems Institute, Iizuka, Japan. He is co-president of the International Conference on Information Processing and Management of Uncertainty, Paris. He is Editor-in-Chief of the *International Journal of Intelligent Systems*, and also serves on the editorial board of a number of other journals including the *Journal of Approximate Reasoning, Fuzzy Sets and Systems*, and the *International Journal of General Systems*. Dr. Yager has published over 300 articles and has edited ten books. His current research interests include multi-criteria decision making, uncertainty management in knowledge-based systems, fuzzy set theory, fuzzy logic control, fuzzy systems modeling, neural modeling, and nonmonotonic reasoning.
Current Address: Machine Intelligence Institute, Iona College, New Rochelle, NY 10801.

Jia Hong Yan: Received her B.S. degree in 1987 from Jillin Institute of Chemical Engineering, Jillin, China, where she worked as an assistant. She received the M.S. degree from Yamagata University, Yonezawa, Japan, in 1992. Her current research interest is in fuzzy control.
Current Address: Yamagata Mitsumi Co., Ltd., 1-1059-5 Tachiyagawa, Yamagata 990-12, Japan.

Shuichi Yoshida: Received the B.S. degree in Electrical Engineering and the M.S. degree in Electronic Engineering from Kyoto Institute of Technology, Kyoto, Japan, in 1980 and 1982, respectively. He joined Matsushita Electric Industrial Co., Ltd. in 1982, and since then he has been engaged in research and development of servo systems for disk drives. He is now an Engineer at the Information Equipment Research Laboratory.
Current Address: Information Equipment Research Laboratory, Matsushita Electric Industrial Co., Ltd., Osaka, 571 Japan.

Xinghu Zhang: Received the B.S. and M.S. degrees in Mathematics from Beijing Normal University, China, in 1983 and 1986, respectively. He was at the China Academy of Electronics and Information Technology, Beijing, from 1986 to 1991 as an associate Researcher. Currently, he is a Research Scholar at the Institute of Systems Science, National University of Singapore, majoring in fuzzy neural networks and their applications. He has published over 20 papers in mathematics, fuzzy logic, and neural networks. His research interests include fuzzy logic, uncertainty reasoning, decision-making theory, knowledge engineering, neural networks, hyperspace theory, and mathematical problems in artificial intelligence.

Current Address: Institute of Systems Science, National University of Singapore, Kent Ridge, Singapore 0511.

INDEX

A

Actors, 25, 27, 28
Adaptive control systems, 22, 32
    hierarchies, 24, 38
Adaptive process control, 295
Afferent activities, 26
Agents, 25, 27, 30
Analogical processor, 64
    learning, 65
    parametric learning, 72
    reasoning, 66

B

Backpropagation, 4, 6

C

Cache memories, 541
    organization, 543
    replacement algorithms, 542
CAM. *See* Computer aided manufacturing
Cart pole system, 181, 185, 191, 431
    See also Inverted pendulum
Cell state space methods, 397
    optimal control, 402

Cell-to-cell mappings, 397, 402
Center of area, 5, 106, 267
    See also Center of gravity
Center of gravity, 5
Chaining of fuzzy rules, 114
Chaotic fuzzy PLL, 574
Chaotic systems, 475, 478
Computer aided manufacturing, 20, 21
Consequence modeling, 306
    linear substructures, 306
    modeling support, 307
Controllability, 452

D

Defuzzification, 5, 78, 101, 167, 168, 260, 266, 268, 437
Direct fuzzy control, 146
Direct rule-based fuzzy controller, 526
Disk drive, 579
    fuzzy control, 583
    seek control, 581

E

Efferent activities, 26

F

Fuzzification, 5, 32, 436
Fuzzy auto-tuning, 147
Fuzzy control neural networks, 432
    implementation, 436
    learning capacity, 438
Fuzzy identification, 36
Fuzzy inference, 167, 168
    constraint-oriented, 181, 183
Fuzzy language understanding, 462
Fuzzy learning, 166, 168
    of control rules, 376
Fuzzy management of cache memories, 541, 544
Fuzzy membership functions, 379
    modification, 379
Fuzzy neural networks, 430
Fuzzy path planning, 464
Fuzzy phase-locked loop, 574, 575
Fuzzy reasoning, 337, 356, 357
    self-tuning of, 337
Fuzzy transportation system, 460

G

Genetic algorithms, 161, 183, 188, 191, 337, 341, 482, 489
Global control law, 253
Gradient descent, 10, 339
Granularity, 259

H

Hardware accelerator, 168
Heterogeneous control laws, 244, 250
Heterogeneous controller, 248
Hierarchical representation of rules, 123
Hybrid control system, 22
Hybrid neural-fuzzy reasoning, 355
    design, 367
    model, 356, 365
Hyperellipsoidal clustering, 295, 301
    algorithm, 303
    criterion, 301
    design parameters, 304
    membership functions, 305

I

Identification, 409
    fuzzy, 36
    parameter, 409
Inference process, 5
Inference rules
    optimization of, 341, 343
Information complexity, 76
Information of fuzzy sets, 83
Interior penalty method, 198, 199, 202

Inverse matching, 57
Inverted pendulum, 411, 431
   four-dimensional, 420, 423
   stability, 414
   three-dimensional, 419, 422
   See also Cart pole system

L

Learning algorithms, 4, 23
   supervised, 4
   using descent method, 339
Least square method, 11
   recursive, 11
Linguistic rule-based fuzzy system, 5
Logic-based neurons, 59

M

Machine-tool control optimization, 523
Mamdani model, 104, 110
Matching, 57
Matrix representation, 316, 318
   reduction of, 322
Max criterion, 5
Maximum uncertainty operations, 89
Mean of maximum, 5
Membership functions
   adjustment, 236
   learning, 444
   modification, 379
   optimization, 12
   uncertainty, 79
Minimization of uncertainty, 94
Min-max-gravity, 275, 276, 278, 282
Min/product-max-gravity, 291
Min-sum-gravity, 291
Model reference adaptive fuzzy control system, 219
Multilevel control, 40
   organization, 43
   strategies, 43

N

Neural-fuzzy reasoning, 356
Neural networks, 4, 6, 328
   fuzzy, 430
Neuro-fuzzy networks, 4, 5
   differentiable, 15
   non-differentiable, 14
Neurons
   logic-based, 59
New fuzzy reasoning method, 356, 361

O

Optimal control, 21, 401
   algorithm, 406
   cell state space, 402
Optimal controller, 23
Optimization
   machine-tool, 523
   of inference rules, 341, 343
   of membership functions, 12
   random, 13

P

Parametric learning, 72
Perceptron, 7
Phase-locked loop, 561, 563
   fuzzy control of, 564
PLL. See Phase-locked loop
Possibility theory, 33
Power sets, 32
Predictive control, 310
Premise modeling, 307
   conditional variables, 307
   model evaluation, 308
Process control, 295
Product-sum-gravity, 275, 280, 282, 285

Q

Quasi-fuzzy inference, 591

R

Reasoning by analogy, 55, 56
   characteristics of, 69
   with input uncertainty, 68
Relational systems, 32
Robotics, 20
Robustness, 225
   of fuzzy controller, 238
Rule generation, 207
   hybrid algorithm, 209
   of fuzzy controller, 232, 234
Rule modification algorithm, 531

S

Self-organizing control, 189, 193
Self-organizing fuzzy controller, 530
Self-tuning method, 198, 199, 202
Specifity, 260
Stability, 142, 452
   of fuzzy PLL, 574
   of inverted pendulum, 414
Superconductors, 499
Superconducting actuators, 499
   fuzzy control, 512
   levitation mechanism, 500
   pump actuator, 515
   radial bearing, 506
Supervised learning, 4

T

Task descriptors, 39
Topology preserving mapping, 7
Traffic control, 331

V

Variable structure systems, 225
VLSI implementation, 161, 172, 551
VSS. *See* Variable structure systems
VSS controller, 227

   fuzzy, 230

W

Weight identification, 9
Weight perturbation algorithm, 12